HYDROLOGY OF THE NILE BASIN

OTHER TITLES IN THIS SERIES

1   G. BUGLIARELLO AND F. GUNTER
COMPUTER SYSTEMS AND WATER RESOURCES

2   H.L. GOLTERMAN
PHYSIOLOGICAL LIMNOLOGY

3   Y.Y. HAIMES, W.A.HALL AND H.T. FREEDMAN
MULTIOBJECTIVE OPTIMIZATION IN WATER RESOURCES SYSTEMS:
THE SURROGATE WORTH TRADE-OFF-METHOD

4   J.J. FRIED
GROUNDWATER POLLUTION

5   N. RAJARATNAM
TURBULENT JETS

6   D. STEPHENSON
PIPELINE DESIGN FOR WATER ENGINEERS

7   V. HÁLEK AND J. ŠVEC
GROUNDWATER HYDRAULICS

8   J. BALEK
HYDROLOGY AND WATER RESOURCES IN TROPICAL AFRICA

9   T.A. McMAHON AND R.G. MEIN
RESERVOIR CAPACITY AND YIELD

10 G. KOVÁCS
SEEPAGE HYDRAULICS

11 W.H. GRAF AND C.H. MORTIMER (EDITORS)
HYDRODYNAMICS OF LAKES: PROCEEDINGS OF A SYMPOSIUM
12–13 OCTOBER 1978, LAUSANNE, SWITZERLAND

12 W. BACK AND D.A. STEPHENSON (EDITORS)
CONTEMPORARY HYDROGEOLOGY: THE GEORGE BURKE MAXEY MEMORIAL VOLUME

13 M.A. MARIÑO AND J.N. LUTHIN
SEEPAGE AND GROUNDWATER

14 D. STEPHENSON
STORMWATER HYDROLOGY AND DRAINAGE

15 D. STEPHENSON
PIPLELINE DESIGN FOR WATER ENGINEERS
(completely revised edition of Vol. 6 in the series)

16 W. BACK AND R. LÉTOLLE (EDITORS)
SYMPOSIUM ON GEOCHEMISTRY OF GROUNDWATER

17 A.H. EL-SHAARAWI (EDITOR) IN COLLABORATION WITH S.R. ESTERBY
TIME SERIES METHODS IN HYDROSCIENCES

18 J. BALEK
HYDROLOGY AND WATER RESOURCES IN TROPICAL REGIONS

19 D. STEPHENSON
PIPEFLOW ANALYSIS

20 I. ZAVOIANU
MORPHOMETRY OF DRAINAGE BASINS

HYDROLOGY OF THE NILE BASIN

MAMDOUH/SHAHIN

International Institute for Hydraulic and Environmental Engineering
Oude Delft 95, 2601 DA Delft, The Netherlands

ELSEVIER

Amsterdam — Oxford — New York — Tokyo   1985

ELSEVIER SCIENCE PUBLISHERS B.V.
Molenwerf 1
P.O. Box 211, 1000 AE Amsterdam, The Netherlands

Distributors for the United States and Canada:

ELSEVIER SCIENCE PUBLISHING COMPANY INC.
52, Vanderbilt Avenue
New York, NY 10017

ISBN 0-444-42433-4 (Vol. 21)
ISBN 0-444-41669-2 (Series)

Printed in The Netherlands

PREFACE

This book aims at describing a number of the hydrological aspects of the basin of the Nile River and the different factors affecting them. With this aim in mind it deals primarily with the inflow-outflow balance of the Nile system from the source up to mouth sub-basin-wise.

The components of the hydrologic cycle which enter in the water balance and which are considered here are the rainfall, evaporation, evapotranspiration and the change of water in storage, both in volume and level. Each of these components is presented as observed in nature, recorded from experiments or found from computations, together with an explanation of the procedures used and the interpretation of the results obtained. Attention is paid to the losses which take place in certain parts of the basin.

The meteorologic and hydrologic data at the key stations on the Nile and its tributaries are analyzed and their basic statistical properties given. A special chapter is devoted to the geohydrology of the basin and to the groundwater situations and potentialities in some of the countries sharing the Nile Basin. Last, but not least, a whole chapter has been left to the storage, control and conservation works, both existing and planned, and to the impacts of such works on the environment.

In attempting to cover here as many of the hydrological aspects of the Nile Basin as possible, it has not been intended that this book shall compete with existing literature on the same subject but rather complete it. Moreover, this book, when added to those describing aspects other than hydrological, such as biological and geological, shall certainly help to provide the reader with a more complete picture of this river basin.

Almost two-thirds of the surface of Africa are drained by seven major rivers, including the Nile. So the knowledge gained of the hydrology of any of them shall, no doubt, contribute to a better understanding of the water resources of a continent with an acute water-shortage problem. Such an understanding is needed by every hydraulic or water resources engineer aiming at a more efficient utilization of the readily available, as well as the potential, water resources.

The author has depended in some parts of the book on his experience and viewpoints, and on the existing literature in the remaining parts. A list of the references and data sources used appears at the end of every chapter.

A sense of gratitude must be expressed here for the many who directly and indirectly, by their constructive criticism and advice, have helped in the preparation of this book. Most of the appreciation goes, in fact, to Prof. ir L.J. Mostertman, Director of the International Institute for Hydraulic and Environmental Engineering, Delft, The Netherlands, whose continuous support and encouragement dates back to as early as 1962. Gratitude is also extended to many of the author's former colleagues at the Nile Control Department, the Ministry of Irrigation, Egypt, and at the Faculty of Engineering, Cairo University. Special mention must be made of Ms P.E. Röell, Librarian of the International Institute for Hydraulic and Environmental Engineering, Delft, for providing the author with an enormous number of references and documents, Ms P. Schott-León who undertook all the typing and Mr W. van Nievelt for preparing the graphic work in this book. The author is greatly indebted to his family, whose patience and tolerance have been his support in the many years spent in compiling the book.

M.M.A. Shahin,
Delft, 1984

CONTENTS

Chapter 1

HISTORICAL INTRODUCTION

1.1   SOURCE OF THE NILE

The known history of the Nile River dates back to just before 5000 B.C.
From then till recently various theories about the source of the Nile and its
rise have been laid down. Some of these theories were so conflicting that it
became customary in ancient Rome to say: quaerere fontes Nile (search for the
Nile) when someone talked about a mysterious or impossible matter (Pierre,
1974).

The name Nile is said to be derived from the Greek Neilos, whose origin is
unknown. Both Neilos and Aiguptos (masculine) were used in Greek drama when
referring to the Nile, whereas Aiguptos (feminine) alone was used when referring to Egypt (Encyclopaedia Britannica, 1969).

It is often claimed that the ancient Egyptians knew nothing of the origin
of their river. This claim is supported on one hand by the naive conviction
of the common ancient Egyptian that the Nile flows out of the full breasts of
the Nile God, Hapi. The priests of ancient Egypt, on the other hand, had
their own theory, which they faithfully founded on theological grounds. Those
priests were convinced that somewhere the course of "The Celestial Nile" was
beset by monstrous rocks and stones and that below this barrier rose Egypt's
Nile or Egypt's heaven-descended stream. After all, the Nile priests were one
of the sources that provided Herodotus, the Ionian, with most of the information that appeared in his written accounts on Egypt and the Nile.

The philosophers and savants of ancient Greece had their views on, and
opinions of, the source of the Nile and of its rhythmic pattern of flow. These
views comprised the role of the Etesian (northerly) winds in the build-up of
the Nile, the origin of the river from Oceanus (the ocean surrounding the
earth), and the rise of the Nile from the peaks of the Lybian mountains and
its supply from the melted snow thereon. Herodotus wrote in his accounts that,
during his visit to Egypt in 457 B.C., one informant said that the Nile rose
from a powerful spring feeding a deep lake situated between the hills of
Mophi and Chrophi. There are two trains of thought: the first is that the
said lake lies between the island of Elephantine and Aswan, which confines
the whole story to the Nile in Egypt, and the other is that the lake is situated far more to the south. In comparison with recent discoveries the latter is
sometimes interpreted as the lake supplying the Semliki River which discharges
its water into Lake Albert (also called Mobutu-Sese Seko). If we are
prepared to accept this interpretation, it is then fair to conclude that

Herodotus can be complimented for throwing some light on the western tributary
of the Nile River. In a paper entitled "The Nile, its Origin and Rise" it is
mentioned that Herodotus believed that the Upper Nile flowed from west to east,
but he confused the Niger with the Nile (Biswas, A., 1966). This idea prevailed
for quite some time. Juba II (20 A.D.), the King of Mauritania, affirmed that
the source of the Nile was in western Africa, thereby supporting the conviction
that the Niger is a branch of the Nile (Biswas, A., 1966, 1970).

The Greek philosopher Aristotle (384-322 B.C.) thought that the river
descended from a mountain of silver (montagne d'argent) and that heavy spring
and summer rains on the highlands of the catchment areas (what we call nowadays the Blue and White Niles) were responsible for the flow in the Nile.

Almost two centuries after Herodotus, came the mathematician and geographer
Eratosthenes (276-194 B.C.) who described the Nile far better than any of his
predecessors. His idea was that two principal streams sprang out of some lakes
situated to the east and encircled Meroe, a considerably large island (see map,
Fig. 1.1.). The eastern tributary was the Astaboras (now called the Atbara) and
the western the Astasobas (now called the Blue Nile). The Astapus (now called
the White Nile) was a different river entirely, which rose from some lakes to
the south and carried the summer rains to form the direct stream of the Nile.

Ptolemy, the Roman, who lived in Alexandria in the second century A.D.,
thought that the main Nile came from the Mountains of the Moon, which were
permanently covered with snow and passed through two lakes. It is probable
that he meant by the Mountains of the Moon, the Ruwenzori range and by the two
lakes, Victoria and Albert Nyanza. The map of the Nile as developed by Ptolemy
is as shown in Fig. 1.2.

From the second half of the second century onwards, for at least thirteen
centuries, there were hardly any discoveries, with the exception of some des-
criptions of the lower reaches of the Nile by the moslim geographers in Egypt.
Examples are Al-Khuwarazmi in 864 A.D. and Al-Masoudy in 957 A.D.

The triumphant wars of Portugal against the Moors in north-west Africa in
the fifteenth century had, no doubt, paved the road to the Portuguese infiltration in Africa, both east and west. Two Portuguese missionaries, Pedro
Paez and Jerome Lobo, visited Ethiopia in the seventeenth century. Father Paez
visited Lake Tana (1618). His successor, Father Lobo wrote an account about
his visit to Tissisat Falls. About 150 years later, in 1770 A.D., a Scottish
explorer named James Bruce, after having journied five years in Ethiopia,
succeeded in discovering the source of the little Abbai (see map, Fig. 1.1.).
A summary of Bruce's views on Paez and Cheesman's on the expedition of Bruce
can be found in the Book on The Blue Nile (Moorehead, A., 1962).

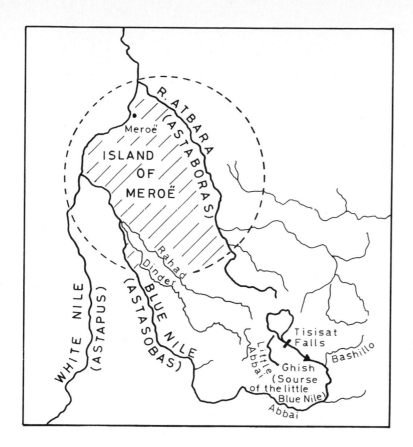

Fig. 1.1. Location of the old island of Mero and tributaries

The journeys made by the Arab traders on the east coast of Africa to the interior, the religious missions to East Africa and the steadily growing connections between the Coptic churches of Alexandria and Ethiopia have all led to more concrete information on the snow-capped mountains (the Mountains of the Moon) described by Ptolemy in 150 A.D.

The history of exploring the Nile river system in the nineteenth century begins with the invasion of the Sudan by Mohammed Ali Pasha and his sons from 1821 onwards. As a result of this, the Blue Nile was explored as far as its exit from the Ethiopian foothills, and the White Nile as far as the Sobat mouth. The Bahr el Ghazal was explored by Petherick in the eighteen fifties.

The considerable interest of the Europeans came shortly after the reports of Knoblecher (1850) about the existence of some large lakes in the south.

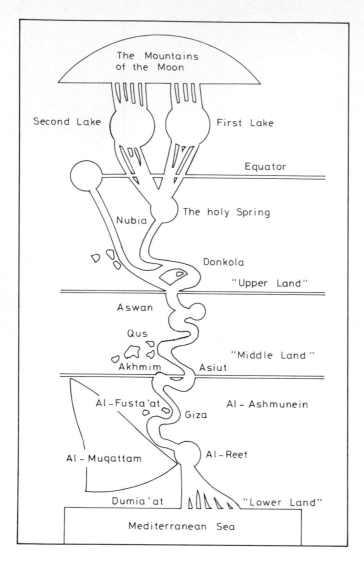

Fig. 1.2. Map of the Nile from the source at the Mountains of the Moon to the mouth in the Mediterranean Sea, as prepared by Ptolemy

From about 1857 onwards for about 30 years, the Equatorial Lakes Plateau and surroundings were traversed by several explorers, all searching for the source of the river that had puzzled the whole world for several centuries. The principal explorers and their expeditions were as follows:

i) Burton and Speke went from Bagamoyo to Tabora to Ujiji on Lake Tanganyika (1857-59). In 1858 Speke alone went on northwards and discovered Lake Victoria.

ii) Speke and Grant set off on an expedition (1860-63) around Lake Victoria and reached Ripon Falls. They thought that the stream flowing out of this lake was the source of the Nile.

iii) Samuel Baker, after exploring the Atbara, went on to discover Lake Albert (1862-64) as well as the Albert Nile and the upper reach of Bahr el Jebel.

iv) Livingstone, like Burton and Speke, set off on his route from Tabora to Ujiji (1872). He thought the Lake Nyasa might be draining into Lake Tanganyika, which could then be linked with the Albert Lake and the Nile.

v) Stanley travelled up from the east coast and circumnavigated Lake Victoria (1872). His attempt to get to Lake Albert was not successful, though he reached as far as the escarpment about Lake George. In a later journey (1889) he reached the Semliki and Lake Edward.

The routes of these leading explorers are indicated on the map, Fig. 1.3. (Stamp and Morgan, 1972).

In spite of all the expeditions already mentioned, a number of tributaries of the Nile remained undiscovered. Exploration missions stopped in the period 1881 up to 1898 as a result of the rebellions of the Mahdis in the Sudan. A historical account of this period can be found in the book entitled "The White Nile" (Moorehead, A., 1960). After the reopening of the Sudan in 1898 and at about the same time the opening up of Kenya and Tanzania (formerly called Tanganyika) and Uganda, irrigation services and survey and geological depart- ments were established in the respective countries. Since then these depart- ments have taken over the exploratory work as well as the collection of hydrologic and other relevant data (Hurst, H., 1952).

At about the end of 1902 the Irrigation Department of Egypt started two expeditions; the one under Mr C.E. Dupuis to visit Lake Tana and the other under Sir W. Garstin to visit the Lakes Victoria and the then Albert and Edward. The expeditions which followed were sent to collect, or report on, hydrological data needed for the Nile projects (Hurst, H.E., 1931). Unfortun- ately, the Macmillan-Jessen expeditions in 1902 and 1905 failed in exploring Lake Tana or finding the source of the Blue Nile. After this there was a long period of inactivity on the river until R.E. Cheesman arrived in 1925 in Ethiopia. For the next eight years Cheesman devoted himself to the exploration of the gorge of the Blue Nile and to circumnavigate Lake Tana. This accomplish- ment was terminated by writing an account on Lake Tana and the Blue Nile (Cheesman, R., 1936).

More recent expeditions have been sent, especially by the Egyptian Govern- ment, to explore more of the tributaries in the catchments of the Equatorial Lakes, the White Nile and the Sobat. However, it was not until 1937 that the

southernmost source of the Nile in the headstreams of the Kagera, largest tributary of Lake Victoria, was located. Further exploration of the Nile tributaries in the Ethiopian Plateau has been for quite some time less fortunate than
to other parts of the Nile Basin. The gorge of the Blue Nile (Great Abbai) was
not fully traversed until a British military and scientific expedition con-
quered it during the flood of 1968 (Blashford-Snell, J.N., 1970).

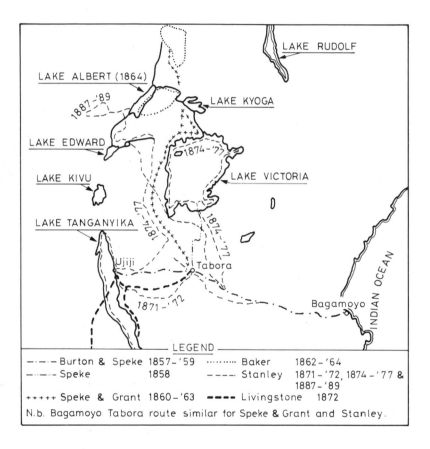

Fig. 1.3. The routes of some leading explorers of the Nile Source (Stamp, L.D.
and Morgan, W.T., 1972)

1.2   HISTORY OF HYDROLOGY OF THE NILE BASIN

1.2.1   From 3200 B.C. to 1900 A.D.

   Whether the ancient Egyptians knew the origin of their river or not,
(see 1.1) what cannot be denied is that they were deeply keen to observe all
astronomical phenomena and terrestrial events associated with, or related to,
the Nile floods.

One of the most ancient records is a drawing of an imperial macehead held by the protodynastic king Scorpion when celebrating the occasion of cutting an irrigation ditch some 3200 years B.C. (Biswas, A., 1970). The same reference mentions, among others: damming off of the Nile and diverting its course by King Menes in 3000 B.C. At the same time the ancient Egyptians began to use the Nilometers to record the fluctuation of the Nile, the construction of Sadd el-Kafara (Pagants dam) some 30 kilometres south of Cairo and its failure in 2850 B.C., the connection of the Nile and the Red Sea by a navigational canal during the reign of Seostris I in about 1950 B.C., the hydraulic works of Amenemhet III (including Lake Moeris) in about 1850, the water codes of King Hammurabi in 1750 B.C., Joseph's well near Cairo (more than 100 metres deep) in about 1700, etc. (Biswas, A., 1970). Another reference states that "hydraulic engineering" at that time reached a high degree of accomplishment; reclamation schemes on the left bank of the Nile were initiated during the floods; dams, dikes and canals were constructed and later water-lifting machinery was invented (Teclaff and Teclaff, 1973).

Probably the first non-religious theory explaining the yearly flooding was that derived by the Greek Thales (600 B.C.), the founder of deductive geometry. His theory was that the Etesian winds drove the sea high against the mouths of the river and thereupon prevented them from discharging their water. The river therefore returned upon itself and, whenever it could, it burst out into forbidden ground. A couple of centuries later another Greek geometer, Democritus, came up with a somewhat different theory. He thought that when snow in the northern parts of the world was melted at the time of the summer solstice and flowed away, clouds were formed by the vapour. When the clouds were driven towards the south and towards Egypt by the Etesian winds, violent storms arose and caused the lakes feeding the River Nile to be filled (Frisinger, H., 1959).

Since the Nile used to rise in flood at about the same time every year, its behaviour had been described as regular. The Bible tells us, however, "....there came seven years of great plenty throughout the land of Egypt. And there shall rise after them seven years of famine .." (Genesis 41, 29-30). The interpretation by Joseph to this dream of the Pharao of Egypt was probably the first indication of the persistence in the hydrologic time series.

The Roman savant, Pliny (23-79 A.D.), thought the two probable theories about the flooding of the Nile were those of Thales and Democritus. After all, in both theories the action of the Etesian winds was the culprit. After Pliny, the English historian and theologian, Beda (674-735 A.D.), compiled and summarized the knowledge then available about the Nile flood. The theory he proposed was very similar to that of Thales. He claimed that the northerly

winds forced the sea waves to pile up sand at the Nile mouths, thus causing
the river to back up upon itself and flood (Frisinger, H., 1959). From that
time onwards until the nineteenth century there was very little done to explore
the sources of the Nile and its tributaries, except for the recordings which
were done whenever possible. The nineteenth century was the century of discovering the Nile source and tributaries rather than collecting and/or inter-
preting its hydrologic data.

The record of the Nile levels dates back to about 3000 to 3500 years B.C.
The river gauge is called Nilometer (in Arabic Miqyas An-Nil). Three types of
Nilometers were used. The first type consisted simply of marking the water
levels on cliffs on the banks of the river, e.g. the second cataract at Semna.
The second type consisted essentially of a scale, usually of marble, on which
the water level was observed. The standard gauge consisted of a series of steps
or pillars built into the river bank to each of which a section of the scale
was fixed. It is claimed that there are 140 of these gauges scattered over the
basin outside Egypt, and many more on the Nile in Egypt. Most of them are
observed daily, and the readings of the more important are telegraphed or tele-
phoned to Cairo (Hurst, H.E., 1952). The catastrophic floods of 1954 and 1958
were measured at 93 gauging points in Egypt only, 36 gauges in upper Egypt and
the remainder in lower Egypt (Hashem and El-Sherbini, 1961). The third and most
accurate Nilometer used to bring water of the Nile to a well and the water
level was marked either on the walls of the well or on a central pillar. The
most notable Nilometer is situated at Roda near Cairo. The recorded water level
there dates back to 641 A.D. The Arab chaliphs (Kings) built a new Nilometer
there in 715 A.D. This was rebuilt in 861 A.D. It consists of a square well
connected to the Nile by means of three conduits. At the centre of the well is
a graded octagonal pillar of white marble divided into 19 cubits (see Fig. 1.4).
It was reported that as a result of poor joining of the lower broken part of
the pillar, the corresponding cubit now measures 31 cm only instead of the original 54 cm (Ghaleb, K., 1935). This example and many others shows that the
records available since 641 A.D. need much adjustment before having them
analyzed. The suitability of the record for statistical analysis has been
argued recently: "Simply, the reliability of flood data for the assessment of
water averages, the reliability of preservation of a long-range constant gauge
datum in the past, and accuracies in observations (all kinds of changes), unfortunately do not permit one to draw dependable conclusions". (Yevjecvich, V.,
1983).

1.2.2   From 1900 A.D. till now

The turn of the last century and the beginning of the twentieth century
witnessed a number of notable accomplishments. When the then Anglo-Egyptian

9

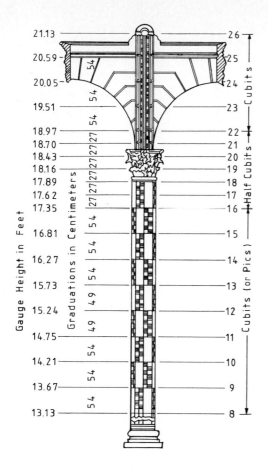

Fig. 1.4. The Nile gauge at Roda, Cairo, Egypt

Sudan was reoccupied in 1898 all swamp rivers were found blocked up. It was not before 1905 that a channel through the Bahr el Jebel had been made clear. The Ghazal was not made navigable to the principal capital Wau, until 1904.

In 1902 the works in the first Aswan dam and some of the Barrages on the Nile in Egypt were completed. The history of scientific study of the hydrology of the Nile begins with the introduction of current metres by Sir H. Lyons in about 1902. Previous to this, flow measurements had been made by floats. The Survey Department of Egypt became charged with the survey of all river gauge records south of Aswan in 1902 and 1903. This was the origin of the Hydrology Service which formed part of the later Physical Department of Egypt (Hurst and Philips, 1931).

The results of the expeditions to Lakes Tana, Victoria, Albert and Edward conducted by Dupuis and Garstin in 1901-1904 (see 1.1) were concluded in the 1904 report by Sir W. Garstin. Shortly after that, in 1906, Sir H. Lyons published his book "The Physiography of the Nile". This book contained the information gathered from travellers and scientific explorers available at that time. In 1905 Sir M. MacDonald introduced a new method of river measurement at Aswan using the flow through the sluices of the dam. A large masonary tank was used to measure the discharge of one type of sluice under all conditions of head and sluice opening. The results were then applied to flow from other sluices.

The Sudan branch of the Egyptian Irrigation Service was formed in 1905 with the object of performing all hydrologic and hydrographic works for the different projects aiming at the improvement of the water supply of Egypt and the development of perennial irrigation in the Sudan. The data collected in the period 1906 to 1913 were useful in the design of the Jebel Aulia dam on the White Nile, about 45 km above the junction of the White and the Blue Niles, and Makwar dam on the Blue Nile, some 360 km above the same junction.

The foundations of the present Meteorological Service of Egypt were laid in 1900. Since then many stations were established and more observations were taken. In 1915 all the work of a physical nature done by the Survey Department, and the hydrological work done by the Irrigation Department, were combined to form the Physical Department of the Ministry of Public Works, Egypt.

The work on the Nile projects stopped during the First World War (1914-1918), though routine observations were carried on. From 1912 up to 1923, especially in the post-war period, most of the progress was directed at establishing permanent discharge sites at a number of important stations where regular observations had been taken. Advances in measuring devices and techniques followed. Data collection and analysis went on and that was a great help in the design and construction of the Makwar dam, which was finally built in 1925 on the

Blue Nile for the benefit of the Sudan.

In 1923 the Ministry of Public Works, Egypt, sent a mission to the Equatorial Lakes with the aim of preparing a programme for investigations in connection with the possible Upper Nile projects. That mission was followed by others in 1924, 1926, 1930 and 1931 to investigate the hydrology of the Lake Plateau, Bahr el Ghazal and the White Nile basins.

The East African Meteorological Service, later Department, EAMD, was formed in 1927. This department was partly financed by the Egyptian Government. It operates over all of East Central Africa and the data it collects are undoubtedly valuable for the hydrology of the Nile Basin.

The original Nile waters agreement was laid down in 1929 and had, for some time, been the basis of the water allocation between Egypt and the Sudan. The most important item in the agreement was "... no works should be constructed or measures taken, on the Nile or its branches or on the lakes from which it flows, in the Sudan or in the territories under British administration, which would affect the flow of the river in such a way as to cause prejudice to the interests of Egypt".

To ensure the continuity of dissemination of the knowledge on the hydrology of the Nile Basin and to present the ever-increasing data in a systematic way, for both scientific and practical purposes, it was decided to issue the volumes and supplements of "The Nile Basin" successively.

The basic data about these references are as follows:

Volume No.	Subject matter	Author(s)	Year of publication
I	General description of the basin; meteorology, topography of the White Nile	H.E. Hurst and P. Philips	1931
II	Discharge and stage measurements of the Nile and its tributaries (with 9 supplements: 1928-32, 33-37, 38-42, 43-47, 48-52, 53-57, 58-62, 63-67 and 1968-72)	H.E. Hurst, P. Philips, Y.M. Simaika, R. Black and Nile Control Staff	from 1932 onward
III	Ten-day mean and monthly mean gauge readings of the Nile and its tributaries (with 9 supplements: 1928-32, 33-37, 38-42, 43-47, 48-52, 53-57, 58-62, 63-67 and 1968-72)	H.E. Hurst, P. Philips, Y.M. Simaika, R.P. Black and Nile Control Staff	from 1932 onward

Volume No.	Subject matter	Author(s)	Year of publication
IV	Ten-day mean and monthly mean discharges of the Nile and its tributaries (with 9 supplements: 1928-32, 33-37, 38-42, 43-47, 48-52, 53-57, 58-62, 63-67, 1968-72)	H.E. Hurst, P. Philips, Y.M. Simaika, R.P. Black and Nile Control Staff	from 1933 onward
V	The hydrology of the Lake Plateau and Bahr el Jebel	H.E. Hurst and P. Philips	1938
VI	Monthly and annual rainfall totals and number of rainy days at stations in and near the Nile Basin for periods: 1938-42, 43-47, 48-52, 53-57, 58-62, 63-67 and 1968-72 (7 supplements)	H.E. Hurst, R.P. Black, Y.M. Simaika and Nile Control Staff	from 1950 onward
VII	The future conservation of the Nile	H.E. Hurst, R.P. Black and Y.M. Simaika	1946
VIII	The hydrology of the Sobat and White Nile and the topography of the Blue Nile and Atbara	H.E. Hurst	1950
IX	The hydrology of the Blue Nile and Atbara and the Main Nile to Aswan with some reference to projects	H.E. Hurst, R.P. Black and Y.M. Simaika	1959
X	The major Nile projects	H.E. Hurst, R.P. Black and Y.M. Simaika	1966

The data and information contained in the above-listed volumes of the Nile Basin, together with those appearing in the other papers of the Physical Department, later the Nile Control Department, have been employed in the design and construction of the major hydraulic works on the Nile and its branches and tributaries. Examples of these are: the heightening of the original Aswan dam in 1912 and in 1937, the Jebel Aulia dam on the White Nile in 1934, the Owen Falls dam at the exit of Lake Victoria in 1950, ... etc.

The idea of constructing a high dam at Aswan led to another agreement between Egypt and the Sudan in 1959 for the full utilization of the Nile waters. It is worthwhile mentioning here that the design of this dam was based on the theory of over-year storage. The earliest thought of this theory goes back to before 1946 (Hurst et al, 1946). The development of the theory marked the birth of modern hydrology, especially the stochastic part of it (Mandelbrot and Wallis, 1968).

The year 1959 also witnessed the first attempt to plan for the ultimate hydraulic development of the Nile Valley using an electronic digital computer

(Morrice and Allan, 1959). Although the Nile has been the best-studied river in the world for a generation, the need for many more investigations and research work is there. The completion of the first phase of the Roseires dam on the Blue Nile, the Khashm el-Girba dam on the Atbara and the High dam on the Main Nile at Aswan are hydrologic highlights in the nineteen hundred and sixties. A very important step which began in 1967, and has continued for quite some years, is the collaboration between Kenya, Tanzania, Uganda, the Sudan and Egypt in a hydrometeorological survey of the catchments of Lakes Victoria, Kyoga and Albert. This project included the upgrading of some of the existing hydrometrical stations and the establishment of new hydrometrical stations and river discharge measurement sites. All these stations have been equipped with modern instruments. The rainfall-runoff relationships were studied in a number of experimental basins and the relevant parameters estimated (WMO, 1974).

The early sixties and the late seventies of this century witnessed an un-usual rise in the surface water levels of the Equatorial Lakes and of other African lakes as well. The level of Lake Victoria rose by over 2.5 metres between 1959 and 1964 (Kite, G., 1981). For the same period, the rise reached 3.3 metres for Lake Albert, 2.6 for Lake Tanganyika and 1.5 metres for Lake Malawi. The second substantial rise began in 1978 and by mid 1979 reached about 1.8 metres for Lake Victoria, 3.0 for Lake Malawi and 1.0 metre for Lake Tanganyika. The considerable rise in the Lake Victoria water level in 1964 led to an excessive flow in the Nile to such an extent that it flooded some parts of Cairo at that time.

Two principal projects have been taking place during the last few years and are probably worth recording here. One is the first phase of the diversion scheme (called Jonglei canal). The canal connects the Bahr el Jebel at Bor straight to about Malakal on the White Nile and conveys 20 million m^3 per day at maximum. The annual volume of water saved by this scheme is 3.8 milliard m^3 estimated at Aswan. Half of this amount will be taken by the Sudan and the other half by Egypt (Executive Organ for the Development Projects in Jonglei Area, 1975). The second event, since 1978, is the joint work of the Ministry of Irrigation, Egypt, represented by its organs (mainly the Master Water Plan and the Research Institute for Water Resources Development), the University of Cairo, Egypt, and the Massachusettes Institute of Technology, U.S.A. in the analysis of the hydrologic data of the Nile Basin. The results so far obtained are available in a series of technical reports. Additionally, once every two years, they organize a sort of conference where problems related to water resources planning, management and development are discussed, together with the possible solutions.

Last but not least, both Egypt and the Sudan are working jointly to esta-
blish a commission of all countries sharing the Nile waters. The road to
realizing this step is, no doubt, rough and full of difficulties. Nevertheless,
such a step is, in the author's opinion, unavoidable if these countries are
keen on having a more efficient utilization of the water resources in the Nile
Basin.

REFERENCES

Bixwas, A.K., 1966. The Nile, its origin and rise. Water and Sewage Works,
    113: 283-292
Biswas, A.K., 1970. History of hydrology. North-Holland, Amsterdam, 336 pp.
Blashford-Snell, J.N., 1970. Conquest of the Blue Nile. Geogr. Journ. 136:
    42-51.
Cheesman, R.E., 1936. Lake Tana and the Blue Nile. Macmillan, London, 400 pp.
Encyclopaedia Britannica, 1969. Nile, Vol. 16: 516-523.
Executive Organ for the Development Projects in Jonglei Area, 1975. Jonglei
    Project (Phase One). Tamaddon P. Press, Khartoum, 99 pp.
Frisinger, H.H., 1959. Early theories on the Nile floods. Weather, Vol. 20:
    206-207.
Ghaleb, K.O., 1935. Discussion of: Flood-stage records of the River Nile, by
    C.S. Jarvis. Trans. ASCE, Paper No. 1944: 1063-1067 (discussion: 1063-1067).
Hashem, A. and El-Sherbini, H., 1961. The hydrologic features of the 1954 and
    1958 floods (in Arabic). The Government Printer, Cairo, 98 pp.
Hurst, H.E. and Philips, P., 1931. The Nile Basin, Vol. I, General description
    of the basin, meteorology and topography of the White Nile Basin. Physical
    Department Paper 26, Government Press, Cairo, 128 pp.
Hurst, H.E., Black, R.P. and Simaika, Y.M., 1946. The Nile Basin, Vol. VII,
    The future conservation of the Nile, Physical Department Paper 51, Eastern
    Press, Cairo, 159 pp.
Hurst, H.E., 1952. The Nile, a general account of the river and the utilization
    of its waters. Constable, London, 326 pp.
Kite, G.W., 1981. Recent changes in the level of Lake Victoria. Bulletin of
    Hydrological Sciences, No. 26, 3: 233-243.
Mandelbrot, B.B. and Wallis, J.R., 1968. Noah, Joseph, and Operational
    Hydrology. Water Resources Research, Vol. 4, No. 5: 909-918.
Moorehead, A., 1960. The White Nile. Hamish Hamilton, London. 385 pp.
Moorehead, A., 1962. The Blue Nile. Hamish Hamilton, London. 308 pp.
Morrice, A.W. and Allan, W.M., 1959. Planning for the ultimate development of
    the Nile Valley. Proc. Inst. Civil Eng. 14, Paper 6372: 101-155.
Pierre, B., 1974. Le Roman du Nil. Librarie Plon, Paris, 480 pp.
Stamp, D.L. and Morgan, W.T., 1972. Africa: A study in tropical development.
    John Wiley and Sons, Inc., New York, 520 pp.
Teclaff, L.A. and Teclaff, E., 1973. A history of water development and water
    quality. In: Environment Quality and Water Development (Editors: Goldman,
    C.R., McEvoy III, J. and Richerson, P.M.). W.H. Freeman and Company, San
    Francisco: 26-77.
World Meteorological Organization, 1974. Hydrometeorological survey of the
    catchments of Lakes Victoria, Kyoga and Albert, RAF 66-025, Tech. Report 1,
    Vols I, II, III and IV.
Yevjevich, V., 1983. The Nile River Basin: hard core and soft core water
    projects. Water International Vol. 8, No. 1: 23-34

Chapter 2

PHYSIOGRAPHY OF THE NILE BASIN

2.1   INTRODUCTION

The Nile Basin covers a surface of about 2.9 million square kilometres,
approximately one-tenth of the surface area of Africa. It extends from 4°S to
31°N latitude and from about 21° 30´E to 40° 30´E longitude. The hydrographic
boundaries of the Nile system are as shown on the map, Fig. 2.1. The highest
and the lowest points in the basin are represented by the top of the Ruwenzori
Range and the trough of El-Quattara depression respectively. They are at elevations of about 5 120 metres above mean sea level (a.m.s.l.) and about 160 metres
below mean sea level (b.m.s.l.), respectively.

The length of the River Nile from its most remote source, at the head of
River Luvironza, near Lake Tanganyika, to its mouth on the Mediterranean Sea, is
about 6 500 kilometres. The river course and its tributaries traverse the territories of Tanzania, Uganda, Rwanda, Burundi, The Congo (Zaire), Kenya, Ethiopia,
the Sudan and the Arab republic of Egypt. This state of affairs has made an
international river of the Nile, whose water is shared by a number of countries.

Although the Nile is an ancient river, the existing hydrologic pattern may be
as young as 10 000 years. If we exclude the drastic changes in the basic nature
of the river, the uninterrupted life of the modern Nile configuration has been a
short one. Additionally, a wide variety of topographic features - climate, geology, soil, plant and vegetal cover and other hydrology-affecting factors -
can be found in the Nile Basin. The integrated effect of such causative factors
on the run-off from a river basin can be represented by the so-called specific
discharge, $\bar{q}$. Specific discharge values of some of the world rivers are listed
in Table 2.1. These values have been calculated using the relationship $\bar{q} = \bar{Q}/A$,
where $\bar{Q}$ is the long-term mean discharge and A = surface area of the river catchment.

It is very clear from Table 2.1 that of all world rivers with drainage basin
areas, each larger than 1 million km^2, the River Nile has the lowest specific
discharge. If the estimated $\bar{q}$ of the Congo Basin, which is geographically the
closest to the Nile Basin, is fairly correct, the specific discharge of the
latter would then be just one tenth the specific discharge of the Congo.

The two causative factors which probably have the biggest effect on the run-
off from a drainage basin are the climate and the topography. The former will be
discussed in Chapter 3, while the topography of the Nile Basin is dealt with in
this chapter.

16

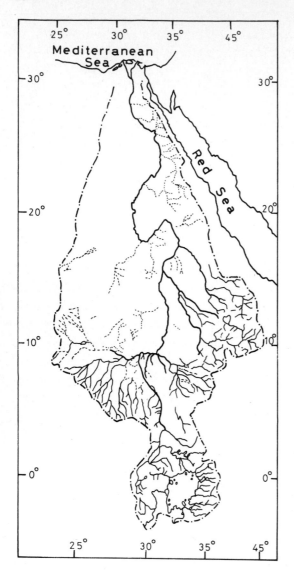

Fig. 2.1. The hydrographic basin of the Nile system (the boundaries of the drainage basin are indicated by a dash-dot line)

TABLE 2.1   Specific discharges of rivers with catchment areas each larger than
1 million square kilometres (Kalinin, G., 1971)

River	Site	Catchment area, A, km^2	Long-term discharge, $\bar{Q}$, m^3/sec	Specific discharge, $\bar{q}$, lit/sec/km^2
Nile	Aswan	2.880.000	2 830	0.98
Missouri	Hermann	1.369.000	2 187	1.69
Mississippi	St. Louis	1.817.000	4 900$^{.}$	2.70
Amur	Khabarovsk	1.620.000	7 300	4.51
Ob	Salekhard	2.450.000	12 460	5.09
Volga	Kuibyshev	1.220.000	7 480	6.13
Lena	Kyusyur	2.430.000	15 900	6.54
Yenisei	Ingarka	2.470.000	18 100	7.33
Congo[+]	River mouth	3.700.000	36 000	9.73
Yangtze	Hankow	1.490.000	23 700	15.91

[+]estimated

The general topographic map, Fig. 2.2., shows that the basin of the Nile is characterized by the existence of two mountainous plateaus rising some thousands of metres above mean sea level. The Lake Plateau in the southern part of the Nile Basin is generally at a level of 1 000 to 2 000 metres. The Ruwenzori mountainous range extending between Lakes Edward and Albert (Mobutu-Sese Seko) at the west of the Lake Plateau has a peak rising more than 5 100 metres whereas the peak of Mt. Elgon north-east of Lake Victoria is at a latitude of 4 300 metres. All the lakes in this plateau, except Lakes Victoria and Kyoga are at levels below 1 000 metres a.s.l. The other mountainous plateau in the basin of the Nile is the Ethiopian or Abyssinian Plateau, which forms the eastern part of the basin. The peaks of this plateau rise to more than 3 500 metres a.m.s.l.

North of the Lake Plateau the basin descends gradually to the Sudan plains where the Nile runs at altitudes lower than 500 metres in its northerly direction. At about 200 kilometres south of the southern frontier of Egypt, the river cuts its channel in a narrow trough bounded from each side by the contour line of 200 metres ground surface level. In general, the width of this trough increases as the river proceeds northwards. Almost two hundred kilometres before discharging into the sea, the river bifurcates and its two branches encompass the Nile Delta. A fairly detailed description of the various parts of the Nile Basin is presented in the following sections.

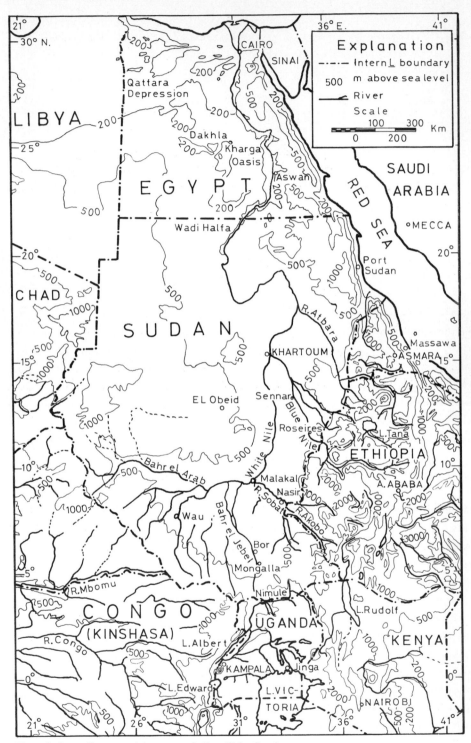

Fig. 2.2.   Topographic map of the Nile Basin

2.2   THE EQUATORIAL LAKES PLATEAU

2.2.1   Lake Victoria

The Great Rift Valley which runs with some interruptions from Zimbabwe to the Jordan Valley, including the Red Sea, is divided into two branches in the southern part of the Nile Basin. The eastern branch of the Rift Valley runs through Kenya and is not included in the Nile Basin. The western branch, however, contains Lakes Tanganyika, Kivu, Edward, George and Albert. It continues north along the Bahr el Jebel. The range of Mufumbiro mountains, with peaks reaching 4 500 metres a.m.s.l., extends between Lakes Edward, Kivu and Tanganyika and separates the latter two lakes from the Nile Basin (see map, Fig. 2.3.: Stamp and Morgan, 1972). The most upstream tributary of the Nile, also the most important feeder of Lake Victoria, is the River Kagera. This tributary has a drainage basin of 63 000 km^2 in an area situated between 1^O and 4OS latitude and between 29° 30´and 31° 40´E longitude as shown on the map, Fig. 2.4. Practically the whole of the Kagera Basin is mountainous country and the greater part of it is situated between the 1 200 and 1 600 metre levels. In the extreme west, the country level is at 2 500 metres a.m.s.l. and rises to about 4 500 metres to form the peaks of the Mufumbiro Range. The Kagera Basin is a complex of streams of varying order which are intercepted and interconnected by lakes and swamps. This complex begins with the River Luvironza in the south-west of the Kagera Basin about 40 km from the eastern shore of Lake Tanganyika. After flowing in a very winding course for about 100 km at levels higher than 1 600 m a.m.s.l., it continues for some 180 km in a relatively straight channel traversing a lower-lying country. There the river name changes to Ruvuvu and it joins the Kagera downstream of the Bugufi Falls. The Ruvuvu draws its supplies from the high land in Burundi. Moreover, this river is joined by a number of seasonal-flowing streams all coming from the east and by the Nyavarongo from the west, a few kilometres upstream of the Bugufi Falls. The River Nyavarongo flows from the high land east of Lake Kivu and receives water from the River Akanyaru in the south and the River Nyaranda in the north-east (see map, Fig. 2.4.). Below the junction of these rivers the main stream traverses an area surrounded by lakes and swamps up to the confluence with the Ruvuvu. Downstream of the Bugufi Falls the Kagera runs to the north then to the north-west in a less winding course for about 170 km, where it is joined by the River Kalangassa from the south-west and by the River Kakitumba from the west. The Kagera then continues its course along the so-called big eastward bend to the village of Bibatura where it enters a relatively low-lying country. About 70 km further to the south-east the Kagera receives some water brought by a stream flowing out of the Muisha swamp in a northerly direction. A few kilometres below the confluence of this stream with the Kagera the latter is joined by the River Ngono which runs west of the coast

20

of Lake Victoria. The Kagera continues its course along the eastward bend for about 20 km before it finally enters Lake Victoria (Hurst, H.E. 1927).

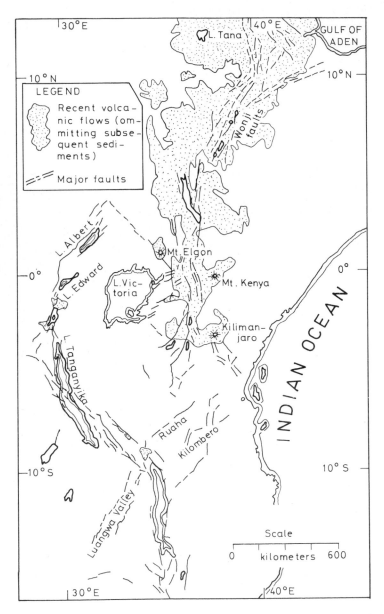

Fig. 2.3.   Rift Valleys and volcanic areas of eastern Africa (Stamp, L.D. and Morgan, W.T., 1972)

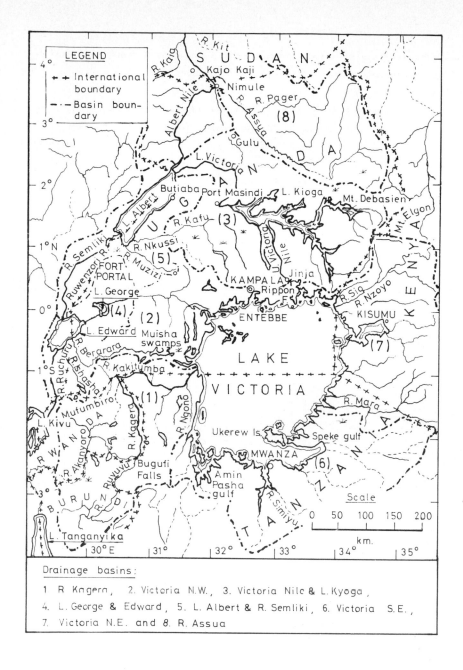

Fig. 2.4. Map showing the drainage basins in the Equatorial Lakes Plateau

The Lakes Plateau is situated between the two branches of the Great Rift Valley. The average elevation of this plateau is about 1 300 metres a.m.s.l. The plateau contains Lakes Victoria, George, Edward and Albert. Lake Victoria is a depression whose surface has an area of about 69 000 km^2, corresponding to a water level of 1 134 metres a.m.s.l. The net water area is about 4% less than the total area, the difference is occupied by the Sese islands in the north-west and the Ukerwe island in the south-east and many other less important islands. The water surface is divided between Kenya, about 5%, Tanzania, about 51%, and Uganda, 44%. The lake has an O-shaped surface which extends from about 3°S to 0° 30´N latitude and from about 31° 40´E to 34° 50´E longitude. The average depth of the lake is 40 m and the maximum depth as far as it has been sounded is 79 m. The bathymetric map of the lake is shown in Fig. 2.5. (Talling, J.F., 1969). The shallow depth of this lake is why there is no stratification in the water temperature. Instead, complete mixing occurs and the water temperature varies between 23.8° C and 26.0° C, depending on the time of the year (Beauchamp, R.A., 1964).

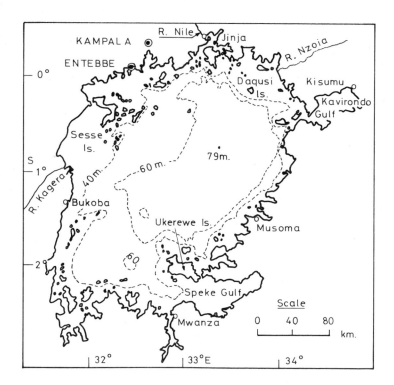

Fig. 2.5.   Bathymetric map of Lake Victoria (Talling, J.F., 1966)

The land portion of the Lake Victoria catchment is about 193 000 km^2. This area is divided between Kenya, 44 000, Tanzania, 84 200, Uganda, 32 100 and Rwanda-Burundi, 33 600 km^2 (Zaghloul, S.S., 1982). An insight into the topography of the catchment surface can be seen from the cross-sections which are presented in Fig. 2.6.

Three sources contribute to the net supply to Lake Victoria. These are: the outflow of the River Kagera, the direct precipitation on the lake surface and the run-off from the land portion of the catchment. The Kagera Basin has already been described. In spite of the fact that it receives more rainfall than the other two sources and the slope of the streams discharging into the Kagera is, generally, not small, the discharge of the Kagera is rather low. The reasons behind it are the swamps and lakes which exist in the basin and the considerable length of streams flowing in it.

The direct precipitation on the Lake Victoria is almost lost by the evaporation from its surface. Although the difference between the average depths of precipitation and evaporation in a year is too small, the corresponding volume is quite big. A yearly excess of the precipitation on the lake surface of 15 mm over the evaporation from the lake surface means a gain to the volume of lake water content of 1 milliard m^3. This therefore constitutes an important source of supply to the lake. The third source of supply to Lake Victoria is formed by the perennial streams in the eastern side of the lake. Of these may be mentioned the Simiyu and the Ruwand which flow into Speke Gulf, the Mara River which enters the lake somewhere about the middle point of the eastern shore, and the Nzoya, Yala and Sio which enter the lake in its north-eastern corner.

2.2.2   The Upper Victoria Nile

The Upper Victoria Nile is the only outlet of Lake Victoria and it connects the latter with Lake Kyoga. The river is about 130 km long and the difference in level between its head and its tail is about 105 m. This difference has been brought by the Owen and the Ripon Ralls. These falls are formed by a reef of rock crossing the stream diagonally. Since 1952 the Nile leaves Lake Victoria through the turbines of the power plant annexed to the Owen Dam which is built at the foot of the Owen Falls. The width of the water surface in the Upper Victoria Nile varies between 300 and 600 m.

The region north of Lake Victoria has been tilted in such a way as to reverse the flow in the upper part of the Kafu River. The valleys at the head of the river therefore became flooded, to form the practically continuous pair of Lakes Kyoga and Kwania (Wayland, E.J., 1934).

24

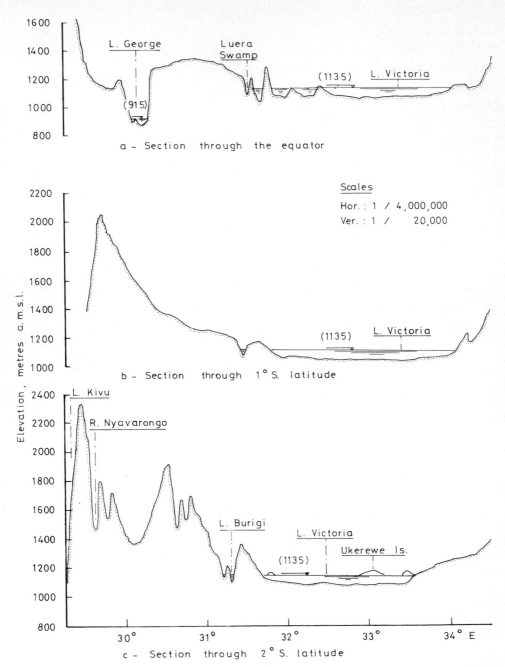

Fig. 2.6.   Sections across Lake Victoria and adjacent country

2.2.3   Lake Kyoga

Lake Kyoga has undergone some changes in its old pattern. The older Kyoga
was a larger lake than the present body of water. It is a shallow depression
consisting of a number of arms, many of which are filled with swamp vegetation.
The lake has a basin 75 000 km^2 in area including 6 270 km^2 which form the
areas occupied by the lake arms and enclosing high land up to an elevation of
1 030 metres a.m.s.l. The depth of the lake at its western end is from 3 to 5 m,
the maximum recorded depth is 7 m. The drainage basin of Kyoga, with the excep-
of the Debasien Mountain and the western half of the Elgon Mountain, is characterized by a series of low hills and flat valleys with impeded drainage (see
cross-sections in Fig. 2.7). In spite of the almost 1 300 mm yearly rainfall,
the excessive evapotranspiration from the swamps covered with cyperus papyrus
and water lilies and the insignificant supply brought by many of the rivers
draining into the lake make Lake Kyoga a source of loss. Heavy rains in the
lake basin are likely to set loose large masses of vegetation which block the
outlet of the Nile from the lake.

2.2.4   The Lower Victoria Nile

The Lower Victoria Nile leaves Kyoga at Port Masindi and runs as a sluggish
swampy river to the north for a distance of about 75 km. Here it bends west-
wards and after a succession of rocks and rapids descends the Marchison Falls
and shortly afterwards enters Lake Albert through a swampy delta. On the west
of the Victoria Nile Basin there is a large system of swamps whose drainage
enters the Nile by the Kafu River (see Fig. 2.4.). The contribution of this
river, except after heavy rains, may be considered negligible. The difference
in the water level between the two ends of the Lower Victoria Nile, i.e. between Lake Kyoga and Albert is almost 410 metres.

2.2.5   Lake Albert (Mobutu-Sese Seko)

As mentioned earlier, the chain formed by the lakes Albert, Edward, and
George, together with their respective drainage basins, forms a part of the
Great Rift Valley. In some places the escarpments of the valley rise directly
from the water surface of Lake Albert, which is at an altitude of about 617
metres a.m.s.l., reaching an elevation of 2 000 metres or higher a short distance inland from the lakes (see the sections in Fig. 2.7.). Lake Albert has a
surface area of 5 300 km^2 corresponding to an elevation of 617 metres a.m.s.l.
The Bathymetric map (Fig. 2.8.) shows that the depth of water reaches 50 metres
at some places in the lake. The run-off from the drainage basin of Lake Albert,
17 000 km^2 in area, plus the direct precipitation on the lake itself, are all
lost by evaporation from the lake surface. The net gain by Albert comes from

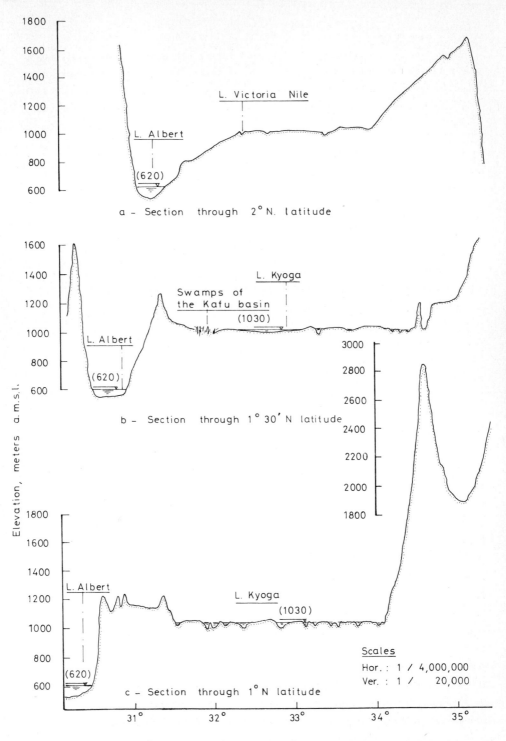

Fig. 2.7.   Sections across Lake Kyoga and adjacent country

the outflow of the River Semliki, which enters the lake from the south-west.
The Semliki connects Lake Edward to Lake Albert, after flowing a distance of
about 250 km down the Rift Valley to the west of the Ruwenzori Mountain. The
drainage basin of the Semliki is 8 000 km^2 in area. It covers the western
slopes of the Ruwenzori Range and is traversed by many streams. The difference
in water level between the two ends of the Semliki is 295 metres. Most of the
drop takes place over the rapids which exist in the upper of the river's
course. In the lower part, the river has a width of 150 m in flood reduced to
50 m at low stage. The average depth of water in these two seasons is 5 m and
3 m respectively.

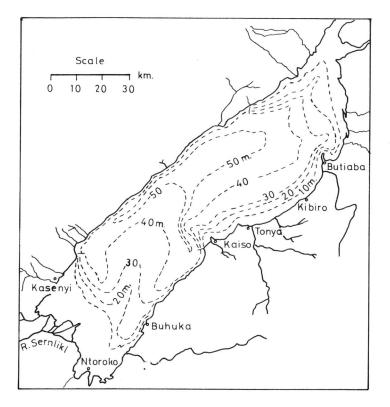

Fig. 2.8.   Bathymetric map of Lake Albert (from Rzóska, J., 1977)

2.2.6   Lakes Edward and George

Lake Edward is connected to Lake George by the Kazinga Channel. Lake George
is situated on the equator and its surface area at an elevation of 915 metres
a.m.s.l., is 300 km^2. This lake has a drainage basin 8 000 km^2 surface area.

It is drained by a number of streams flowing down from the Ruwenzori into the swamps at the northern end of the lake. The principal tributary, the Mbuku, carries a considerable flow during the flood. The outflow from Lake George runs through the Kazinga Channel, which is practically nothing but a carrier. Lake Edward lies in the western Rift Valley and at an altitude of about 915 metres a.m.s.l. has a surface area of about 2 200 km^2. The cross-section (Fig. 2.9.) shows how the escarpment of the Rift Valley rises steeply from the water surface level of about 915 m to more than 2 500 m on the western side of Lake Edward. This is, however, not the case for the other sides of the lake, though at the north-east corner the outlying hills of the Ruwenzori Range come down within a few kilometres of the lake. The lake has a basin 12 000 km^2 in area, which is traversed by a number of streams often fringed by thick forest at their low ends. The principal streams debouching their waters into Lake Edward are: Nyamgasani flowing down the Ruwenzori Range north-east of the lake, the Rivers Berarara and Ishasha flowing from the east in a northerly direction towards the lake, a system of rivers pouring into the main stream, the Ruchuru, running down the Mufumbiro mountains towards the lake in a northerly direction and the River Ruindi reaching Lake Edward at its south-west corner.

From the above description, it is clear that the River Semliki supplies Lake Albert with the run-off from a total catchment of about 30 500 km^2 in area, including the surfaces covered by Lakes George and Edward. The Nile flows out of Lake Albert at the extreme north corner of the lake under the name of the Upper White Nile or Bahr el-Jebel.

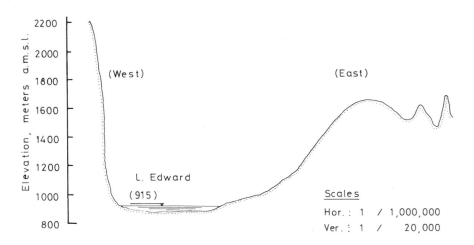

Fig. 2.9.   Section across Lake Edward and adjacent country through 0° 30′S latitude

2.3   THE BAHR EL JEBEL BASIN

The major affluents of the Bahr el Jebel or the Upper White Nile and the
swamps and lakes already discussed in the preceding section, are summarized by
the drawing presented in Fig. 2.10.(Thompson, K., 1975).
From the outlet of Lake Albert down to Nimule, 225 km downstream, the river is
a rather broad, sluggish, stream fringed with swamps and lagoons. It meanders
east and west through a narrow flood plain between hilly country on either side
so that the area of the swamp is well defined. The area occupied by swamps and
open water is estimated at about 380 km^2. The Bahr el Jebel from the outlet of
Lake Albert to Nimule is a placid stream with an average slope of only about
2.2 cm/km or 2.2 x 10^{-5}. It is not a deep river and its width varies from 100
to 300 m. A number of small streams join the Bahr el Jebel from both sides in
this reach (see Fig. 2.4.).

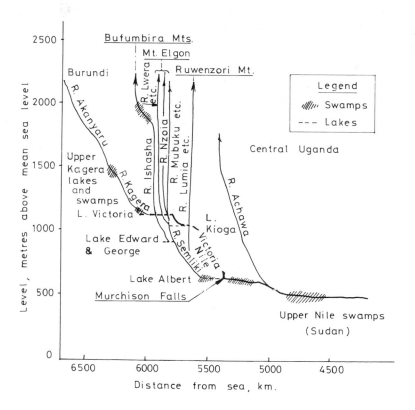

Fig. 2.10.   Altitudinal map of major affluents of the Upper White Nile and the
occurrence of swamps (Thompson, K., 1975)

At Nimule the river course is twisted in a sharp bend and its direction changes suddenly to the west then to the north and north-east up to Mongalla.

In the reach between Nimule and Rejaf, a distance of about 156 km, the river is a narrow and fast stream interrupted by such rocky rapids as the Fola and Bedan. The river flows into a narrow valley cut through hilly country and descends about 150 m. The average slope is nearly 1 m/km or 1×10^{-3}.

In the reach from Nimule to Mongalla the Bahr el Jebel receives a number of small but torrential streams which run full after heavy rains. Of these streams the Assua, the Kaia and the Kit are the largest. They carry some flow even in the dry season. The River Assua joins the Bahr el Jebel at its right bank almost 20 km below Nimule. The former rises in the vicinity of the Moroto Mountain and is joined by some streams descending from the north and from the east not far from the Morongole Mountain. The Assua thus drains a vast area east of the Bahr el Jebel, whereas the River Kaia drains part of the country west of it. The Kaia is joined by the River Kijo a few kilometres before its junction with the Bahr el Jebel. The River Kit rises from the high land between the Tereteinia and Imatong Mountains and after flowing a distance of about 160 km, it joins the Bahr el Jebel at the right bank a few kilometres upstream of Rejaf, where it enters the Sudan Plains. Nevertheless, the slope in the 57 km from Rejaf to Mongalla is rather steep, on average 0.3 m/km, falling off gradually as the Mongalla is approached. From Rejaf to some distance northwards the valley is well-defined, though shallow, and the river winds about in the plain forming the valley floor (Hurst, H.E., 1931).

In the reach from Rejaf to Malakal on the White Nile, the river is not confined to a single channel except at Mongalla where it is in one channel at low stage. Between Rejaf and Bor, a distance of about 180 km, the valley is wide and flat and there is usually a channel on either side along the higher ground while occasional channels cross the swampy valley floor (see Fig. 2.11.).

The distribution of the swamp vegetation on the flood plain in the reach between Juba and Bor was investigated. It was found that this distribution is controlled by seasonal flooding and the form of control can be deduced by relating the distribution to elevation and thus to hydrological conditions (Sutcliffe, J.V., 1974). Furthermore, it seems that the results of that investigation apply to the river reach north of Bor.

North of Bor the Valley widens and becomes more swampy, while the sides are less defined. Extensive swamps spread out on either side of the river and continue down to Lake No. This region is known as the Sudd. North of Kenisa, about 85 km downstream of Bor, the dry land can hardly be seen, except in a few places. The river flows northwards between walls of papyrus and tall grasses reaching 4 or 5 m in height. These plants have their roots in water and the

31

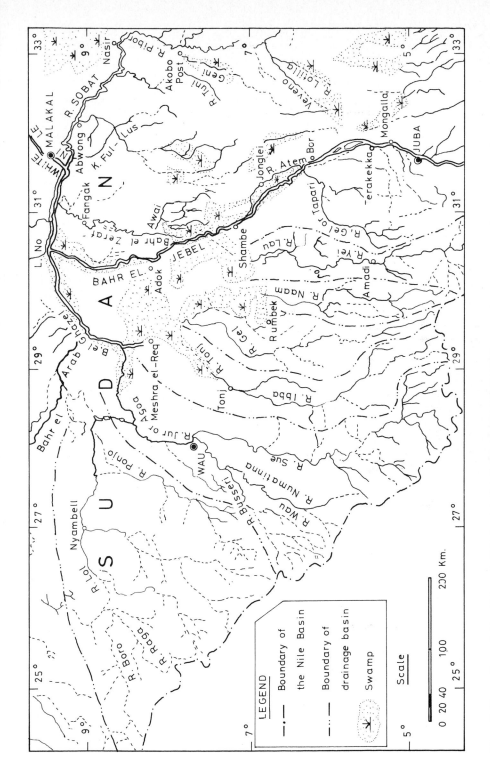

Fig. 2.11. Map of the Bahr el Jebel and the Bahr el Ghazal Basins

river bank is partly formed of masses of roots of former vegetation. Furthermore, there are many patches of open water alongside the river north of Bor,
many of which are connected directly with the river or with the side channels.
Of the latter the Awai and Atem Rivers and Gage's and Peake's Channels may be
mentioned (see map, Fig. 2.11.). North of Ghaba Shambe, some 140 km from Bor,
the swamps are wide and the plain is full of vegetation and lagoons. A plan
and cross-section of a typical lagoon in this area is shown in Fig. 1.12.
(Hurst, H.E. 1931). Because of the high rate of loss of water from the Sudd
region and the vast area from which this loss takes place, the total loss in
an average year amounts to approximately one half of the total flow at
Mongalla. In an attempt to transport the water in this region with less loss,
the Bahr el Jebel was joined to the Bahr el Zaraf by two cuts at distances of
106 and 112 km from Shambe. Unfortunately these two cuts and the channel which
runs between them are so heavily blocked with vegetation that their efficiency
in reducing the transmission losses in the swamps is questionable. Between the
two cuts and Lake No in the north there are occasional isolated spots of high
ground compared to the surrounding swamps. At Lake No the Bahr el Jebel is
joined by the Bahr el Ghazal and the combined stream turns abruptly to the
east, bearing the name "The White Nile". Here the swamps end and the White
Nile flows northwards in a fairly well-defined valley of moderate width.

The Bahr el Zaraf starts somewhere about latitude 7° 20′N in the swamps
east of the mouth of the River Awai. It is probable that there are some
channels connecting the Jebel, the Awai, the Atem and the Zaraf through which
the latter derives its supply of water. The Bahr el Zaraf has a winding course
of about 280 km in length to its mouth on the White Nile some 80 km from Lake
No. Along this course there is not so much papyrus along the Jebel and the
principal swamp plant is um soof, with reeds and bulrushes.

Higher ground exists not very far to the east of the Upper Zaraf so that
swamps reach their definite limit in the east. Some of this high ground, however, becomes swampy after heavy rain. In the neighbourhood of the Jebel-Zaraf
cuts and for a long way north, the Zaraf flows through swamp, winding about
forming lagoons in its bends like the Bahr el Jebel.

The edges of the Bahr el Zaraf are swampy in places as far north as kilometre 100 (measured from the mouth) and there is always a fringe of um soof.
The banks gradually become high as one goes northwards until they form definite
boundaries limiting the Zaraf to a narrow channel.

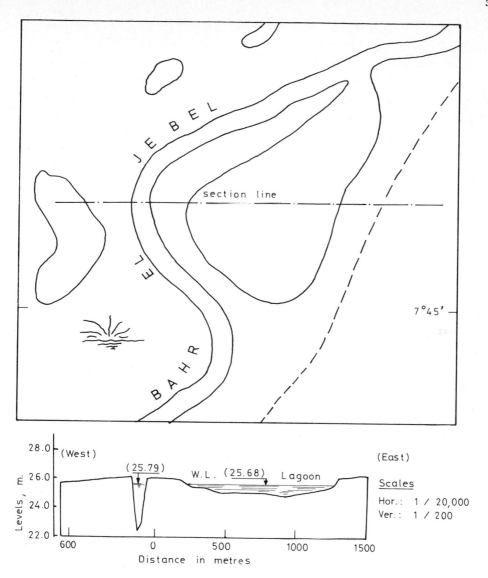

Fig. 2.12.   Plan and cross-section of a lagoon in the Bahr el Jebel Basin
(reproduced from the Nile Basin Vol. I: Hurst, H.E. and Philips, P., 1931)

2.4   THE BAHR EL GHAZAL BASIN

The Bahr el Ghazal is the name given to the waterway from Meshra el Req to Lake No (see map, Fig. 2.11.). Though the length of this stream does not exceed 160 km, the size of its basin is approximately 526 000 km^2 which is by far the largest of any of the sub-basins of the tributaries of the Nile River. The annual rainfall on the basin is estimated at 500 x 10^9 m^3. Of this amount only 0.6 x 10^9 m^3/year reaches the basin outlet at Lake No.

All along the Bahr el Ghazal and to the south and east of it are large areas of swamp which are fed by a number of streams. The country where the upper courses of these streams flow is entirely covered by a sort of savannah forest. Nevertheless, in the ravines formed by the streams, there is a thick forest similar to the tropical rain forest of parts of the Lake Plateau and the Congo Basin. On the lower courses of all the tributaries of the Bahr el Ghazal and along the Ghazal itself are large areas of swamps. Unfortunately, most of the flow carried by the tributaries is lost in the swamps. Near the Bahr el Arab, the forest is of the thorny savannah type and this gradually changes to shrub steppe as one goes northwards.

On either side of the Uganda-Sudan boundary that coincides with the divide between the Nile and the Congo Basins, numerous streams arise. Most of them descend to a large swampy plain in which they wind and finally spread and cease to exist as streams with definite courses, except for the Jur, which preserves its channel and joins the Bahr el Ghazal.

The tributaries of the Bahr el Ghazal from east to west are: the Gel or Tapari, the Yei or Lau, the Naam, the Meridi or Gel, the Ibba or Tonj, the Jur, the Lol and the Bahr el Arab. Some of the data belonging to these rivers are included in Table 2.2. A map illustrating the drainage basin of the Bahr el Arab and surroundings is shown in Fig. 2.13.

The data presented in Table 2.2 may help to show that the Jur is the most important tributary of the Ghazal. The former has two main tributaries, namely, the Sueh and the Busseri. Both are relatively large streams.

Lake Ambadi is about 10 km long by 1 km wide and mostly less than 3 m deep. From this lake down the Ghazal to the mouth of the Bahr el Arab, the country remains swampy and the river does not have a defined bank. The vegetation bordering the river is um soof with little papyrus. The lower Ghazal is fringed by papyrus, though its growth is stunted and less luxuriant than on the Jebel. The many temporary streams which join the Ghazal on both sides are usually blocked and therefore cannot contribute much water. As Lake No is approached, the dry land nears the river on the north and the river loses its defined banks. Lake No is nothing but a large shallow lagoon. Here the sluggish Bahr el Ghazal

TABLE 2.2   Some data of the tributaries of the Bahr el Ghazal (Hurst, H.E.
             and Philips, P., 1938)

Item	Tributary of the Bahr el Ghazal							
	River[+] Tapari	River[+] Yei	River Naam	River Meridi	River Tonj	River Jur	River Lol	Bahr el Arab
Basin area, km^2	12 800	25 000	16 000	22 000	27 000	64 000	82 000	209 000[++]
Mean rain-fall, mm/yr	1 050	1 250	1 200	1 200	1 220	1 200	1 100	700
Trough width, m	50	100	90	–	70	130	270	35
Max. depth, m	4	5	4	–	3	6	3	5
Max. dis-charge, m^3/s		400	160	–	110	600	500	
Place of observation	a	near Amadi	near Rumbek	–	Tonj	Wau	b	c

[+]usually considered as a tributary of the Bahr el Jebel and not of the Bahr
el Ghazal

[++]Excluding the swamps

[a]At the road crossing between Amadi and Terrakekka

[b]Intersection with road from Nyamlell to Bahr el Arab

[c]Safaha due north of Nyamlell

joins the Bahr el Jebel after having a tremendous volume of water wasted in the
vast swamps. A schematic cross-section through the Upper Nile swamps from Bahr
el Ghazal to the Bahr el Jebel can be seen in Fig. 2.14. (Jonglei Report,
1954).

2.5   THE SOBAT BASIN

The Sobat Basin, approximately 225 000 km^2 in area, includes most of the
plain east of the Bahr el Jebel and Bahr el Zaraf and parts of the Abyssinian
Mountains and the Lakes Plateau (see Fig. 2.15.). In view of the large size of
the basin area and the diversity in its topography, the annual rainfall varies
from about 650 mm near the mouth of the Sobat, to about 2 000 mm in the most
elevated parts of the basin eastwards. The Sobat is formed by the junction of
its two main tributaries, the Baro and the Pibor. The Baro is claimed to be
the principal feeder of the Sobat, though its basin has a surface area of
41 400 km^2, whereas the surface area of the Pibor is 10 900 km^2.

36

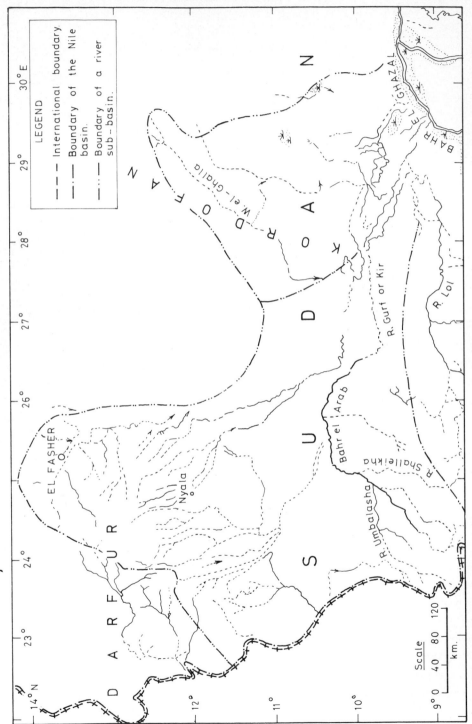

Fig. 2.13. Map showing the drainage basin of the Bahr el Arab and surroundings

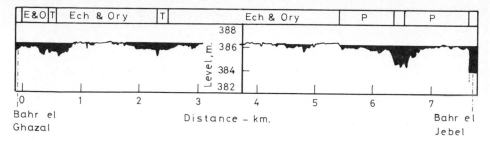

Fig. 2.14.   Schematic cross-section through Upper Nile swamps from Bahr el
Ghazal to the Bahr el Jebel (Jonglei Report, 1954)

The Baro is formed by a number of streams which in some places flow through
deep gorges in their descent from the plateau. A good deal of the mountainous
part of the basin lies above 1 500 m with portions even higher than 2 000
metres a.m.s.l. Soon after leaving the mountains, the Baro reaches Gambeila,
which is on the plain, almost 520 metres a.m.s.l. Here the Baro does not
receive tributaries, but rather shallow swampy khors[*], the principals of which
are Khor Jokau coming from the north, the Atura branch, and the Mokwai.

Down of Gambeila up to the junction of the Baro with the Pibor, is the Baro,
with an average width of 100 metres, increasing to 250 metres in some places
and an average depth of more than 6 metres during the flood. Unfortunately,
some 40 km upstream of the junction some water leaves the Baro through Khor
Machar to feed a large swampy area north and east of El-Nasir. This swamp is
fed by other streams flowing from the Abyssinian Plateau (see map, Fig. 2.15.).
The part flowing to Khor Machar constitutes, however, a permanent source of
loss from the Baro (Hurst et al, 1966).

The River Pibor runs in a northerly direction. It draws the greater part of
its supply from Abyssinia and the rest comes from the northern slopes of the
lake plateau and from the Sudan plains. The basin area of the Pibor has already
been mentioned as being larger than that of the Baro and the annual rainfall is
probably more, since the mountainous portion in Abyssinia is larger than the
corresponding portion of the Baro Basin. In spite of these two factors, the
flow of the Pibor is quite inferior to that of the Baro. This is because the

---

[*]A khor is a temporary stream which runs full during and after rainfall. It
could be of a torrential nature

38

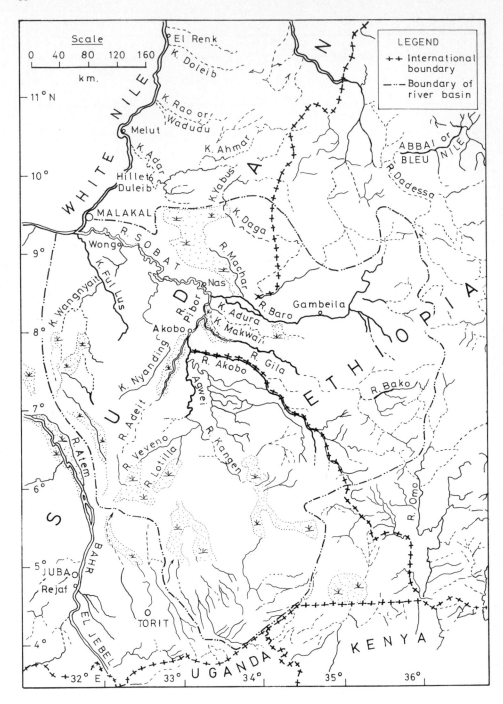

Fig. 2.15. Map showing the approximate boundaries of the drainage basin of the River Sobat

slope of the Pibor is very flat compared to the Baro and has consequently more chance of forming large swamps and evaporating the water thereof.

The Pibor is formed by the junction of the Veveno, Lotilla and the Kangen (see map, Fig. 2.15.). None of these streams carries much water and all are reduced to pools in the dry season. Cross-sections of these streams can be found in Vol. VIII of the Nile Basin (Hurst et al, 1950). Some of these cross-sections are reproduced in Fig. 2.16. The Pibor, like many of its tributaries, becomes narrower and deeper in section as it goes down. Near its mouth, the clear width of the river channel drops from 150 to 60 m, whereas the depth increases from about 4 m in the upper reaches to about 6.5 m.

Downstream of the junction of the Kangen and Lotilla the Pibor runs northwards in a winding course and receives the Agwei, Akobo, Gila and Khor Makwai on the east bank. On the west bank, it receives the Khor Adeit and several smaller khors.

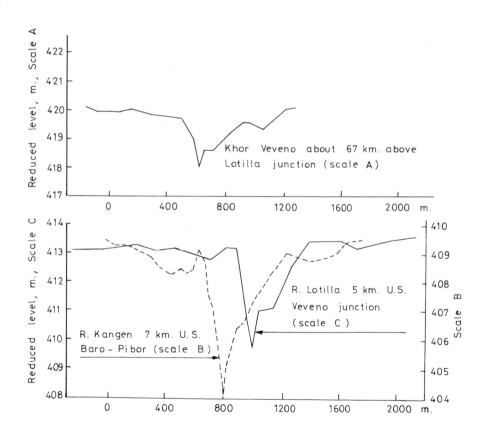

Fig. 2.16. Cross-sections of some of the tributaries of the River Pibor (taken from the Nile Basin Vol. VIII: Hurst, H.E., 1950)

From the Baro-Pibor junction to the mouth of the Sobat the country is a flat plain intersected by swampy depressions or khors. The Sobat has a winding course and its surface width near the mouth varies from 100 m or less in the low-flow period to more than 150 m in flood time. For these two seasons, the depth of water is about 3.5 and 6.5 m respectively. The principal khors joining the Sobat in this reach are the Nyanding, Wangnyait and Fullus. These khors, in spite of the conveyance loss in the Sobat, help to increase the flow in the Sobat in a normal year from 12.4×10^9 m^3 at Nasir to 13.6×10^9 m^3 at Hillet Doleib near the mouth.

2.6   THE WHITE NILE

The stretch of the White Nile (An Nil El Abiad) from Lake No down to its junction with the Blue Nile is known as the White Nile. The remarkable feature about this river is its extremely flat slope. In the upper 120 km, i.e. from Lake No to the mouth of the Sobat, there are several swamps, khors and lagoons. In the remaining 800 km, i.e. from Malakal to just upstream of Khartoum, the channel of the White Nile is almost free of swamps. As the average slope of the river in this stretch is about 1.4×10^{-5}, it is, for a large part, sluggish.

The drainage basin of the White Nile extends from the foothills of the lake plateau in the south to the junction of the White and Blue Niles up north and from the foothills of the Abyssinian Plateau in the east to the Nile-Congo divide in the south-west and Nuba Mountains in the west (see map, Fig. 2.17.).

In the distance from Lake No to the mouth of the Sobat, the river flows in an undefined vegetated course in a plain with a width of 1 km at Malakal. From the mouth of the Sobat to north of Kosti the width of the depression between the river banks increases from 3 to 4 km, whereas the width of the river channel itself is between 300 and 400 m. North of Malakal up to its junction with the Blue Nile, flows the White Nile in a well-defined channel or channels.

From Lake No to just south of Melut the main feeders of the White Nile are the Khor Lolle, Bahr el Zaraf, the Sobat, Khor Wol and Khor Adar. Many swampy khors join the river reach between Melut and El-Renk. Of these khors, the Rao or Wadudu and the Doleib are the largest. North of Jebelein, swampy khors are fewer and the country gradually becomes more arid.

2.7   THE ETHIOPIAN OR ABYSSINIAN PLATEAU

2.7.1   The Blue Nile

The Blue Nile (An Nil el Azraq or Abbai) and its tributaries all rise on the Ethiopian Plateau at an elevation of 2 000 to 3 000 metres a.m.s.l. Volume VIII of the Nile Basin (Hurst, H.E., 1950) mentions that the source of the Blue Nile is a small spring at a height of 2 900 m and at about 100 km south of Lake Tana.

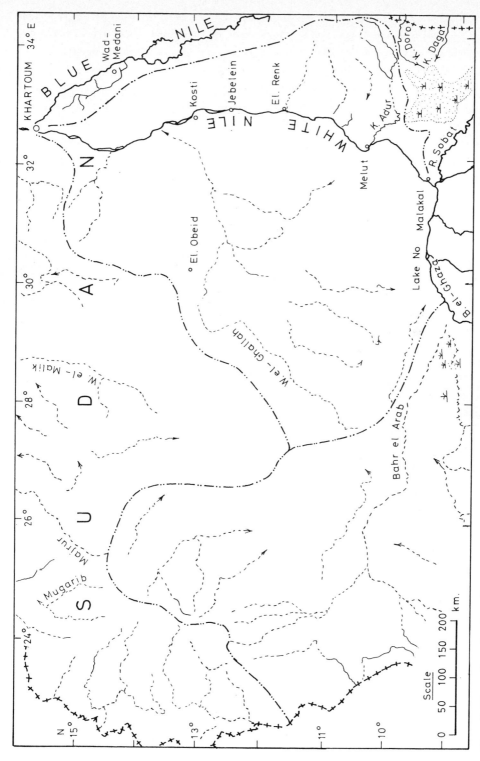

Fig. 2.17. Map showing the part of the drainage basin of the White Nile in the reach from Lake No to Khartoum

From this spring the Little Abbai flows down to Lake Tana (1 829 metres a.m.s.l.). There are 60 affluents of Lake Tana, of which the Little Abbai is usually regarded as the most important.

The Ethiopian Plateau country cannot be described as flat. Most of it is hilly with grassy downs, swamp valleys and scattered trees. The high country is cut up by deep ravines or canyons in which the rivers flow, the greatest of which is that of the Blue Nile. In some places this river flows in a channel that is about 1 200 metres below the level of the country on either side. The drop of the plateau to the Sudan Plain is, in most places, steep. However, there are many outlying hills, some of which are as high as the plateau itself. The plain is largely covered with thin Savannah forest, but north of latitude 13°N there is a good deal of open grassland.

The Blue Nile Basin, including Lake Tana and its basin, has an area of 324 530 km^2. This area covers most of Ethiopia west of longitudinal 40°E and between latitudes 9° and 12°N (see Fig. 2.18.). This area can be divided into a number of sub-basins as follows (Hurst et al, 1959):

Lake Tana Basin, including the lake	17 500 km^2
Khor Didessa Basin	25 800
Khor Balas Basin	15 200
Khor Dabus or Yabus Basin	14 000
Khor Tumat Basin	4 370
Khor Bashilo Basin	13 900
Khor Jamma Basin	19 800
Khor Mugor Basin	7 270
Khor Guder Basin	6 390
River Rahad Basin	35 600
River Dinder Basin	34 700
Blue Nile Basin, excluding above areas	130 000

Lake Tana is a fresh-water body situated in north-central Ethiopia. Its maximum length is 78 km, width 67 km and depth 14 m. A bathymetric map of the lake is shown in Fig. 2.19. Relatively important feeders to Lake Tana other than the Little Abbai are the Rivers Reb, Gumara, Magetch, Gelda and Unfraz. From this lake the Greater Abbai flows in a large loop first south-east, then south and then west. About 35 km from the exit of Lake Tana, the river drops approximately fifty metres into the Tissisat Falls. Fifty kilometres further downstream, it begins to cut a deep gorge through the plateau which, as already mentioned is, in some places, 1 200 m below the country level on either side.

Numerous rock-outcrops occur in the river bed, the last of which are a few kilometres south of Roseires, some 1 000 km from its source beyond Tana, and known as the Damazin rapids. The Blue Nile emerges from the plateau close to

43

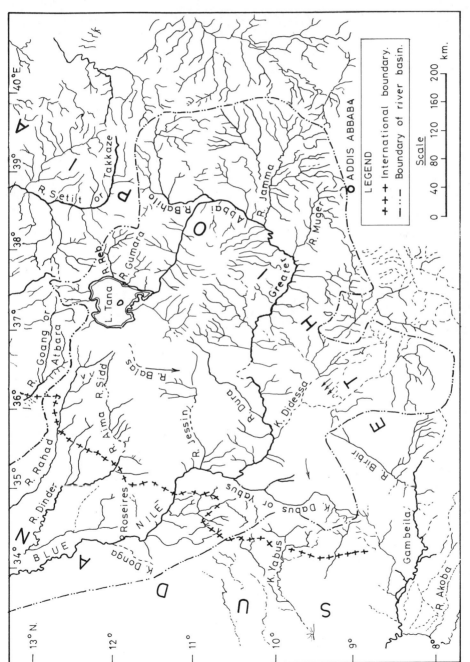

Fig. 2.18. Map showing the drainage basin of the Blue Nile up to latitude 13°N

Fig. 2.19.   Bathymetric map of Lake Tana (Moradini, G., 1940)

the western border of Ethiopia, where it turns north-west and enters the Sudan
at an altitude of 490 metres a.m.s.l. (see Fig. 2.20.). Just before crossing the
frontier, the river enters the clay plain, through which it flows to Khartoum.
At this point the Blue Nile joins the White Nile to form the main stem of the
Nile River. The area bounded by these two rivers is known as the Gezira Plain
(formerly called Meroe Island, see Fig. 1.1.). The physiography of this area can
be seen in Fig. 2.21. (Berry L. and Whiteman, A.J., 1968).

The Dinder and the Rahad join the Blue Nile in the reach between Sennar and
Wad-Medani. The head streams both of the Dinder and the Rahad rise on the
Ethiopian Plateau about 30 km west of Lake Tana. These two rivers are seasonal
streams, and in the dry season they are reduced to pools separated by stretches
of dry sandy bed. They are nearly equally long, each about 750 or 800 km. The
effective catchment areas are about 16 000 and 8 200 km^2 for the Dinder and the
Rahad respectively. The two rivers are very similar to each other but the Rahad

has a flow volume in a normal year equal to 1.1 x 10^9 m^3, which is about one third of the annual volume of flow of the Dinder.

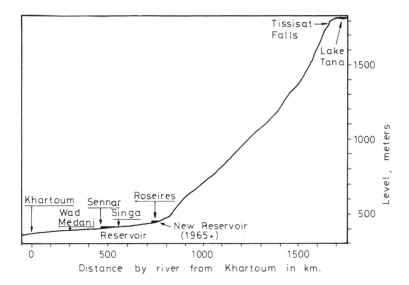

Fig. 2.20.   Longitudinal profile of the Blue Nile

2.7.2   The Atbara

The River Atbara, which is the last tributary of the Nile, enters the Main Nile at about 320 km downstream of Khartoum. It is 880 km long and the greater part of its catchment is situated in Ethiopia. The highest points in the catchment reach more than 3 500 metres a.m.s.l., whereas the eastern watershed of the Atbara is, for the most part, more than 2 500 m high.

The Atbara does not spring from a lake and relies totally on many small tributaries, of which the Takazze or the Setit is the principal one. The latter has a catchment area of 68 800 km^2 out of 112 400 km^2 (the total catchment area of the Atbara). Above the Setit junction the Atbara receives a number of tributaries of which the Bahr el Salam is the principal.

The Atbara is more strongly seasonal in its flow, compared to the Blue Nile. Moreover, the big drop in elevation between the head and the junction of the Salam River is responsible for the excessive sediment load of the Atbara in proportion to its flow volume. In this reach the Atbara has a slope of about 5 x 10^{-3} (see Fig. 2.21.). The longitudinal section of the Blue Nile shows that this river has a slope of only 2 x 10^{-3} in the upper 300 km.

An important place in the basin of the Atbara is Khashm el Girba. There the river was provided with a gauging station at the beginning of the century.

North-east of Khashm el Girba lies Kassala on the River Gash (see Fig. 2.22.).
This is, like the Atbara, a torrential stream, but much smaller and flows for a
shorter time. Below Kassala the Gash spreads out into a Delta and its waters
ultimately disappear into the soil. It is quite possible that the Gash was a
tributary of the Atbara in the past.

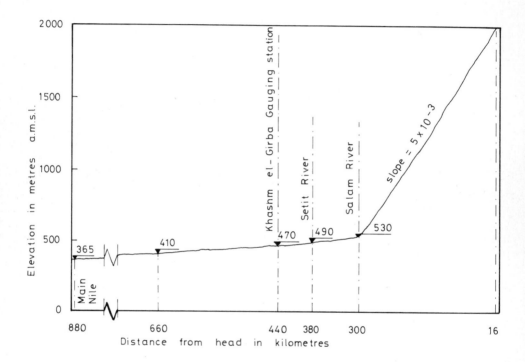

Fig. 2.21.   Longitudinal profile of the River Atbara

2.8   THE MAIN NILE

A description of the physiography and the topography of the basin of the
Main Nile from Khartoum up to the Mediterranean Sea is given in Vol. IX of the
Nile Basin (Hurst et al, 1959). A summary of the marked features of the Nile
Basin in this stretch is as follows:

2.8.1   From Khartoum to Aswan

At Khartoum the Blue Nile joins the White Nile and the combined waters flow
for 1 885 km to Aswan through a region of Nubian sandstone overlying an old
eroded land surface of crystalline rocks which has been laid bare at places in
the course of the still incomplete degradation of the river bed. These

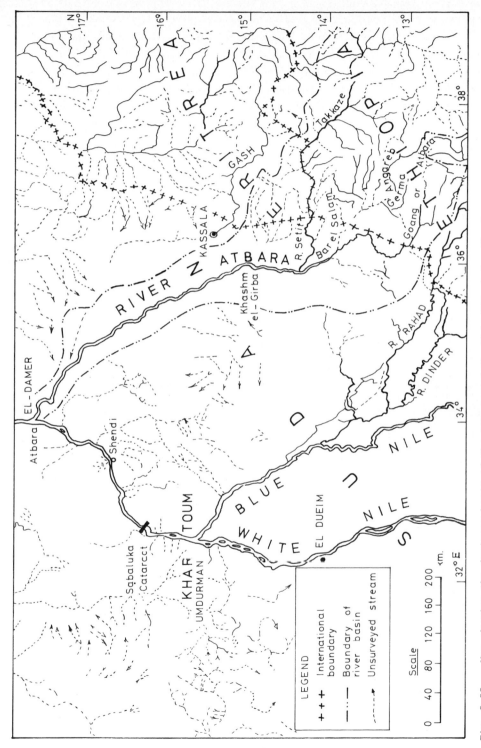

Fig. 2.22. Map showing the drainage basin of the Atbara and the Main Nile below Khartoum

crystalline rocks offer a much greater resistance to the river's action than does the softer Nubian sandstone. Upstream, therefore, in the place where the former rocks are exposed, degradation ceases for a while, while the river cuts its way through the rocky obstacle. The river's course thus consists of a series of placid reaches of mild slope separated by rocky rapids, called the Cataracts, where the slope is greater and the flow more turbulent. The rapids themselves are caused by bars of hard rock crossing the course of the river. These rocks are more slowly eroded than the neighbouring rocks and so form sills or steps. The approximate cross-sectional area of the Main Nile at the locations of the Cataracts and in between them is included in Table 2.3. Fig. 2.23. also shows the longitudinal profile of the water surface together with the water surface width at the different stretches of the river.

TABLE 2.3   Approximate sectional areas, in square metres of some stretches of the Main Nile between Atbara and Halfa

Location	Sectional area, m^2, for months of the year											
	Jan.	Feb.	Mar.	Apr.	May	Jun.	Jul.	Aug.	Sep.	Oct.	Nov.	Dec.
Atbara to 5th Cataract	2600	1900	1500	1250	1160	1800	2800	4500	6750	6100	4600	3300
5th Cataract	1800	1400	1100	900	800	1000	2000	4000	4850	4500	3500	2500
5th Cataract to 4th Cataract	2500	1800	1300	1100	1000	1300	2500	5700	6950	6400	5000	3600
4th Cataract	1800	1400	1100	900	800	1000	2000	4000	4850	4500	3500	2500
4th Cataract to 3rd Cataract	2300	1700	1300	1100	1000	1300	2400	5400	6550	6100	4700	3300
3rd Cataract	2600	2000	1500	1100	860	900	1900	4100	5450	5000	3900	3300
3rd Cataract to 2nd Cataract	2600	2000	1500	1100	860	900	1900	4100	5450	5000	3900	3300
2nd Cataract	1800	1400	1100	900	860	1000	2000	4000	4850	4500	3500	2500

For the first 80 km north of Khartoum, the river flows northwards, thence to Berber (km 387 from Khartoum). The course of the river runs successively east, north-east, and north. North of Berber the river turns north-west to Abu-Hamed (km 578), where it abruptly turns south-west to Korti (km 872). From Korti the course swings around a bend back to the north at Kerma (km 1 145), where it proceeds north and north-east past Wadi-Halfa (km 1 435) to El-Derr (km 1 671). From El-Derr the river, after a short right-hand loop to the south, flows in a northerly direction to Aswan (see Figs. 2.22., 2.24., and 2.25.).

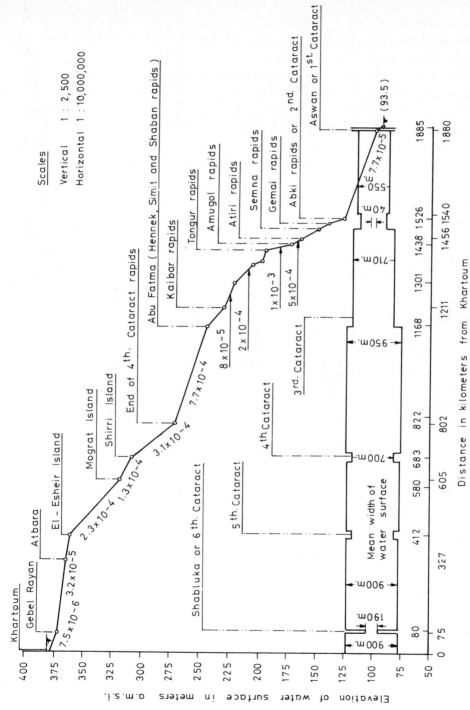

Fig. 2.23.   Longtidunal profile of the water surface in the Main Nile from Khartoum to Aswan

50

As degradation is still in progress throughout the reach described, the
river deposits no flood plains. Cultivation, therefore, is confined to those
few stretches where natural conditions permit irrigation.

The first storage work in the Nile Valley, the old Aswan Dam, was built in
1902 at the foot of the Aswan Cataract.

2.8.2   From Aswan to the Mediterranean Sea

The old Aswan Dam has been heightened twice; once in 1912 and the second
time in 1934. This dam, together with the other storage works on the Blue and
White Niles have changed the Nile from Aswan to the sea into a partially regulated river instead of a naturally flowing one. Full regulation has almost been
achieved as a result of the formation of the Nasser Lake upstream of the high
dam at Aswan in 1965. This huge artificial impoundment of the Nile water
extends from Aswan to a little south of the Dal Cataract between the 23° 58'N
and 20° 27'N latitude and 30° 35'E and 33° 15'E longitude (see Fig. 2.25.).
Table 2.4 shows the development of the reservoir level and capacity in the
period from 1964-65 till 1975-76.

TABLE 2.4   The gradual filling of the reservoir formed by the High Dam at
Aswan (Abu el-Ataa, A., 1978)

Year	Highest level, m a.m.s.l.	Date of occurrence	Maximum storage 10^9 m^3	Lowest level, m a.m.s.l.	Date of occurrence	Minimum storage 10^9 m^3
1964/65	127.60	18.01.1965	9.620	111.89	1.08.1964	-
1965/66	132.86	4.11.1966	13.590	119.02	29.07.1966	4.650
1966/67	142.48	4.02.1967	24.320	113.48	26.07.1967	14.130
1967/68	151.21	10.10.1967	39.640	145.27	21.07.1968	28.516
1968/69	156.55	21.11.1968	39.005	150.85	22.07.1969	39.005
1969/70	161.30	25.10.1969	62.400	153.81	3.08.1970	42.280
1970/71	164.88	26.11.1970	77.468	159.65	23.08.1971	60.450
1971/72	167.64	4.12.1971	87.757	162.49	28.07.1972	68.774
1972/73	167.52	1.01.1973	87.320	158.20	8.07.1973	56.960
1973/74	166.32	9.11.1973	82.716	161.00	16.07.1974	64.500
1974/75	170.63	5.11.1974	100.309	165.60	30.07.1975	80.060
1975/76	175.71	10.12.1975	124.990	172.42	26.07.1976	108.370

This table shows clearly that the water level upstream of Aswan has risen in
the last ten years by a minimum of 40 metres compared to the flood level in the
pre-High Dam period (121.0 metres a.s.l.). This considerable rise in the water

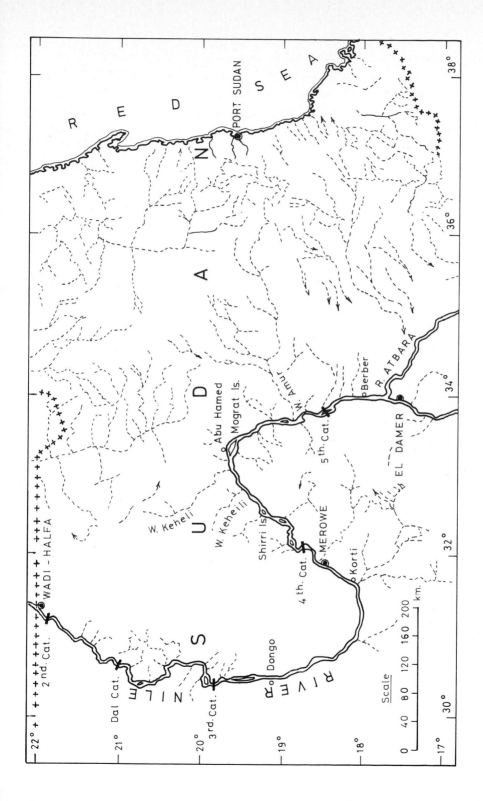

Fig. 2.24. Map showing the Main Nile in the reach from the Atbara junction to Wadi-Halfa

level has resulted in the inundation of some parts of Nubia.

In its natural condition, the length of the river from Aswan to the Delta Barrages was 968 km in the low-flow season and 923 km in the flood season and the slope was 7.7×10^{-5} and 8.5×10^{-5} respectively. The cross-sectional area during the flood was about 7 500 m^2, the mean width about 900 m and the mean velocity between 1.0 m/s and 2.0 m/s.

From Cairo to a little south of Luxor the cultivated land is usually several kilometres wide but towards Aswan it narrows to about one kilometre, and in places the desert hills are close to the river. These conditions persist for a long way south of Wadi-Halfa.

Perennial irrigation in Egypt has become possible only after the construction of a number of barrages on the Nile and its branches. A barrage - sometimes called an open-type weir - is different from a dam as its function is not to form a storage reservoir, but merely to raise the level of the water upstream of it so as to divert some of it into the canals whose entrances are above the barrage. The old Delta barrages were completed in 1861 and the new ones in 1939. Other barrages were built at Esna, Nag-Hammadi, Assiut and Zifta.

The surface of the cultivated area in both the Nile Valley and the Nile Delta amounts to only 3% of the total surface area of Egypt. The eastern and western deserts occupy 23 and 74% of the surface area of Egypt, respectively. The eastern desert is rugged and mountainous and is much cut up by deep valleys (Wadis), down which occasional heavy rains cause torrents to flow. There are no Wadis in the western desert. This desert is lower and more undulating, but is nevertheless sharply divided from the Nile Valley, because cultivation ceases as soon as the ground begins to rise above the level which can be flooded by the Nile water.

There are a number of oases in the western desert. These are simply depressions where the ground level is near the water level, which is easily reached by wells. A cross-section extending from the coast of the Red Sea in the east to the western boundary of Egypt passing through the El-Khargah Oasis is shown in Fig. 2.26.

The Fayum is a depression situated about 70 kilometres south of Cairo and separated from the Nile Valley by a narrow strip of desert. This depression is quite different both from the oases in the desert and the cultivated area in the Nile Valley or the Delta. On one hand the Fayum gets its water via a canal from the Nile, whereas the oases are supplied by groundwater. On the other hand the land in El-Fayum has a considerable slope compared to the land in the Valley or in the Delta.

The bottom of the El-Fayum depression is filled by Lake Qarun (in ancient times called Lake Moeris) and most of the remainder is cultivated. The lake has

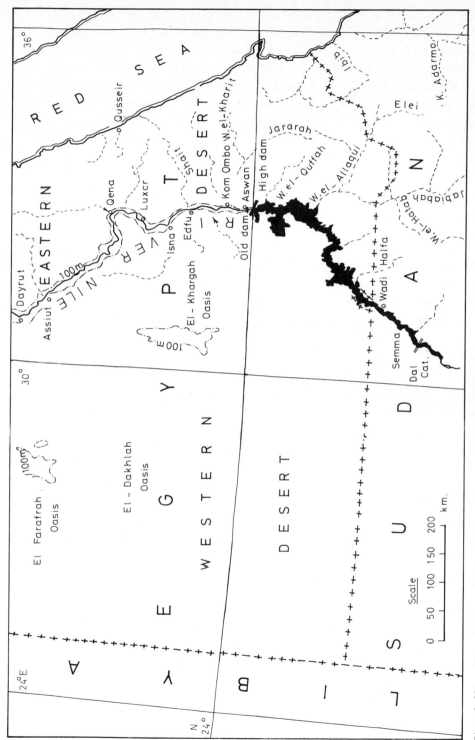

Fig. 2.25.   Map showing the Nile River from south of Wadi-Halfa to a little north of Assiut

no outlet and receives the drainage water from the cultivated land. Its level
is kept fairly constant by evaporation balancing the inflow, so that the lake
water steadily becomes more saline. About 50 kilometres south-west of El-Fayum
town and 80 kilometres west of Beni-Suef lies another depression, known as
Wadi el Rayyan. This Wadi had often been considered as a possible solution for
side storage of the flood water. The Wadi is a depression whose maximum depth
is 50 metres b.s.l. as compared with the 45 metres of the present Lake Qarum
level.

The largest, and at present the deepest, depression in the Egyptian part of
the western desert, is the Qattarah depression. This depression has a surface
area corresponding to the mean sea level of about 4 million feddans or about
50% larger than the area contained between the two branches of the Nile (see
Fig. 2.27.). The deepest point in the depression is at a level of 159 metres
b.m.s.l. The Qattarah depression has, for the last thirty years or more, been
considered as a possible scheme for generating electric power. This can be
achieved by connecting the depression with the Mediterranean Sea by an intake
at, or about, El-Alamein. The yearly inflow to the depression will be balanced
by the yearly evaporation, so the surface water level in the depression remains
constant.

The Nile north of Cairo bifurcates into the Rosetta and Damietta branches.
Very close to the mouth of each branch is the site where an earth bank used to
be constructed each year. This was completed when the flow into the river was
shut off at the Delta Barrage and all the water diverted to the canals because
it was needed for cultivation. The banks prevented salt water from penetrating
inland, and also enabled seepage and drainage back into the river trough from
the Delta Barrage northwards to be collected and used for irrigation of small
areas further north along the river. The earth bank on the Rosetta branch was
replaced in 1951 by the Edfina Barrage, whereas the planned barrage at
Faraskour for the Damietta branch has never been executed. After the construction of the High Dam at Aswan in 1964, the flow of the Nile from Aswan to the
Mediterranean Sea has been under different regulation.

In the pre-High Dam time, it was claimed that the Damietta branch was
gradually silting up and therefore decreasing in size, whereas the Rosetta
branch was scouring in high floods. The mean width of the Rosetta branch is
500 m and the mean sectional area during the flood was 4 000 m^2. The maximum,
minimum, and mean discharges were about 6 500, 2 600 and 4 000 m^3/sec., respectively. The mean width of the Damietta branch is 270 m and its mean
sectional area during the flood was 2 700 m^2. The maximum, minimum, and mean
discharges were about 4 600, 1 300, and 2 300 m^3/sec., respectively.

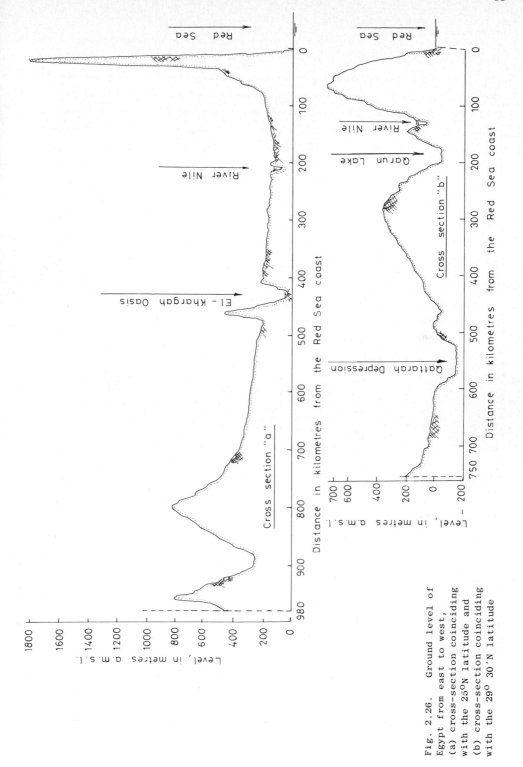

Fig. 2.26. Ground level of
Egypt from east to west,
(a) cross-section coinciding
with the 25°N latitude and
(b) cross-section coinciding
with the 29° 30´N latitude

The two branches of the Nile have their mouths situated at the coast of the Mediterranean Sea. The coastline from Alexandria to Port Said is an undulating line that bears the features of an advancing delta (Fig. 2.27.). Three shallow lakes occupy a great part of the northern section of the Delta. These are: Lake Idku in the west, Lake Burullus in the middle and Lake Manzala in the east. These lakes receive a considerable amount of the drainage wager from the Delta, and are separated from the sea by narrow strips of land and all have outlets to the sea.

The coastal line of the Nile Delta has, for some time, been undergoing a rather active process of retreat. The High Dam at Aswan has brought the slow but continuous process of building the Nile Delta to an end. This means that the coastal line is left to the erosive action of the shore currents. A detailed discussion of this matter is presented in a later chapter.

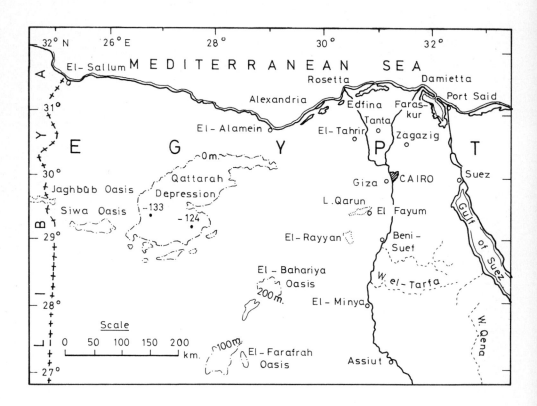

Fig. 2.27.   Map showing the Nile and its branches from Assiut to the Mediterranean Sea coast

REFERENCES

Abu el Ataa, A., 1978. Egypt and the Nile after the High Dam (text is in Arabic), Ministry of Irrigation, Cairo, Egypt, 145 pp (with 18 plates).

Beauchamp, R.S., 1964. The Rift Valley Lakes of Africa. Verh. Int. Ver. Theor. Angew. Limnol. 15: 91-99.

Berry, L., and Whiteman, A.J., 1968. The Nile in the Sudan. Geogr. Journ. 134 I: 1-37.

Hurst, H.E., 1927. The Lake Plateau Basin of the Nile, 2nd part, Phys. Dept. Paper 23, Government Press, Cairo, Egypt, 66 pp.

Hurst, H.E., and Philips, P., 1931. The Nile Basin, Vol. I: General description of the Basin, meteorology and topography of the White Nile Basin, Phys. Dept. Paper 26, Government Press, Cairo, Egypt, 128 pp.

Hurst, H.E., 1950. The Nile Basin, Vol. VIII: The hydrology of the Sobat and White Nile and the topography of the Blue Nile and Atbara, Phys. Dept. Paper 55, Government Press, Cairo, Egypt, 125 pp.

Hurst, H.E., Black, R.P., and Simaika, Y.M., 1959. The Nile Basin, Vol. IX: The hydrology of the Blue Nile and Atbara and the Main Nile To Aswan, with some reference to projects, Nile Control Dept. Paper 12, Government Printing Office, Cairo, Egypt, 207 pp.

Hurst, H.E., Black, R.P., and Simaika, Y.M., 1966. The Major Nile Projects, Nile Control Dept. Paper 23, Government Printing Office, Cairo, Egypt, 159 pp.

Jonglei Investigation Team, 1954. The Equatorial Nile project and its effect in the Anglo-Egyptian Sudan. Sudan Government, London.

Kalinin, G.P., 1971. Global hydrology (translated from Russian, Israel Program for Scientific Translations Ltd.). U.S. Dept. of Comm., National Technical Information Service, Springfield Va 22151, U.S.A.

Morandini, G., 1940. Ricerche Limnologiche. Geografia-Fisica, Vol. III, 1. Missione di studio al Lago Tana, 319 pp.

Rzóska, J. (editor), 1976. The Nile, biology of an ancient river. Dr W. Junk B.V. Publishers, The Hague, The Netherlands, 417 pp.

Stamp, D.L., and Morgan, W.T., 1972. Africa: a study in tropical development. John Wiley and Sons Inc., New York, 520 pp.

Sutcliffe, J.V., 1974. A hydrological study of the southern Sudd region of the Upper Nile, Bull. Hydro. Sci. 19: 237-255.

Talling, J.F., 1966. The annual cycle of stratification and phyto-plankton growth in Lake Victoria (East Africa), Int. Rev. Hydrob. 51: 545-621.

Talling, J.F. and Rzóska, J., 1967. The development of plankton in relation to hydrological regime of the Blue Nile, Journ. Ecol. 55: 637-662.

Thompson, K., 1975. Productivity of Cyperus papyrus L., In: Photo-synthesis and productivity in different environments. (Ed.) J.P. Cooper; I.B.P. Synthesis Series 3; Cambridge University Press.

Wayland, E.J., 1934. Rifts, rivers, rains and early man in Uganda. Journ. Roy. Anthropol. Inst. 64: 333-352. II.

Zaghloul, S.S., 1982. Water balance of Lake Victoria and the effect of gravity. M.Sc. Thesis, Fac. Engrg., Cairo University, Giza, Egypt, 239 pp. (excluding annexes).

Chapter 3

CLIMATE OF THE NILE BASIN

3.1   HISTORICAL INTRODUCTION

There is evidence of some climatic changes in the Nile Basin, especially in
Egypt and the Sudan, in the last 25 000 years.

In Egypt the wet phase that terminated some 25 000 years BP (before present)
was followed by a dry phase that lasted about 7 000 years. The subsequent period
from 18 000 to, say, 5 000 years BP was characterized by its heavy winter rain
and by increased flow in the Nile coming from the Ethiopian Plateau. The gradual
aridity which swept over Egypt since then was interrupted by some wet, though
short, intervals. These were from about 10 000 to 8 000 years BP and from about
6 000 to 4 500 BP. The moist intervals have been terminated since about 2 500
years B.C. (Butzer, K.W., 1966 and 1971).

In the Sudan the period from 20 000 to 15 000 years BP was very arid. This
was followed by a wet period that lasted from 12 000 to 7 000 years BP and by
somewhat fluctuating intervals from 7 000 to 6 000 years BP. The climate in the
interval from 6 000 to 3 000 years BP can be described as fairly wet. From 3 000
years ago up to the present, the climate in the Sudan, like that in Egypt, has
gradually been becoming arid (Wickens, G., 1975).

A detailed description of these changes are beyond the scope of this book.
Our interest here is focused on the climate as it has been in the past 100 years
or so.

3.2   CLIMATIC REGIONS

A short description of the climate in the area occupied by the Nile Basin is
given in Vol. I of the Nile Basin (Hurst, H.E. and Philips, P., 1931). In this
description the climate has been divided into three main types. These are:

Type 1 - Mediterranean climate covering the area from the sea coast to a little
         south of Cairo. The annual rainfall decreases from 150 to 200 mm/yr on
         the coast to about 25 to 30 mm/yr at Cairo.

Type 2 - Desert or Saharan climate covering the area from a little south of
         Cairo to Atbara. There is practically no rainfall in this area.

Type 3 - Tropical climate covering the area south of Atbara. This type has
         further been sub-divided into:

         3a - The Sudan Plains - There is a steady increase in rainfall south
              of the rainless region (type 2). An annual depth of, say, 1 000 mm
              is reached in the south.

60

3b - The Highlands of Abyssinia - This could be a region of relatively
       heavy rainfall, since an annual depth of 1 600 mm is reached in
       some places.

3c - The Highlands of the Lake Plateau - The average annual rainfall
       could be in the order of 1 250 mm.

The climate of the Sudan was described by Ireland in "Agriculture in the
Sudan" (edited by Tothill, J.D., 1948). The Sudan, he mentioned, lay wholly
within the tropics between latitudes 22° and 3^ON. It is almost entirely land-
locked and has a predominantly continental climate. The effect of the Red Sea is
quite limited and there are no lakes or inland water surfaces large enough to
produce even local climatic effects. Broadly speaking, therefore, the Sudan is
one vast plain, interrupted only by the Marra Mountains of Darfur and the Nuba
Mountains of southern Kordofan.

The climate of the Sudan may be divided into 3 regions.
Region 1 is situated north of about latitude 19^ON. In this desertic region the
dry northerlies prevail throughout the year and rain is infrequent. It experi-
ences large diurnal and annual variations in temperature, characteristics of a
desert climate.
Region 2 is situated south of latitude 19^ON to latitude 3^ON. Here the climate
is typical of a tropical continent, though the northern part is semi-arid.
Region 3 comprises the areas along the Red Sea coast and the eastern slopes of
the Red Sea hills. Like region 1, the northerlies prevail throughout the year,
except that the climate is profoundly affected by the maritime influence of the
Red Sea. The rain is partly orographic and partly convectional.

The climate of Africa was classified (Trewartha, G.T., 1962), based on a
simplified form of the classification that had originally been developed by
W. Köppen. The results of that work have been used to derive the map of the
climate of Africa (Stamp, D. and Morgan, W., 1972). The part of the map covering the Nile Basin is shown in Fig. 3.1. The categories included in it are:
1) The equatorial or tropical rain forest climate - This is characterized by
   almost constant heat, constant humidity and constant rainfall. Plant growth
   takes place throughout the year, so that luxuriant vegetation is the rule.
   The sun heat causes evaporation from lakes and moist land surfaces. The
   heated, saturated, air rises and is cooled by convection so that rain falls
   in almost the same area from which the moisture originates (convective
   rains).
2) The tropical savannah - This region extends from the tropical rain forest on
   one side to the desert margins on the other. Accordingly, the savannah
   climate varies greatly between these two limits. The variation is primarily
   in the annual depth of precipitation, commonly 1 500 mm or more in the

equatorial margins to 400 mm in the semi-arid part. The season from November
to February is relatively cool and dry. This is followed by a hot, dry,
season where the hottest month is April or May. The coming of rains from
June to October, causes the lowering of the temperature. The amount of the
lowering depends on the amount of rainfall. The seasonal depth of rain has a
wide relative range but that of the monthly depth is much wider.

3,4) The semi-arid and arid climate or the steppe and desert - The savannah
limit may be taken as the 400 mm/yr isohyet. If less than this depth, the
climate may be described as steppe. It is rather difficult to say where the
steppe changes into desert. On the poleward margins, a low rainfall limit
may be taken to indicate where the steppelands fade into the Mediterranean.
Along the southern margins, the rainy season is the hot season, whereas
along the Mediterranean margins it is the winter.

5) The Highlands - There are two areas whose climate belongs to this category;
around the Equatorial Lakes and a good part of the Ethiopian Plateau (see
map, Fig. 3.1.). The climate here is very much modified by the elevation.
Some parts rise to the snow line. The annual precipitation easily reaches
1 500 mm.

The moisture index method has been used in the hydrometeorological survey of
the catchments of Lakes Victoria, Kyoga and Albert for the purpose of classify-
ing the climate of the Lake Plateau area. The index has been computed for
several points within the area from the expression

$$I_m = \frac{100 \ (S - 0.6D)}{PET} \qquad (3.1)$$

where

I_m = moisture index,

S = cumulated monthly surplus, $(R - PET)_m$,

D = cumulated monthly deficit, $(PET - R)_m$,

R = rainfall,

m = subscript referring to month, and

PET = annual potential evapotranspiration

The different types of climate correspond to the following I_m values

Type	I_m value
Perhumid	100 and above
Humid	20 to 100
Moist sub-humid	0 to 20
Dry sub-humid	- 20 to 0
Semi-arid	- 40 to -20
Arid	-100 to -40

62

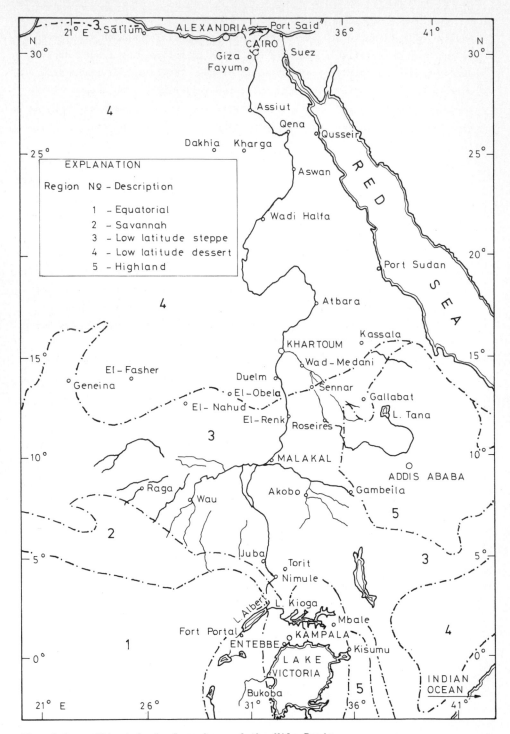

Fig. 3.1.   Climatological regions of the Nile Basin

The contour lines of equal moisture index values in the Equatorial Lake
Plateau are as shown in Fig. 3.2. From this map one can easily see that the
greater part of the plateau area falls in the class of the dry sub-humid climate
and only a few small parts fall in the moist sub-humid class. The information
presented on the map in Fig. 3.2. does not fully agree with the corresponding
part of the map in Fig. 3.1. They differ mainly in that part situated west of
Lake Victoria, in the basin of the Kagera and north of it.

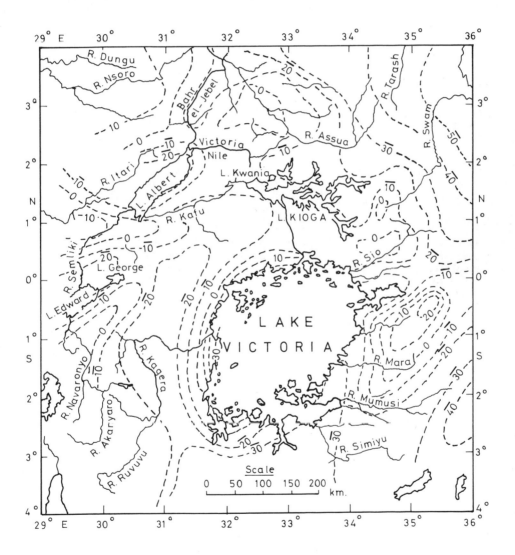

Fig. 3.2.   Contour lines of equal moisture index values for the Equatorial
Lake Plateau area (WMO, 1974)

3.3 NETWORK OF METEOROLOGICAL STATIONS

A description of some of the climatic features of the Nile Basin area is included in the following sections of this chapter. These features include the temperature, humidity, radiation, sunshine, cloudiness, wind and general circulation of the air masses. This description is based on the data observed at some, or all, of the meteorological stations listed in Table 3.1.

TABLE 3.1   Data of some of the meteorological stations in the Nile Basin
(Ireland, 1948; Ministry of War and Marine, Egypt, 1950; WMO, 1974)

Station	Country	Latitude N		Longitude E		Altitude metres
Sidi Barrani	Egypt	31°	38′	25°	58′	22
Salum (Observatory)		31	33	25	11	4
Damietta		31	25	31	49	3
Rosetta		31	24	30	25	2
Mersa Matruh		31	22	27	14	7
Edfina		31	18	30	31	3
Port Said (Airport)		31	17	32	15	1
Sirw		31	14	31	39	2
Alexandria		31	12	29	53	32
Atf		31	11	30	31	10
Arish		31	07	33	46	10
Sakha		31	07	30	57	6
Mansura		31	03	31	23	7
Damanhur		31	02	30	28	6
Qurashiya		30	51	31	07	8
Gemmeiza		30	48	31	07	9
Tanta		30	47	31	00	14
Zagazig		30	35	31	30	13
Benha		30	28	31	11	14
Delta Barrage		30	11	31	08	20
Cairo (Ezbekiya)		30	03	31	15	20
Giza		30	02	31	13	21
Suez (Port Tewfik)		29	56	32	33	10
Helwan		29	52	31	20	116
Fayum		29	18	30	51	30
Siwa		29	12	25	29	-15
Beni Suef		29	04	31	06	28
Minya (Airport)		28	05	30	44	39
Hurghada		27	14	33	51	3
Assiut		27	11	31	13	55
Qena		26	10	32	43	75
Qusseir		26	08	34	18	7
Nag-Hammadi		26	03	32	15	70
Luxor		25	39	32	39	78
Dakhla		25	29	29	00	122
Kharga		25	26	30	34	72
Esna		25	18	32	34	82
Deadalus		24	55	35	52	4
Kom Ombo		24	29	32	56	102
Aswan		24	02	32	53	111

TABLE 3.1   (continued)

Station	Country	Latitude N		Longitude E		Altitude metres
Wadi Halfa	Sudan	21°	55′	31°	20′	125
Port Sudan		19	37	37	13	5
Dongola		19	10	30	29	225
Karima		18	33	31	51	250
Tokar		18	26	37	44	20
Atbara		17	42	33	58	345
Khartoum		15	37	32	32	380
Kassala		15	28	36	24	500
Jebel Aulia		15	14	32	30	380
Wad-Medani		14	24	33	29	405
El-Dueim		14	00	32	30	380
El Fasher		13	38	25	21	740
Sennar		13	33	33	37	420
Geneina		13	29	22	27	805
El-Obeid		13	11	30	14	565
Singa		13	09	33	57	430
Gallabat		12	58	36	10	760
El-Nahud		12	42	28	26	565
El-Roseires		11	51	34	23	465
Renk		11	45	32	47	380
Malakal		09	33	31	39	390
Addis Ababa	Ethiopia	09	02	38	45	2450
Raga	Sudan	08	28	25	41	460
Gambeila		08	15	34	35	450
Akobo		07	47	33	01	400
Wau		07	42	28	01	435
Jimma	Ethiopia	07	39	36	51	1750
Juba	Sudan	04	51	31	37	460
Torit		04	25	32	33	625
Loka		04	22	30	57	965
Gulu	Uganda	02	45	32	20	926
Moroto		02	33	34	36	1524
Lira		02	18	32	56	1095
Butiaba		01	50	31	20	621
Soroti		01	43	33	37	1127
Masindi		01	41	31	43	1146
Fort Portal		00	40	30	17	1539
Mubende		00	35	31	22	1542
Namulonge		00	32	32	37	1148
Eldoret	Kenya	00	31	35	17	2084
Kampala	Uganda	00	20	32	36	1230
Entebbe		00	03	32	27	1146
Kitale	Kenya	00	01	35	00	1896
			S			
Equator	Kenya	00	01	35	33	2762
Kisumu		00	06	34	35	1146
Kericho		00	21	35	20	2070
Mbarara	Uganda	00	37	30	39	1443
Kabale		01	15	29	59	1868
Bukoba	Tanzania	01	20	31	49	1137
Musoma		01	30	33	48	1147
Mwanza		02	28	32	55	1140

The data used, part of which is included in the climatic tables available in
this book, are extracted from a number of references. Examples of these references are: the publications of the East African Meteorological Department
(E.A.M.D.), the Climatological Normals for Egypt, the Climates of Africa and
some volumes of the Nile Basin.

Each of the countries sharing the basin of the Nile has its own national
network of meteorological stations. However, the hydrometeorological project of
the Equatorial Lake area had among its objectives the strengthening of the
previously existing networks in Uganda, Kenya and Tanzania, and later in Rwanda
and Burundi. To fulfill this objective, twenty-five meteorological stations
have been established and thirty existing stations have already been up-graded
by the provision of additional instruments. Moreover, 200 ordinary rain gauges
have been installed and 23 totalizers placed in remote places and islands. For
the estimation of evaporation from the Equatorial Lakes, a network of six
solarimetres for radiation measurement, seven stations with wind masts for
measurement of wind speed, and eight stations for the measurement of lake surface water temperature, have been installed.

For the estimation of evapotranspiration, a network of Thornthwaite tanks
and some special lysimetres have been installed at a number of locations.

An automatic weather station had been established on the Nabiyongo Island
in Lake Victoria, with an auxiliary station at Entebbe. The principal station
at the latter has among its equipment the Russian 20 m^2 and GGI 3000 evaporation pans (WMO, 1974).

The majority of the meteorological stations in Egypt and the Sudan are
stations of the second order, where observations are taken every day at 08.00,
14.00 and 20.00 hours standard local time. At first-order stations, the observations are usually taken eight times a day in the synoptic hours of observations,
and at third-order stations, observations are taken at 08.00 hours local time
only.

Screen observations usually comprise the air temperature, maximum and minimum temperatures, all in degrees centigrade, the barometric pressure in milli-
bars, the humidity as obtained from the wet and dry bulb thermometers, and the
evaporative capacity of the air as measured with a Piche evaporimeter. The
duration of the bright sunshine is measured mostly by a Campbell-Stokes
recorder. The wind is expressed by a number on the Beaufort Scale when the wind
force is estimated, or in kilometres per hour when its speed is measured by
means of an anemometer. The soil temperature is measured at a few locations
only, and at a depth varying from 0.60 m to 2.10 m, whereas the grass minimum
temperature is measured at a height of 0.10 m above the ground level. The river
and sea temperatures are measured at some selected sites in Egypt and the Sudan.

The rainfall is measured not only at the meteorological stations, but also at many other locations. Most of the rain gauges installed are cylinderical in form with a catch of 200 cm^2 in surface area and a rim of about 1.0 m height from the ground surface. The analysis of the rainfall data is presented in Chapter 4.

3.4   TEMPERATURE

The mean daily temperature for the months of the year at a number of stations in the Nile Basin is listed in Table 3.2. These data are based on the daily mean temperature which is calculated on the basis of the number of observations taken every day. So, for all stations in Egypt and Sudan observing thrice daily, the mean temperature is one-fourth the sum of the temperatures at 08.00, 14.00, and 20.00 hours plus the minimum temperature. For stations observing twice daily, the temperature is the mean of the 08.00 hour and 20.00 hour observations, and for stations observing once daily the temperature is simply the mean of the maximum and minimum temperatures.

The meteorological stations installed by the hydrometeorological project in the Equatorial Lake Plateau as well as the up-graded stations in East Africa are spot-read visually during the synoptic hours: 06.00 Z[*], 09.00 Z and 12.00 Z.

The mean daily temperature shows a distinct pattern characteristic of each part in the Nile Basin. Generally speaking, the coolest month in Egypt is January and the warmest is July, except along the coasts of the Mediterranean and the Red Seas, where August is the warmest month. For the greater part of the Sudan, January is the coolest month. The month with the highest mean daily temperature changes rapidly with latitude from July in Wadi Halfa, similar to Egypt, to May in Wad Medani, April in Malakal and February in Juba down south. This main cycle is followed, in many places, by a less pronounced cycle where the second minimum falls in August and the second maximum in September or October. In the Equatorial Lake Plateau the wave of the mean daily temperature is quite similar to that in the southern part of the Sudan. The primary maximum occurs in February and the lowest temperature in July. The secondary maximum takes place in October and is followed by a secondary minimum in November. The ratio of the mean daily temperature in the warmest month to the mean daily temperature in the coolest month, $(\bar{T}_{mx}/\bar{T}_{mn})$, has been computed for all stations given in Table 3.1 and plotted versus the latitude, ϕ. Three curves are obtained as shown in Fig. 3.3. In the very northern latitude, about N 32°, the grouping of the stations is not clear. South of this latitude one can easily

---

[*]Z = G.M.T. = Greenwich Meridian Time
$\bar{T}_{mx}$ = mean daily temperature in the warmest month, and
$\bar{T}_{mn}$ = mean daily temperature in the coolest month

distinguish one curve for the stations located west of the Nile, another curve for the stations on the Nile and its tributaries and between the branches and a third curve for those stations on the Red Sea coast and east of the Nile. In each case the curve consists of a very short rising limb followed by a long falling tail. The peak occurs at approximate latitudes of 28°, 27° and 30° north for the three curves in the order described above. The corresponding $(\bar{T}_{mx}/\bar{T}_{mn})$ ratio is about 2.65, 2.5 and 2.05 respectively.

The annual mean daily temperature at those stations listed in Table 3.2 have been used for plotting the mean annual isotherm over the Nile Basin. The map in Fig. 3.4. shows that the mean of the annual mean daily temperature increases almost steadily from about $19^{\circ}C$ on the Mediterranean Sea coast in the north to almost $29^{\circ}C$ in Atbara down south. A mean temperature of 29 to $29.5^{\circ}C$ covers the belt from Atbara to Khartoum. South of Khartoum the temperature falls, but slowly, to reach $26^{\circ}C$ along the southern frontier of the Sudan. North-west of Lake Victoria the temperature drops rather rapidly to reach about $21^{\circ}C$ in Entebbe and $20^{\circ}C$ in Fort Portal. The topography of the highlands in the eastern part of the Nile Basin causes the cooling of the mean temperature to about $17^{\circ}C$ as shown in Fig. 3.4. The mean annual temperature has a small range of variation. This range varies from about $3^{\circ}C$ in the major part of the Equatorial Lake Plateau to a maximum of less than $6^{\circ}C$ in the central plains of the Sudan. This narrow range is produced by the relatively small annual variation of radiation. In contrast, the diurnal range of temperature is quite large. The figures presented in Table 3.3 show the range to be largest in the northern part of the Sudan and the southern part of Egypt, and smallest in the Mediterranean and Red Sea areas and the Lake Plateau area.

TABLE 3.2 Mean daily temperature at certain stations in the Nile Basin (data are from Griffiths, 1972; Ireland, 1948; Ministry of War and Marine, Egypt, 1950; WMO, 1974)

Mean daily temperature in °C for

Station	Jan.	Feb.	Mar.	Apr.	May	Jun.	Jul.	Aug.	Sep.	Oct.	Nov.	Dec.	Year
Sidi Barrani	12.3	13.2	14.2	16.6	19.3	22.0	23.8	24.5	23.5	20.9	17.4	13.6	18.4
Salum (Observatory)	11.3	12.3	14.1	16.8	19.4	22.3	24.3	24.3	23.0	21.2	17.8	13.4	18.4
Damietta	13.2	14.0	15.3	18.3	22.2	24.2	26.0	26.2	24.6	23.2	19.8	15.4	20.2
Rosetta	15.2	15.2	16.7	19.0	22.0	24.5	26.3	27.2	26.3	24.5	21.2	17.2	21.3
Mersa Matruh	12.4	12.9	14.5	17.2	19.9	22.9	24.7	25.5	24.4	22.2	18.7	14.4	19.1
Port Said	13.7	14.3	16.2	18.7	21.8	24.6	26.4	26.9	25.8	23.9	20.4	15.6	20.7
Alexandria	13.7	14.1	15.8	18.1	21.0	23.6	25.4	26.2	25.3	23.3	19.9	15.7	20.2
Mansura	13.4	14.0	16.2	19.8	23.8	26.3	27.8	27.8	26.1	24.0	20.3	15.4	21.2
Damanhur	13.6	14.2	16.2	19.4	22.9	25.2	26.4	26.6	25.2	23.5	19.8	15.5	20.7
Tanta	11.6	12.3	14.9	18.7	22.9	25.4	26.5	26.4	24.4	22.1	18.3	13.6	19.8
Zagazig	11.5	12.6	15.2	18.9	23.0	25.7	26.8	26.6	24.5	22.4	18.4	13.4	19.9
Delta Barrage	13.0	14.0	16.0	19.8	23.7	26.2	27.7	27.4	25.2	23.2	19.2	14.8	20.8
Cairo (Ezbekiya)	12.3	13.5	16.3	20.2	24.2	26.8	27.7	27.6	25.3	22.7	18.7	14.0	20.8
Giza	11.2	12.5	15.4	19.2	23.3	26.0	26.9	26.7	24.3	22.0	18.0	13.2	19.9
Suez (Port Tewfik)	13.8	14.6	17.1	20.5	24.4	26.9	28.4	28.5	26.3	24.0	20.0	15.4	21.7
Helwan	12.3	13.5	16.4	20.4	24.3	26.6	27.5	27.4	25.4	23.3	19.0	14.1	20.8
Fayum	11.6	13.2	16.1	20.4	25.1	27.2	28.1	28.0	25.6	23.2	18.7	13.5	20.9
Siwa	10.7	12.6	15.8	20.3	25.3	27.9	28.9	28.5	26.0	22.4	17.4	12.3	20.7
Beni Suef	12.4	14.1	16.9	20.2	23.7	26.6	27.4	27.6	24.6	22.6	18.5	13.5	20.7
Minya	12.2	14.1	17.1	21.4	26.1	28.0	29.0	28.7	26.1	23.8	19.2	14.0	21.6
Hurghada	15.8	16.2	18.7	22.2	26.2	28.5	29.5	30.0	28.0	25.2	21.2	17.7	23.3
Assiut	11.7	13.3	17.1	22.2	26.6	28.8	29.4	29.1	26.5	23.8	18.6	13.6	21.7
Qena	13.2	15.0	19.4	24.6	29.8	31.7	32.0	32.1	29.1	26.0	20.3	15.0	24.0
Qusseir	17.8	18.4	20.7	23.4	26.8	28.9	29.8	30.3	28.7	26.7	23.4	19.6	24.5
Nag Hammadi	12.0	12.7	16.6	21.5	27.3	28.5	29.1	29.4	26.9	24.6	19.6	14.6	21.9
Luxor	13.0	15.4	19.4	25.0	30.2	31.4	32.3	32.1	29.7	26.8	20.5	15.1	24.2
Dakhla	12.3	14.1	18.1	23.4	28.2	30.4	30.8	30.5	28.1	24.9	19.2	13.8	22.8
Kharga	13.1	14.9	19.1	23.9	29.0	31.0	31.4	31.1	28.6	26.0	20.2	15.0	23.6

TABLE 3.2 (continued)

Mean daily temperature in °C for

Station	Jan.	Feb.	Mar.	Apr.	May	Jun.	Jul.	Aug.	Sep.	Oct.	Nov.	Dec.	Year
Esna	13.1	14.4	18.5	23.3	27.6	29.7	30.1	30.1	27.9	24.2	19.1	14.6	22.7
Deadalus	21.8	21.5	22.7	24.6	27.2	28.3	29.7	30.3	29.1	28.0	26.1	23.3	26.0
Kom Ombo	14.9	16.3	20.4	25.0	29.6	31.2	31.4	31.4	29.4	27.2	21.6	16.8	24.6
Aswan	15.5	17.2	21.3	26.2	30.5	32.9	33.2	33.0	30.9	28.3	22.6	17.4	25.8
Wadi Halfa	15.9	17.5	21.9	26.7	30.5	32.2	32.2	32.2	30.5	28.2	22.6	17.6	25.7
Port Sudan	23.5	23.2	24.2	26.6	29.4	32.3	34.5	34.8	32.2	29.4	27.4	25.0	28.6
Tokar	24.4	24.3	26.0	28.0	30.9	33.4	35.0	35.0	33.6	30.6	28.0	25.6	29.6
Atbara	22.2	23.4	26.6	30.4	33.4	34.8	33.6	32.7	33.6	31.6	27.4	28.2	29.5
Khartoum	23.6	25.0	28.2	31.4	33.6	33.6	31.7	30.6	32.2	32.1	28.4	25.0	29.6
Kassala	25.0	26.1	28.8	31.6	33.2	32.1	29.1	28.0	29.6	31.2	29.4	26.2	29.2
Wad-Medani	24.2	25.2	28.2	31.0	32.5	31.9	29.1	27.8	28.9	30.2	27.7	25.0	28.5
El-Dueim	23.7	25.0	27.8	30.9	32.1	31.8	29.6	28.4	29.5	30.8	28.2	24.8	28.6
Fasher	20.6	22.2	25.3	28.4	30.1	30.5	28.7	27.0	28.1	27.8	23.8	21.1	26.1
Sennar	25.0	26.1	29.1	32.0	32.4	31.2	28.4	27.4	28.2	29.9	28.4	26.0	28.7
Geneina	22.4	24.5	25.8	29.1	29.6	29.4	27.1	25.4	26.1	26.1	24.2	23.2	26.1
El-Obeid	21.0	22.6	25.8	29.3	30.7	30.2	28.0	27.0	27.9	28.7	25.3	22.0	26.6
Singa	25.7	26.8	29.6	32.0	32.1	30.3	27.8	26.8	27.6	29.4	28.5	26.4	28.6
Gallabat	26.0	27.5	29.4	30.6	29.7	26.6	24.0	23.8	24.6	26.0	26.0	25.6	26.6
El-Nahud	23.4	24.2	25.6	30.4	30.4	30.0	27.4	26.6	27.6	28.4	24.8	23.7	26.9
El Roseires	26.2	27.5	29.8	31.6	31.0	28.6	26.8	26.4	26.9	28.0	27.8	26.5	28.1
Renk	24.7	26.5	28.1	31.9	31.2	29.3	27.1	26.3	26.9	28.5	27.3	26.5	27.9
Malakal	27.0	28.4	30.4	31.0	29.4	27.4	26.3	26.2	27.0	27.8	27.6	27.0	28.0
Addis Ababa	15.6	16.9	18.2	18.0	18.7	17.5	16.5	16.1	16.3	16.5	15.1	14.8	16.7
Raga	24.2	25.8	27.4	28.8	28.0	26.6	25.5	25.2	25.8	26.6	25.3	24.2	26.1
Gambeila	27.5	28.6	29.9	29.4	27.5	26.2	25.6	25.6	25.9	26.5	26.8	27.0	27.2
Akobo	28.4	29.8	30.9	30.2	28.0	27.1	26.1	26.1	26.9	27.4	28.2	27.8	28.1
Wau	26.7	28.0	29.5	29.8	28.4	27.2	26.2	26.0	26.6	27.4	27.5	26.8	27.5
Jimma	19.5	20.0	20.0	20.0	19.5	19.0	18.0	18.0	18.5	18.0	18.0	18.0	19.0
Juba	28.8	29.6	29.5	28.8	27.4	26.5	25.5	25.6	26.4	27.2	27.7	28.1	27.6
Torit	28.3	28.9	28.8	28.1	26.8	25.9	24.8	24.8	25.8	26.4	27.0	26.9	26.9

TABLE 3.2   (continued)

Mean daily temperature in °C for

Station	Jan.	Feb.	Mar.	Apr.	May	Jun.	Jul.	Aug.	Sep.	Oct.	Nov.	Dec.	Year
Gulu	24.2	24.8	24.5	23.6	22.5	22.2	21.6	21.8	22.3	22.8	23.6	23.4	23.2
Lira	24.5	25.1	24.9	23.8	22.9	22.9	21.7	21.9	22.5	22.9	23.3	23.5	23.3
Butiaba	26.1	26.5	26.5	25.9	25.7	25.3	24.8	24.5	25.1	25.5	25.6	25.7	25.6
Soroti	25.5	25.9	25.7	24.4	23.5	23.1	22.5	22.6	23.3	24.1	24.5	24.7	24.1
Masindi	23.8	24.1	24.0	23.3	22.9	22.3	21.6	21.5	21.9	22.5	22.9	22.9	22.8
Fort Portal	19.9	20.2	20.1	20.0	19.6	19.2	19.0	19.2	19.4	19.0	19.2	19.6	19.5
Entebbe	21.7	21.7	21.8	21.3	21.0	20.8	20.4	20.4	20.8	21.2	21.2	21.3	21.1
Kisumu	25.1	24.3	24.0	23.6	23.2	22.7	22.2	22.6	23.3	23.8	24.6	24.9	23.5
Bukoba	20.2	20.3	20.4	20.3	20.2	20.3	20.2	20.4	20.5	20.6	20.4	20.2	20.4
Mwanza	22.6	22.5	22.2	22.0	22.2	22.0	21.8	22.2	23.0	23.4	23.1	22.8	22.5

72

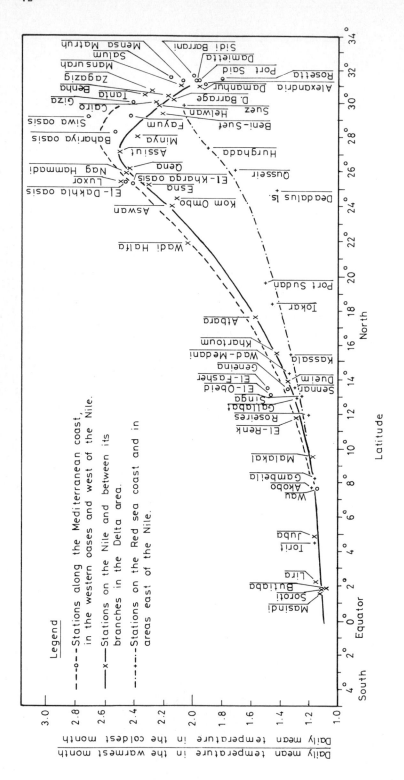

Fig. 3.3. Variation of the ratio of the daily mean temperature in the warmest month to the daily mean temperature in the coldest month with decreasing latitude in the Nile Basin

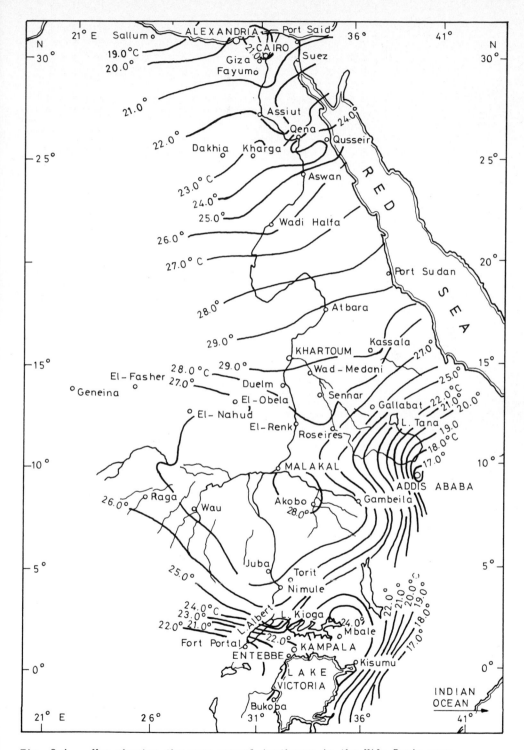

Fig. 3.4.   Map showing the mean annual isotherms in the Nile Basin area

TABLE 3.3   Mean daily range of temperature at a number of stations in the Nile Basin (Ireland, 1948; Ministry of War and Marine, Egypt, 1950; WMO 1974)

Mean daily range, in $^{\circ}$C, for

Station	Jan.	Feb.	Mar.	Apr.	May	Jun.	Jul.	Aug.	Sep.	Oct.	Nov.	Dec.	Year
Sidi Barrani	10.6	10.5	10.7	11.0	11.6	10.0	8.4	8.5	9.9	10.4	10.7	10.5	10.3
Salum(Observatory)	9.8	9.9	10.2	10.9	9.8	10.4	10.1	9.7	9.5	10.6	9.4	9.8	10.0
Damietta	9.8	9.6	8.8	8.8	8.8	8.3	9.0	8.7	8.6	8.3	7.9	9.3	8.8
Rosetta	5.8	6.5	6.2	6.5	5.7	4.6	4.2	4.5	5.0	5.8	5.7	6.2	5.5
Mersa Matruh	9.2	9.1	9.0	9.4	8.8	7.6	6.4	6.6	7.1	8.4	8.4	9.1	8.2
Port Said	8.2	8.6	8.0	7.9	7.9	7.8	8.1	8.0	7.8	7.6	7.5	8.1	7.9
Alexandria	7.9	8.1	8.4	8.5	8.2	7.4	6.6	6.8	7.3	8.2	8.1	8.0	7.7
Mansura	12.6	13.1	13.7	15.3	15.7	15.2	14.7	14.0	13.6	12.9	11.8	12.2	13.7
Damanhur	12.1	12.5	13.2	14.7	14.8	13.9	12.7	12.4	12.6	13.0	11.9	11.8	13.0
Tanta	13.6	14.3	15.5	17.2	17.4	16.8	15.5	15.2	15.1	14.7	13.5	13.1	15.1
Zagazig	14.0	14.3	15.2	16.6	16.8	16.4	15.3	14.8	14.5	14.4	13.4	13.4	14.9
Delta Barrage	13.3	14.3	15.3	17.3	17.6	16.9	16.2	14.6	13.5	13.7	13.1	13.1	14.9
Cairo (Ezbekiya)	12.1	13.0	13.9	15.2	15.7	15.4	14.2	13.3	12.7	13.0	12.4	11.8	13.5
Giza	13.8	14.9	16.1	17.8	17.9	17.3	16.1	15.0	14.2	14.4	13.7	13.3	15.4
Suez (Port Tewfik)	10.5	11.0	11.9	13.5	13.9	14.0	13.6	13.1	12.2	11.8	11.0	10.9	12.3
Helwan	10.5	11.5	12.8	14.4	15.0	14.9	14.4	13.5	12.4	11.7	10.7	10.2	12.7
Fayum	14.6	14.9	15.6	17.0	17.0	16.3	15.5	14.8	14.1	14.3	13.8	13.9	15.1
Siwa	16.1	16.4	17.1	17.9	18.0	18.5	17.6	17.7	17.3	17.6	16.7	16.1	17.2
Beni Suef	14.3	15.4	16.8	18.1	18.2	17.7	16.3	15.1	15.1	14.7	14.6	13.8	15.8
Minya	13.8	14.5	15.8	17.1	17.0	16.3	15.2	14.4	13.0	12.7	12.8	12.9	14.6
Hurghada	11.5	12.4	12.4	12.5	12.1	11.0	10.4	10.0	10.0	11.0	11.1	11.4	11.3
Assiut	14.0	15.3	16.5	17.0	16.6	16.7	14.8	13.8	12.8	12.1	13.7	13.7	14.7
Qena	15.8	17.5	19.1	19.2	18.0	17.8	16.9	16.5	16.1	16.3	16.5	15.5	17.1
Qusseir	8.8	8.7	8.3	7.9	7.6	7.1	7.2	7.1	6.9	7.3	7.9	8.4	7.7
Nag Hammadi	15.0	16.8	17.4	18.5	17.5	17.1	16.8	17.7	14.6	15.3	16.6	15.8	16.5
Luxor	18.0	19.4	19.7	20.0	19.5	19.7	18.9	18.4	17.8	18.2	18.3	18.0	18.8
Dakhla	16.7	17.7	18.3	19.1	18.3	16.9	16.1	15.9	15.5	16.0	16.1	16.2	16.9
Kharga	16.1	17.1	17.3	17.9	16.8	15.9	16.1	16.6	15.4	15.6	15.8	15.8	16.4

TABLE 3.3 (continued)

Mean daily range, in °C, for

Station	Jan.	Feb.	Mar.	Apr.	May	Jun.	Jul.	Aug.	Sep.	Oct.	Nov.	Dec.	Year
Esna	16.4	17.1	18.1	18.4	19.0	18.0	17.6	16.5	14.3	14.5	16.4	15.7	16.8
Deadalus Island	3.8	3.8	4.1	4.4	4.2	4.5	5.1	5.1	4.2	4.2	4.2	3.7	4.3
Kom Ombo	16.2	17.4	18.9	19.7	18.7	18.7	18.4	17.8	17.5	17.4	16.3	15.7	17.8
Aswan	13.4	14.3	15.9	16.3	15.9	15.9	14.9	14.7	15.4	15.2	14.1	13.5	15.0
Wadi Halfa	16.2	17.6	18.9	19.5	18.8	18.2	17.9	16.8	16.2	17.0	16.5	16.3	17.5
Port Sudan	7.2	8.0	9.0	10.1	11.3	13.2	12.8	11.9	11.4	9.1	7.4	7.0	9.9
Tokar	8.7	9.2	10.2	11.5	15.0	17.4	15.1	13.1	14.7	11.3	10.1	8.9	12.1
Atbara	16.3	17.2	18.1	18.7	16.9	16.1	14.1	14.0	14.9	15.6	15.5	15.9	16.1
Khartoum	16.9	17.6	19.0	18.7	16.7	15.0	13.6	12.2	14.2	16.2	16.3	16.7	16.1
Kassala	18.1	19.0	19.1	18.1	16.0	14.6	12.2	11.3	13.5	15.4	16.3	17.3	15.9
Wad Medani	19.8	20.4	20.8	20.3	17.4	15.2	13.0	11.5	13.8	16.7	18.6	19.3	17.2
El Dueim	18.6	19.6	20.3	20.0	17.6	15.4	12.9	11.7	13.6	16.2	17.3	17.9	16.7
El Fasher	21.5	21.8	22.2	21.3	18.8	17.2	14.2	13.1	16.0	19.1	21.3	21.8	19.0
Sennar	19.8	20.6	21.2	19.9	17.3	15.6	13.1	11.9	13.7	16.8	18.7	19.3	17.3
Geneina	22.9	22.6	21.2	20.4	19.3	17.3	13.0	11.6	14.8	20.0	20.5	22.1	18.8
El-Obeid	18.8	19.5	19.9	18.8	16.8	14.6	12.1	11.3	13.4	15.8	17.9	18.8	16.5
Singa	18.8	19.4	20.0	18.8	16.0	14.6	11.9	10.8	12.9	16.5	18.6	18.5	16.4
Gallabat	20.0	19.1	18.7	17.7	15.8	13.7	11.1	11.0	12.5	16.1	19.5	10.2	16.3
El-Nahud	16.9	18.7	18.3	17.7	16.2	13.5	10.8	10.4	12.5	15.1	16.3	18.2	15.4
El Roseires	20.6	20.6	20.0	18.4	15.4	13.1	11.2	11.1	12.6	15.9	17.4	20.4	16.6
Renk	19.1	19.0	19.5	16.7	14.6	12.3	10.1	8.9	11.5	15.4	18.3	18.8	15.3
Malakal	17.5	17.5	17.4	15.3	13.1	11.3	10.8	9.7	10.7	12.3	16.3	17.7	14.1
Addis Ababa	17.0	17.0	16.0	15.0	16.0	13.0	9.0	9.0	11.0	15.0	18.0	17.0	14.4
Raga	22.8	22.4	21.2	16.9	14.4	12.9	11.2	11.1	12.3	14.0	19.6	22.3	16.8
Gambeila	18.6	17.9	17.2	14.8	12.2	11.1	10.3	10.5	11.6	13.8	15.7	17.5	14.2
Akobo	16.4	15.7	15.2	13.7	11.9	11.4	10.2	9.8	9.8	11.3	13.8	15.7	13.0
Wau	18.2	17.2	16.8	14.7	13.2	12.1	11.0	10.9	12.2	13.5	16.0	17.9	14.6
Juba	17.2	16.2	15.1	13.6	12.1	12.0	11.2	11.4	12.8	14.0	15.6	16.6	14.0
Gulu	15.5	15.1	13.6	11.6	10.5	10.5	9.8	10.1	11.3	12.0	13.2	14.1	12.3
Moroto	15.0	14.8	13.4	12.2	12.1	12.3	11.6	12.1	14.3	13.6	13.8	14.0	13.3

TABLE 3.3 (continued)

Station	Jan.	Feb.	Mar.	Apr.	May	Jun.	Jul.	Aug.	Sep.	Oct.	Nov.	Dec.	Year
						Mean daily range, in °C, for							
Lira	17.2	16.5	14.9	12.8	11.7	11.5	11.2	11.7	12.7	13.8	14.7	15.5	13.7
Butiaba	7.9	7.5	7.2	7.3	7.2	7.3	7.0	6.5	7.4	7.3	7.4	7.8	7.3
Masindi	14.2	14.1	12.8	11.5	10.7	11.2	10.6	10.7	11.3	11.7	12.2	12.9	12.0
Fort Portal	14.6	14.6	13.5	11.9	11.2	12.1	12.2	11.7	12.2	12.1	11.8	12.8	12.6
Mubende	10.4	10.6	10.0	9.0	8.2	8.3	8.7	8.9	9.0	8.8	9.1	9.1	9.2
Eldoret	16.0	16.8	15.9	13.3	12.9	13.3	11.7	12.1	14.0	13.8	13.5	14.0	13.9
Entebbe	9.8	9.7	8.8	7.8	7.7	8.3	8.8	8.8	9.4	9.2	8.9	9.0	8.9
Kitale	10.0	10.4	11.5	12.7	12.4	11.4	11.5	11.2	10.7	11.1	10.7	10.6	11.0
Kericho	15.2	15.4	14.6	11.8	11.8	10.6	11.1	12.0	12.9	13.5	13.7	12.5	12.9
Mbarara	12.6	12.5	11.8	11.0	10.8	12.1	13.1	12.2	11.6	11.2	11.0	11.5	11.8
Kabale	14.1	13.8	13.1	11.5	11.0	13.2	14.5	13.6	13.6	12.9	12.5	13.0	13.1
Bukoba	10.7	10.5	10.3	9.4	9.1	10.1	10.3	10.3	10.4	10.1	9.9	10.3	10.1
Musoma	8.8	9.0	9.2	9.0	9.4	9.9	10.2	9.5	9.4	9.4	8.8	8.8	9.3
Mwanza	8.8	9.3	9.4	9.2	9.8	11.8	12.4	11.2	10.5	9.7	9.2	8.8	10.0

3.5   ATMOSPHERIC HUMIDITY

The atmospheric humidity in the Nile Basin area is expressed mostly in terms
of the relative humidity. In Egypt and the Sudan this measurement is made once
or twice a day, except, of course, at the first-order stations. For stations
observing twice or more a day, the mean relative humidity is the mean of the
relative humidities measured at 08 00 hours and 20 00 hours, and for those
observing once daily, it is simply that measurement made at 08 00 hours. For the
up-graded stations in the catchments of the Equatorial Lakes, the measurements
of the relative humidity are made at 03 00, 06 00 and 12 00 Z hours, and the
others are at 06 00 and 12 00 Z hours only. The deviations from the mean daily
relative humidity averaged for January, April, July and October for Alexandria,
Egypt, Khartoum and the Sudan are shown in Figs. 3.5(a) and 3.5(b) respectively.
The graphs for Alexandria, and similarly for many stations, show that the daily
mean is reached twice every day; once between 07 30 hours and 10 30 hours (fore-
noon) and another time between 17 30 hours and 22 00 hours (afternoon), depending on the location and month of the year. This is nearly so everywhere in the
Nile Basin, excluding Khartoum (Fig. 3.5(b)). Generally speaking, therefore, in
the absence of continuous, or frequent, measurements of the relative humidity,
the average of the observations made at 08 00 hours and 20 00 hours, or simply
the observation at 08 00 hours will not be too far from the true mean (Olivier,
H., 1961). The mean daily relative humidity for some stations in Egypt, the
humidity at 08 00 hours in the Sudan and at 06 00 hours elsewhere, are included
in Table 3.4.

The relative humidity at noon, H_n, seems to bear a certain relationship to
the daily mean humidity, H_m, or to the humidity at 08 00 hours or any other
reference hour[+]. This relationship is shown graphically in Figs. 3.6(a) and
thru' (d) for a number of combination of stations from some parts of the Nile
Basin.

Of interest is Fig. 3.6(c), which shows two distinct relationships; one for
stations located along the Mediterranean Sea coast and the Suez Canal, and the
other for those stations located along the Red Sea coast. For values of H_m less
than 80%, the relative humidity at noon, H_n, is much more for the stations on
the Red Sea than for the stations on the coast of the Mediterranean Sea. The
plot of the H_{06} humidity versus the H_{12} relative humidity for the Equatorial
Lake Plateau area does not exhibit a single relationship for all the stations in
the area. Among the factors influencing the relationship are the geographic
location of the station, the altitude, and the distance from the station to the
nearest lake. The graphic plot for the stations east and south-east of Lake

[+]Relative humidity at 06 00 Z hours

78

Victoria presents a much wider scatter than does the plot for the stations in
the north, north-west and west of the same lake.

The times are not uniform across the Nile Basin area and the early observation of humidity is taken at a time when the relative humidity is changing
rapidly. In some cases this is done without having a readily available simul-
taneous temperature to specify the climate completely, as is the case with the
noon humidity. In this connection, Griffiths, F.J. (1972) recommends the use of
other parameters to describe the atmospheric humidity as the dew point or the
absolute humidity. These two parameters do not show the large diurnal fluctuation typical of the relative humidity curve.

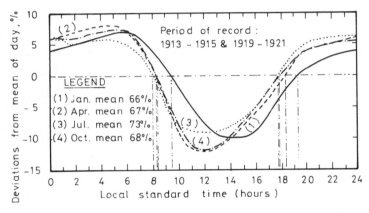

Fig. 3.5(a)   Deviations from mean daily relative humidities for Alexandria,
Egypt (Olivier, H., 1961)

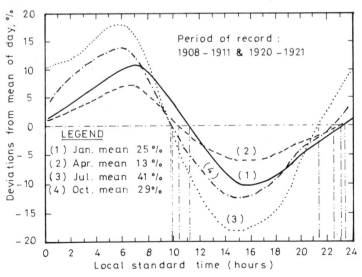

Fig. 3.5(b)   Deviations from mean daily relative humidities for Khartoum,
Sudan (Olivier, H., 1961)

TABLE 3.4   The mean daily relative humidity at a number of stations in the Nile Basin (Griffiths, 1972; Ireland, 1948; Ministry of War and Marine, Egypt, 1950; WMO, 1974)

Station	Mean daily relative humidity, in percent, for												
	Jan.	Feb.	Mar.	Apr.	May	Jun.	Jul.	Aug.	Sep.	Oct.	Nov.	Dec.	Year
Salum (Observatory)	71	72	69	66	70	67	70	76	74	74	73	70	71
Mersa Matruh	76	73	74	72	74	77	81	80	77	75	75	74	76
Edfina	79	78	74	70	68	70	73	73	71	74	78	79	74
Port Said	76	75	73	73	73	75	77	76	73	72	73	76	74
Alexandria	69	68	68	69	72	74	77	75	69	69	70	70	71
Tanta	81	78	76	68	61	63	70	74	77	79	81	80	74
Zagazig	83	79	75	67	60	62	68	73	78	80	82	83	74
Cairo (Ezbekiya)	74	68	65	58	52	55	61	65	69	72	74	76	66
Giza	79	72	67	60	53	56	62	68	72	73	78	81	68
Suez (Port Tewfik)	68	66	63	60	59	61	62	65	67	68	69	68	64
Helwan	61	56	52	45	41	44	51	54	58	59	62	62	54
Fayum	68	63	58	50	42	46	51	57	62	64	69	72	58
Siwa	70	64	61	56	52	53	55	58	60	63	66	70	61
Minya	64	58	52	43	39	42	46	51	58	61	65	68	54
Assiut	69	62	54	41	36	37	42	46	55	62	67	69	53
Qena	63	56	44	31	27	28	31	32	44	53	59	63	44
Qusseir	56	54	52	52	52	50	52	51	53	56	58	57	54
Nag Hammadi	69	59	48	40	38	40	44	45	54	58	64	71	52
Luxor	68	58	46	34	30	32	34	37	46	53	60	66	47
Dakhla	51	47	41	35	32	30	30	33	37	43	47	53	40
Kharga	58	54	46	40	38	38	38	38	44	48	55	60	46
Esna	61	54	41	32	27	25	27	31	42	49	52	59	42
Deadalus Island	68	70	74	70	77	78	77	76	78	79	73	68	74
Aswan	45	40	32	27	27	25	26	29	33	35	40	45	34
Wadi Halfa	48	40	31	24	22	21	23	26	30	34	39	42	32
Merowe	31	26	20	14	15	18	23	22	23	22	24	26	22
Atbara	38	34	27	18	19	24	32	33	30	28	31	34	29
Khartoum	29	24	16	15	22	34	47	52	46	31	30	29	31

TABLE 3.4   (continued)

Mean daily relative humidity, in percent, for

Station	Jan.	Feb.	Mar.	Apr.	May	Jun.	Jul.	Aug.	Sep.	Oct.	Nov.	Dec.	Year
Gallabat^x	45	43	36	38	53	68	78	80	77	67	54	49	57
Kassala^x	62	56	48	40	40	49	66	72	65	48	52	60	55
Port Sudan^x	66	65	64	56	45	37	39	41	47	66	68	69	55
El Roseires^x	41	34	27	31	48	66	79	83	80	70	49	42	54
Wad Medani^x	36	26	21	20	31	48	67	77	70	50	36	39	43
El Obeid^x	37	28	23	26	41	56	73	79	69	48	33	36	46
El Fasher^x	35	28	24	21	31	47	65	74	61	37	31	34	41
Malakal	30	24	36	46	58	75	84	87	85	80	63	34	58
Addis Ababa^x	61	64	58	65	63	76	86	86	76	56	59	62	68
Akobo^x	43	43	45	63	75	79	84	85	84	78	71	58	67
Wau	35	29	36	45	64	73	78	80	76	71	55	46	57
Juba	43	41	50	63	85	77	82	83	77	71	63	52	66
Torit^x	39	45	53	67	73	75	79	79	74	72	62	53	64
Mongalla	50	52	55	68	75	81	84	83	81	78	73	57	70
Gulu^+	63	66	73	83	85	84	87	88	85	85	74	70	79
Moroto^+	57	55	62	72	75	75	79	76	69	66	63	62	68
Lira^+	70	73	78	84	89	87	90	89	86	82	79	74	78
Butiaba^+	68	67	70	73	74	75	79	80	78	77	74	71	74
Masindi^+	73	73	75	80	81	83	86	86	83	81	78	77	79
Fort Portal^+	87	87	89	90	95	88	90	90	90	89	88	88	89
Mubende^+	77	77	81	86	88	80	82	84	83	94	81	84	83
Eldoret^+	62	58	62	73	76	79	82	82	72	66	67	67	71
Entebbe	84	88	90	85	78	87	91	92	88	85	88	85	87
Kitale^+	70	76	73	80	83	84	85	84	79	74	73	72	78
Kericho^+	66	70	78	85	83	84	85	81	78	74	72	75	78
Mbarara^+	85	84	85	87	86	82	79	80	82	84	85	85	84
Kabale^+	94	95	96	97	95	95	95	92	91	93	94	95	96
Bukoba^+	85	86	86	88	85	79	76	82	82	81	82	83	83
Musoma^+	74	74	76	80	79	73	70	69	70	69	67	64	72
Mwanza^+	77	78	79	81	77	68	66	63	61	66	68	76	72

x = relative humidity at 08 hr 30   + = relative humidity at 06 00 Z hours

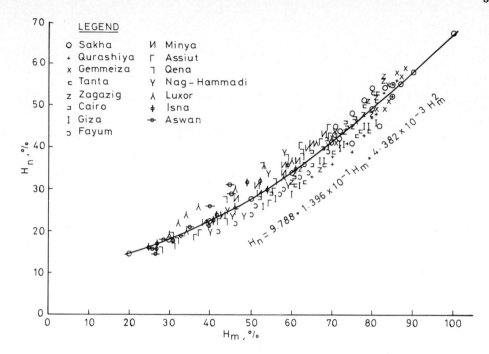

Fig. 3.6(a)   The mean daily relative humidity, H_m, versus the relative humidity at noon, H_n, for the inland stations in Egypt.

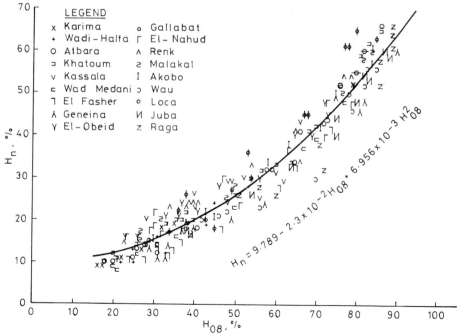

Fig. 3.6(b)   The relative humidity at 08 00 hours, H_{08}, versus the relative humidity at noon, H_n, for the inland stations in the Sudan.

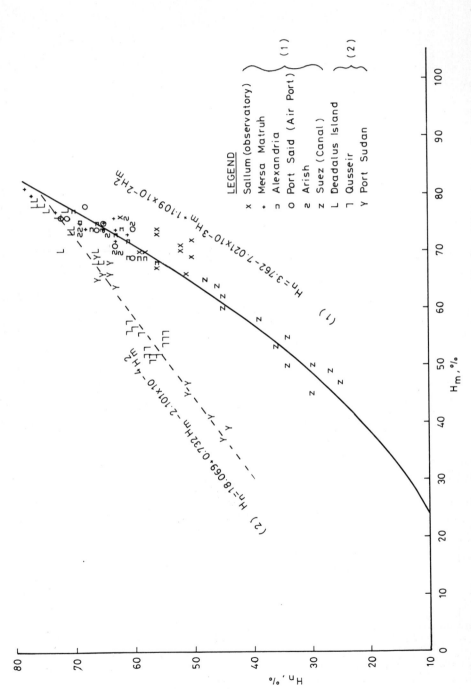

LEGEND

x Sallum (observatory)
+ Mersa Matruh
п Alexandria
o Port Said (Air Port)
я Arish
z Suez (Canal)
L Deadalus Island
٦ Qusseir
Y Port Sudan

(1)

(2)

$H_n = 3.762 - 7.021 \times 10 - 3 H_m + 1.109 \times 10 - 2 H_m^2$

(1)

$H_n = 18.069 + 0.732 H_m - 2.101 \times 10 - 4 H_m^2$

(2)

$H_m, \%$

$H_n, \%$

Fig. 3.6(c)   The mean daily relative humidity, H_m, versus the relative humidity at noon, H_n, for (1) stations along the Mediterranean coast and the Suez Canal and (2) stations along the Red Sea coast

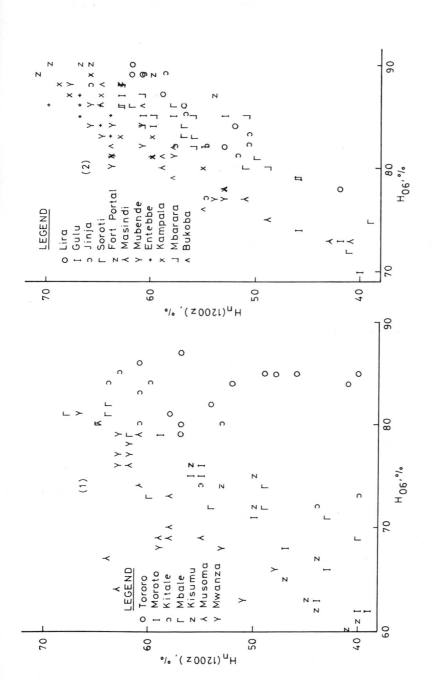

Fig. 3.6(d)   The relative humidity at 06 00 Z, H_{06}, versus the relative humidity at 12 00 Z, H_n, for stations (1) east and south of Lake Victoria, and (2) west and north of Lake Victoria

84

3.6   RADIATION, SUNSHINE AND CLOUDINESS

All atmospheric processes such as temperature, pattern of barometric pressure, wind flow, rainfall, humidity and evaporation, are influenced directly or indirectly by the flux of the short wave radiation. Unfortunately, the density of the network of radiation stations in the Nile Basin is still unsatisfactory. This is the state of affairs in spite of the network strengthening and the updating of some of the previously existing stations in the catchments of the Equatorial Lakes. Many more sunshine measuring stations are, however, available than for radiation.

The global radiation (solar and sky) at the top of the earth's atmosphere depends on the geographic latitude and the time of the year. This radiation, expressed as Angot's value, can be computed from astronomical considerations, and is available in some references in meteorology in tabular or graphical form. The monthly Angots for latitudes from 30°N up to 10°S with 10° latitude intervals have been quoted and included in Table 3.5 just for comparison with the actual global radiation at the earth's surface measured at some locations in the Nile Basin area.

TABLE 3.5   Solar radiation at the top of the atmosphere expressed in Langleys/day[*]

Latitude	Jan.	Feb.	Mar.	Apr.	May	Jun.	Jul.	Aug.	Sep.	Oct.	Nov.	Dec.
30°N	520	630	775	895	975	1000	990	925	820	685	560	490
20°N	660	750	850	920	960	965	960	935	875	785	685	630
10°N	780	840	900	925	915	900	905	915	905	865	800	760
Equator	885	915	925	900	850	820	830	870	905	910	890	875
10°S	965	960	915	840	755	710	730	795	875	935	955	960

The actual radiation received on the earth's surface depends mainly on the theoretical radiation, the cloudiness and the atmospheric turbidity. The estimate of the actual radiation, R, from the relative duration of sunshine, n/N, was first suggested by Ångström, A. (1924). Since then the expression in common use is

$$R = R_A (a + b \cdot n/N) \tag{3.2}$$

where R_A = total radiation received if the atmosphere were perfectly transparent.

---

[*]Langley/day = 1 gm cal/cm^2/day

The overall regression equation developed by Black, J., Bonython, C., and
Prescott, J. (1954), for 32 stations scattered over a range of latitudes from
about $7^{\circ}S$ to about $65^{\circ}N$ is

$$R = R_A \ (0.23 + 0.48 \ n/N) \tag{3.3}$$

The same authors upon grouping the 32 stations into seven groups found that
the regression constant, a, varies from 0.19 to 0.40 and the slope of the
regression line, b, from 0.274 to 0.613.

Darlot and Le Carpentier found a and b at 0.18 and 0.62 respectively,
whereas the data analyzed by Woodhead for 15 stations in East Africa gave 0.23
for a and 0.53 for b.

The equation developed by Glover, J., and McCulloch, J.S. (1958), includes
the effect of the geographical latitude of the station, ϕ, on it. It reads

$$R = R_A \ (0.29 \cos \phi + 0.52 \ n/N) \tag{3.4}$$

For East Africa we may take N = 12.1 (maximum error $\pm$ 0.6) and $\phi = 0$, so
$\cos \phi = 1$ (maximum error - 0.015) so that eq. 3.4 becomes

$$R = R_A \ (0.29 + 0.043 \ n) \tag{3.5}$$

For annual values R_A = 850 gm $cal/cm^2/day$ (maximum error 1 percent), so that
R = 250 + 37 n.

The global radiation at the earth's surface, R, is usually measured by the
Gunn-Bellani instrument and the duration of the sunshine is recorded by the
Campbell-Stokes sunshine recorder. The data available for these two meteorologic parameters in the Nile Basin area are summarized and given in Tables 3.6
and 3.7 respectively. From the data in Tables 3.5 and 3.6 the ratio R/R_A was
computed for 14 stations scattered over the range from about $2^{\circ}S$ to about $30^{\circ}N$
latitude. The relative duration of bright sunshine was computed from Table 3.7
for those stations having N equal to 12 hours. For the remaining stations, n/N
was taken directly from literature by Olivier, H. (1961) and WMO (1974).

The plotted points of n/N versus R/R_A are shown in Fig. 3.7. It is clear
that the scatter of the points is so large that a single regression relation
between n/N and R/R_A applying to all the stations used here will not be statistically justified. This scatter is partly due to the difference in the time
period and the number of years each set of points belonging to a station represents. Another reason could be that not all the sunshine data were measured by
the same instrument. If there is a need, however, to link the two variables n/N

TABLE 3.6   The monthly and annual mean solar radiation at the earth's surface for some radiation stations in the Nile Basin

Station	Average global radiation in cal/cm² day for												
	Jan.	Feb.	Mar.	Apr.	May	Jun.	Jul.	Aug.	Sep.	Oct.	Nov.	Dec.	Year
Tahrir	293	401	489	568	659	684	682	627	538	413	325	282	497
Giza	290	375	498	576	635	667	663	610	533	420	319	266	488
Wadi Halfa	457	524	588	639	667	655	638	609	582	525	461	426	564
Port Sudan	354	449	539	612	625	577	564	552	547	486	383	340	502
Khartoum	483	557	612	648	622	587	571	563	561	522	488	455	556
Wad-Medani	483	433	576	602	584	459	474	484	562	533	489	466	512
El-Fasher	457	534	579	594	592	550	532	534	563	529	486	456	534
Tozi	422	474	516	528	529	499	466	470	504	476	415	399	475
Malakal	471	521	527	536	510	448	448	486	484	487	487	476	490
Juba	458	464	462	454	479	459	415	466	509	480	457	445	462
Gulu	474	468	397	428	413	408	378	409	461	477	486	483	436
Lira	405	493	437	459	474	452	409	423	498	528	512	485	464
Masindi	393	423	421	415	400	376	415	416	444	490	477	511	432
Kabale	334	410	378	353	316	383	380	325	399	417	347	351	366
Tororo	456	501	473	501	444	445	426	409	452	483	503	477	464
Namulonge	403	397	415	423	405	408	366	379	416	431	421	411	403
Jinja	405	450	448	457	423	435	390	419	453	452	447	432	434
Entebbe	474	459	445	461	402	432	392	421	433	445	460	421	438
Kericho	446	481	483	470	397	437	380	397	435	482	404	450	437
Mbarara	375	408	376	386	377	420	411	418	444	416	404	419	404
Kisumu	490	542	462	510	463	490	443	470	500	527	521	533	499
Bukoba	339	305	386	381	338	403	427	448	355	433	395	386	383
Mwanza	433	502	477	491	470	494	487	486	492	496	445	430	475

TABLE 3.7   The monthly and annual means of the daily sunshine hours at certain stations in the Nile Basin (Griffiths, 1972; Ministry of War and Marine, Egypt, 1950; WMO, 1974)

Station	Jan.	Feb.	Mar.	Apr.	May	Jun.	Jul.	Aug.	Sep.	Oct.	Nov.	Dec.	Year
Sallum	7.0	7.5	9.0	9.9	9.9	12.3	12.7	12.2	10.9	9.7	7.3	6.8	9.6
Mersa Matruh	5.5	7.3	7.8	10.0	10.7	11.9	12.2	11.9	10.4	8.8	8.3	6.6	9.4
Edfina	5.9	7.7	8.7	10.4	10.7	12.2	12.2	11.9	11.0	9.5	7.9	6.7	9.7
Port Said	6.9	7.3	8.6	9.8	10.9	12.0	12.2	11.8	11.0	10.0	8.7	6.6	9.6
Alexandria	7.0	7.8	9.0	10.6	10.9	11.9	12.0	11.9	11.1	9.9	8.2	6.7	9.8
Tanta	7.0	7.5	9.2	10.1	11.0	12.2	12.1	11.6	10.7	9.4	8.0	6.6	9.6
Cairo	7.4	8.1	9.0	10.2	10.7	12.1	12.0	11.5	10.6	9.5	8.2	6.8	9.7
Giza	7.3	8.1	8.9	10.2	11.0	12.4	12.1	11.6	10.6	9.3	8.2	7.2	9.7
Helwan	7.6	8.5	9.4	10.6	11.4	12.8	12.6	12.1	11.1	9.8	8.6	7.6	10.2
Wadi Halfa	10.2	10.4	10.3	10.0	11.3	11.9	11.0	11.1	10.0	10.7	10.4	9.8	10.6
Karima	9.9	10.6	10.6	11.0	11.4	10.7	11.1	9.6	9.6	10.2	10.7	10.5	10.5
Atbara	10.1	10.6	10.5	12.0	11.1	10.4	9.9	9.5	9.8	10.2	10.3	10.1	10.4
Khartoum	10.6	10.7	9.9	10.4	10.3	9.9	8.6	8.7	10.0	10.3	10.8	10.6	10.1
Kassala	10.1	10.4	10.2	10.8	10.5	10.2	8.4	8.4	9.7	10.3	10.1	10.0	9.9
Port Sudan	6.9	8.2	9.1	10.4	10.9	10.3	9.9	9.6	10.0	9.9	8.3	7.6	9.3
Wad Medani	10.6	10.6	10.5	10.8	10.2	9.4	7.7	7.8	8.8	10.0	10.6	10.5	9.8
El Obeid	10.3	10.6	9.9	10.3	9.8	8.5	7.3	6.8	8.3	9.3	10.6	10.5	9.3
El Fasher	10.3	10.7	10.0	10.1	9.9	9.2	7.7	7.4	8.6	9.9	10.9	10.5	9.6
Malakal	9.7	9.3	9.0	8.3	7.7	5.5	5.3	5.9	6.0	7.3	8.6	10.1	7.7
Addis Ababa	8.7	8.5	8.0	7.1	7.0	5.2	2.2	2.7	4.6	8.6	8.9	8.6	6.4
Jimma	8.0	6.4	6.7	6.3	6.3	5.1	3.4	4.0	5.7	6.6	6.2	7.3	6.0
Wau	9.7	9.3	8.4	7.6	8.2	7.9	5.7	6.2	7.0	7.4	8.9	9.8	8.0
Juba	9.3	8.2	7.1	6.5	7.8	7.5	5.9	6.9	7.5	7.6	7.9	8.1	7.5
Gulu	8.9	8.7	8.0	7.4	7.8	7.7	6.2	6.4	8.1	8.1	8.2	8.6	7.8
Soroti	7.4	8.4	8.0	7.5	7.9	7.4	6.2	6.9	7.9	8.5	8.1	8.3	7.7
Namulonge	6.1	6.0	6.1	5.4	5.7	5.6	5.0	4.9	5.2	5.4	5.7	5.7	5.5

TABLE 3.7 (continued)

Station	Jan.	Feb.	Mar.	Apr.	May	Jun.	Jul.	Aug.	Sep.	Oct.	Nov.	Dec.	Year
Kampala	7.7	7.9	6.6	6.0	6.2	6.3	5.6	5.8	6.1	6.3	6.3	7.1	6.5
Entebbe	7.5	7.2	6.6	6.0	6.2	6.2	6.4	6.3	6.5	6.5	6.6	6.8	6.6
Eldoret	8.9	9.7	8.6	8.1	7.5	7.6	5.9	5.9	7.5	7.8	7.4	8.8	7.8
Equator	8.9	9.3	8.6	7.2	7.7	6.7	5.4	5.4	7.4	8.0	7.5	8.3	7.5
Kericho	8.0	8.1	7.6	5.7	6.4	6.8	5.8	5.6	6.0	5.7	5.6	6.9	6.5
Kabale	4.8	5.6	5.0	4.3	3.7	5.7	5.5	4.8	5.2	5.1	4.5	4.4	4.9
Kisumu	8.6	8.8	8.5	7.7	7.8	7.5	6.9	6.9	7.6	7.7	7.3	8.2	7.8
Mwanza	7.4	7.7	7.5	8.0	8.2	9.4	9.6	9.0	8.4	7.8	7.0	7.2	8.1

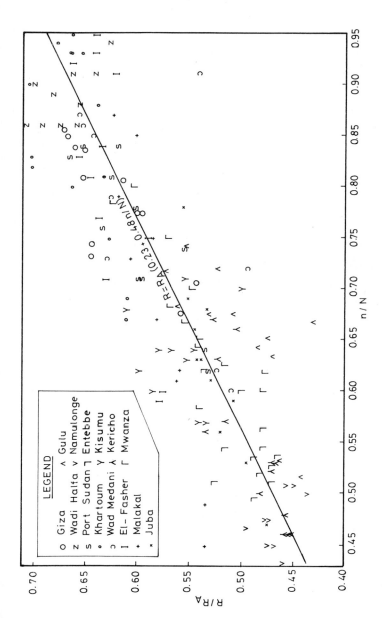

Fig. 3.7. The relative duration of bright sunshine, n/N, versus the ratio of the short wave radiation at the earth's surface to the short wave radiation at the top boundary of the earth's atmosphere for 14 stations in the Nile Basin area

and R/R$_A$ for the whole basin area by a single regression relation, the line best fitted by eye will be that represented by eq. 3.2 (see Fig. 3.7).

In the absence of observed sunshine data, one may try to estimate the ratio n/N from the amount of sky cloudiness that could be available at the place in question. The monthly and annual mean cloudiness at a number of stations spread in the Nile Basin area are summarized and presented in Table 3.8. The monthly values, S$_c$, are plotted versus the corresponding monthly mean relative duration of the bright sunshine, n/N, as shown in Fig. 3.8. From this figure it is clear that the points are so widely scattered that a fairly accurate relationship between S$_c$ and n/N for all stations combined can hardly be developed. Among the reasons behind this wide scatter is the inconsistency of the sky cloudiness produced by differences in the length of record and frequency and precision with which measurement is taken.

Doorenbos, J., and Pruitt, W.O. (1977), suggested the use of Table 3.9 as a rough guide for converting cloudiness into equivalent values of n/N. Palayasoot, P. (1965), upon investigating the relationship between S$_c$ and n/N, suggested the use of the equation

$$n/N = 100 - 1.6 \ S_c - 0.84 \ S_c^2 \tag{3.6}$$

where S$_c$ is in tenths on the scale 0 - 10.

From the data included in Table 3.9 and the results obtained from eq. 3.6, two lines have been drawn, as shown in Fig. 3.8.

The line given by Doorenbos and Pruitt is the approximate lower envelope for nearly all plotted points, whereas the curve described by eq. 3.6 forms the upper envelope of the n/N ratio from cloudiness from zero up to slightly less than 4 oktas (1 okta = 1.25 times 1 tenth).

The thick solid line drawn in Fig. 3.8. and which can be described by the equation

$$n/N = 88 + 0.43 \ S_c - 1.19 \ S_c^2 \tag{3.7}$$

where S$_c$ is in oktas, is the curve fitted by the method of least squares to the annual mean values (not shown in Fig. 3.8.). This curve as judged by eye seems to fit the monthly mean values with S$_c$ above 1 okta much better than any of the enveloping lines. In the lower half of the graph it falls inside the band formed by the boundary lines, and thereupon it passes halfway through the plotted points. For oktas more than 4, the n/N ratio obtained from eq. 3.6 is closest to the measurement than those estimated from the other methods. The major drawback of eq. 3.6 is that it gives for S$_c$ = 0, n/N = 88% instead of 100% and for S$_c$ = 8, n/N = 15% instead of 0%.

TABLE 3.8 Monthly and annual mean sky cloudiness at certain stations in the Nile Basin area (Griffiths, 1972; Ministry of War and Marine, Egypt, 1950; WMO, 1974)

Station	Amount of cloudiness, in oktas, for												
	Jan.	Feb.	Mar.	Apr.	May	Jun.	Jul.	Aug.	Sep.	Oct.	Nov.	Dec.	Year
Sidi Barrani*	3.4	3.5	2.6	2.7	2.2	1.4	1.8	1.9	1.5	2.2	2.4	2.8	2.4
Sallum (Observatory)	3.8	3.8	3.1	2.5	3.3	1.0	0.5	0.6	1.3	2.4	3.8	3.5	2.4
Damietta*	3.3	2.7	2.9	2.5	2.1	1.1	1.4	1.7	1.7	2.1	3.0	3.4	2.3
Rosetta*	4.0	3.5	3.4	2.7	2.3	1.6	1.7	2.1	2.2	2.3	3.4	3.7	2.7
Mersa Matruh	3.8	3.4	2.9	2.0	2.2	1.2	1.0	1.2	1.4	2.1	3.4	3.5	2.3
Port Said	3.2	2.9	2.7	2.3	1.9	1.1	1.2	1.4	1.5	2.0	2.6	3.2	2.2
Alexandria	4.1	3.7	3.2	2.6	2.4	1.3	1.4	1.6	1.9	2.4	3.4	4.1	2.7
Mansura*	2.7	2.4	2.2	1.6	1.5	1.0	1.5	1.3	0.9	1.4	2.4	2.7	1.8
Damanhur*	3.4	2.9	2.6	2.2	1.9	0.9	1.3	1.4	1.1	1.6	2.7	3.2	2.1
Tanta	3.3	3.3	2.7	2.2	2.0	0.6	1.0	1.1	1.3	1.8	2.8	3.4	2.1
Zagazig	2.4	2.3	1.8	1.4	1.0	0.5	0.7	0.9	0.9	1.2	2.0	2.3	1.4
Delta Barrage*	2.8	2.3	2.1	1.8	1.5	0.9	1.6	2.1	1.7	1.8	2.3	2.6	2.0
Cairo (Ezbekiya)	3.2	2.8	2.4	1.8	1.7	0.6	0.9	1.1	1.1	1.7	2.5	3.0	1.9
Giza	3.4	3.1	2.8	2.2	2.2	0.9	1.0	1.2	1.6	2.0	3.0	3.4	2.2
Suez (Port Tewfik)	2.7	2.4	2.0	1.6	1.5	0.4	0.3	0.5	0.6	1.3	2.0	2.6	1.5
Helwan	3.2	2.9	2.5	2.2	1.9	0.5	0.5	0.6	0.6	1.5	2.4	3.1	1.8
Fayum	2.8	2.5	2.1	1.8	1.9	0.4	0.3	0.5	0.5	1.3	2.3	3.0	1.6
Siwa	1.4	1.5	1.0	0.9	1.2	0.3	0.1	0.1	0.2	0.6	1.2	1.7	0.8
Beni Suef*	1.2	1.1	0.7	0.9	0.4	0.2	0.5	0.3	0.4	0.6	0.6	1.2	0.7
Minya	1.9	1.8	1.6	1.2	1.4	0.2	0.2	0.2	0.3	1.1	1.7	2.3	1.2
Hurghada*	1.6	1.4	1.3	0.8	0.9	0.1	0.0	0.1	0.2	0.7	1.1	1.9	0.8
Assiut*	1.3	1.0	0.8	0.7	0.8	0.2	0.1	0.0	0.1	0.4	0.8	1.3	0.6
Qena*	1.9	1.5	1.6	1.4	1.5	0.3	0.2	0.2	0.2	0.9	1.4	2.0	1.1
Qusseir	1.4	1.1	0.9	0.7	0.7	0.1	0.0	0.1	0.0	0.6	0.8	1.1	0.6
Nag-Hammadi	1.6	0.7	1.5	0.8	0.8	0.0	0.0	0.0	0.0	0.4	0.8	1.6	0.7
Luxor	2.3	1.8	1.8	1.8	1.5	0.2	0.3	0.2	0.2	1.0	1.2	2.1	1.2
Dakhla	1.4	1.0	0.9	0.8	0.9	0.2	0.1	0.0	0.0	0.3	0.6	1.1	0.6
Kharga	1.2	0.9	0.7	0.6	0.7	0.2	0.0	0.0	0.0	0.3	0.6	1.0	0.5
Esna	1.1	0.9	0.9	0.6	0.6	0.0	0.1	0.2	0.1	0.4	0.5	0.9	0.5

* = observed at 08 00 hours local time

TABLE 3.8 (continued)

Amount of cloudiness, in oktas, for

Station	Jan.	Feb.	Mar.	Apr.	May	Jun.	Jul.	Aug.	Sep.	Oct.	Nov.	Dec.	Year
Deadalus Island	2.3	2.0	1.7	1.6	1.6	0.6	0.8	1.0	1.1	1.6	2.1	2.5	1.6
Kom Ombo	1.2	0.8	0.7	0.5	0.9	0.1	0.1	0.1	0.1	0.3	0.4	0.9	0.5
Aswan	1.2	1	0.9	0.9	0.8	0.2	0.2	0.3	0.2	0.4	0.9	1.0	0.7
Wadi Halfa+	1	1	1	1	1	<1	1	1	1	<1	1	1	1
Port Sudan+	3	3	2	1	1	1	2	2	2	2	3	2	2
Dongola+	<1	<1	1	1	1	<1	1	1	1	<1	<1	<1	1
Khartoum+	1	1	2	2	2	3	4	5	4	2	1	1	2.3
Kassala+	1	1	3	2	2	3	5	5	3	2	1	1	2.0
El Roseires+	1	1	1	2	3	4	5	5	4	3	1	1	3.0
Malakal+	2	3	3	4	5	6	6	6	6	5	3	2	4.3
Addis Ababa	3	4	5	5	5	6	7	7	6	4	2	2	4.7
Wau+	2	3	4	5	5	5	6	6	6	5	4	3	4.5
Jimma++	3	3	4	5	4	5	7	6	5	3	2	2	4.1
Juba+	3	4	5	6	6	5	6	5	5	5	5	4	4.9
Gulu	5	5	6	6	6	6	6	6	6	6	5	5	5.7
Entebbe	6	6	6	6	6	6	6	6	6	6	6	6	6.0
Equator	6	5	6	7	6	6	6	6	5	6	6	6	5.9
Kisumu	5	5	6	6	5	5	5	5	5	5	6	5	5.2
Kabale	7	7	7	7	7	6	6	6	7	7	7	7	6.8

+ = mean of 06 00, 12 00 and 18 00 hours local time
++ = mean of 08 00 and 14 00 hours local time

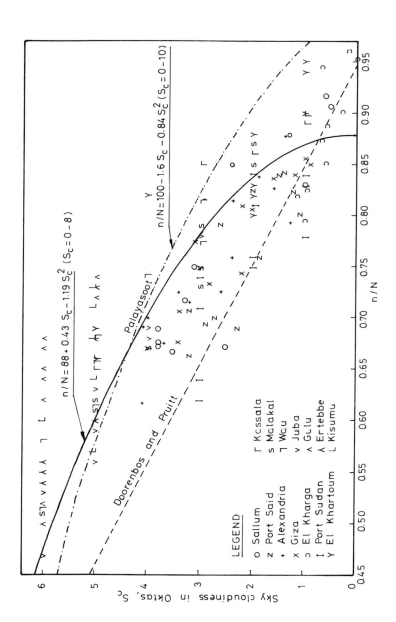

Fig. 3.8. The amount of sky cloudiness versus the relative duration of bright sunshine for 14 stations in the Nile Basin area

94

TABLE 3.9   Conversion of sky cloudiness, S_c, into relative duration of bright
sunshine, n/N, as suggested by Doorenbos and Pruitt (1977)

S_c, oktas (0-8)	0	1	2	3	4	5	6	7	8		
n/N ratio		.95	.85	.75	.65	.55	.45	.30	.15	-	
S_c, tenths (0-10)	0	1	2	3	4	5	6	7	8	9	10
n/N ratio		.95	.85	.80	.75	.65	.55	.50	.40	.30	.15	-

3.7 WIND

In this section we shall try to draw a sketch for the surface winds blowing in the Nile Basin area.

Wind is usually measured at about 2 metres from the earth's surface by means of atmometers. The wind characteristics at a height of 10 metres above the surface are automatically recorded at a limited number of places by means of mechanical recording anemographs.

The wind force is given in terms of numbers on the Beaufort Scale when the force is estimated, or in units of velocity as m/sec or km/hr, when the velocity is measured by an anemometer. The frequency with which this velocity is measured varies from one country to another, and for the same country from one group of stations to another. In Egypt, the wind force is estimated once at 08 00 hours local time, whereas the velocity measurements are taken at 08 00, 14 00 and 20 00 hours local time and the mean given. In the Sudan, the wind speed and direction are measured at 07 00 and 13 00 hours local time. In east Africa, wind measurements are taken at 06 00 Z and 12 00 Z (09 00 and 15 00 hours local time). A summary of the wind speed, prevailing direction and percentage frequency of calm conditions in Egypt, Sudan and Ethiopia, Uganda and Kenya, is given in Tables 3.10(a), (b) and (c) respectively. In East Africa, the velocity measured at a height of 10 metres above the surface does not compare to the fast blowing wind often noted in the temperate zones. The monthly mean value hardly reaches 10 m/sec. The maximum gust recorded is about 30 m/sec at Kisumu, while values above 18 m/sec are rare and seldom last for more than a minute or two (Griffiths, J.F., 1971).

The seasonal variation in the wind speed is not strong everywhere. In the highly elevated places such as Addis Ababa, Jimma and Equator, the ratio of the maximum to the minimum monthly mean is between 2.7 to 1 and 2.0 to 1. Low elevated stations such as Gulu, Entebbe, Kisumu and Kabale show the maximum to minimum ratio of between 1.5 to 1 and 1.15 to 1. In general, the wind blows faster in the early spring and in autumn than in the remaining seasons of the year.

For all stations included in Table 3.10(c) and for other East African stations which are not included in this table, the wind velocity measured at 12 00 Z (second measurement) is more than that measured at 06 00 Z (first measurement). The ratio between these two measurements varies from 1.05:1 for the annual mean at the Equator to 2.72 at Kabale. The magnitude of this ratio depends, among other factors, on the altitude of the point in question and its distance to the nearest lake shore. The prevailing wind along the shore of Lake Victoria changes its main direction from south-east at the southern part of the western shore to south at the northern part of the same shore, and to south-west at the northern part of the eastern shore. North of Lake Victoria, the direction of the prevailing wind swings between the east and the north. A more detailed review of the characteristics of the surface winds in the Equatorial Lake Plateau can be found in the report on the Hydrometeorological Survey of the Catchments of Lakes Victoria, Kyoga and Albert ((WMO, 1974).

The wind blowing on the Ethiopian highlands is characterized by the considerable variation in the percentage of calm conditions during the year with the local conditions (compare Addis Ababa to Jimma, Table 3.10(c)). Generally speaking, however, the percentage of calm condition is greatest at the early morning reading (06 00 hours). At Addis Ababa it reaches the maximum in May and the minimum in November, both at the early morning reading and in July and October respectively, both at the midday reading. At Jimma it is autumn and winter which enjoy the greatest percentage of the calm condition.

Wind directions are generally dominantly east to south-easterly during the dry season but veer around to westerly or north-westerly in the rainy season.

The annual mean velocity of the surface wind over the Sudan is generally not high. It increases slightly from Wadi Halfa to Khartoum and from there decreases rather rapidly in the direction to the south. To the east, except on the Red Sea coast, and to the west, the wind speed is, on average, less than at those stations on, or near, the Nile and its main tributaries.

From the data listed in Table 3.10(b) it is clear that the maximum average speed occurs in the winter time, except for Wadi Halfa and Kasssala as it occurs in the spring and summer, respectively. The minimum occurs in the summer everywhere, except at Kassala where late autumn is the time of occurrence of the minimum of the average wind speed. The seasonal effect is quite small, except for Malakal and Kassala where the ratio of the maximum to the minimum, both at average speed, is about 2.7 or 2.8 to 1.

The prevailing wind changes its direction from north at Halfa to north-west at Dongola and then back to north at Khartoum except during the summer season where the direction becomes mostly south and south-south-west. In this period the "haboob" (Arabic word meaning "blowing"), which refers mostly to duststorm or sandstorm, occurs rather frequently. This wind applies to a rapidly-moving and unstable type of air mass associated with thunder activity and related to advancing or early monsoon. The average number of duststorms reaches a maximum around Khartoum with averages 4, 6, 4 for May, June and July, respectively. At Atbara the maximum of 5 occurs in August.

Along the diagonal joining Kassala and Malakal and on both sides of its extension, i.e. Juba and Wau, the north-north-east wind prevails in the winter time. During the spring, summer and part of the autumn, the southerlies and the south-westerlies prevail.

The unusual nature of the diurnal wind variation, i.e. the decrease of the wind speed from morning to midday was investigated by Farquharson, J.S. (1939). He concluded that this type of variation is a characteristic of tropical Africa and not only of Central Sudan. In his comment on that work, Sutton presented the two maps which are shown here in Figs. 3.9(a) and 3.9(b). These maps show that the unusual type of diurnal variation occurs in an extensive area between latitudes 20°N and S. Within these latitudes the usual type of diurnal variation covers, however, large regions (Sutton, L.J., 1939). Sutton added further that in some parts of the Sudan there are large seasonal variations. For example, in the north (Merowe) the morning wind is stronger than that in the afternoon only from August to October during the SW monsoon. In the south (Malakal, Nasser), the morning wind is stronger than the afternoon wind only from December to March during the period of NE winds. But in the south-west, Bahr el Ghazal Province (Aweil, Raga, Wau) the morning wind is weaker than the afternoon wind all the year round. The contrast in the behaviour of the wind in these south and south-west regions, which form a continuous plain in the same latitudes, does not seem attributable to local variations of topography.

In the winter season the stations situated along the coast of the Mediterranean Sea in Egypt and the stations in the Nile Delta are strongly affected by the so-called Mediterranean depressions. At Sallum, which is located at the rear of the track of these depressions, the prevailing wind direction is west to north-west. Those places located in front of the depressions, like Alexandria and Qurashiya receive south-westerly winds (see Table 3.10(a)). Upper Egypt and the northern Red Sea area are practically unaffected by the Mediterranean depressions and are, for most of the time, under the eastern flank of the sub-tropical high pressure cell that covers the western desert of Egypt (Stamp, D.L. and Morgan, W.T.W., 1972). The prevailing winds are, therefore, northerlies (see data of Helwan, Minya and Luxor in Table 3.10(a)).

In the summer season the northerly winds prevail over most of Egypt. The northern part of the Nile Delta area and west of it up to Alexandria are an exception to this rule. The north-west direction usually prevails.

The direction of the prevailing wind over Lower Egypt changes in the spring and autumn to north-east because of the passage of the centre of the Khamasin depressions (duststorms blowing from the western quadrant and lasting fifty days) south of the Mediterranean coastline.

The mean annual wind speed decreases along the Mediterranean coast from west to east. It also decreases from north to south. The northern part of the Red Sea coast is a relatively windy area. The wind conditions inland are adjusted, however, by the local topography.

Late winter and early spring are the seasons in which gales and strong winds take place in Lower Egypt. The frequency of these gales decreases considerably along the north-south direction. In general, autumn, especially October, is the time of lowest wind speed.

TABLE 3.10(a)   Wind speed and direction at certain stations in Egypt (Ministry of War and Marine, Egypt, 1950)

Month and Year	Sallum (Observatory)			Port Said (Airport)			Alexandria			Qurashiya		
	Avg. speed, km/hr	Prev. direction	Calm, %	Avg. speed, km/hr	Prev. direction	Calm, %	Avg. speed, km/hr	Prev. direction	Calm, %	Avg. speed, km/hr	Prev. direction	Calm, %
January	19.9	W	5.5	15.6	W	8.4	15.8	SW	11.6	4.9	SW	53.0
February	20.4	NW	4.8	14.2	W	7.8	16.4	NW	9.0	5.7	SW	45.6
March	20.3	N	5.4	17.9	N	1.7	16.3	NE	7.1	5.5	SE	49.5
April	20.0	N	4.6	15.3	N	2.5	15.5	NE	4.6	5.9	SE	43.1
May	17.1	NE	4.2	14.3	N	3.2	14.0	NE	6.1	5.4	SE	49.7
June	19.8	N	3.2	14.7	N	2.4	15.3	NW	4.6	5.6	SE	42.6
July	24.9	N	1.3	14.5	N	1.3	16.2	NW	1.3	5.4	NW	44.7
August	22.6	N	1.7	11.8	N	3.2	14.9	NW	3.8	4.0	SW	57.2
September	19.3	N	2.4	11.3	N	4.6	13.1	N	5.2	2.9	NW	71.0
October	16.1	N	6.7	11.8	N	5.8	11.1	NE	9.9	2.3	SE	76.0
November	16.6	NW	6.1	13.7	N	5.3	12.0	NE	11.5	3.0	SW	71.3
December	19.6	NW	4.6	11.7	W	10.1	12.8	SW	15.6	4.2	SW	59.4
Year	19.7	N	4.2	13.9	N	4.7	14.4	NW	7.5	4.6	SW	55.3

Month and Year	Cairo (Almaza Airport)			Giza			Helwan			Minya (Airport)		
	Avg. speed, km/hr	Prev. direction	Calm, %	Avg. speed, km/hr	Prev. direction	Calm, %	Avg. speed, km/hr	Prev. direction	Calm, %	Avg. speed, km/hr	Prev. direction	Calm, %
January	12.3	SW	8.8	7.6	NW	26.4	13.1	N	2.5	6.4	N	23.9
February	11.5	NE	7.3	8.6	NW	22.6	14.4	N	3.5	6.8	N	20.8
March	14.2	NE	5.0	9.8	N	16.9	17.2	N	2.3	9.4	N	11.0
April	13.5	NE	2.5	9.9	N	14.4	19.0	N	1.2	9.6	N	13.7
May	13.0	NE	3.7	10.5	N	15.6	20.0	N	0.3	10.7	N	11.1
June	11.3	N	2.9	10.5	N	10.5	20.2	N	0.2	12.7	N	7.1
July	7.7	N	6.2	9.7	NW	11.0	18.1	NW	0.2	9.5	N	16.8
August	8.5	N	6.9	9.6	N	14.1	18.0	N	0.5	7.7	N	23.2
September	8.7	N	12.4	9.4	N	15.6	19.0	N	0.7	8.3	N	17.1
October	9.0	NE	15.7	8.2	N	19.3	18.4	N	2.7	6.9	N	23.6
November	9.8	NE	14.4	7.4	N	24.4	15.5	N	4.6	7.1	N	27.3
December	9.4	NE	13.1	6.8	N	30.7	12.6	N	3.7	5.6	N	21.5
Year	10.9	NE	8.2	9.0	N	18.5	17.1	N	1.9	8.4	N	18.1

TABLE 3.10(a)   (continued)

Month and Year	Qusseir Avg. speed, km/hr	Prev. direction	Calm, %	Luxor Avg. speed, km/hr	Prev. direction	Calm, %	Dakhla Avg. speed, km/hr	Prev. direction	Calm, %	Aswan Avg. speed, km/hr	Prev. direction	Calm, %
January	18.0	W	4.2	6.8	N	28.2	6.3	NW	23.8	7.7	N	48.0
February	18.0	W	5.3	7.2	N	27.6	5.9	NW	23.0	7.7	N	43.0
March	18.0	N	6.1	7.9	NW	14.3	6.8	NW	15.7	8.6	N	36.4
April	18.0	N	9.3	7.2	NW	14.1	7.2	NW	12.8	8.6	N	35.9
May	18.0	N	9.4	7.6	NW	13.9	6.8	NW	10.6	8.1	N	42.8
June	14.4	N	7.2	7.2	NW	12.7	7.2	NW	9.4	8.5	N	36.1
July	14.4	N	10.8	7.6	NW	13.0	7.7	NW	6.4	7.2	N	52.3
August	18.0	N	13.2	7.6	NW	13.2	7.7	NW	6.9	7.2	N	41.0
September	14.4	N	8.9	6.9	N	18.6	8.1	NW	5.8	8.1	N	29.6
October	18.0	N	7.2	5.8	N	19.0	6.8	NW	11.1	7.7	N	37.9
November	14.4	N/W	2.8	6.1	N	19.3	5.4	NW	16.5	6.8	N	46.0
December	14.4	W	2.1	6.5	N	27.5	5.9	NW	19.8	7.2	N	45.6
Year	16.5	N	7.5	7.0	N/NW	18.5	6.8	NW	13.5	7.8	N	41.2

TABLE 3.10(b) Wind speed and direction at certain stations in the Sudan (Griffiths, 1972)

Month and Year	Wadi Halfa			Port Sudan					Dongola			
	Avg. speed, km/hr	Prev. direction	Calm, %	Avg. speed, km/hr	Prevailing direction		Calm, %		percentage frequency of avg. wind speed		Prev. direction	Calm, %
					07 hr	13 hr	07 hr	13 hr	1-3	4-7		
January	14	N	5	18.5	NW	NE	2	0	97	2	NE	1
February	15	N	4	18.5	NW	NE	3	0	96	3	NE	<1
March	16	N	3	16.7	NW	NE	9	0	92	8	NE	<1
April	16	N	6	16.7	NW	NE	15	0	93	5	NE	2
May	16	N	5	14.8	NW	NE	23	0	94	4	NE	<1
June	15	N	6	13.0	NW	E	33	1	95	4	NE	1
July	14	N	10	14.7	SW	E	30	1	94	4	NE	2
August	13	N	9	13.0	SW	NE	23	0	94	4	NE	2
September	14	N	4	13.0	NW	NE	33	0	95	5	NE	<1
October	14	N	3	13.0	NW	NE	6	0	96	4	NE	<1
November	13	N	4	14.9	NW	NE	1	0	94	6	NE	0
December	11	N	6	16.7	NW	NE	4	0	96	4	NE	<1
Year	14	N	5	15.3	NW	NE	15	0	95	5	NE	1

TABLE 3.10(b) (continued)

Month and Year	Khartoum Avg. speed km/hr	Khartoum Prev. direction	Khartoum Calm %	Kassala Avg. speed km/hr	Kassala Prev. direction	Kassala Calm %	Malakal Avg. speed km/hr	Malakal Prev. direction	Malakal Calm %	Wau Avg. speed km/hr	Wau Prev. direction	Wau Calm %
January	16	N	<1	5	NNE	6	14	NNE	2	5	NNE	34
February	16	N	<1	5	NNE	6	14	NNE	5	3	NNE	29
March	15	N	1	5	NNE	6	10	N	6	5	V	28
April	14	N	1	5	N	7	10	S	8	5	S	23
May	16	S	2	5	S	12	10	S	4	5	S	23
June	14	S	2	6	S	9	8	S	8	3	SW	28
July	16	SSW	1	8	S	3	8	S	10	3	SW	33
August	15	S	2	6	S	5	5	V	13	3	SW	35
September	11	N	3	5	S	9	5	V	15	3	V	38
October	13	N	1	3	S	12	5	N	14	3	V	43
November	14	N	0	3	NNE	8	8	N	9	3	NNE	34
December	16	N	0	5	NNE	7	12	NNE	5	3	NNE	32
Year	15	N	1	5	-	7	9	-	8	4	-	

Month and Year	Juba Avg. speed km/hr	Juba Prev. direction	Juba Calm %
January	5	NNE	36
February	5	NNE	29
March	5	S	21
April	5	S	17
May	5	S	19
June	3	S	30
July	3	S	31
August	3	S	26
September	3	S	27
October	3	S	23
November	3	S	29
December	3	NNE	32
Year	4	S	27

TABLE 3.10(c) Wind speed and direction at certain stations in Ethiopia, Kenya and Uganda (Griffiths, 1972; WMO, 1974)

Month and Year	Addis Ababa				Jimma				Gulu			
	Avg. speed, km/hr	Prev. direction	Calm, % 06 hr	Calm, % 12 hr	Avg. speed, km/hr	Prev. direction	Calm, % 06 hr	Calm, % 12 hr	Avg. speed, km/hr 09 hr	Avg. speed, km/hr 15 hr	Prev. direction	Calm, %
January	11.1	SSE	56	12	7.4	E	97	53	13.0	18.5	N/E	0
February	9.3	E	51	10	9.3	SE	92	42	13.0	18.5	N	0
March	11.1	E	40	11	7.4	E	87	56	13.0	18.5	E	0
April	13.0	E	41	10	5.6	E	87	47	14.8	14.8	S	0
May	11.1	S	64	16	7.4	SE	88	49	13.0	14.8	S	0
June	9.3	S	52	23	5.6	SSE	95	54	11.1	13.0	S	0
July	7.4	S	53	39	7.4	SE	87	63	11.1	13.0	N	1
August	7.4	S	51	13	5.6	E	83	54	11.1	14.8	N	0
September	9.3	S	57	21	3.7	E	95	50	13.0	14.8	N	0
October	13.0	E	33	6	5.6	SSE	97	64	13.0	14.8	E	0
November	14.8	SE	29	7	3.7	E	96	76	13.0	16.7	E	0
December	13.0	E	40	8								
Year	10.9	-	47	15	6.2	-	92	56	12.5	15.4	-	0

TABLE 3.10(c)   (continued)

Month and Year	Avg. speed, km/hr 09 hr	Avg. speed, km/hr 15 hr	Prev. direction	Calm, %
		Entebbe		
January	7.4	18.5	S	0
February	7.4	22.2	S	0
March	9.3	22.2	S	0
April	11.1	18.5	S	0
May	11.1	18.5	S	0
June	11.1	22.2	S	0
July	11.1	18.5	S	1
August	9.3	18.5	S	1
September	11.1	18.5	S	0
October	9.3	18.5	S	1
November	7.4	18.5	S	1
December	7.4	16.7	S	1
Year	9.4	19.3	S	<1

Month and Year	Avg. speed, km/hr 09 hr	Avg. speed, km/hr 15 hr	Prev. direction	Calm, %
		Equator		
January	31.5	27.8	E	0
February	29.6	29.6	E	0
March	31.5	31.5	E	0
April	25.9	27.8	E	0
May	20.4	25.9	E	1
June	14.8	22.2	E	1
July	13.0	18.5	W	4
August	13.0	16.7	W	2
September	16.7	20.4	E	1
October	25.9	24.1	E	0
November	33.3	29.6	E	0
December	35.2	29.6	E	0
Year	24.2	25.3	–	1

Month and Year	Avg. speed, km/hr 09 hr	Avg. speed, km/hr 15 hr	Prev. direction	Calm, %
		Kisumu		
January	5.6	22.2	SW	1
February	5.6	24.1	SW	0
March	5.6	20.4	SW	1
April	5.6	18.5	SW	1
May	5.6	14.8	SW	1
June	7.4	14.8	SW	0
July	7.4	14.8	SW	1
August	7.4	16.7	SW	1
September	7.4	18.5	S	0
October	7.4	18.5	SW	0
November	5.6	18.5	SW	1
December	5.6	18.5	SW	1
Year	6.4	18.4	SW	1

Month and Year	Avg. speed, km/hr 09 hr	Avg. speed, km/hr 15 hr	Prev. direction	Calm, %
		Kabale		
January	5.6	13.0	NE/SE	1
February	3.7	13.0	NE/SE	0
March	3.7	13.0	E/SE	0
April	3.7	13.0	SE	0
May	5.6	13.0	SE	0
June	5.6	14.8	SE	0
July	5.6	14.8	SE	0
August	5.6	14.8	SE	0
September	5.6	14.8	SE	0
October	5.6	14.8	SE	0
November	5.6	14.8	E/SE	0
December	5.6	13.0	E/SE	0
Year	5.1	13.9	–	0

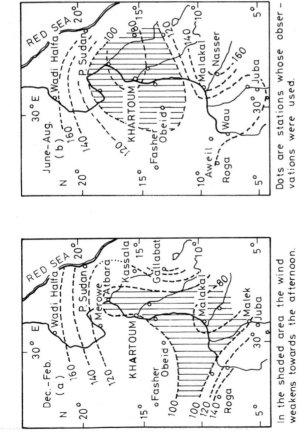

Fig. 3.9.   Wind force at 14 00 hours expressed as percentage of force at 08 00 hours (Sutton, 1939)

3.8.   CIRCULATION OF AIR MASSES

The general circulation of the atmosphere in Africa is dominated by two sub-tropical high-pressure belts and the equatorial low-pressure belt between. These belts move northwards in the northern summer, lagging behind the overhead sun, and southwards in the southern summer. The circulation of the air masses in the Nile Basin area is part of the general circulation over the continent, modified by the local conditions. The two hemispheric anticyclones which extend from latitudes 30° north and south to the Equator, control the wind systems in the equatorial region and the tropics. The north and south anticyclones are most intense during the northern winter and summer respectively.

The Equatorial Lake Plateau is covered by a part of the equatorial trough which extends over Central Africa and where winds are generally light and variable. The easterlies, north-east and south-east monsoons, prevail over the Lake Plateau area. The resulting weather pattern, however, is significantly modified by the westerlies. The latter are known to be a rain-bringing agent.

The north-east monsoon originates from the north sub-tropical anticyclone. It flows towards lower latitudes when the sun is south of the Equator, from October to March. It is that segment which originates over Saudi Arabia while travelling over the Red Sea and part of the Indian Ocean that acquires some moisture as it goes down to lower latitudes. This air mass prevails over the Lake Plateau.

The south-east monsoon takes place from April to September and originates as an easterly wind from the south-tropical anticyclone over the south Indian Ocean. By the time this air mass enters the eastern seaboard of Africa it already has a maritime trajectory.

The westerlies cover a narrow belt stretching from the equatorial regions of South America to equatorial Africa. They intrude into the circulation established by the easterlies and play a role in the change of the weather patterns. The north-east monsoon crosses the Equator during the winter in the northern hemisphere and under the influence of the change in the direction of the Criolis force, the stream blowing south of the Equator appears as if coming from the west. The westerly flow is supported by the pressure gradient caused by the Lake Victoria low. A segment of the south-east monsoon mentioned above and which has continued its trajectory to the west splits before leaving the west coast and invades a narrow belt north of the Equator. Here the stream moving to the east is deflected by the Criolis force and caused to blow over the Zaire Basin as a westerly wind which then intrudes in the established circulation. The zone of separation of the northern from the southern hemispheric air masses is known as the intertropical convergence zone, ITCZ. This zone migrates in the course of the year between the two tropics, i.e. between latitude $23^\circ30'$N and

$23^{o}30'$S. The winds are generally south-east to south from 5°S to the Equator
and north to north-east from the Equator to 5°N. This extent covers the Lake
Plateau and the extreme south of the Sudan. Part of the rain falling on this
area comes from the ITCZ as it follows the polar migration of the sun and the
thereupon lagged monsoons. Since the movement and position of ITCZ are
influenced by the strength of the monsoons, the rainfall thus derived is
variable in amount and intensity.

A detailed summary of the work of Trewartha (1962) and Flohn and Struning
(1969) concerning the atmospheric circulation in the Lake Plateau area appears
in the report on the hydrometeorological survey of the catchments of Lake
Victoria, Kyoga and Albert (WMO, 1974).

In the northern summer (mainly June to August) a trough of low pressure
covers the northern Sudan. An oscillatory barometric minimum is located over
the north-east Sudan, while a barometric maximum (of the Indian Ocean anticyclone) extends over the greater part of Abyssinia and south-east Sudan. This
maximum suffers fluctuations in its strength and northward extent. The general
wind circulation meets, therefore, with some disturbances. Four types of disturbances have been distinguished by El-Fandy (1949). These are:

(i)   Sand storms (haboobs) with a modification of the general circulation by
        the onset of strong southernly winds. These storms are common over the
        desert areas of the central and northern Sudan. Marked strengthening and
        deepening of the westerlies usually accompany the northward extension of
        the barometric maximum of the south-east Sudan together with a northward
        oscillation of ITCZ. The modification manifests itself by a rise of
        pressure and fall of temperature in the southern Sudan-Uganda area and
        spreads northwards rapidly.

(ii)   Duststorms (haboobs) with "induced cold fronts" or line squalls set up by
        thunderstorms.

(iii)   Low-level thunderstorms, which are formed under favourable conditions of
        temperature and humidity, by direct convection within the south to south-
        east monsoon. These generally occur when the ITCZ is at its farthest
        north, the moisture content showing a marked increase within the lowest
        layers.

(iv)   High-level thunderstorms followed by outbreaks of rain covering a wide
        area and possibly persisting for about 24 hours in some places.

El-Fandy (1948) also found that thundery conditions in Egypt occur when the
Sudan monsoon low intensifies towards the north and supplies the eastern
Mediterranean with a warm south-easterly current. Quite often shallow depressions form over the south-eastern Mediterranean, or the adjacent land areas, as
a result of the meeting of the above current with the relatively cold north-

easterly air of anticyclonic distribution in Asia Minor. In a few cases only, they form over the western desert. Once fully developed, these depressions move eastwards or north-eastwards, while the inflow of the cold air in the rear gives thundery showers. Such thunderstorms are specially marked over the sea, Syria and Israel. Over Egypt, however, high-level instability frequently exists between the warm current and the cool air advancing above. Owing to the stability of the surface layers, this high-level instability cannot pass into the thunderstorm stage without sufficient convergence of damp air. With a high degree of latent instability, severe thunderstorms are then produced.

The above discussion of the atmospheric circulation might help us form an idea about the rainy seasons in the Nile Basin area. Generally speaking, the rainfall over the Equatorial Lake Plateau, the extreme south of the Sudan up to Juba and the Ethiopian Highlands can occur in every month of the year. Over the Sudan, away from the Red Sea coast, rainfall is effectively confined to a definite season. This season lasts from March to November along the line connecting Wau and Malakal. Further north, the season becomes progressively shorter; from May to October on the line passing through Geneina, El Nahoud, El-Obeid, Khartoum and Kassala, from June to September at Atbara, and July and August for places situated more to the north up to Wadi Halfa. The rainfall on Lower Egypt and the Mediterranean coast can occur in all months of the year except June, July, and August.

The peak rainfall over the Equatorial Lake Plateau occurs in April with the possibility of a less pronounced second peak around November in some places. The rainfall over the Ethiopian Plateau and the Sudan shows that July and August are the months of peak rain. The month of maximum rainfall over the Mediterranean coast and Lower Egypt is January, and over Middle Egypt, December. Though the rainfall over Upper Egypt is extremely low, yet its distribution among the months of the year clearly shows two marked maxima, one in October and the other in May.

The diurnal variation of the incidence of rainfall on the Lake Plateau was investigated and discussed by Thompson, B.W. (1957). A few of his many fundamental findings is that the maximum of rainfall in a day over western Lake Victoria during November to May, i.e. in the north-eastern monsoon, occurs in the early morning (00 - 06 hours local time) and over north-western Lake Victoria during May to September, i.e. in the south-east monsoon, in the morning (06 - 12 hours local time). The heavy rainfalls east of Lake Victoria occur between 12 and 18 hours and over north-eastern Victoria in the late evening (18 - 24 hours). There is a lag in time between the incidence of heavy rains inland and on the lake. A detailed investigation and discussion of the diurnal variation of the incidence of monsoon rains over the Sudan and of rain and thunder at Asmara and Addis Ababa, Ethiopia were given by Pedgley, D.E., (in 1969 a and b, and 1971 respectively).

A similar investigation on the rainfall over Egypt, with emphasis on the diurnal variation of the fall on Alexandria, Port Said and Cairo (Almaza) was carried out and reported by Soliman, K. (1953).

A summary of these results is presented in Fig. 3.10. Since July and August are the two months with the most frequent occurrence of rain in the Sudan and the Ethiopian Plateau, the results obtained by Pedgley have been averaged over July and August. The results obtained by Soliman for Egypt have been left as they are for the whole duration of the rainy season.

The patterns of diurnal variation may be grouped under four types with the following characteristics:

(i)   Rainfall reasonably uniformly distributed throughout the day, with a weak maximum in the afternoon or early evening. This is the case at Juba, Malakal, El Obeid, Addis Ababa, etc.

(ii)   Similar to (i) except that the weak maximum takes place during the early morning. This pattern can be found at Kosti and Tozi in the Sudan and at Alexandria in Egypt.

(iii) A well-defined peaked distribution where the peak takes place in the late afternoon or early evening. Examples of this pattern can be found at Wadi Halfa, Abu Hamed, Port Sudan, Geneina, etc.

(iv)   A well-defined peaked distribution where the peak takes place during the late evening or early morning and with infrequent rains during the middle of the day. This is the case at Wad Medani and El-Khartoum.

Pedgley concluded that the distribution of the four types of stations forms a definite pattern over the Sudan. Type (iii) is to be found in the middle of the country, between the extensive highland massif of Ethiopia and the smaller Marra Mountains which are situated west of El-Fasher. Except for El-Fasher and Geneina, the remaining stations of types (i) and (iv) lie within 700 km of the Ethiopian Highlands. On the diurnal variation of precipitation in Egypt, generally speaking, the maritime type has its maximum in the early morning or during the night, and the continental type, which has its maximum in the afternoon.

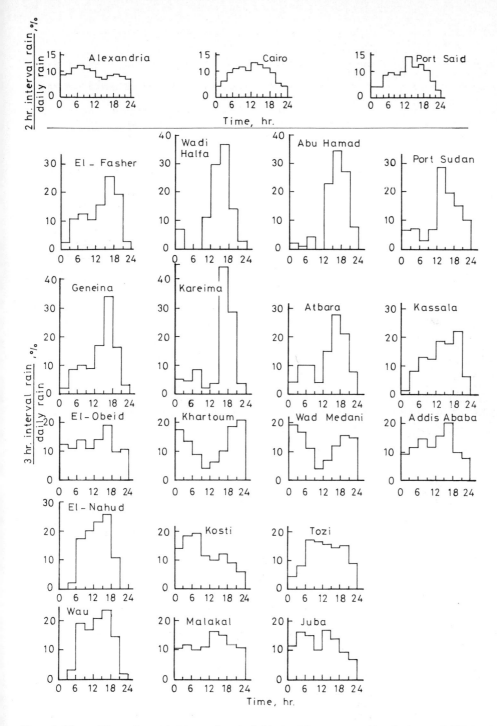

Fig. 3.10. Diurnal variation of rainfall incidence over the Sudan and Ethiopia, average of July and August, and over Egypt average for rainy season

REFERENCES

Angström, A., 1924. Solar and atmospheric radiation, 1924. Quart. Journ. Roy.
    Meteo. Soc., 50: 121-125.
Black, J.N., Bonython, C.W., and Prescott, J.A., 1954. Solar radiation and the
    duration of sunshine. Quart. Journ. Roy. Meteo. Soc., 80: 231-235.
Butzer, K.W., 1966. Climatic changes in the arid zones of Africa during early to
    mid Holocene times. Roy. Meteo. Soc.: World Climate from 8000 to 0 BC.
    Proc. 72-83.
Butzer, K.W., 1971. Environment and Archaeology, 2nd edition. Methuen & Co.;
    Aldine Atherton, Chicago, New York.
Doorenbos, J., and Pruitt, O.W., 1977. Crop water requirements. FAO Irrigation
    and drainage paper 24 (Revised), FAO, Rome.
El-Fandy, M.G., 1948. The effect of the Sudan monsoon low and the development of
    thundery conditions in Egypt, Palestine, and Syria. Quart. Journ. Roy. Meteo.
    Soc., 74: 31-38.
El-Fandy, M.G., 1949. Forecasting the summer weather of the Sudan and the rains
    that lead to the Nile Floods. Quart. Journ. Roy. Meteo. Soc., 75: 375-398.
Farquharson, J.S., 1939. The diurnal variation of wind over tropical Africa.
    Quart. Journ. Roy. Meteo. Soc., 65: 165-180.
Flohn, H., and Struning, J.O., 1969. Investigations on the atmospheric circula-
    tion above Africa. Meteo. Inst. Univ. Bonn (W. Germany), Bonner Meteo.
    Abhandlungen: 10-55.
Glover, J., and McCulloch, J.S., 1958. The empirical relation between solar
    radiation and hours of sunshine. Quart. Journ. Roy. Meteo. Soc., 84: 172-175.
Griffiths, J.F., 1971: East Africa: Its peoples and resources. Chapter 9:
    Climate, 107-118, edited by W.T. Morgan, Oxford University Press, Nairobi,
    London, New York.
Griffiths, J.F., 1972: Climates of Africa, Chapter 10: World Survey of
    Climatology, edited by J.F. Griffiths, Elsev. Pub. Co., Amsterdam, New York.
Hurst, H.E., and Philips, P., 1931. The Nile Basin, Vol. I, General description
    of the basin, meteorology and topography of the White Nile Basin, Physical
    Department Paper 26, Government Press. Cairo, 128 pp.
Ireland, A.W., 1948. Agriculture in the Sudan, Chapter V: The climate of the
    Sudan, 62-83, edited by Tothill, J.D., Oxford University Press, London.
Ministry of War and Marine, Egypt, 1950. Climatological normals for Egypt.
    Meteorological Department of Egypt, C. Tsoumas & Co. Press, Cairo.
Olivier, H., 1961. Irrigation and climate. Edward Arnold (Publishers) Ltd.,
    London, 250 pp.
Palayasoot, P., 1965. Estimation of pan evaporation and potential evapotranspi-
    ration of rice in the central plain of Thailand by using various formulas
    based on climatological data. M.Sc Thesis, College of Engineering, Utah State
    University, Logan.
Pedgley, D.E., 1969a. Diurnal variation of the incidence of monsoon rainfall
    over the Sudan, I. Quart. Journ. Roy. Meteo. Soc., 98: 97-107.
Pedgley, D.E., 1969b. Diurnal variation of the incidence of monsoon rainfall
    over the Sudan, II. Quart. Journ. Roy. Meteo. Soc., 98: 129-134.
Pedgley, D.E., 1971. Diurnal incidence of rain and thunder at Asmara and Addis
    Ababa, Ethiopia. Quart. Journ. Roy. Meteo. Soc., 100: 66-71.
Soliman, H.K., 1953. Rainfall over Egypt. Quart. Journ. Roy. Meteo. Soc., 79:
    389-397.
Soliman, H.K., 1972. Climates of Africa, the climate of the United Arab
    Republic (Egypt), Chapter 3: The northern desert, edited by J.F. Griffiths,
    Elsev. Pub. Co., Amsterdam, New York.
Stamp, D.L. and Morgan, W.T., 1972. Africa: A study in tropical development, 3rd
    edition, John Wiley and Sons, Inc., New York, London, 520 pp.
Sutton, L.J., 1939. Discussion of the paper: The diurnal variation of wind over
    tropical Africa. Quart. Journ. Roy. Meteo. Soc., 65: 181-182.

Thompson, B.W., 1957. The diurnal variation of precipitation in British East
    Africa. Techn. Memor. 8, East African Meteo. Dept., Nairobi.
Trewartha, G.T., 1962. The earth's problem climates, Univ. of Wisconsin Press,
    Madison, and Methuen & Co. Ltd., London, 334 pp.
Wickens, G., 1975. Changes in the climate and vegetation of the Sudan since
    20 000 BP. Proceedings 8th Plenary session AETFAT. Geneva, 1974, Boissiera 24.
WMO, 1974. Hydrometeorological survey of the catchments of Lakes Victoria, Kyoga
    and Albert, Vol. I: Meteorology and hydrology of the basin, Geneva.

Chapter 4

ANALYSIS OF RAINFALL ON THE NILE BASIN

4.1   NETWORK OF RAIN-GAUGING STATIONS AND MEASUREMENT OF RAINFALL

Measurement of rainfall on the Nile Basin commenced about the beginning of
the century at a few places only. In 1924, it was reported that a few records
from East Africa were available for the years previous to 1900 (Brooks, C.E.,
1924). The total number of rain-gauging stations in East Africa (Uganda, Kenya
and Tanzania) was estimated in 1960 at about 850 (Johnson, D.H., 1962). The
number of rain-gauging stations that existed up to 1969, prior to the hydrometeorological survey of the catchments of Lakes Victoria, Kyoga and Albert was
estimated at 721.

The number of stations whose records are included in Supplements I thru' VI
of the Nile Basin Volume VI is much less than the above figures. A survey of the
number of the rain-gauging stations observed in the period from 1938 up to and
including 1967 is given in Table 4.1. The geographic co-ordinates, altitude and
annual rainfall depth at each station are tabulated in Appendix A1, whereas the
location of each rain-gauging station is shown on the map included in Appendix
A2. It is unfortunate that up to the time of writing this chapter, none of
the Supplements VII or VIII, which are supposed to include the rainfall data
for the periods 1968-72 and 1973-77 respectively, has yet appeared, making it
impossible to include them in the analysis presented here.

It is quite startling that whereas the number of stations in the first 5-year
section of the span of record was 337, it fell to 250 stations (about 74%) in
the last 5-year section of the available record. The military operations which
took place in Egypt, the Sudan and east of the Mediterranean during the Second
World War, the wars between Egypt and Israel in 1947-48, 1956, 1967 and 1973 and
the rebels and fights in the southern Sudan, Ethiopia and Uganda, all have
affected the operation of some of the rain-gauging stations here and there in
one way or another.

The rain gauge in use in Egypt and the Sudan is made of sheet zinc in the
form of a cylinder 45 cm in height. It consists of an upper part which has a
sharp rim and funnel for collecting the rain, and a lower part to hold the
vessel which is to contain the rain. The rim is made of a brass ring of which
the upper edge is bevelled. The ring is 15.9 cm in diameter. The catch thus has
an area of 200 cm^2. The zinc gauge is supported on a wooden or an iron post. The
details of this gauge are shown in Fig. 4.1.

TABLE 4.1   Number of rain-gauging stations whose records are available in the
             Supplements of Vol. VI of the Nile Basin (Hurst, H.E. et al, 1950,
             1955 and 1957; Nile Control Staff, 1963, 1969 and 1972)

Country	Number of stations recorded in the period						
	1938-42	1943-47	1948-52	1953-57	1958-62	1963-67	1938-67[*]
Egypt	62	64	59	69	67	65	65
Sudan	97	86	80	77	77	76	77
Ethiopia	02	01	01	01	01	01	01
Uganda	108	98	94	95	79	64	64
Kenya	43	42	36	31	29	25	25
Tanzania	25	25	25	25	21	18	18
Total	337	316	295	298	274	249	250

[*]Period of observation taken for the analysis of rainfall data

One of the items included in the project of the hydrometeorological Survey of
the Catchments of Lakes Victoria, Kyoga and Albert was to improve the network
density in these catchments. This was done by installing an additional 200
ordinary rain gauges in the project land area and 13 automatic Rimco-type
recording rain gauges in Lake Victoria. Manually read rain gauges were installed
inside the Lake Victoria in some of the accessible and inhabited islands.
Details about the network of ordinary rain gauges before and during the project,
in 1971, are presented in Table 4.2. In part I of Vo. I of the project report
(WMO, 1974), it was concluded that the already achieved network density of about
396 km^2 per station for the land area can be considered as quite satisfactory $\vphantom{j}$
for the purpose of water balance studies. Rainfall observations are generally
made with standard 5-inch rain gauges, the standard exposure being with the rim
of the gauge at 12 inches above the ground. The rain gauge is placed inside a
5-metre square lot in which grass is periodically cut low to ground level.

Rainfall is generally measured at 09 hours East African Standard Time (06 GMT)
for the 24 hours ending at 09 hours on the day of observation and entered against
the previous date prior to the date of observation. The Rimco recording gauge
works on the tilting-bucket principle and it can operate unattended for six
months. The annual rainfall depths in the period from 1938 up to and including
1967 for 250 stations are given in Appendix A_1. As has already been mentioned,
these data are available in Vol. VI of the Nile Basin, Supplements I thru' VI
(Hurst et al, 1950, 1955 and 1957 and the Nile Control Staff, 1960, 1969 and
1972). A look at these data will show that not every station has a complete

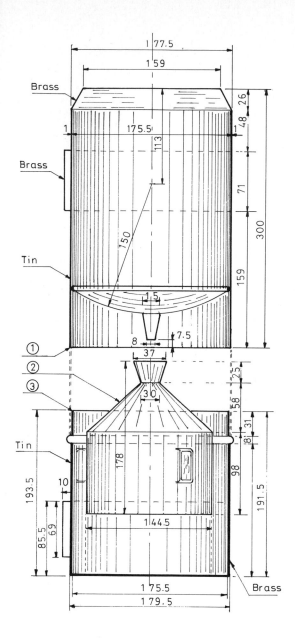

N.B. - The gauge consists of three main parts shown
separated in the drawing. All dimensions are
in millimetres

Fig. 4.1.   The normal type rain gauge in use in Egypt and the Sudan (Hurst,
H.E., and Black, R.P., 1950)

TABLE 4.2   Network of ordinary rain gauges in the catchments of Lakes Victoria, Kyoga and Albert (WMO, 1974)

Name	Area km^2	Existing No. of rain-gauging stns. (pre-project)	Density in area per stn. km^2	Stns. added	Network total in 1971	Density in area per stn. (1971) km^2
Lake Albert						
Semiliki	1 800	1	1 800	–	1	1 800
Muzizi	3 800	6	633	2	8	475
Nkusi	3 500	6	583	2	8	437
Kyoga Nile from Masindi	11 100	24	462	2	26	396
East Albert Shore	6 100	19	321	4	23	265
Lake Kyoga						
Kafu Mayanja	16 700	32	522	8	40	417
Kyoga South Shore	4 200	20	210	10	30	200
Victoria Nile (Jinja)	3 500	18	194	–	18	194
Mpologoma Malaba	14 100	52	271	–	52	271
Salisbury Basin	24 000	36	666	9	45	544
Kyoga North Shore	13 000	21	619	3	24	542
Lake Victoria						
Sio	1 450	9	866	–	9	866
Nzoia	12 400	85	146	–	85	146
Yala	3 500	26	135	–	26	135
Nyando	3 600	45	80	–	45	80
Sondu	3 600	19	189	–	19	189
Gucha-Migori	6 600	25	264	2	27	244
Kavirondo Gulf	5 500	30	183	1	31	177
Mara Kenya	7 800	9	866	3	12	600
Mara Tanzania	5 200	3	1 733	3	6	867
Mori Tanzania	1 600	3	533	1	4	400
Suguti	1 100	2	550	1	3	366
Ruwana-Gurumeti	11 150	3	3 716	4	7	1 010
Mbalagati	3 350	–	–	0	0	670
Simiyu Duma	10 800	9	1 200	3	12	900
Magogo Moame	3 300	5	660	3	8	412
Isanga	4 700	4	1 175	4	8	470
Victoria South Shore	21 500	26	827	11	37	581
Kagera Tanzania	21 500	39	577	10	49	425
Kagera Uganda	4 500	11	409	2	13	423
Ruizi Kibale	8 400	18	466	2	20	442
Katonga	14 700	29	506	6	35	420
Victoria North Shore	5 300	23	230	3	26	203
Mara Total	13 000	12	1 083	6	18	
Kagera in Tanz/Uganda	26 000	50	520	12	62	
Vict. Is. Stns. & Lake						
Uganda		7		9	16	
Kenya	70 260	–		2	2	
Tanzania		5		10	15	
Total	333 610	721	466	120	841	396

record covering the entire period used. As a matter of fact, about one-third of
the stations has a record of less than 90% of the whole period, and about 5% of
the stations have records covering less than 60% of the same period. The frequency of the length of record in years for the different stations is illustrated by the histogram in Fig. 4.2. The serial number, name, geographic co-
ordinates and altitude of each rainfall station is included in the tables,
Appendix A1.

In order to enable the analysis of the annual rainfall to cover as many of
the locations in the Nile Basin as possible, the missing data for 110 stations,
from serial number 97 up to and including serial number 206, have been supple-
mented. The majority of these stations enjoys a reasonably long record (see
Fig. 4.2.), and each is surrounded by stations with more years of observation,
which makes interpolation possible. To supplement the missing data at those
stations in Uganda, the adjacent rainfall stations with long records, even those
situated outside the Nile Basin, e.g. in the Congo, have been used. The rainfall
stations with a serial number less than 97 are located in Egypt and the northern part of the Sudan where the annual rainfall depth is small. Supplementing
the missing rainfall for this group of stations was felt unnecessary. A computer programme for supplementing the missing values by means of interpolation
based on the squared inverse of the distances to nearby stations and adjusted
for differences in altitude has been developed and applied here. Supplementing
the missing data bring the percentage of rain-gauging stations with 30 years of
record to about 70 and of those with more than 27 years of record (90% of the
full length of record) to 91 of the total number of stations. The completed
rainfall data of stations 97 to 206 are included in Appendix B. These data have
been employed conjunctively with the data in Supplements I thru' VI of the Nile
Basin Vol. VI for preparing the monthly rainfall and its percentage to the
annual rain depth for all stations. The results are given in Appendix C.

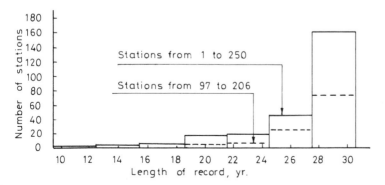

Fig. 4.2.   Frequency histograms for the length of rainfall record at stations
from 1 to 250 and from 97 to 206

4.2   RAINFALL VARIATION

4.2.1   Annual rainfall

The basic statistical descriptors of the annual rainfall have been computed for the 250 stations and the results obtained are listed in Table 4.3. These descriptors are the arithmetic mean, $\bar{X}$, the standard deviation, s, the coefficients of variation, C_v, skewness, C_s, and kurtosis, C_k, and the first five serial correlation coefficients r_1 thru' r_5. The serial coefficients have been computed only for the stations having uninterrupted records of a minimum of 27 years.

It may be worthwhile mentioning that the analysis of the variation of rainfall on the catchments of Lakes Victoria, Kyoga and Albert (WMO, 1974), comprised 22 stations, 19 of which are included here. The earliest years of measurement at any of these stations is 1926 and the last is 1970. Earlier observations at some stations in Uganda and Tanzania were analyzed and discussed (Brooks, C.E., 1924). Moreover, the rainfall data up to 1940 at a number of gauging stations in the Sudan were analyzed and the results reported by Ireland, A.W. (1948). The station serial number and the period of observation appearing in those works are summarized in Table 4.4. In this section we shall present the methods and equations which were used to get the results included in Tables 4.3.

Denote the annual rainfall by X_i (i = 1, 2, ..., n) and the years of record by n. The arithmetic mean, $\bar{X}$, is therefore given by the equation

$$\bar{X} = \frac{1}{n} \sum_{i=1}^{n} X_i \tag{4.1}$$

the standard deviation, s, and the coefficients of variation, C_v, skew, C_s, and kurtosis, C_k, by the equations

$$s = \sqrt{\frac{\sum\limits_{i=1}^{n} (X_i - \bar{X})^2}{n - 1}} \tag{4.2}$$

$$C_v = \frac{s}{\bar{X}} \tag{4.3}$$

$$C_s = \frac{\frac{1}{n} \sum\limits_{i=1}^{n} (X_i - \bar{X})^3}{s^3} \tag{4.4}$$

$$C_k = \frac{\frac{1}{n} \sum\limits_{i=1}^{n} (X_i - \bar{X})^4}{s^4} \tag{4.5}$$

respectively.

The results obtained from equations 4.4 and 4.5 have been corrected for the bias by multiplying them by $\dfrac{n^2}{(n-1)(n-2)}$ and $\dfrac{n^3}{(n-1)(n-2)(n-3)}$ respectively (Haan, C.T., 1977).

The serial correlation coefficient, r_L, corresponding to any lag, L, $(L = 1, 2, \ldots\ldots, \frac{n}{4})$ has been computed using the formula

$$r_L = \frac{\sum\limits_{i=1}^{i=n-L} (X_i - \bar{X})(X_{i+L} - \bar{X})}{\sum\limits_{i=1}^{i=n} (X_i - \bar{X})^2} \tag{4.6}$$

The 95% confidence limits for r_L can be computed from the equation

$$\frac{-1 \pm 1.96\,(n-L-2)^{\frac{1}{2}}}{(n-L-1)} \tag{4.7}$$

The mean annual rainfall listed in Table 4.3 has been used for constructing the isohyetal map, Fig. 4.3. This map compares fairly well with the map in Fig. 4.4., which is available in Supplement VI, Vol. VI of the Nile Basin (Nile Control Staff, 1972). Both give the annual precipitation slightly less than that in the map given in the Water Resources of the Earth (UNESCO, 1978), which is shown in Fig. 4.5.

Furthermore, to get an idea about the possible change in the mean annual rainfall with the time and number of years of record, the values of the mean $\bar{X}$ in Table 4.3 for the 30-year period 1938-1967 have been plotted versus the mean $\bar{X}'$ available for the stations listed in Table 4.4. This plot, together with the line $\bar{X} = \bar{X}'$, is shown graphically in Fig. 4.6. It is quite possible that the position of the years of record on the time scale plays a role in the value of the mean annual rainfall. However, it is clear from the information in Table 4.4 and the plot in Fig. 4.6., that those stations with length of record of 25 years and more give almost the same mean regardless of their positions on the time scale. Stations with fewer years of record such as 142, 232, 202, ... show a considerable deviation from the 1:1 line.

A picture of the geographic distribution of the coefficient of variation, C_v, of the annual rainfall can be seen from Fig. 4.7., where it can be immediately noticed that C_v exercises a steep rise from less than 0.4 along the Mediterranean Coast to about 4.0 in southern Egypt and around the oases in the Western Desert. This is quite understandable as the mean rainfall there is extremely low and occasionally comes a year with relatively intensive rainfall, thereby producing a considerable variation. The gradual improvement of the mean depth of annual rainfall, in a southerly direction, results in a rapid reduction in C_v, so that a value as low as 0.4 is reached around Khartoum. Further increase in the annual rainfall and its uniformity with time brings C_v up to 0.2 along the fringes of southern Sudan. Lower values can be found in some parts of the Equatorial Lakes Plateau.

A graphical plot of the coefficient of variation versus the mean of the annual rainfall for the 250 gauging stations in the Nile Basin shows a certain amount of scatter. Nevertheless, the decline in C_v with increasing $\bar{X}$ is definitely unmistakable. This can be clearly seen from Fig. 4.8.

The distribution of the skewness coefficient over the Nile Basin assumes the pattern shown in Fig. 4.9. This pattern is not similar to that of the coefficient of variation, at least in some parts of the basin. Most of these parts are concentrated in the Equatorial Lakes Plateau, where the skewness appears to be rather high whereas the coefficient of variation is quite small. This conclusion is confirmed by the results appearing in Part I, Vol. 1 of the Hydrometeorological Survey of the Equatorial Lakes. The skewness of the annual rainfall at most of the stations in Uganda and Tanzania which are listed in Table 4.4 seems to be in close agreement with our results, which are presented in Table 4.3.

The annual rainfall depths composing the series whose serial correlation coefficients are not significantly different from zero at the specified level of confidence (95%) have been termed as independent, I. Those series whose serial correlation coefficients lie outside the confidence limits have been termed as dependent, D. A number of statistical models have been tried to the independent series, and the distribution function of best fit to the observations in each of them has been used for estimating the annual rain depth corresponding to return periods of 1.01, 2, 5, 10, 20, 50 and 100 years. These results, together with the distribution function used in estimating them, are included in Table 4.5.

Of the 146 independent series analyzed here, 96 series, or 65.8% have been found to be best fitted by Pearson III distribution function, 34 series, or 26.0%, by the normal distribution and the rest, or 8.2% by the two-parameter lognormal distribution function. Examples of rainfall series fitted by each of these distribution functions can be seen in Figs.4.10a thru' c. The general formula used for estimating the rainfall depth, X_T, that corresponds to any return period T, is

$$X_T = \bar{X} + K_T \cdot s \tag{4.8}$$

The frequency factor, K_T, depends on the distribution function used and the coefficient of variation, C_v, or the skewness, C_s, except for the normal distribution. The table of K_T, given by Chow (1964) is satisfactory for fitting the lognormal function, whereas the tables presented by Kite (1977) are fairly satisfactory for the purpose of fitting the normal and the lognormal functions.

The frequency factors, K_T, used in connection with Pearson III have been taken from the "Guidelines for Determining Flood Flow Frequency" by the Water Resources Council of the United States of America (1977).

In the analysis of the annual rainfall which is presented in the report of the hydrometeorological survey of the catchments of Lakes Victoria, Kyoga and Albert, the two-parameter Gamma distribution is reported as the distribution function of best fit to the data of the 22 stations included in that investigation. The same function has been reported as the function best fitting the distribution of the monthly rainfall. The Pearson III found by us here as the distribution function of the annual rainfall at two-thirds of the stations is simply a Gamma function, but with three parameters.

It may be of interest to mention that the lognormal and the Gamma functions have proven to be functions of best fit to the distribution of rainfall depths in many parts of the world (Markovic, R., 1965; Huynh Ngoc Phien et al, 1980; Kottegoda, N.T., 1980).

The rain-gauging stations with complete record and not appearing in Table 4.5 have rainfall series whose elements experience some dependence. These have already been termed by the letter D in Table 4.3. In order to describe the dependent component in these series, we tried to fit an autoregressive model with lag 1, 2, 3 or 4 to the observed data. The basic equation can be written as

$$X_t - \bar{X} = \alpha_1 (X_{t-1} - \bar{X}) + \alpha_2 (X_{t-2} - \bar{X}) + \alpha_3 (X_{t-3} - \bar{X}) + \\ + \alpha_4 (X_{t-4} - \bar{X}) + \varepsilon_t \tag{4.9}$$

where,

X_t = annual rainfall in any year t,

X_{t-L} (L = 1, 2, 3, 4) = annual rainfall in year t-L,

α_L = coefficient to be found from the serial correlation coefficients, for
example in the 1st order model, $X_t - \bar{X} = \alpha_1 (X_{t-1} - \bar{X}) + \varepsilon_t$, $\alpha_1 = r_1$,
and in the 2nd order model, $X_t - \bar{X} = \alpha_1 (X_{t-1} - \bar{X}) + \alpha_2 (X_{t-2} - \bar{X}) + \varepsilon_t$
$\alpha_1 = \dfrac{r_1 - r_1 r_2}{1 - r_1^2}$ and $\alpha_2 = \dfrac{r_2 - r_1^2}{1 - r_1^2}$ .... and so on.

Expressions for α_1, α_2, and α_3 in a 3rd order model and for α_1, α_2, α_3, and α_4 in a 4th order model can be found in some of the literature on stochastic hydrology (Yevjevich, V., 1973 and Kottegoda, N.T., 1980). The computer programme developed by R. Clarke (1973) has been worked out to obtain the residuals left after fitting the 1st, 2nd, 3rd and 4th order models to every rainfall sequence and to obtain the serial correlation coefficients of each set of residuals. The values of these coefficients form the criterion to judge the suitability of the model as a good fit to the sequence in question. When the first order model is suitable, the higher order models usually prove to be equally suitable. In such a case the selection of the model order is based on the variance of the residuals. The order whose residuals gave the smallest values of the serial correlation coefficients and which showed the least variance has been selected. From the 37 sequences investigated the autoregressive model failed in 4 cases. For the remaining 33 sequences, 22 could be fitted by lag 1 model, 9 by lag 2 model, 0 by lag 3 model and 2 by lag 4 model. The basic properties of these models and the values of the serial correlations of the residuals are given in Table 4.6. Once a model is established, one can generate any number of annual rainfalls which can be used for determining the rain depth corresponding to any given return period.

TABLE 4.3 Basic statistical descriptors and serial correlation coefficients of the annual rainfall data (Shahin, M.A., 1983)

Station		Basic statistical descriptors					Serial correlation coefficients					I/D[4]
No.	Name	$\bar{x}$[1]	s[1]	C_v	C_s[2]	C_k[3]	r_1	r_2	r_3	r_4	r_5	
1	Sidi Barrani	167.1	70.38	.4211	1.5715	5.5214						
2	Borollos L.H.	181.4	63.34	.3492	-.0417	2.7740	-.0968	-.2124	.0927	-.1129	.1545	I
3	Sallum	108.8	75.70	.6959	1.7371	6.3878						
4	Damietta L.H.	123.9	46.37	.3740	-.1755	2.1746	.0857	-.3077	-.1742	.0724	.2927	I
5	Damietta (Town)	109.9	48.61	.4423	.3652	2.9334	.1708	.1309	-.0976	.2278	-.0716	I
6	Rosetta L.H.	190.7	97.40	.5108	.8184	2.6121						
7	Rosetta (Town)	212.7	93.23	.4382	.6060	2.8067						
8	Mersa Matruh	136.2	59.43	.4360	.9754	3.5703						
9	Tolombat el-Boseili	157.1	65.47	.4167	.8627	3.5773	-.0129	-.1539	.0813	-.1644	.0783	I
10	Tolombat el-Tolombat	184.7	90.41	.4895	.3964	2.3954	-.0501	-.1023	-.1053	.1118	.2318	I
11	Edfina	184.1	65.40	.3553	.2611	2.4923						
12	Port Said	73.0	36.60	.5014	1.1094	5.1014	-.0714	-.0945	-.2482	.4593	-.2212	D
13	Sirw	72.4	30.86	.4262	.6298	3.4871	.3044	-.2430	-.2107	-.0334	-.0632	I
14	Kom el-Tarfaia	146.2	75.18	.5142	.8091	2.7095	-.1534	-.1168	-.0258	-.1422	.2246	I
15	Alexandria	189.6	62.21	.3281	.2372	2.5448	-.1025	-.1567	-.0810	-.0706	.2438	I
16	Mex	186.7	84.46	.4524	.4635	2.5417	-.0124	-.0328	.0371	.0420	.3593	I
17	Kafr el-Dawar	155.4	76.59	.4929	.7069	3.0097	.0616	-.3214	-.2894	-.1748	.1595	I
18	Kafr el-Sheikh	80.1	36.40	.4544	.6722	2.5902	.0481	-.2842	.1429	-.0265	.1519	I
19	Al-Arish	105.3	44.06	.4185	.5187	3.4684						
20	Sakha	66.6	30.74	.4616	.6396	2.6866	.0262	-.1750	.1450	-.0408	.2236	I
21	Ras el-Dabaa	135.4	54.17	.4001	.3805	3.1654						
22	Mansura	57.1	22.48	.3937	.3026	2.2199	-.1203	-.5192	.2824	.2194	-.4325	D
23	Fuka	113.5	60.16	.5302	.9237	3.4447						
24	Damanhur	89.3	39.61	.4436	.6187	2.8923	.0657	-.0692	-.0653	.1063	-.0921	I
25	Amria	141.3	84.07	.5950	1.1690	5.5110						

(1) in mm/yr; (2), (3) corrected for bias in estimation from sample; (4) I = independent and D = dependent

TABLE 4.3   (continued)

No.	Name	$\bar{x}$ (1)	s (1)	C_v	C_s (2)	C_k (3)	r_1	r_2	r_3	r_4	r_5	I/D (4)
26	Borg el-Arab	163.2	74.77	.4581	.3886	3.2379						
27	Hamman	114.0	61.95	.5434	1.0013	4.4936						
28	Kafr el-Zayyat	36.4	24.35	.6690	.9544	2.9680	.2118	-.0341	-.0412	.0449	-.3710	I
29	Tanta	50.8	22.84	.4496	.2651	3.2317	-.1575	-.1537	.0942	-.0761	-.0726	I
30	Faqus	39.3	18.14	.4616	.2650	2.7047	.3570	.3703	-.0779	.3025	.0637	D
31	El-Quseima	54.4	29.09	.5346	.6022	3.4053						
32	Zagazig	36.0	30.87	.8575	4.1461	23.0186	-.1769	-.1649	-.1720	.3406	-.0036	I
33	Shebin el-Kom	36.3	14.96	.4125	.1326	3.3464						
34	El-Hassana	27.5	17.77	.6470	.1586	3.3306						
35	Benha	22.6	14.63	.6473	.6621	2.5589						
36	Wadi el Natrun	42.6	60.33	1.4161	3.5945	16.8848						
37	Fayed	22.5	16.48	.7308	1.0947	4.4302						
38	Delta Barrage	20.3	12.97	.6391	1.1863	5.7927						
39	Giza	23.1	13.04	.5645	.7487	3.5886	.1110	.0603	.0512	.1437	.3273	I
40	Kuntella	22.8	23.15	1.0155	1.4161	4.3525						
41	Suez	25.6	19.20	.7500	.8542	3.1395	.0677	-.0177	-.0648	.0556	-.1254	I
42	El-Nekhl	19.0	18.14	.9545	1.1977	4.5985						
43	Attaqua	14.7	12.46	.8498	1.9521	7.5117						
44	Helwan	25.2	20.85	.8274	1.4937	4.7287	.0206	.1379	-.2855	.2219	.1551	I
45	El-Themed	18.9	19.39	1.0286	1.6645	5.3242						
46	Shakshuk	09.3	06.97	.7495	1.0465	3.8441						
47	Fayum	13.1	14.16	1.0809	2.0732	8.2254	.3269	.1039	.2002	.3238	.2423	I
48	Ras el Negb	27.4	23.18	.8457	.9401	3.2016						
49	Siwa	09.3	10.14	1.0904	1.5194	5.2160	-.1651	-.3424	.2247	.0281	-.2488	I
50	Beni-Suef	07.0	07.29	1.0408	1.8590	7.1406						

(1) in mm/yr; (2), (3) corrected for bias in estimation from sample; (4) I = independent and D = dependent

TABLE 4.3 (continued)

No.	Name	Basic statistical descriptors					Serial correlation coefficients					I/D(4)
		$\bar{X}$(1)	s(1)	C_v	C_s(2)	C_k(3)	r_1	r_2	r_3	r_4	r_5	
51	Baharia	03.4	04.80	1.4245	1.6708	5.4642	-.2050	-.2568	.0828	.0698	.3962	I
52	Tor	10.1	11.91	1.1818	1.4658	4.6035						
53	Minya	03.1	05.18	1.6716	2.2887	7.4406	-.0611	-.0610	.1428	-.0423	-.1106	I
54	Hurghada	04.4	09.86	2.2306	3.6922	17.3111						
55	Assiut	02.9	04.64	1.6200	1.7751	5.2604	.3655	.1201	.0968	-.0132	-.0310	D
56	Farafra	01.3	02.66	1.9978	2.7785	11.6421						
57	Qena	04.7	11.26	2.4197	3.9016	20.0833						
58	Qusseir	03.5	08.26	2.3820	3.5151	16.4855	-.0387	-.1556	-.1707	-.1081	.1488	I
59	Nag-Hammadi	01.2	04.03	3.3618	4.2048	20.7059						
60	Luxor	00.9	01.48	1.6480	1.4795	4.1012						
61	Dakhla	00.5	02.20	4.5850	4.9269	26.8735						
62	Kharga	01.4	02.45	1.7059	1.777	5.2374	.1298	.1924	.0453	-.4267	-.3828	I
63	Deadalus Island	08.6	12.99	1.5049	2.1236	6.9631	.1651	-.1164	-.0902	-.1262	-.0819	I
64	Kom Ombo	00.7	01.37	2.0560	1.9957	6.3826						
65	Aswan	01.2	02.01	1.6205	1.9584	6.7914						
66	Wadi-Halfa	03.6	07.48	2.0970	2.6956	10.3612						
67	Port Sudan	99.0	57.81	.5839	.5666	2.4869						
68	Abu-Hamed	15.9	27.37	1.7214	3.4710	17.1297	-.0680	-.1359	.2053	-.1552	-.1613	I
69	Gebeit	123.4	67.59	.5477	1.1927	4.5434	-.4458	.0121	.1827	-.4202	.1053	D
70	Sinkat	128.8	93.81	.7283	1.3823	5.2346	-.1943	-.0523	-.0954	-.2365	-.1297	I
71	Kareima	39.0	40.64	1.0421	1.2130	3.8206	-.0398	-.1670	.3272	.0090	-.3053	I
72	Tokar	92.6	66.76	.7210	.8518	3.2578						
73	Tahamiyam	79.4	61.65	.7768	1.5774	6.8951						
74	Talguharia	79.7	66.06	.8289	1.9146	8.4202						
75	Atbara	65.0	43.06	.6625	.7587	3.0301	-.2239	-.0734	.2181	-.1229	-.1151	I

(1) in mm/yr; (2), (3) corrected for bias in estimation from sample; (4) I = independent and D = dependent

TABLE 4.3 (continued)

No.	Name	$\bar{X}$(1)	s(1)	C_v	C_s(2)	C_k(3)	r_1	r_2	r_3	r_4	r_5	I/D(4)
		Basic statistical descriptors					Serial correlation coefficients					
76	Zediab	55.4	50.99	.9198	.7291	2.9178	.3149	.2011	.3298	-.0123	.0135	I
77	Abu Deleig	224.7	97.29	.4330	.3946	3.4076						
78	Khartoum G.C.	175.1	79.61	.4547	.7576	3.7292						
79	Khartoum	171.5	80.81	.4712	.4881	2.8969	-.0941	-.0984	.0941	.0772	.0464	I
80	Kassala	321.4	85.88	.2672	-.1322	2.7130	-.0113	-.2618	.0862	.2377	.0364	I
81	Jebel Aulia	187.0	71.31	.3813	.1588	2.6616						
82	Wadi-Turabi	249.5	98.14	.3933	.4414	3.6382	-.1674	-.1228	.0244	.0854	-.2872	I
83	Kamlin	262.9	88.97	.3384	.4268	3.2821	.0160	.0525	.2779	.0285	.1043	I
84	Khashm el-Girba	338.1	113.49	.3357	1.2700	4.3005						
85	Geteina	203.3	73.52	.3616	1.3528	5.3100	-.0833	-.0869	.1708	-.2786	-.2972	I
86	Rufaa	296.9	89.22	.3005	.6753	4.2637	-.1728	.2431	.0392	.1177	-.1082	I
87	Wadi-Shair	301.7	82.93	.2748	.4573	3.3238	-.2495	.0331	.2043	-.2345	-.1463	I
88	Wad Medani	365.9	107.99	.2951	-.4088	3.6197						
89	Managil	298.5	102.19	.3423	.6710	4.1381	-.1253	.0580	.2781	-.3380	-.1175	I
90	Gedaref	580.7	99.19	.1708	.5930	4.3354	-.2958	-.1041	-.0493	.1317	-.1468	I
91	Dueim	303.4	90.59	.2986	.0949	2.8073	-.0800	-.1412	.1020	.0391	-.0822	I
92	Haj Abdalla	435.5	83.32	.1913	.4697	4.1929	-.2868	.0337	.1489	-.0030	-.0956	I
93	Wad-Haddad	402.4	89.74	.2230	-.2188	6.3673						
94	Bara	284.1	72.94	.2567	.1031	2.1945	-.1179	.1990	.0741	-.0526	.0134	I
95	El Fasher A.P.	302.8	118.15	.3902	2.1606	8.6072						
96	Mfaza	591.4	146.62	.2479	.2778	2.6046						
97	Sennar D.S.	481.7	97.50	.2024	-.1249	2.6373	-.0134	-.1728	-.1485	-.5053	.1323	D
98	Geneina	553.0	123.21	.2228	.6456	2.6435	.2899	-.0726	-.1742	-.2550	-.1181	I
99	El-Obeid Town	404.2	116.40	.2880	.8694	3.9926	-.0141	-.0186	-.0314	-.2981	-.0666	I
100	Kosti/Rabak	401.9	80.13	.1994	.5504	2.8227	-.1426	.2102	.3030	-.1959	.1824	I

(1) in mm/yr; (2), (3) corrected for bias in estimation from sample; (4) I = independent and D = dependent

TABLE 4.3   (continued)

No.	Name	Basic statistical descriptors					Serial correlation coefficients					I/D(4)
		$\bar{X}$(1)	s(1)	C_v	C_s(2)	C_k(3)	r_1	r_2	r_3	r_4	r_5	
101	Singa	591.7	89.90	.1519	-.2641	4.4685	-.1611	.2280	.0777	-.0601	-.1570	I
102	Tendelti	369.5	96.77	.2619	.1071	3.3796	.3092	.0922	.2310	.2712	-.2435	I
103	Um Rwaba	417.2	133.77	.3206	2.6676	13.6276	-.0758	-.1982	-.2502	-.1288	.1265	I
104	Rahad	460.3	121.95	.2649	1.9179	8.3518	.1274	-.0580	-.2671	-.2101	-.0082	I
105	El Nahud	417.9	92.59	.2216	.2382	3.2438	.0202	-.2251	-.2180	-.1106	-.1354	I
106	Jebelein	423.4	82.12	.1940	-.0374	2.1020	-.0923	.0004	-.0470	-.1739	.1658	I
107	Nyala	478.4	110.00	.2293	-.1455	3.4023	.1613	-.3771	-.1112	-.0372	-.2266	I
108	Dilling	640.8	145.69	.2274	-.0256	3.1763	.3073	.0579	-.1459	-.1265	.1582	I
109	Rashad	777.2	110.38	.1420	.5692	3.0244	-.1108	.0452	.007	.0302	-.2050	I
110	Roseires	732.0	109.74	.1499	.5775	4.0764	.3017	-.0140	-.1302	.3542	-.4355	I
111	Renk	567.0	152.52	.2690	.7772	2.8204	.4138	.2894	-.0229	-.0835	-.1324	D
112	Abri	692.2	96.76	.1398	-.0416	2.0987	.1318	-.0656	.0016	-.2339	-.3173	I
113	Kadugli	787.6	146.56	.1861	.5740	2.8731	.1801	-.1782	.1876	.0319	-.4827	D
114	Talodi	821.2	138.58	.1688	.0416	3.1320	-.2100	-.2407	.4540	-.1811	-.4615	D
115	Kurmuk	932.1	206.18	.2212	-.4560	2.6780	.3062	.2424	-.0487	-.0702	-.1635	I
116	Melut	642.1	166.17	.2588	1.7961	10.3313	-.0895	.1557	-.4233	-.3719	-.1797	D
117	Kodok	734.7	132.61	.1805	.4365	4.4227	.1484	-.3941	-.2610	.0506	.1035	I
118	Malakal	794.5	140.98	.1774	.2079	4.4517	.0985	-.0968	-.1419	-.0181	.0120	I
119	Tonga	888.8	188.23	.2118	.2548	3.3455	.3979	.2688	.1822	-.0851	.0225	D
120	Abwong	756.7	183.81	.2429	-1.0775	4.9807	.4598	.2165	.1878	.1535	.1373	D
121	Fangak	887.7	273.53	.3081	.1225	2.6662	.5700	.5555	.4935	.5935	.3040	D
122	Aweil	897.5	217.78	.2345	.1236	3.0943	.2047	-.0821	-.0925	-.0954	-.1966	I
123	Nasser	798.2	146.48	.1835	-.1761	4.4046	-.3231	-.1666	-.0139	.1301	.0744	I
124	Raga	1228.7	168.54	.1372	.5852	3.8970	-.0618	-.1259	.0115	.2992	-.1686	I
125	Meshra er-Rek	919.4	209.88	.2283	.6294	3.9399	.2019	-.1199	-.1887	-.0893	.0237	I

(1) in mm/yr; (2), (3) corrected for bias in estimation from sample; (4) I = independent and D = dependent

TABLE 4.3 (continued)

No.	Name	$\bar{X}^{(1)}$	$s^{(1)}$	C_v	$C_s^{(2)}$	$C_k^{(3)}$	r_1	r_2	r_3	r_4	r_5	I/D$^{(4)}$
		Basic statistical descriptors					Serial correlation coefficients					
126	Gambeila	1316.1	287.91	.2188	- .4554	3.8904	.2865	.3175	.1536	-.2162	.0614	I
127	Akobo	984.3	194.54	.1976	- .4969	5.0569	.3280	-.0043	.1084	.1519	-.2753	I
128	Wau	1182.3	159.28	.1347	.0969	2.3407	-.3088	-.0209	.2650	-.1343	-.0958	I
129	Tonj	1072.7	206.78	.1928	.7399	3.0459	-.1313	-.0793	.0001	.3950	-.2327	D
130	Ghabe Shambe	758.7	185.81	.2449	.1526	2.4769	.1939	-.0834	.0935	.0728	-.1658	I
131	Rumbek	962.4	209.45	.2176	- .6624	3.1784	.2892	.1854	-.0114	-.1217	-.0952	I
132	Pibor Post	909.3	242.14	.2663	.2813	3.0006	.2645	.2788	-.1676	.3462	.2534	I
133	Bor	892.2	153.04	.1715	.7445	3.6988	.2597	.2667	-.0235	-.1918	.0565	I
134	Amadi	1186.8	197.12	.1661	- .1097	2.9269	.2438	.2953	.1075	-.1450	-.0033	I
135	Terakekka	979.3	202.88	.2072	1.0746	4.6239	.4272	.3382	.0550	-.0448	-.0349	D
136	Maridi	1357.5	220.43	.1624	- .0698	3.1662	.3203	.0292	.0586	.2159	.0997	I
137	Juba	999.3	182.54	.1827	.2843	2.8288	.2096	-.1039	.2892	-.3982	-.1390	I
138	Yambio	1472.8	156.64	.1064	.1874	3.333	-.1431	-.0358	.1330	.0056	-.2692	I
139	Loka	1256.9	176.99	.1408	.5380	3.8270	.4648	-.0126	-.2258	-.2957	-.0587	D
140	Yei	1335.4	209.92	.1572	.1908	3.2537	.1936	.0671	-.0795	-.4616	-.1318	D
141	Kajo Kaji	1319.0	303.18	.2299	.5955	5.4390	.0640	-.4164	.0018	.0261	-.0996	D
142	Nimule	1275.8	247.11	.1937	.8785	4.7248	.1270	-.0458	.4609	-.1195	-.0844	D
143	Kitgum	1252.2	192.84	.1540	.0954	2.4929	-.2613	.1748	-.0200	-.2731	.1196	I
144	Atua	1325.3	329.93	.2489	- .8324	3.9897	.4452	.4139	.0744	.1851	.0084	D
145	Gulu	1549.8	230.41	.1487	- .2915	3.2422	.4105	-.1323	-.1322	-.0749	-.1518	D
146	Moroto	910.0	218.12	.2397	.2456	2.8725	-.1696	-.2104	-.1200	.2180	.1354	I
147	Negetta Farm/Lira	1451.9	180.79	.1245	.3509	4.3080	.1881	.1309	.0493	.0447	-.1401	I
148	Aduku dispensary	1285.9	179.63	.1397	.2741	2.9989	.2752	-.1182	-.0393	-.0743	.2558	I
149	Katakawi	1135.7	172.29	.1517	.5121	3.1972	-.0913	-.2879	.1181	-.0517	.0116	I
150	Butiaba	755.2	250.10	.3312	2.3766	10.1242	.1950	-.0433	-.1525	-.2043	-.0424	I

(1) in mm/yr; (2), (3) corrected for bias in estimation from sample; (4) I = independent and D = dependent

TABLE 4.3 (continued)

No.	Name	Basic statistical descriptors					Serial correlation coefficients					$I/D^{(4)}$
		$\bar{X}^{(1)}$	$s^{(1)}$	C_v	$C_s^{(2)}$	$C_k^{(3)}$	r_1	r_2	r_3	r_4	r_5	
151	Soroti	1290.7	240.18	.1861	.3967	3.1071	.1578	-.4590	-.4060	-.0582	.2878	D
152	Masindi	1254.6	190.49	.1518	-.1435	3.8712	.4310	.0041	.0020	-.2062	-.2841	D
153	Ongino	1242.8	238.23	.1917	.3728	6.2210	.1993	-.1922	-.2533	.1495	.4191	I
154	Serere Agr. St.	1345.1	260.25	.1935	-1.0758	6.4170	.2157	-.1086	.0998	.0138	.3602	I
155	Kyere	1290.4	176.46	.1367	-.1350	2.3608	-.0548	-.4277	-.1811	.0824	.1743	D
156	Bulindi	1301.8	197.27	.1515	.4610	2.7500	.2566	.4008	.0925	-.2046	-.0285	D
157	Ngora	1353.7	176.76	.1306	.9742	5.0904	.0206	-.3792	.0853	.0371	-.1121	I
158	Nakasangola	1034.1	174.02	.1683	.3596	4.1195	-.0749	-.0657	.0488	-.3314	.3444	I
159	Bukedia	1229.7	194.80	.1584	.1227	2.7029	.2648	.0485	-.1952	.1758	.1201	I
160	Kachimbala	1196.2	218.38	.1826	.6776	3.6600	.4413	-.0238	-.2806	.0102	.1323	D
161	Kibale	1385.7	242.17	.1748	1.0664	4.9411	.0261	.0277	.0642	.0355	-.0680	I
162	Bugaya	1315.1	228.16	.1735	1.1410	4.9904	-.1176	-.0569	-.0919	.0684	-.1312	I
163	Mbale	1169.0	245.18	.2097	1.0882	3.4210	.0767	-.0937	.0233	-.1366	-.1533	I
164	Namasagali	1266.7	214.28	.1692	1.0764	4.6595	-.0130	-.1530	.1805	.0218	.1525	I
165	Vukula	1363.5	255.04	.1870	1.1595	4.6978	.2600	.1374	-.0266	-.0529	-.2891	I
166	Kiboga	1199.6	271.09	.2260	1.7211	8.5143	-.1057	-.1885	.0989	-.3239	-.0845	I
167	Bulopa	1352.9	217.49	.1608	-.0591	2.7189	-.1033	.0697	-.0552	.2753	-.2286	I
168	Ntenjeru	1221.2	203.26	.1664	.8284	3.8008	.0258	.1835	.1721	.0642	-.0115	I
169	Bukalasa	1298.7	241.24	.1858	1.4703	7.2773	.1450	-.0852	.0659	.2379	.0711	I
170	Kahangi Estate	1367.7	199.10	.1456	1.2295	6.1086	-.0062	-.0446	.0851	.0156	.0170	I
171	Tororo	1426.2	222.77	.1562	.4018	3.1733	.4811	-.0125	-.0824	.1745	.3642	D
172	Fort Portal	1517.9	226.46	.1492	.8182	3.8060	-.0005	.0152	-.0947	-.1325	-.0531	I
173	Kalagala Agr. St.	1243.0	257.56	.2072	1.2027	5.7374	.0700	.0739	-.0670	.1150	-.4210	I
174	Inganga	1279.7	240.62	.1880	.6356	4.7638	-.1001	.2004	-.1762	.0378	-.0786	I
175	Mubende	1189.2	175.07	.1472	.5325	4.2978	-.1224	-.0453	-.0435	-.0735	.0172	I

(1) in mm/yr; (2), (3) corrected for bias in estimation from sample; (4) I = independent and D = dependent

TABLE 4.3   (continued)

No.	Name	Basic statistical descriptors					Serial correlation coefficients					I/D(4)
		$\bar{X}^{(1)}$	$s^{(1)}$	C_v	$C_s^{(2)}$	$C_k^{(3)}$	r_1	r_2	r_3	r_4	r_5	
176	Nawanzu	1324.8	295.00	.2227	2.2355	11.5580	-.0403	.2294	-.0271	.1503	-.1745	I
177	Dabani	1538.8	275.11	.1737	.6494	2.6743	.0135	-.0303	-.1007	.2360	.0498	I
178	Nagoje	1403.5	349.79	.2492	.9408	3.9994	.3490	.4027	.2635	.3290	.0319	D
179	Masafu Dispensary	1539.5	326.86	.2123	1.3474	5.1022	-.0802	-.0615	-.1083	.0121	.1097	I
180	Lugala	1396.8	264.75	.1895	1.2182	4.5157	.0721	-.0090	-.0384	.1040	-.1681	I
181	Moniko Estate	1500.4	233.00	.1553	.7977	4.1457	.2662	.2340	.0091	.1521	.0332	I
182	Namanve	1345.6	274.72	.2042	1.4886	6.2496	.2433	-.1000	-.1365	-.0679	.0887	I
183	Mukono Agr. Stat.	1466.3	338.02	.2305	.6838	3.1562	.6001	.3714	.1404	.0305	-.2430	D
184	Bwavu	1370.4	244.17	.1782	1.1412	4.2658	-.0386	.1743	-.1105	.1690	-.1744	I
185	Nsyamba	1149.9	256.20	.2228	.3913	3.6206	.3099	.0286	.0996	.2682	.2089	I
186	Kabasanda	1323.5	261.31	.1974	.6975	3.5811	.0727	-.0741	-.0775	.2185	-.0798	I
187	Budo King's School	1264.6	196.48	.1554	.6483	2.8198	-.0021	-.2642	-.2072	.1392	-.0554	I
188	Ngogwe Coffee Nursery	1540.8	294.49	.1911	.6414	3.2720	.1976	.1899	-.1480	-.0788	-.1121	I
189	Buvuma Isl.	1578.0	306.64	.1943	.1443	2.2876	.0486	-.1886	-.2205	.0749	.2469	I
190	Kisubi	1424.4	258.48	.1815	.1862	4.2016	.1261	-.2260	-.2878	-.2967	.0312	I
191	Entebbe	1557.7	268.05	.1721	-.4874	2.8591	.2737	-.2850	-.3438	-.2210	.0994	I
192	Nkozi	1019.0	232.93	.2286	.3978	2.7988	.1831	.1060	.0673	.0778	-.1258	I
193	Kalungu	1054.8	210.60	.1997	.2004	3.1832	-.0195	.0044	.1480	-.0209	-.0356	I
194	Katigondo	1084.5	206.18	.1901	.4274	2.9364	-.2628	.0171	.1786	-.1437	-.0367	I
195	Lyantonde Dispensary	910.7	249.56	.2740	1.0695	3.7603	.0509	-.2181	.1746	.0945	-.1031	I
196	Masaka	1076.4	219.34	.2038	.7233	3.2775	.1389	-.0348	.3062	-.0958	-.3267	I
197	Kiwala Estate	1171.8	274.08	.2339	1.0211	3.7067	.0390	-.0468	.0116	-.0806	-.1740	I
198	Kalangala Didpensary											
199	Buunga	1142.7	248.57	.2175	.9646	3.4779	-.1652	-.1251	.0093	.1093	-.1004	I
200	Kyanawkaka	1155.2	218.11	.1888	1.2631	4.3958	-.1606	-.0712	.3984	.0216	-.1622	D

(1) in mm/yr; (2), (3) corrected for bias in estimation from sample; (4)   I = independent and D = dependent

TABLE 4.3   (continued)

No.	Name	$\bar{X}$[1]	s[1]	C_v	C_s[2]	C_k[3]	r_1	r_2	r_3	r_4	r_5	I/D[4]
201	Busenyi	1258.8	262.83	.2088	.3081	2.8557	.1746	.0198	-.4592	-.2776	-.1452	D
202	Mbarara	921.9	186.00	.2018	1.2704	5.3994	-.1921	.0497	-.2075	-.0773	.0311	I
203	Bikira	1054.4	162.41	.1540	.5790	2.6892	-.0231	-.2690	.1971	-.0966	-.3954	I
204	Lwasamaire	1064.7	182.07	.1710	.4334	2.8380	.2224	-.0274	-.1781	-.3396	-.3965	I
205	Katera	1165.2	176.82	.1518	-.1584	3.6911	-.2846	.0602	-.1151	.1248	-.1532	I
206	Kabale	1003.8	173.45	.1728	.6260	3.8610	.0459	-.2995	-.0609	.2305	.1351	I
207	Kapenguria	1264.1	322.50	.2551	.9150	4.5000						
208	Endebess (Mt. Elgon)	1251.7	210.60	.1682	.2659	2.5443	.0673	-.3649	-.0230	-.0380	.0361	I
209	Kitale Agr. Stat.	1151.3	218.05	.1894	.5045	3.3466	.0761	-.0526	.1386	-.0170	.0959	I
210	Turbo	1267.1	274.08	.2163	1.2210	4.8852	.3869	-.0018	.3475	.3732	-.0348	D
211	Tambach	1206.5	276.10	.2288	.2147	2.6351	-.0967	-.2702	.0772	-.0525	.0029	I
212	Myanga	1358.4	483.71	.3561	2.2958	10.8581						
213	Bungoma V.S.	1560.9	260.67	.1670	.6982	2.8679	.1344	-.3544	-.0740	-.1328	-.1452	I
214	Mumais	1738.8	317.15	.1824	.0118	2.1907	.3288	.2014	-.0274	-.1630	-.2435	I
215	Tororo/Nangina	1396.1	205.38	.1471	1.2721	4.5792						
216	Kakamega	1982.2	284.08	.1433	.4644	2.5430						
217	Kapasabet	1546.3	290.40	.1878	.6539	3.2129						
218	Rangala	1563.8	245.96	.1573	.2767	3.0669						
219	Maseno, V.S.	1590.3	269.68	.1696	.7141	3.4504	.2201	.0177	-.2055	.1201	.2390	I
220	Equator	1552.5	286.95	.1848	.0208	3.4451	.1396	-.2559	-.1300	-.0638	.0765	I
221	Miwani	1122.7	184.87	.1646	.0556	2.3785						
222	Kisumu P.C.	1086.6	228.08	.2099	.5947	2.7984	.0976	-.1288	-.2725	-.2582	.1285	I
223	Chemelil	1443.5	398.99	.2764	2.1485	9.5738						
224	Muhoroni	1385.5	302.63	.2184	.4448	2.9430	.0858	-.2311	-.2736	-.0039	.0509	I
225	Fort Ternan	1249.8	207.97	.1664	.1972	2.0881	.1719	-.0886	.1626	.2596	.0151	I

(1) in mm/yr; (2), (3) corrected for bias in estimation from sample; (4) I = independent and D = dependent

TABLE 4.3 (continued)

No.	Name	$\bar{X}^{(1)}$	$s^{(1)}$	C_v	$C_s^{(2)}$	$C_k^{(3)}$	r_1	r_2	r_3	r_4	r_5	$I/D^{(4)}$
							Basic statistical descriptors →				Serial correlation coefficients →	
226	Londiani	1189.1	217.73	.1831	-.2540	2.3639	.1137	-.3093	-.2416	-.1701	.0671	I
227	Lumbwa	1108.8	176.03	.1588	-.3474	3.0842	.1809	-.0054	-.0991	-.2053	.1017	I
228	Molo	1073.6	278.44	.2594	.2092	3.0331	.3645	-.0631	-.0602	-.1013	-.2829	D
229	Kericho	1887.0	284.73	.1509	.8191	3.8312	.1299	-.4536	-.0226	-.2896	.0383	D
230	Sotik	1350.4	204.41	.1514	-.4772	4.7605	.1829	.0366	.1777	.3211	.0616	I
231	Kisu	1842.4	561.75	.3048	2.7433	13.3440	-.1106	.1655	.0259	-.1947	-.2996	I
232	Bukoba	2090.1	292.51	.1399	.6911	3.0121	.6742	.3719	.1884	.1795	.1710	D
233	Tarime	1442.0	323.32	.2242	1.7789	5.6362	-.0507	-.0099	.3681	.2939	-.2571	D
234	Musoma	825.9	152.73	.1849	-.1904	2.9022	-.0981	.0499	-.3245	.1646	.0742	I
235	Kagondo	1714.8	339.72	.1981	.9061	4.4254						
236	Kawalinda	1650.4	290.02	.1757	.5554	2.8975	.3398	.3561	.2380	.1792	-.0020	D
237	Rubya	1284.9	263.54	.2051	-.0248	2.9343	-.2521	.1056	.2387	.1757	-.6647	D
238	Igabiro	1078.4	229.91	.2133	.2966	3.1706						
239	Ikizu	967.8	269.75	.2787	.5966	3.3395						
240	Kome	1062.1	263.76	.2483	.6855	2.6430						
241	Ngara	1017.7	167.76	.1648	.9066	4.1991	-.0042	-.1116	.1189	-.1564	.0089	I
242	Mwanza	1026.9	253.30	.2467	.9969	3.8617						
243	Biharamulo	957.4	178.31	.1862	1.7031	8.7233						
244	Ukiriguru	837.4	215.67	.2576	.3235	3.0355	.2880	-.2941	-.1091	.2635	.3133	I
245	Samwe Mission	957.4	480.77	.5022	1.6237	5.9419						
246	Geita	1007.6	209.02	.2074	1.2053	5.0079	.1073	-.1720	-.0615	.1557	.2733	I
247	Ngudu	794.0	210.99	.2657	.3888	3.3287						
248	Kijima Mission	755.3	166.32	.2202	1.3145	6.4833						
249	Shanwa	769.7	194.94	.2533	1.2028	4.8015	.1245	-.1265	-.3445	-.0146	-.0663	I
250	Addis Ababa	1162.5	165.26	.1422	.6765	3.3761	.2440	-.0110	.2779	-.1897	-.1489	I

(1) in mm/yr; (2), (3) corrected for bias in estimation from sample; (4) I = independent and D = dependent

TABLE 4.4   Periods of rainfall measurement at some of the gauging stations in
the Sudan, Uganda and Tanzania as reported in a number of
references

Station No.	Period(a)*	Station No.	Period(b)*	Period(c)*
66	1902-40	137	1901-13	
67	1905-40	142	1904-13	
69	1916-40	145	1911-20	1926-70
71	1905-40	146		1933-67
72	1913-40	150	1904-20	
75	1902-40	151	1914-20	
79	1900-40	152	1906-20	1926-70
80	1901-40	156	1909-20	
81	1920-40	157	1909-20	
82	1930-40	163	1907-20	1926-70
87	1929-40	167	1908-20	
88	1919-40	172	1901-20	1926-70
91	1902-40	175	1909-20	
92	1930-40	176	1908-20	
95	1918-40	181	1911-20	
97	1922-40	185	(1893-98,	
98	1938-40		1907-20)	
99	1901-40	187	1908-20	
100	1938-40	191	1896-1920	1934-70
101	1912-40	196	1902-20	1926-70
105	1938-40	202	1907-20	1926-70
110	1904-40	206	1918-20	
111	1938-40	214	1896-1915	1933-70
118	1915-40	222	(1899-1903,	1926-70
124	1928-40		1913-1920)	
126	1909-40	229		1930-70
127	1932-40	231		1926-70
128	1902-40	232	(1893-98,	1926-68
133	1931-39		1901-11)	
139	1929-40	233		1932-69
		234		1926-70
		238		1932-70
		242		1926-68
		243		1928-69
		249		1928-60

*(a) Ireland, The Climate of the Sudan, edited by Tothill (1948)

(b) Brooks, The Distribution of Rainfall over Uganda, with a note on Kenya
Colony (1924)

(c) UNDP & WMO, Vol. I, Part I, Meteorology and Hydrology of the Basin,
Hydrometeorological Survey of the Catchments of Lakes Victoria, Kyoga
and Albert (1974)

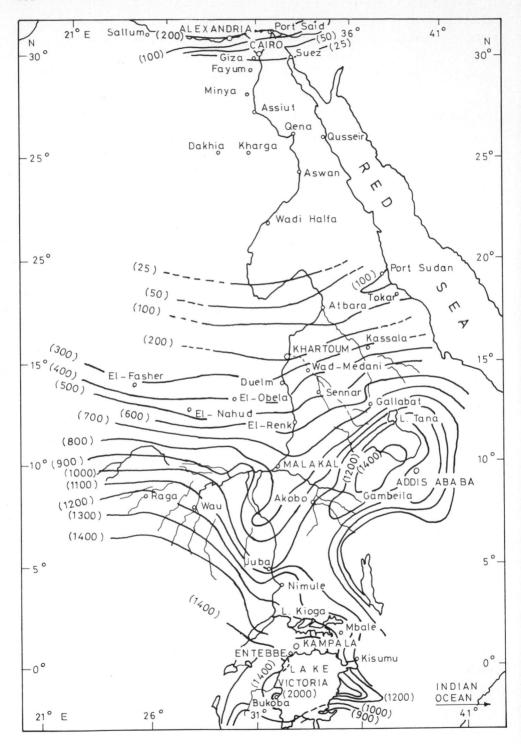

Fig. 4.3.   Isohyetal map of the annual rainfall on the Nile Basin

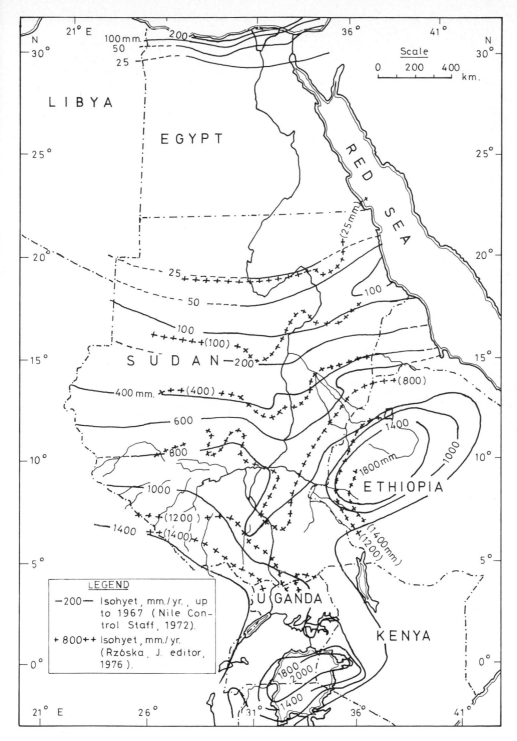

Fig. 4.4.   Isohyetal map of the mean annual rainfall on the Nile Basin till 1967

136

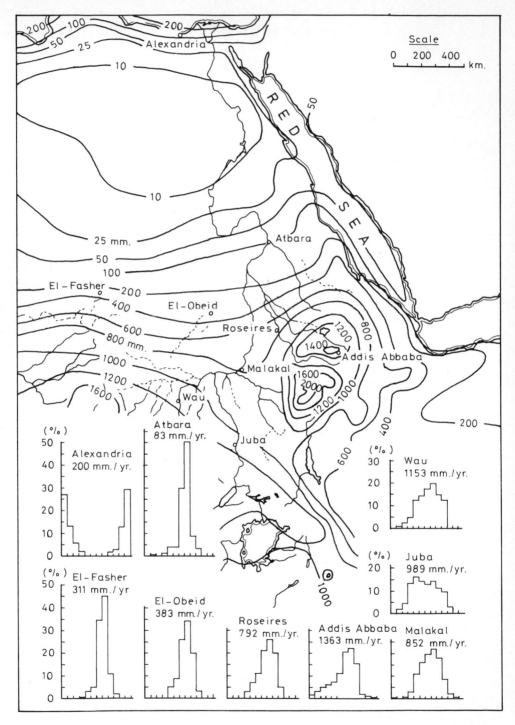

Fig. 4.5. Map of the mean annual rainfall on the Nile Basin and the distribution diagrams of monthly rainfall (the World Water-Balance, UNESCO, 1978)

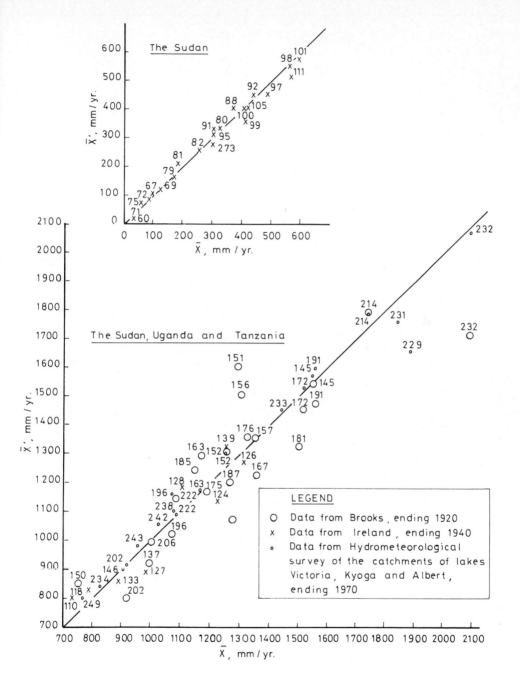

Fig. 4.6. The mean annual rainfall over the 30-year period 1938-1967, $\bar{X}$ versus the mean annual rainfall over other periods of record, X'

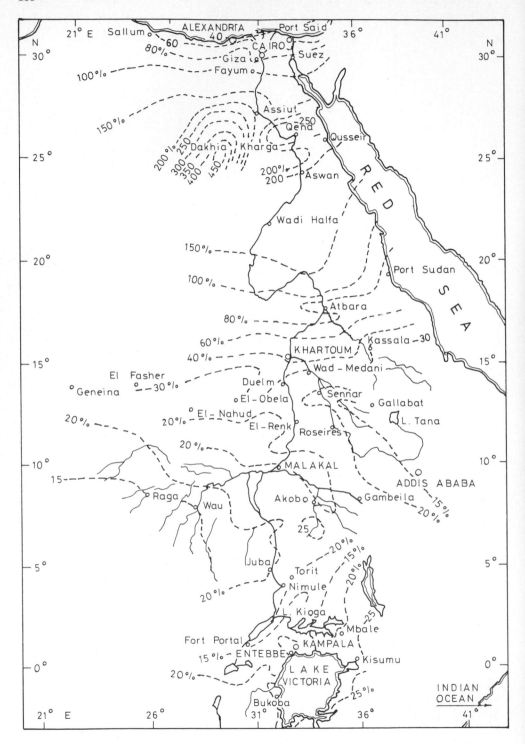

Fig. 4.7.   Contour lines of equal coefficient of variation of annual rainfall

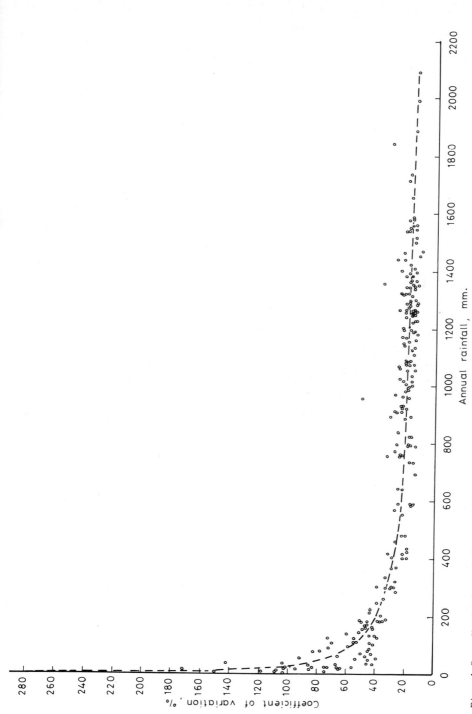

Fig. 4.8. Variation of the coefficient of variation of the annual rainfall with the mean of the annual rainfall at 250 stations in the Nile Basin

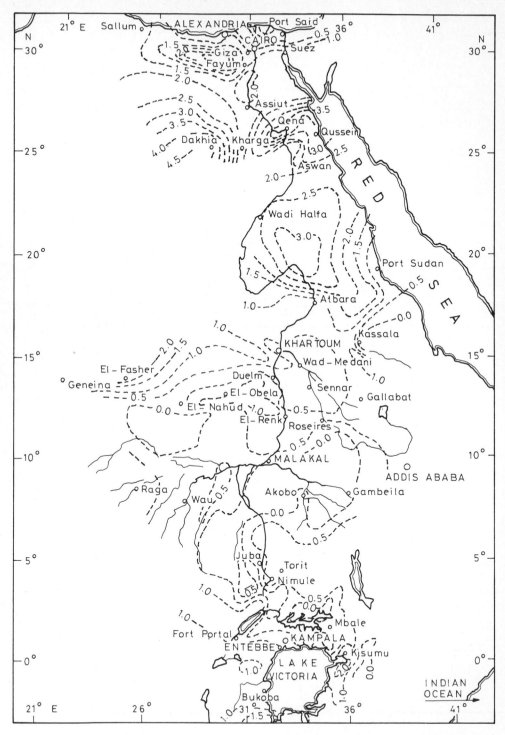

Fig. 4.9.   Contour lines of equal coefficient of skewness of annual rainfall

TABLE 4.5   Estimated annual rainfall for given return periods at some of the rain-gauging stations in the Nile Basin (Shahin, M.M., 1983)

Station Serial Number	Distribution Function	Rainfall depth mm/yr, for return period, yr						
		1.01	2	5	10	20	50	100
2	Normal	34	181	235	263	286	311	329
4	Normal	16	124	163	183	200	219	232
5	Pearson III	06	107	150	174	195	219	236
9	Pearson III	47	148	206	245	278	320	349
10	Pearson III	02	179	262	304	343	389	421
13	Pearson III	15	69	97	114	128	146	158
14	Pearson III	17	136	204	247	285	331	346
15	Normal	45	190	242	269	292	317	334
16	Pearson III	06	180	255	298	336	380	411
17	Pearson III	18	146	216	258	295	340	372
18	Pearson III	15	76	109	129	146	167	182
20	Pearson III	10	63	91	107	122	140	152
24	Pearson III	15	85	121	142	161	183	199
28	Pearson III	-	33	55	69	82	98	110
29	Normal	-	51	70	80	88	98	104
32	Pearson III	21	23	42	66	95	137	172
39	Pearson III	00	22	33	41	47	55	60
41	Pearson III	-	23	41	51	61	72	81
44	Pearson III	-	20	40	53	66	82	95
47	Pearson III	00	09	22	32	41	55	65
49	Pearson III	-	07	16	23	29	37	43
51	Pearson III	-	02	06	10	13	17	20
53	Pearson III	-	01	06	10	14	19	22
58	Pearson III	-	00	06	13	20	30	38
62	Pearson III	-	01	03	05	06	08	10
63	Pearson III	-	09	21	25	35	47	56
68	Pearson III	00	05	25	46	70	104	131
70	Pearson III	05	108	195	254	310	383	436
71	Pearson III	00	31	69	93	117	146	167
75	Pearson III	-	60	99	123	144	170	189
77	Pearson III	27	218	304	353	395	445	479
79	Pearson III	08	165	237	278	315	357	388
80	Normal	122	321	394	431	463	498	521
82	Pearson III	53	242	329	378	422	473	509
83	Pearson III	83	256	334	379	418	465	497
85	2-Par. Lognormal	85	191	257	294	340	388	432
86	Pearson III	134	288	308	410	459	511	517
87	Pearson III	137	295	369	411	448	492	521
89	Pearson III	112	287	379	434	484	543	585
90	2-Par. Lognormal	386	572	660	712	757	809	850
91	Normal	93	303	380	419	452	489	514
92	Pearson III	271	430	503	546	583	627	658
94	Normal	115	284	346	378	404	434	454
98	2-Par. Lognormal	323	540	650	719	775	851	901
99	2-Par. Lognormal	200	388	493	556	618	690	748
100	2-Par. Lognormal	246	394	466	507	544	589	621
101	Normal	383	592	667	707	740	776	801
102	Normal	144	370	451	494	529	568	595
103	Pearson III	317	367	482	582	686	830	940
104	Pearson III	338	424	536	620	704	812	895
105	Normal	203	418	496	537	570	608	633

TABLE 4.5   (continued)

Station Serial Number	Distribution Function	Rainfall depth mm/yr, for return period, yr						
		1.01	2	5	10	20	50	100
106	Normal	232	423	493	529	559	592	615
107	Normal	223	478	571	619	659	704	734
108	Normal	302	641	763	828	881	940	980
109	Pearson III	567	767	866	924	975	1037	1080
110	Pearson III	524	722	820	878	928	990	1033
112	Normal	467	692	774	816	851	891	917
115	Pearson III	385	917	1109	1184	1242	1304	1342
117	Pearson III	470	725	843	910	968	1037	1085
118	Normal	466	795	913	975	1026	1084	1123
122	Normal	391	897	1082	1177	1256	1345	1405
123	Normal	457	798	922	986	1039	1098	1139
124	Pearson III	910	1212	1364	1453	1531	1625	1691
125	Pearson III	529	898	1087	1198	1298	1417	1503
126	Pearson III	553	1294	1563	1669	1748	1838	1888
127	Pearson III	462	968	1151	1221	1274	1330	1365
128	Normal	812	1182	1316	1387	1444	1509	1553
130	Normal	327	759	915	997	1064	1140	1191
131	Pearson III	377	940	1142	1209	1264	1315	1347
132	Normal	346	909	1113	1220	1308	1407	1472
133	Pearson III	621	873	1012	1096	1172	1264	1330
134	Normal	728	1187	1353	1439	1511	1592	1645
136	Normal	845	1358	1543	1640	1720	1810	1870
137	Normal	575	999	1153	1233	1300	1374	1424
138	Normal	1108	1473	1605	1674	1730	1794	1837
143	Normal	804	1252	1415	1499	1569	1648	1701
146	Normal	403	910	1094	1189	1269	1358	1417
147	2-Par. Lognormal	1077	1440	1599	1690	1766	1872	1919
148	Normal	868	1286	1437	1516	1581	1655	1704
149	2-Par. Lognormal	795	1120	1273	1365	1441	1532	1601
150	Pearson III	544	669	891	1072	1258	1511	1702
153	Pearson III	625	1258	1446	1540	1616	1697	1749
154	Pearson III	547	1391	1566	1635	1681	1722	1745
157	Pearson III	1066	1325	1488	1591	1685	1801	1886
158	Pearson III	876	1024	1177	1263	1363	1424	1487
159	Normal	777	1230	1394	1479	1550	1630	1683
161	Pearson III	1013	1344	1568	1710	1843	2008	2128
162	Pearson III	975	1273	1483	1619	1747	1909	2013
163	Pearson III	794	1126	1353	1498	1633	1801	1923
164	Pearson III	938	1229	1428	1554	1672	1818	1925
165	Pearson III	987	1315	1551	1705	1847	2029	2160
166	Pearson III	893	1126	1377	1558	1734	1965	2136
167	Normal	847	1353	1536	1632	1711	1802	1859
168	Pearson III	872	1194	1379	1493	1596	1721	1812
169	Pearson III	991	1242	1466	1621	1768	1957	2096
170	Pearson III	1083	1328	1513	1635	1749	1893	1999
172	Pearson III	1131	1488	1694	1820	1935	2074	2177
173	Pearson	870	1193	1432	1588	1735	1875	1919
174	2-Par. Lognormal	820	1256	1470	1596	1710	1853	1946
175	2-Par. Lognormal	843	1173	1329	1420	1499	1597	1662
176	Pearson III	1061	1227	1492	1703	1917	2203	2422
177	Pearson III	1075	1555	1803	1950	2080	2238	2349

TABLE 4.5   (continued)

Station Serial Number	Distribution Function	Rainfall depth mm/yr, for return period, yr						
		1.01	2	5	10	20	50	100
179	Pearson III	1098	1468	1772	1977	2171	2417	2599
180	Pearson III	1016	1345	1588	1752	1904	2096	2234
181	Pearson III	1097	1470	1682	1812	1929	2072	2174
182	Pearson III	995	1281	1537	1712	1880	2096	2255
184	Pearson III	1010	1324	1551	1698	1834	2008	2133
185	Pearson III	630	1133	1359	1487	1598	1729	1820
186	Pearson III	852	1293	1529	1671	1798	1953	2063
187	Pearson III	901	1244	1421	1526	1619	1732	1811
188	Pearson III	996	1510	1775	1933	2072	2241	2366
189	Normal	865	1578	1835	1971	2082	2208	2292
190	Normal	823	1424	1642	1756	1850	1955	2026
191	Pearson III	841	1575	1787	1884	1958	2084	2126
192	Pearson III	548	1004	1209	1326	1427	1546	1627
193	Normal	565	1055	1232	1325	1401	1487	1545
194	Pearson III	671	1070	1252	1356	1447	1556	1629
195	Pearson III	526	867	1099	1245	1382	1552	1676
196	2-Par. Lognormal	673	1051	1247	1366	1474	1601	1702
197	Pearson III	740	1126	1379	1539	1687	1871	2002
199	Pearson III	739	1103	1332	1476	1608	1772	1891
202	Pearson III	661	884	1054	1171	1278	1415	1518
203	Pearson III	747	1039	1185	1270	1345	1436	1500
204	Pearson III	700	1052	1213	1305	1385	1481	1545
205	Normal	754	1165	1314	1392	1456	1528	1577
206	Pearson III	680	986	1142	1234	1316	1415	1485
208	Pearson III	801	1243	1426	1526	1613	1712	1779
209	2-Par. Lognormal	720	1132	1326	1439	1537	1653	1740
211	Normal	589	1207	1439	1560	1661	1775	1824
213	Pearson III	1090	1531	1767	1908	2035	2188	2297
214	Normal	1030	1739	2006	2145	2260	2390	2448
219	Pearson III	1107	1558	1802	1950	2082	2241	2354
220	Normal	885	1553	1794	1920	2025	2142	2220
220	2-Par. Lognormal	651	1064	1267	1383	1495	1604	1718
224	Pearson III	737	1154	1631	1785	1920	2078	2190
225	Normal	766	1250	1425	1516	1592	1677	1734
226	Pearson III	643	1197	1374	1461	1531	1607	1655
228	Normal	426	1074	1308	1431	1532	1646	1722
231	Pearson III	661	1814	2305	2578	2812	3084	3312
232	Pearson III	1562	2056	2321	2480	2622	2794	2916
235	Pearson III	1151	1666	1976	2170	2348	2565	2721
242	Pearson III	625	985	1219	1366	1502	1671	1793
244	Pearson III	388	826	1015	1120	1211	1316	1391
246	Pearson III	705	967	1161	1288	1407	1557	1666
249	Pearson III	487	732	913	1031	1142	1282	1384
250	Pearson III	858	1146	1294	1383	1463	1558	1628

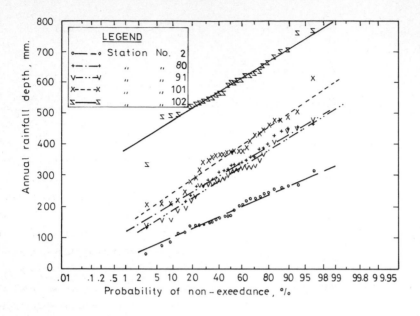

Fig. 4.10a.   Fitting the normal distribution to the annual rainfall depth at some stations in the Nile Basin

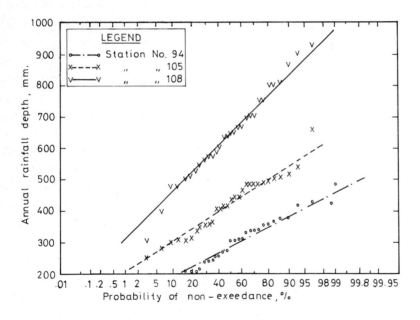

Fig. 4.10a.   Fitting the normal distribution to the annual rainfall depth at some stations in the Nile Basin

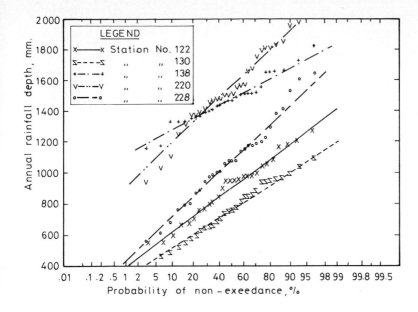

Fig. 4.10a. Fitting the normal distribution to the annual rainfall depth at some stations in the Nile Basin

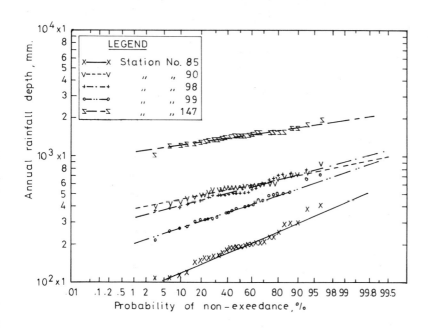

Fig. 4.10b. Fitting the lognormal distribution to the annual rainfall depth at some stations in the Nile Basin

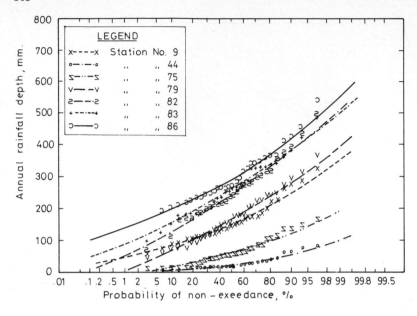

Fig. 4.10c.   Fitting the Pearson III distribution to the annual rainfall depth at some stations in the Nile Basin

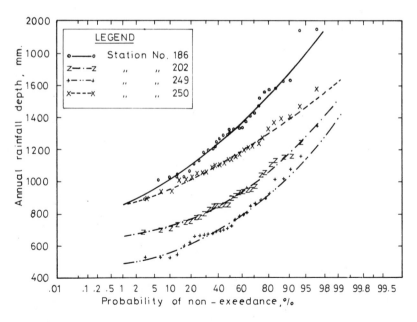

Fig. 4.10c.   Fitting the Pearson III distribution to the annual rainfall depth at some stations in the Nile Basin

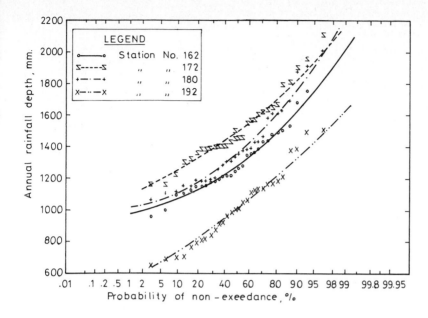

Fig. 4.10c. Fitting the Pearson III distribution to the annual depth at some stations in the Nile Basin

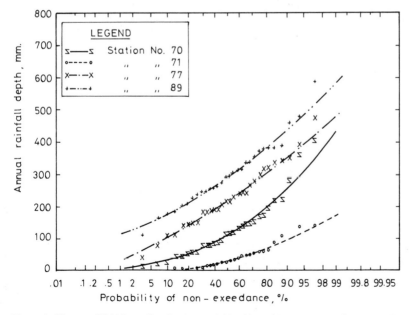

Fig. 4.10c. Fitting the Pearson III distribution to the annual rainfall depth at some stations in the Nile Basin

TABLE 4.6   The basic properties of the autoregressive models fitted to the rainfall sequences of the listed stations (Shahin, M.M., 1983)

Station No.	Model parameters				Serial correlation of residuals					Model
	α_1	α_2	α_3	α_4	r_1	r_2	r_3	r_4	r_5	
12	-0.0718	-0.0051	-	-	-0.039	-0.114	-0.173	0.352	-0.156	2nd order
22	-0.1221	-0.0147	-	-	-0.200	-0.409	0.180	0.194	-0.250	2nd order
30	0.3570	-	-	-	-0.095	0.333	-0.346	0.347	-0.169	1st order
69	-0.4458	-	-	-	-0.038	-0.129	0.023	-0.370	-0.088	1st order
97	-0.0134	-	-	-	0.050	-0.221	-0.101	-0.360	0.135	1st order
111	0.4518	-	-	-	-0.074	0.159	-0.104	-0.028	-0.079	1st order
113	0.1864	-0.0347	0.0065	-0.0012	0.017	-0.224	0.183	0.295	-0.432	4th order
114	-0.2197	-0.0461	-	-	-0.007	-0.098	0.340	-0.149	-0.429	2nd order
116	-0.0895	-	-	-	0.008	0.121	-0.399	-0.374	-0.180	1st order
119	0.3979	-	-	-	-0.040	-0.027	0.254	-0.170	0.171	1st order
120	0.4598	-	-	-	-0.019	-0.035	0.062	0.014	0.145	1st order
121	-	-	-	-	-	-	-	-	-	unsuitable
129	-0.1313	-	-	-	-0.001	-0.062	0.029	0.330	-0.117	1st order
135	0.4272	-	-	-	-0.087	0.219	-0.068	-0.031	-0.044	1st order
139	0.4648	-	-	-	0.070	-0.030	-0.100	-0.274	0.007	1st order
140	0.1936	-	-	-	0.001	0.128	-0.021	-0.376	0.022	1st order
141	0.0640	-	-	-	0.033	-0.388	0.058	0.004	-0.066	1st order
142	-	-	-	-	-	-	-	-	-	unsuitable
144	-	-	-	-	-	-	-	-	-	unsuitable
145	0.4105	-	-	-	0.074	-0.258	-0.016	-0.036	-0.361	1st order
151	0.1618	-0.0255	-	-	0.045	-0.178	-0.203	-0.151	0.154	2nd order
152	0.4310	-	-	-	0.070	-0.168	0.129	-0.171	-0.127	1st order
155	-0.0550	-0.0030	-	-	-0.044	-0.380	-0.181	0.064	0.100	2nd order
156	0.2566	-	-	-	0.025	0.286	0.190	-0.306	0.072	1st order
160	0.5480	-0.2418	-	-	0.033	0.166	-0.245	0.124	-0.010	2nd order
171	0.6260	-0.3012	-	-	-0.008	0.121	-0.047	0.089	0.211	2nd order
178	0.3490	-	-	-	-0.094	0.216	0.043	0.218	-0.277	1st order

TABLE 4.6 (continued)

Station No.	Model parameters				Serial correlation of residuals					Model
	α_1	α_2	α_3	α_4	r_1	r_2	r_3	r_4	r_5	
178	0.3490	–	–	–	-0.094	0.216	0.043	0.218	-0.277	1st order
183	0.6001	–	–	–	-0.001	0.107	0.018	0.120	-0.272	1st order
200	-0.1649	-0.0265	–	–	0.061	0.187	0.271	0.047	0.082	2nd order
201	0.1801	-0.0314	–	–	0.107	0.196	-0.241	-0.013	-0.210	2nd order
210	0.3869	–	–	–	0.042	-0.306	0.289	0.290	-0.117	1st order
229	0.3645	–	–	–	0.070	-0.203	-0.029	0.006	-0.224	1st order
230	–	–	–	–	–	–	–	–	–	unsuitable
233	0.6742	–	–	–	0.121	-0.064	-0.207	-0.035	0.193	1st order
234	-0.0507	–	–	–	-0.053	-0.036	0.338	0.259	-0.265	1st order
236	0.3398	–	–	–	-0.095	0.197	0.085	0.113	-0.020	1st order
237	-0.2705	-0.0732	-0.0197	-0.0050	0.133	-0.097	-0.100	0.195	-0.325	4th order

4.2.2   Seasonal and monthly rainfall

The distribution of the rainfall among the months of the year assumes different patterns. Those covering the 250 gauging stations which are investigated
here have been grouped into principal patterns. To make the comparison between
one group and the other feasible, the monthly rainfall expressed in percent of
the annual rainfall (see Appendix C) has been used as a basis. The nine principal patterns which are shown in Fig. 4.11. show the following characteristics:
Group I comprises stations 1 to 29. These are along the Mediterranean and southwards up to the central part of the Nile Delta area. The rainfall lasts from
October to April or May and the maximum depth falls in December or January.
Group II comprises the stations in the eastern part of Egypt and along the Red
Sea coast up to Port Sudan. This group is characterized by a high peak in
November and the winter rainfall is rather scanty. Typical stations in this
group are No. 45 in Egypt and 67 in the Sudan.
Group III comprises those stations in the northern and central parts of the
Sudan from No. 79 to 109. The maximum rainfall here takes place in August. The
total rainfall in July and August expressed as percentage of the annual rainfall
varies from about 80% in the north to, say, 60% at Jebelein halfway between
Khartoum and Malakal on the White Nile.
Group IV is similar to Group III, except that the base width is larger and the
peak is shorter. So the rain falls on this group from February to March till
November. The August peak is nearly 30% of the annual rain, as at station 110,
and falls to about 21% at Meshra Er-Req, station 125, in the southern part of
the Sudan.
Group V comprises the gauging stations from No. 126 to 139. Here the rain falls
in all months of the year, and the maximum rain falls in August with a depth of
20% or less of the annual rainfall.
Group VI comprises the stations from No. 140 to 162. The rainfall over this
group of stations covers the whole year except that the monthly rain at the
tails is more than that over the stations in the previous group. In other words,
the pattern here tends to be more uniform than in the previous group, and this
uniformity improves gradually as one moves to the south (compare station No. 140
to station No. 161, Fig. 4.11.).
Group VII comprises the stations 163 to 218. The major feature of this group is
the considerable decline of the rainfall in July compared to the previous groups.
Another feature is the development of two peaks; the first and higher peak
occurs in April or May, and the second and lower peak occurs in November.
Group VIII comprises the gauging stations 219 up to and including 231. The main
difference between this group and Group VII is that the rainfall in the second
half of the year is more uniformly distributed.

151

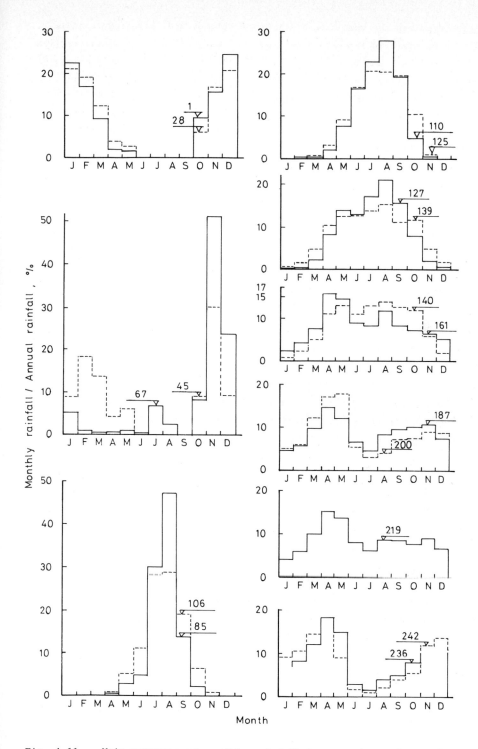

Fig. 4.11.   Main patterns of monthly rainfall in percentage of annual rainfall

Group IX comprises the remaining stations except No. 250, situated at Addis Ababa. The rainfall in the second half of the year for this group differs from that for Groups VII and VIII in that it is distributed nearly like a triangle.

The monthly rainfall pattern at station No. 250 is quite similar to that of Group III, except that the tails here extend to cover the whole year.

The month to month variation, sometimes referred to as seasonal variation, can be formulated, among others, by a harmonic function.

The expression used in the hydrometeorological survey of the catchments of Lakes Victoria, Kyoga and Albert, reads

$$R_t = A_o + \sum_{r=1}^{6} A_r . \sin (\frac{2 \Pi r}{12} . t + \phi_r) \tag{4.10}$$

where,

R_t = rainfall at any month t (t = 1, 2, ..., 12)

A_o = constant equal to the annual rainfall divided by 12,

A_r = amplitude of the rth harmonic (r = 1, 2, ..., 6), and

ϕ_r = phase angle of the rth harmonic (r = 1, 2, ..., 6)

For the above-mentioned hydrometeorological survey, the parameters A_r and ϕ_r were calculated using the rainfall in each year from 1931 up to and including 1970 for each of the 22 gauging stations included in that survey.

Since there are 12 monthly values, R_t, for each year, the maximum number of parameters cannot exceed 12; 6 for A_r and 6 for ϕ_r (ϕ_6 = zero). The values of A_r and ϕ_r (r = 1, 2, ..., 6) averaged over the period 1931-1970 are given in Table 4.7. More detailed information about these harmonic parameters, as well as their maximum and minimum values, dates of occurrence and distribution in space can be found in Part 1, Vol. I of the survey report (WMO, 1974).

From Table 4.7 it is apparent that the first two pairs of the harmonic parameters, i.e. A_1, ϕ_1 and A_2, ϕ_2, are by far the most important of all six pairs in the series. The relative importance of a certain oscillation can be measured by the square of the amplitude of the corresponding sine term. The explainable portion of the variance in a series is the sum of $\frac{1}{2} (A_r)^2$, whereas the total variance can be expressed as

$$Var (R_t) = \frac{1}{2} \sum_{r=1}^{6} A_r^2 + \epsilon \tag{4.11}$$

where ϵ is the residual left after fitting the six harmonics, which is sometimes referred to as the unexplainable portion of the total variance.

TABLE 4.7 The values of the harmonic parameters in the model describing the seasonal variation of rainfall at the listed stations (WMO, 1974)

No.	Name	A_1	ϕ_1	A_2	ϕ_2	A_3	ϕ_3	A_4	ϕ_4	A_5	ϕ_5	A_6
–	El-Doret*	43.9	275.4	15.9	182.4	27.3	116.1	21.2	226.9	8.0	164.0	3.8
–	Bugondo**	46.9	250.5	50.4	212.3	20.0	57.2	11.1	267.8	7.7	117.1	5.3
145	Gulu	84.5	235.1	34.1	233.4	16.5	78.9	13.9	244.8	9.5	358.0	15.3
146	Moroto	54.4	270.2	9.6	188.0	19.8	126.1	14.0	266.5	7.4	88.7	5.3
152	Masindi	35.3	241.4	37.8	217.7	10.8	112.6	6.3	260.8	2.9	20.5	5.8
163	Mbale	50.9	260.5	23.4	205.8	13.9	80.3	7.2	256.1	5.0	74.2	0.7
172	Fort Portal	18.5	177.8	72.4	214.8	5.8	90.0	3.2	291.0	1.2	241.9	6.2
191	Entebbe	60.4	331.5	69.8	181.8	28.4	81.9	17.0	229.3	4.7	79.4	8.5
196	Masaka	25.9	344.0	55.7	194.8	21.9	61.9	6.8	240.7	11.3	88.0	4.0
198	Kalangala	76.7	329.9	74.4	170.3	30.2	89.7	12.4	178.5	7.4	83.7	4.0
202	Mbarara	18.2	71.1	40.0	213.1	10.2	96.7	6.1	205.9	4.1	221.5	3.8
214	Mumias	62.3	282.4	50.9	196.1	18.2	53.5	10.6	225.9	5.5	106.9	2.2
222	Kisumu	41.0	336.6	33.8	197.2	23.1	101.2	6.6	181.4	3.1	177.3	3.5
229	Kericho	48.6	301.6	41.3	196.0	25.1	81.6	5.7	234.1	1.7	23.3	3.8
231	Kisii	36.9	290.4	57.5	206.3	24.4	87.6	5.9	161.5	2.9	230.4	5.3
232	Bukoba	85.8	0.8	86.1	190.7	47.2	66.5	17.5	261.5	12.8	115.3	5.5
233	Tarime	39.0	8.1	52.7	196.5	23.1	103.3	4.3	289.8	3.3	153.4	2.5
234	Musoma	49.5	2.7	36.0	201.5	22.1	103.1	8.0	291.1	5.9	203.1	5.2
238	Igabiro	56.3	30.5	47.2	206.0	21.0	100.1	4.4	328.1	5.2	171.1	0.7
242	Mwanza	68.5	37.3	30.4	190.6	27.9	105.6	6.7	344.9	4.3	241.9	0.3
243	Biharamulo	62.0	32.3	39.9	209.8	23.3	115.8	5.4	292.0	9.7	2o8.6	1.2
249	Shanwa	72.6	35.9	13.2	193.5	25.3	109.6	5.4	9.7	8.0	189.8	7.7

Gauging station — Values of harmonic parameters

* El-Doret station is situated at $0^\circ 36'$N and $35^\circ 26'$E. Its altitude is 2287 m.

** Bugondo station is situated at $1^\circ 37'$N and $33^\circ 17'$E. Its altitude is 1067 m.

The harmonic term r = 1 means that the period of the oscillation is 12 months and r = 2 means that the corresponding oscillation has a period of 6 months. The latter is the bi-annual oscillation responsible for the two peaks in April and November. This is visible in the pattern Groups VI to IX, especially in Groups VII and IX (see Fig. 4.11.).

The hydrometeorological survey of the catchments of Lakes Victoria, Kyoga and Albert included the probabilistic modelling of the monthly rainfall at the chosen stations. It has been reported that the distribution function that serves as a good fit to the observed data is the two-parameter Gamma function (WMO, 1974). This function can be written as

$$P(X) = \frac{\int_{o}^{X} X^{\gamma-1} e^{-X/\beta} \, dX}{\beta^{\gamma} \, \Gamma(\gamma)} \tag{4.12}$$

where

$P(X)$ = cumulative probability

$\Gamma(\gamma)$ = gamma function, and

γ and β = parameters of the distribution function

The monthly values of the γ and β parameters for each of the 22 stations in the catchments of the above-mentioned lakes are included in Table 4.8. An accurate estimate of these parameters is necessary for determining the monthly rain depth corresponding to a certain return period. For this purpose one needs a table giving the cumulative gamma distribution like the one compiled from E.S. Pearson and H.O. Hartley ((1954) and presented by C.T. Haan in his books as Table E.8 (Haan, C.T., 1977). The procedure for determining the magnitude of an event for any given return period using this table is explained in the same reference. We have used this procedure for estimating the monthly rainfall with return periods of 1.01, 1.11, 2, 5, 10, 20, 50 and 100 years at Bukoba and the results obtained are included in Table 4.9. Needless to say, this same procedure can be applied to any other station whose data can be well fitted by the gamma distribution function.

TABLE 4.8 Values of the parameters of the Gamma distribution function fitted to the monthly rainfall at 22 stations in the catchments of Lakes Victoria, Kyoga and Albert (WMO, 1974)

Month	El-Doret		Bugondo		Gulu (145)		Moroto (146)		Masindi (152)		Mbale (163)	
	γ	β	γ	β	γ	β	γ	β	γ	β	γ	β
Jan.	1.094	28.60	1.152	20.30	1.017	21.69	0.809	19.43	1.202	33.17	0.879	40.43
Feb.	0.844	70.99	1.428	39.13	0.913	52.48	0.994	40.35	1.303	42.83	1.418	37.78
Mar.	1.684	45.16	1.438	64.66	3.386	28.29	1.235	64.58	3.428	34.52	2.262	44.78
Apr.	2.485	67.35	4.310	44.67	8.440	20.38	1.671	75.25	7.252	21.11	5.340	29.75
May	3.012	43.56	7.492	24.24	9.228	19.71	2.264	61.62	0.963	29.41	7.532	22.29
June	2.175	38.16	3.241	29.37	5.751	26.33	1.478	59.23	3.632	26.52	5.982	18.96
July	8.586	15.55	4.866	21.65	6.256	27.08	2.795	49.74	4.092	28.18	8.058	15.18
Aug.	6.102	19.82	5.458	25.93	10.060	24.52	1.928	57.84	7.109	19.29	4.898	25.97
Sep.	1.623	36.60	3.815	40.20	4.358	41.39	1.144	47.67	6.925	19.99	4.580	22.22
Oct.	2.146	26.72	3.473	41.23	7.021	24.22	1.346	36.46	6.353	21.63	4.499	19.27
Nov.	1.443	57.67	1.109	81.96	1.476	64.50	0.649	100.78	1.945	58.30	1.182	55.45
Dec.	1.079	48.23	1.196	49.41	1.280	36.49	0.987	32.49	1.382	42.90	1.516	33.08

Month	F. Portal (172)		Entebbe (191)		Masaka (196)		Kalangala (198)		Mbarara (202)		Mumias (214)	
	γ	β	γ	β	γ	β	γ	β	γ	β	γ	β
Jan.	1.314	33.26	1.817	41.65	1.284	42.69	–	–	1.595	33.40	1.068	55.73
Feb.	1.819	43.58	2.910	31.07	2.756	23.68	–	–	2.313	29.34	1.809	52.32
Mar.	5.897	24.40	7.287	24.30	4.581	26.56	–	–	7.224	13.93	3.672	42.31
Apr.	10.518	18.62	9.829	27.73	5.902	31.85	–	–	5.758	21.11	6.222	40.56
May	8.360	17.35	6.394	42.66	3.041	60.19	–	–	2.877	29.71	15.310	17.45
June	3.208	25.16	3.240	33.88	1.392	37.49	–	–	0.862	34.28	6.920	24.49
July	2.306	27.37	2.499	28.49	1.669	24.03	–	–	0.931	22.58	4.441	30.56
Aug.	6.778	18.04	2.928	27.81	1.774	31.76	–	–	2.447	23.38	7.274	21.22
Sep.	7.686	23.70	2.021	38.08	3.144	29.27	–	–	3.407	27.27	6.160	24.91
Oct.	12.869	16.74	3.229	31.59	5.398	20.31	–	–	4.249	23.52	5.176	27.16
Nov.	7.314	22.83	3.904	39.29	4.538	23.17	–	–	3.689	32.17	2.109	56.50
Dec.	3.493	26.37	2.469	46.55	2.211	44.49	–	–	2.359	31.49	2.034	43.81

TABLE 4.8 (continued)

Month	Kisumu (222)		Kericho (229)		Kisii (231)		Bukoba (232)		Tarime (233)	
	Y	β	Y	β	Y	β	Y	β	Y	β
Jan.	1.011	52.42	1.571	50.47	1.442	41.67	4.318	34.52	3.331	26.62
Feb.	2.079	40.67	1.995	47.64	1.750	53.51	3.245	48.95	2.529	41.45
Mar.	4.075	34.00	2.635	62.95	2.689	68.04	9.632	24.27	5.311	32.52
Apr.	5.568	33.37	5.965	41.50	6.343	40.36	12.596	29.64	10.664	21.81
May	4.873	29.87	11.248	19.89	12.457	17.15	7.986	40.14	6.005	27.77
June	2.307	32.85	9.600	14.44	6.205	22.22	2.523	34.18	1.846	40.50
July	2.722	19.85	4.295	26.43	3.795	26.92	1.418	38.41	1.501	39.04
Aug.	2.303	31.76	8.684	16.01	6.942	21.71	2.776	25.98	3.297	22.07
Sep.	3.226	18.63	4.870	24.33	3.541	45.43	3.754	29.40	4.266	21.22
Oct.	1.964	27.19	4.525	24.89	4.047	35.88	5.408	26.44	3.565	34.46
Nov.	1.219	75.05	2.464	47.65	2.214	65.72	6.196	29.42	3.177	46.89
Dec.	2.189	43.72	2.395	45.84	1.966	59.73	4.544	42.92	3.714	36.19

Month	Musoma (234)		Igabiro (238)		Mwanza (242)		Biharamulo (243)		Shanwa (249)	
	Y	β	Y	β	Y	β	Y	β	Y	β
Jan.	1.378	46.55	3.173	28.57	4.168	24.27	3.758	24.63	3.438	32.00
Feb.	1.409	51.14	2.708	38.05	1.987	51.64	5.022	22.26	3.039	36.63
Mar.	2.033	62.58	5.296	29.55	2.514	61.17	5.163	29.07	4.593	27.75
Apr.	6.207	29.13	7.866	24.75	7.592	23.97	5.667	32.76	4.027	37.08
May	3.947	27.55	2.740	40.24	2.516	35.03	1.899	42.58	1.319	35.24
June	1.372	18.02	0.886	25.63	0.953	25.39	0.775	17.20	1.298	4.65
July	0.680	40.54	0.896	16.80	0.746	27.08	0.897	7.24	18.848	0.28
Aug.	1.222	17.66	1.785	13.13	0.183	24.86	1.054	20.34	0.848	12.31
Sep.	1.142	25.09	2.499	24.99	1.461	25.90	1.141	36.64	1.545	8.11
Oct.	1.520	26.43	5.661	15.82	1.193	57.09	2.897	23.23	1.037	28.79
Nov.	2.185	38.21	3.555	34.22	2.275	56.89	4.620	26.73	1.759	50.76
Dec.	1.791	38.89	4.199	28.14	2.595	56.56	5.339	19.94	4.046	33.01

TABLE 4.9   Statistical descriptors of monthly rainfall at Bukoba in the period
            1931-1970, and the estimated monthly rain depth for various return
            periods

	Statistical descriptor			Rainfall, mm/month for return period, yr							
	$\bar{X}$	s	C_v	C_s	1.01	1.11	2	5	10	20	50	100
Jan.	149	72.11	0.4838	0.83	31	66	135	200	243	276	328	366
Feb.	159	81.82	0.5151	0.15	25	59	142	225	279	328	387	436
Mar.	234	73.37	0.3139	0.07	94	144	227	296	331	366	413	439
Apr.	373	102.74	0.2752	0.21	172	245	359	451	507	557	619	656
May	321	109.11	0.3404	0.17	117	187	307	409	470	528	596	642
June	86	56.24	0.6523	1.88	09	28	74	125	158	190	232	258
July	54	42.97	0.7887	1.08	02	07	40	85	115	148	186	215
Aug.	72	37.97	0.5264	0.26	09	24	63	103	129	155	186	208
Sep.	110	56.81	0.5148	0.71	21	46	100	153	187	218	257	285
Oct.	143	62.18	0.4349	0.58	38	69	134	190	225	258	294	320
Nov.	182	81.89	0.4492	1.49	59	97	174	238	279	321	365	385
Dec.	195	94.96	0.4869	1.40	40	79	161	237	284	327	381	420

4.2.3   Daily variation

The day-to-day evolution of rain in East Africa was investigated by D.H.
Johnson, using the so-called regional index of raininess (1962). Daily rainfalls
were plotted on maps for each day from November 1958 to May 1960. The wet areas
were located and each was given an index number. This number, sometimes referred to as concentration, is the percentage, in tens of percent, of stations
reporting rain to the total number of stations confined to the located area. All
daily maps were so analyzed. A sample of the results obtained is shown in
Fig. 4.12.

This kind of analysis is quite subjective and certainly of limited value for
design purposes. Moreover, the extent of the short-period variations in rainfall distribution is so large that corresponding months and seasons in successive years would sometimes exhibit very different characters. The original paper
by Johnson supports this conclusion very strongly (1962).

The hour-to-hour, or the diurnal variation of rainfall on the Lake Plateau,
the Sudan, and Egypt has already been discussed in Chapter 3.

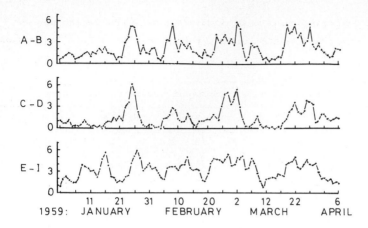

Fig. 4.12. Sequences of daily rainfall indexes for the combined regions A and B (Lake Victoria and Uganda), C and D (most of Kenya), and E to I (mainly Tanganyika). January to early April 1959 (Johnson, D.H., 1962)

4.3   EXTREME RAINFALL INTENSITY FOR DURATIONS OF ONE DAY AND SHORTER

The maximum rainfall in a day during the years of observation of each rainfall station in Egypt and the Sudan up to 1967 is listed in Table 4.10. Since the period of observation at these stations varies in a wide range, from a minimum of about 20 years to a maximum of 70 years or more, it is not possible to reduce the observed extreme intensities to a single return period.

The procedure used in the hydrometeorological survey of the catchments of Lakes Victoria, Kyoga and Albert is based on fitting the Gumbel Type I extremal function to the distribution of the estimated continuous 24-hour maximum, R_x. This can be obtained from the relation (WMO, 1974)

$$R_x = (R_x)_{obs} + \tfrac{1}{2} R' \tag{4.13}$$

where $(R_x)_{obs}$ is the observed fixed day maximum and R' is the highest daily rainfall on the preceding or following day.

This procedure made it possible to develop, among others, generalized charts of 1-day extreme rainfall of 2 and 100 years return periods and a nomogram for estimation of 1-day extreme rainfall corresponding to any given return period.

A comparison between the two maps in Figs. 4.13a. and 4.13b. shows that, all over the area covered by the survey, the 100-year 1-day extreme rainfall is nearly twice as much as the 2-year 1-day extreme rainfall. If this ratio holds everywhere in the Nile Basin, one may use the nomogram shown in Fig. 4.13c., together with Table 4.10, after correcting the tabulated rainfalls, for the estimation of the rain depth corresponding to any return period. A fixed 24-hour rainfall can be corrected to a continuous 24-hour rainfall simply by increasing the former by about 10-13% of its value.

As an example, consider the 1-day extreme rainfall that occurred in the 50-year period 1918-67 at stations 27 in Egypt and 88 in the Sudan. The observed depths, $(R_x)_{obs}$, were 72 and 118 mm respectively. These can be corrected to give R_x of 80 mm for station 27 and 130 mm for station 88. By plotting these two values on the 50-year return period line of the nomogram, Fig. 4.13c., and drawing a straight line passing through each point such that the 100-year return period line reads a 1-day rainfall twice as much as the 1-day rainfall to be read on the 2-year return period line. One arrives at the following results:

Station No.	Rainfall, mm/day, for return periods of						
	2-yr	5-yr	10-yr	20-yr	25-yr	50-yr	100-yr
27	44	55	64	71	73	80	88
88	71	88	103	114	118	130	142

It goes without saying that the above results should be regarded as approximate only.

The East African Meteorological Department, EAMD, has the practice of expressing the extreme rainfall intensity for durations shorter than 1 day, $(I_x)_t$, by the relation

$$(I_x)_t = I_o \cdot (t + a)^{-n} \qquad (4.14)$$

where I_o, a, and n are parameters varying with time and space, and t is the duration of the rain storm for which the maximum intensity, (I_x) is computed. The parameter n has been found to vary between 0.5 and 1.0.

The strange thing in eq. 4.14 is that it does not include any term connected to the return period. On the other hand, the observed extreme 1-hour monthly and annual rainfall at a few stations, after being corrected, have been fitted by Gumbel Type-I extremal function and extreme hourly rainfall corresponding to any given return period determined. The correction of a fixed 1-hour intensity to a continuous 1-hour intensity is done, similar to the 1-day rainfall, by multiplying the fixed intensity by about 1.13.

Further investigation of the intensity-duration-frequency-relationships of the rainfall on the catchments of the Equatorial Lakes is essential before such inconsistency is resolved. Moreover, it is certainly useful to extend such an investigation to cover other parts of the Nile Basin area.

TABLE 4.10 Maximum rainfall in a day as observed in Egypt and the Sudan (Nile Control Staff, 1972)

No.	Station Period of observation	Maximum rainfall in a day, mm	No.	Station Period of observation	Maximum rainfall in a day, mm	No.	Station Period of observation	Maximum rainfall in a day, mm
1	1948-67	54	48	1941-67	20	95	1942-67	128
2	1912-67	75	49	1931-67	23	96	1910-67	115
3	1946-67	121	50	1946-67	17	97	1914-67	105
4	1912-67	67	51	1931-67	16	98	1928-67	87
5	1931-67	55	52	1921-67	37	99	1942-67	97
6	1912-67	85	53	1941-67	10	100	1943-67	193
7	1931-67	56	54	1943-67	25	101	1905-67	130
8	1947-67	76	55	1946-67	03	102	1929-67	105
9	1928-67	60	56	1955-67	11	103	1911-67	99
10	1928-67	85	57	1935-67	53	104	1908-67	110
11	-	-	58	1931-67	34	105	1911-67	170
12	1941-67	48	59	1942-67	39	106	1908-67	108
13	1931-67	41	60	1948-67	06	107	1918-67	171
14	1928-66	49	61	1931-67	08	108	1915-67	97
15	1942-67	65	62	1931-67	08	109	1915-67	107
16	1903-67	56	63	1931-67	39	110	1900-67	116
17	1928-67	58	64	1954-67	04	111	1906-67	145
18	1910-67	68	65	1935-67	06	112	-	-
19	1936-67	59	66	1941-64	19	113	1910-67	127
20	1931-67	40	67	1906-67	112	114	1915-67	178
21	1948-67	49	68	1908-67	53	115	1913-67	114
22	1931-67	48	69	1908-67	71	116	1906-67	140
23	1933-67	51	70	1919-67	101	117	1900-67	144
24	1931-67	37	71	1917-67	63	118	1915-67	126
25	1918-66	59	72	1913-67	82	119	1906-67	130
26	1921-67	102	73	1909-67	130	120	1919-64	160
27	1918-67	72	74	1908-67	77	121	1922-67	140
28	1910-67	56	75	1907-67	98	122	1932-67	164

TABLE 4.10   (continued)

Station			Station			Station		
No.	Period of observation	Maximum rainfall in a day, mm	No.	Period of observation	Maximum rainfall in a day, mm	No.	Period of observation	Maximum rainfall in a day, mm
29	1931-67	39	76	1914-67	76	123	1922-67	150
30	1914-67	75	77	1905-67	120	124	1909-67	160
31	1938-64	72	78	1908-67	100	125	1906-63	148
32	1926-67	24	79	1946-67	84	126	1905-67	129
33	1946-67	28	80	1894-1967	107	127	1913-67	123
34	1938-67	32	81	1920-67	102	128	1904-67	125
35	1921-67	28	82	1930-67	147	129	1944-67	108
36	1946-67	70	83	1905-67	87	130	1903-65	150
37	1941-67	27	84	1903-67	139	131	1907-67	140
38	1936-67	28	85	1905-67	126	132	1913-67	131
39	1931-67	53	86	1905-67	112	133	1905-67	162
40	1932-67	32	87	1929-67	89	134	1924-64	128
41	1931-67	50	88	1918-67	118	135	1925-67	156
42	1907-67	32	89	1905-67	143	136	1908-67	121
43	1943-67	49	90	1903-67	127	137	1924-67	116
44	1931-67	33	91	1902-67	130	138	1921-67	136
45	1921-67	142	92	1920-67	127	139	1929-64	107
46	1928-67	16	93	1914-67	125	140	1914-67	125
47	1931-67	44	94	1908-67	150	141	1916-65	137

162

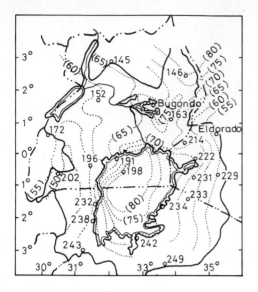

Fig. 4.13a. Generalized chart of 2-year 1-day extreme rainfall (WMO, 1974)

Fig. 4.13b. Generalized chart of 100-year 1-day extreme rainfall (WMO, 1974)

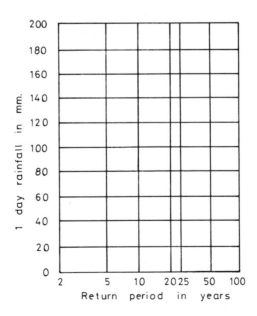

Fig. 4.13c. Nomogram for extimation of 1-day extreme rainfalls (WMO, 1974)

REFERENCES

Brook, C.E.P., 1924. The distribution of rainfall over Uganda, with a note on Kenya Colony. Quart. Journ. Roy. Meteo. Soc., 50: 325-338.

Chow, V.T., 1964. Handbook of applied hydrology. McGraw Hill Book Company, New York, 1453 pp.

Clarke, R.T., 1973. Mathematical models in hydrology. FAO Irrigation and drainage paper 19, FAO, Rome.

Haan, C.T., 1977. Statistical methods in hydrology. The Iowa State University Press, Ames, USA, 378 pp.

Hurst, H.E. and Black, R.P., 1950. The Nile Basin, first supplement to Vol. VI, Physical Department Paper 49, S.O.P. Press, Cairo, 228 pp.

Hurst, H.E., Simaika, Y.M., and Black, R.P., 1955. The Nile Basin, second supplement to Vol. VI, Nile Control Department Paper 4, Government Press, Cairo, 206 pp.

Hurst, H.E., Simaika, Y.M., and Black, R.P., 1957. The Nile Basin, third supplement to Vol. VI, Nile Control Department Paper 9, Government Press, Cairo, 198 pp.

Huynh Ngoc Phien et al, 1980. Rainfall distribution in north-eastern Thailand. Hydro. Sci. Bul. (edited by R.T. Clarke), 25.2: 167-182.

Ireland, A.W., 1948. Agriculture in the Sudan (edited by J.D. Tothill). Chapter V: The Climate of the Sudan, Oxford University Press, London.

Johnson, D.H., 1962. Rain in East Africa. Quart. Journ. Roy. Meteo. Soc., 88.375: 1-19.

Kite, G.W., 1977. Frequency and risk analyses in Hydrology. Water Resources Publications, Fort Collins, Colorado, 224 pp.

Kottegoda, N.T., 1980. Stochastic water resources technology. The McMillan Press Ltd., London, 384 pp

Markovič, R., 1965. Probability functions of best fit to distributions of annual precipitation and runoff. Hydrology Paper 8, Colorado State University, Fort Collins, Colorado, 34 pp.

Nile Control Staff, 1963. The Nile Basin, fourth supplement to Vol. VI, Nile Control Department Paper 18, General Organization for Government Printing Offices, Cairo, 192 pp.

Nile Control Staff, 1969. The Nile Basin, fifth supplement to Vol. VI, Nile Control Department Paper 25, General Organization for Government Printing Offices, Cairo, 169 pp.

Nile Control Staff, 1972. The Nile Basin, sixth supplement to Vol. VI, Nile Control Department Paper 29, General Organization for Government Printing Offices, Cairo, 160 pp.

Pearson, E.S., and Hartley, H.O. (editors), 1954. Biometrika Tables for Statisticians. Vol. 1, Cambridge University Press.

Rzóska, J. (editor), 1976. The Nile, biology of an ancient river. Dr W. Junk B.V. Publishers, The Hague, 417 pp.

Shahin, M.M., 1983. Statistical modelling of rainfall data on the Nile Basin. Paper presented to the conference on water resources development in Egypt (under publication), Cairo.

UNESCO, 1978. World water balance and water resources of the earth. UNESCO, Paris.

United States Water Resources Council, 1977. Guidelines for determining flood flow frequency. Revised edition, Washington, D.C.

Yevjevich, V., 1972. Stochastic processes in hydrology. Water Resources Publications, Fort Collins, Colorado, 276 pp.

WMO, 1974. Hydrometeorological survey of the catchments of Lakes Victoria, Kyoga and Albert, Vol. I: Meteorology and hydrology of the basin, Part 1, Geneva.

Chapter 5

FREE WATER SURFACE EVAPORATION

Evaporation is defined as the transfer of moisture into the atmosphere from
an open or free water surface, a bare soil or interception on a vegetal cover.
The water-resources engineer usually regards evaporation of water as a loss,
whether it occurs from reservoirs, from natural lakes, from bare soil, or from
land-carrying crops.

Since the rainfall on many parts of the Nile Basin is quite scanty, it is,
therefore, necessary to have reliable information about evaporation losses. One
should not forget that the mere existence of some of the countries sharing the
Nile water is almost entirely dependent on the various irrigation schemes and on
the storage works on the river.

In this chapter the evaporation from the free water surfaces such as lakes
and reservoirs in the basin is treated. Evapotranspiration from vegetated and
cropped surfaces is the subject matter of the next chapter.

It is known that evaporation can be measured by atmometers and containers of
various shapes and dimensions. It can also be estimated from the water-balance
of the body of water or the catchment area in question, provided that sufficiently accurate data covering all the terms in the balance equation other than
evaporation are available. In the absence of actual measurements, or where
adequate data of the balance items are missing, one usually resorts to evaporation estimates, using one formula or another.

Our interest here does not extend to the hourly evaporation and its correlation with the climate. We are not even concerned with the daily values, but we
are concerned with the summation of the daily values on a monthly average basis,
and with annual values.

Measurement of evaporation at a number of stations in the Nile Basin dates
back to the beginning of this century. In Egypt and the Sudan the Piche and the
floating tank are still in use, whereas the use of the Wild evaporimeter has
stopped since 1920. The floating tank has a shape of a cube of 1 m side. It is
constructed of iron and floated on a river or a lake by means of a wooden raft.
The rim of the tank remains a few centimetres above the level of the outside
water. The tank is filled up to the same level of the outside water, and the
evaporation loss is measured every day by a gauge which indicates the water
level in the tank. The difficulty with type of device is caused mainly by waves
splashing into it, especially on windy days.

The early experimentation of evaporation measurement using different devices
was described in Vol. I of the Nile Basin (Hurst, H.E., and Philips, P., 1931).
The conclusions that can be drawn for their description are:

i)   the evaporation from a floating tank 2 m square can be considered
      practically as the best available approximation to evaporation from an
      extended surface of water,

ii)   the ratio between evaporation from a 2 m square floating tank and evapora-
      tion from a 1 m square floating tank is 0.88, and

iii) the ratio between evaporation from a 2 m square floating tank and evapora-
      tion from a Piche tube is close to 0.50.

The above conclusions combined with those reported by Keeling at an earlier
time were presented in Vol. I of the Nile Basin with the aim of transorming the
readings of all the measuring devices to free water surface evaporation (Hurst,
H.E., and Philips, P., 1931). This is listed in Table 5.1. The climatological
normals for Egypt (Ministry of War and Marine, Egypt, 1950) show some difference
between the normal evaporation over open water and the original values given in
Table 5.1 for a large number of meteorological screens. A few years later, a
different set of values appeared in "The Nile" (Hurst, H.E., 1952). A summary
of the data available in these two references are also included in Table 5.1.
The data in this table have been used in preparing the map shown in Fig. 5.1.,
so as to get an overall picture of the annual evaporation as deduced from the
Piche readings.

As already mentioned, the water-balance method can be used for estimating the
evaporation loss from an open water where direct measurements are not available.
The balance equation for a body of water, a lake for example, can be set up for
a certain period of time as

$$P_1 + R + I - E_1 - O = \Delta S \qquad\qquad (5.1)$$

where

P_1 = direct precipitation on the lake,

R = run-off from the lake catchment area to the lake itself,

I = inflow to the lake from river tributaries,

E_1 = evaporation from the lake,

O = outflow from the lake, and

ΔS = change in the volume of water stored in the lake.

TABLE 5.1   Normal evaporation over open water in the Nile Basin, in mm/day

Location	Period of observation	Nile Basin Vol. I (1931)					Climatological Normals (1950) Year	The Nile (1952) Year
		Jan.	Apr.	Jul.	Oct.	Year		
Mediterranean Coast								3.0
Mersa Matruh	1920-29	3.2	3.8	3.5	4.2	3.6		
Alexandria	1920-29	1.8	2.1	3.0	2.2	2.0	2.5	
Port Said	1920-29	1.5	2.5	2.3	2.6	2.2		
Nile Delta								2.3
Qurashiya	1907-29	1.0	2.5	2.4	1.6	2.2		
Sakha	1907-29	1.0	2.3	2.8	1.7	1.9	2.3	
Cairo and neighbourhood								2.8
Cairo (Ezbekiya)	1909-29	1.2	2.7	3.7	1.9	2.3	2.3	
Giza	1920-27	1.5	3.8	4.1	2.3	2.8		
Helwan	1920-29	2.5	6.5	7.6	5.0	5.4		
Tor	1905-29	3.2	4.4	5.3	3.7	4.2		
Fayum							3.8	4.0
Qasr el-Gebali	1920-29	1.9	4.8	6.4	3.4	4.1		
Minya (Central Egypt)	1920-29	1.4	4.2	5.0	2.8	3.3	3.8	
Upper Egypt								4.5
Assiut (Upper Egypt)	1920-29	1.8	5.5	7.1	3.4	4.5	4.6	
Aswan	1920-29	3.8	8.7	10.0	7.8	7.5	7.1	
Oases								6.5
Northern Sudan								7.6
(from Halfa to Atbara)								
Wadi-Halfa	1905-29	4.5	9.3	9.9	8.1	7.9		
Merowe	1905-29	5.8	9.8	9.3	8.7	8.4		
Atbara	1905-29	6.8	10.3	9.3	8.3	8.6		
Port Sudan	1905-29	3.7	4.6	7.2	3.5	4.9		
Khartoum and neighbourhood								7.8
Khartoum	1905-29	6.5	10.0	6.7	7.3	7.5		
Kassala	1905-29	4.7	7.7	4.3	5.5	5.4		
Gallabat	1905-29	6.5	8.5	2.3	3.0	5.1		
Wad Medani	1905-29	6.3	9.1	4.9	5.7	6.5		

TABLE 5.1 (continued)

Location	Period of observation	Nile Basin Vol. I (1931)					Climatological Normals (1950) Year	The Nile (1952) Year
		Jan.	Apr.	Jul.	Oct.	Year		
Central Sudan (Dueim to Roseires)								6.3
Roseires	1905-29	6.8	8.1	2.5	3.4	5.3		
Dueim	1905-29	7.6	9.5	4.9	5.8	6.9		
Lake Tana	1921-24	4.1	5.1	1.2	2.2	3.0		
El-Obeid	1907-29	7.2	9.2	4.7	6.3	6.7		
El-Fasher	1918-29	5.2	7.9	4.3	6.0	5.8		
Southern Sudan (Malakal and south outside the swamps)								3.4
Malakal	1915-29	8.5	5.4	1.4	1.8	4.5		
Wau	1906-29	5.9	4.5	1.8	2.3	3.7		
Mongalla	1906-29	5.5	3.0	1.4	2.2	3.0		
Lake Albert								3.9
Lake Edward								3.9
Lake Victoria		3.5	2.9	4.4	3.7	3.6		3.8

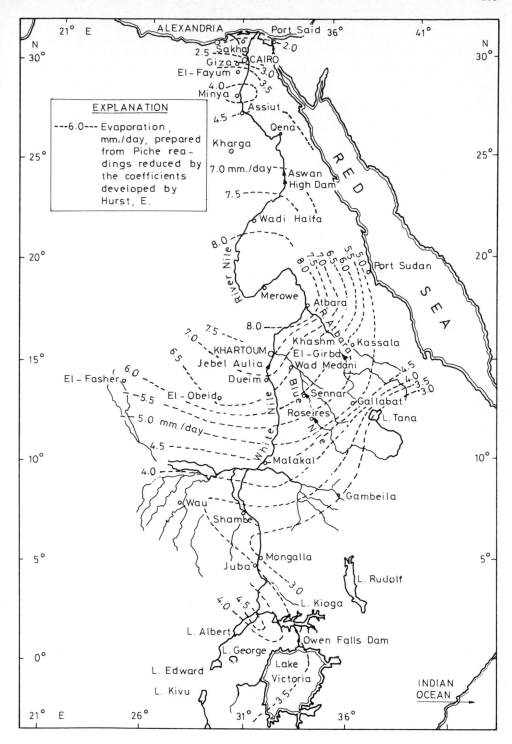

Fig. 5.1.   Mean annual evaporation from an open water surface in the Nile Basin

It is customary among the hydrologists to neglect the change in storage when the period over which the gains are balanced with the losses is sufficiently long, say 1 year or more. When this is the case, ΔS is put equal to zero and eq. 5.1 can be rewritten as

$$E_1 = P_1 + R + I - O \qquad\qquad (5.2)$$

The evaporation from the Great Lakes in the Nile Basin has been estimated by eq. 5.2 using the year as a balance period.

A balance equation quite similar to that given by eq. 5.2 was used in preparing the world water balance (Baumgartner, A., and Reichel, E., 1975). In this work the so-called actual evaporation over the successive 5 degree latitude zones is included. It is not common to use the term actual in connection with evaporation. Instead, this term is used jointly with evapotranspiration from a vegetated or a cropped surface to distinguish it from potential evapotranspiration. Therefore, the discussion of the results obtained by Reichel and Baumgartner concerning what they have called actual evaporation will be presented in the next chapter.

In addition to all that is mentioned above, there has been a number of detailed experiments on evaporation and evapotranspiration from the Nile Basin. Since the bulk of the available information is based, however, on the readings of the Piche atmometer, it is logical to begin by mentioning some remarks about the kind of data one should expect to obtain from such an atmometer. It is important to know this before any detailed investigation can be attempted.

i)   The Piche, like any other atmometer, is mainly used for climatological rather than hydrological purposes, to characterize the drying ability of the air under a given set of conditions.

ii)   The process and circumstances of evaporation from a Piche tube (surface area from which evaporation takes place, two sides = 13 cm^2) are not strictly the same as those from a tank or from a free water surface. The extent of the area affected by humidity caused by evaporation from a nearby water surface is a strong factor influencing the relationship between the evaporation from a Piche instrument and the evaporation from an open water surface. Another strong factor influencing this relationship is the degree of exposure of the atmometer to the surrounding climate.

In his book, "The Nile" (1952), Hurst discussed the use of an evaporimeter in the lakes area saying "... The Great Lakes, however, produce a local climate, so that a Piche evaporimeter in a town on the shores of Lake Victoria will not indicate twice the evaporation from the lake itself, but something appreciably less. This is because the damp atmosphere over the lake extends some distance inland". This is obviously not the case when the evaporimeter is placed at a station next to a river of comparatively much smaller width.

iii) As mentioned in Vol. I of the Nile Basin (Hurst, H.E., and Philips, P., 1931), it is reasonable to suppose that the evaporation from a large open water surface is a little less than that from the largest tanks used in the experiments (here 2 m x 2 m). The ratio between evaporation from an evaporimeter and evaporation from a free water surface was reviewed by Sleight (1927), Hickox (1946) and later by Olivier (1961). The results obtained are summarized in Table 5.2.

TABLE 5.2   Relative rates of evaporation showing effect of pan size on evaporation

Description	Diameter of pan, in feet							
	12	9	6	4	3.39	2.26	2	1
Compiled by R.B. Sleight	1.00	1.009	1.089	1.175	1.202	1.260	1.284	1.589
Compared with 1800-acre reservoir as unity	1.099	1.108	1.196	1.290	1.320	1.383	1.410	1.745

After this historical review we shall deal in the remaining sections of this chapter with evaporation from open water at a number of stations in the different parts of the Nile Basin.

5.1   LAKE PLATEAU AREA

5.1.1   Lake Victoria

The mean annual evaporation from Lake Victoria over the period from 1902 up to and including 1923 can be found from Table 5.1 as 3.6 mm/day. This figure was obtained from the hydrologic balance of the lake for the same period. The distribution throughout the year was developed by comparison with the readings of the wet bulb thermometer. When the balance period extends up to 1936, the average values of P_1, R + I and O become 1151, 276 and 311 mm/yr respectively, thus giving an annual mean evaporation of 3.06 mm/day only (Hurst, H.E., and Philips, P., 1938). A third evaporation rate which was given later by Hurst (1952) is 3.8 mm/day. This is based on P_1, R + I and O of 1463, 239 and 314 mm/yr respectively. A fourth value can still be found in Vol. X of the Nile Basin (Hurst, H.E., Black, R.P., and Simaika, Y.M., 1966). This value of 1150 mm/yr or 3.15 mm/day is based on P_1 = 1260, R + I = 190 and O = 300 mm/yr respectively. From these figures one can easily see that the evaporation rate from Lake Victoria varies from one balance period to another. Each term in the balance equation undergoes variation with time. Next to this, there are definitely some sampling errors. An estimate of the annual rainfall on such an enormous surface of water as Lake Victoria can hardly be accurate. An example of this, though from outside the Nile Basin and to a smaller extent as compared to any of the Equatorial Lakes, is the Lake Hefner in the United States. During the 1950-51 evaporation study from this lake there were 21 eight-inch diameter gauges around the periphery of the 2200-acre lake and one gauge on a raft in the centre. During one storm, rainfall ranged from 0.1 inch on one side of the lake to nearly two inches on the other. During summer convective showers, the areal variability of rainfall may be so great that the average rainfall value on the lake surface is sufficiently inaccurate to invalidate the computed figure of evaporation (US Geological Survey, 1954).

The water-balance of Lake Victoria was recently reviewed in connection with the hydrometeorological network project in the Equatorial Lakes area (Krishnamurthy, K.V., and Ibrahim, A.M., 1973). Based on the information gathered from the meagre network of rain gauges before the commencement of the project in 1967, the best available estimate of P_1 can be taken as 1420 mm/yr $\pm$ 10%. The inflow plus run-off to the lake varied, as reported, from 15 to 18 milliard m^3/yr. This figure is somewhat different from the 12.6 milliard m^3/yr adopted by Hurst and his co-workers in preparing Vol. X of the Nile Basin (Hurst, H.E., Black, R.P., and Simaika, Y.M., 1966). In the balance prepared by Krishnamurthy and Ibrahim the yearly volumes of 18×10^9 m^3 $\pm$ 5% and 23.6×10^9 m^3 $\pm$ 5% were used to represent the run-off to and the outflow from Lake Victoria respectively. The evaporation estimated from this balance lies in the range of 3.65 and 4.50 mm/day. The lower limit of this range does not depart sensibly from the average of the evaporation rates obtained for different balance rates by Hurst and his co-workers. It is also in reasonable agreement with the average of the readings of the Piche evaporimeters installed by the East African Meteorological Service at a number of places near the lake shores. The old work of Keeling based on the wet-bulb depression as an estimate of the daily evaporation over a certain year from a standard Wild instrument was extended by Olivier (1961) to give the mean daily evaporation over a certain month. Keeling, as a result of the experiments made in Egypt and the Sudan, recommended a factor of 1.42 to convert the reading of a Wild instrument to Piche evaporation.

The evaporation formula developed by Olivier reads

$$M_p = \frac{C}{L\phi N} \quad (\text{ or } \frac{C}{L\phi S}) \tag{5.3}$$

where

M_p = free water surface evaporation, in mm/day, from a standard tank at a latitude ϕ,

c = average depression of wet bulb in $^{\circ}$C for a particular month, and

$L\phi N$ = ratio $\dfrac{L}{L_o}$ for latitude ϕN as taken from tables (for latitude $L\phi S$ read 'six months on').

The yearly values of c at Jinja and Entebbe are 36.1°C and 30.1°C respectively. These two palces are located almost on the equator. Assuming that c is uniformly distributed over the 12 months of the year, and using the monthly values $L(\phi = o)$ given by Olivier, the free water evaporation from a standard tank can be calculated from eq. 5.3. The results obtained from calculation together with the average Piche evaporation at the same places are given in Table 5.3. From this table it appears that the computed tank evaporation at both Jinja and Entebbe, which are situated close to the shores of Lake Victoria, is 20 to 30% less than evaporation from the lake as obtained from the water-balance method. Additionally, the ratio Piche evaporation to tank evaporation needs to be changed, at least for the places considered, from 1.42 to 1.21.

TABLE 5.3   Estimate of tank evaporation and Piche evaporation, for Jinja and Entebbe

Place	Method	Evaporation, in mm/day, for												
	*	Jan.	Feb.	Mar.	Apr.	May	June	July	Aug.	Sep.	Oct.	Nov.	Dec.	Year
Jinja	a	2.9	3.0	3.2	3.1	2.9	2.9	2.9	3.0	3.1	3.1	3.0	2.9	3.0
	b	4.5	4.1	3.8	3.2	3.0	3.3	3.3	3.2	3.7	4.1	4.0	3.9	3.7
Entebbe	a	2.5	2.5	2.6	2.6	2.5	2.4	2.4	2.5	2.6	2.6	2.5	2.4	2.5
	c	3.2	3.5	3.1	2.3	2.3	2.6	2.6	2.6	2.9	3.0	2.8	2.7	3.0

*a = Eq. 5.3; b = Piche, 1924-27 and 1932-34; c = Piche, 1924-30 and 1933-34

A series of evaporation experiments using different types of atmometers were carried out for a number of years in East African territories, some of which are located within the boundaries of the catchment area of Lake Victoria. At eleven stations, expressing the annual evaporation in inches, it was found that

$$\text{Class A pan} = 0.87 \text{ Piche} + 24 \qquad (5.4)$$

with a correlation coefficient = 0.95. It was concluded that the Piche readings in a Stevenson screen give a satisfactory estimate of annual evaporation as measured by a class A pan.

Comparison between the so-called Kenya pan and a class A pan were carried out at Dagoretti headquarters, a few kilometres west of Nairobi. Using four years of record (1384 days), a correlation between daily values in inches yielded the result

Class A pan (unscreened) = 1.3 Kenya pan (screened) + 0.02   (5.5)

with a correlation coefficient = 0.95.

Comparisons of monthly means over two years, assuming linear relationship, provided the following results:

Class A pan (unscreened)   = 100
Class A pan (screened)   = 86 (range 84-88)
Class A pan (painted)   = 84 (range 82-86)
Kenya pan (unscreened)   = 95 (only few months)
Kenya pan (screened)   = 71 (range 69-74)
Kenya pan (painted and screened) = 69 (range 65-72)

"Screen" is a 1-inch chicken wire mesh. "Painted" means that the inside is painted dull black with bituminous paint and the outside with aluminium paint.

Further description of those experiments is given in Technical Note No. 83 of the World Meteorological Organization (WMO, 1966). The two stations which are of direct interest to us are Kisumu and Entebbe. The Piche (screened) and Kenya pan (screened) evaporation at Kisumu were 96 and 106 in/yr and at Entebbe 36 and 89 in/yr, respectively. The yearly ratios of the class A pan (unscreened) to Kenya pan (screened) at Kisumu and Entebbe were 1.28 and 1.12 respectively. Assuming the ratio lake evaporation to class A pan evaporation at 0.7, the free water surface evaporation at Kisumu and Entebbe becomes 6.6 mm/day and 4.87 mm/day respectively. The mean annual Piche evaporation is 6.67 mm/day for Kisumu and 2.51 mm/day for Entebbe. The latter figure emphasizes the abnormality in the readings of the Piche atmometer at Entebbe.

Another approach to estimating evaporation from a free water surface is that of applying one or another of the evaporation formulas based on those meteorological factors which are strongly correlated with evaporation. Rijks (1969) has estiamted the evaporation from the meteorological data measured over a swamp at the northern boundary of the Cotton Research Station at Namulonge ($0^{\circ}32'$N, $32^{\circ}37'$E and 1100 m altitude), Uganda. The results he obtained from Penman's formula averaged over the 5-day periods; 8-12 March, 22-26 March and 7-11 April, 1965, are 6.2, 3.8 and 5.5 mm/day respectively. Dagg, M., on Water Requirements of Crops in East Africa (1972) has mentioned that direct and reliable measurements of open water evaporation are deceptively difficult to obtain. Using the estimates of Rijks and Owen, 1965, and of Woodhead, 1967, from the Penman formula, Dagg was able to construct a map of the evaporative demand for East Africa with moderate confidence (1972). Part of this map has been incorporated in the map of the evaporative demand for the Nile Basin prepared by us using the Penman method. This method yields an average rate of 4.65 mm/day for the northern and north-western parts of the catchment area of Lake Victoria. The mean annual Piche evaporation at Jinja, Kampala and Entebbe, which represent those parts of the catchment, is 3.7, 4.5 and 3.0 mm/day respectively.

The above discussion of the evaporation from Lake Victoria and its catchment can be concluded with the following remarks.

i) The water-balance of the lake averaged over different periods gives a mean annual evaporation in the range of from slightly less than 3.1 mm/day to 3.8 mm/day with an overall average of 3.4 mm/day. Substituting higher annual rainfall depths in the balance equation has led to evaporation ranging from 3.65 to 4.50 mm/day.

ii) The evaporative demand for the catchment area of the Lake Victoria varies from one part of the catchment to the other. The range of values obtained from the Penman formula based on the available meteorological observations is from about 4.4 mm/day to slightly more than 6.0 mm/day.

iii) The Piche readings at the stations near the shores of the lake need to be multiplied by a factor ranging from less than 1.0 to more than 1.5 to bring them to their equivalents of free water surface evaporation.

iv) The conversion factor of the Wild instrument to the Piche tube reading seems to be less than the 1.42 originally proposed by Keeling.

Last but not least, it is worthwhile to mention that the hydro-meteorological survey project of Lakes Victoria, Kyoga and Albert has set up five class A evaporation pans on five islands in Lake Victoria as well as nine class A evaporation pans on the shore at some of the first order meteorological screens (see map, Fig. 5.2.). Besides the standard class A pans, a Russian evaporation pan 20 m^2 in size has also been set up to help in the determination of evaporation by the pan method. Furthermore, it has been proposed that meteorological data be collected for estimation of free water surface evaporation by the mass-transfer and energy budget techniques.

In the author's opinion the results that can be expected from these methods still need to be supported, and probably adjusted, by new figures to be obtained from the water-balance method. The meteorological network will help in obtaining more accurate figures about the rainfall not only on the catchment but on the lake itself. Prior to the project, 58% of the total inflow to the lake was measured and the flow from the ungauged areas was estimated on the basis of rainfall, characteristics of the catchment, and similarity with other

176

rivers. The improvement of the network by the project has brought the measured inflow to about 90% of the total inflow (WMO, 1974).

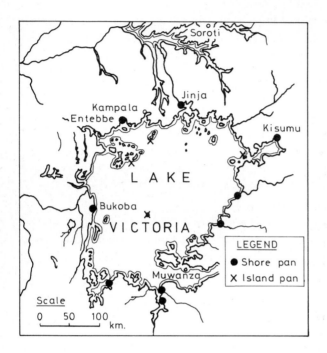

Fig. 5.2.   Map showing the locations of evaporation pans set up by the hydro-meteorological network project on the islands and along the shores of Lake Victoria.

5.1.2   Victoria Nile Basin

The water-balance of Lake Kyoga was drawn by Hurst (1952) considering the inflow from the Upper Victoria Nile as 20.9 mlrd m^3/yr, the inflow brought by the other tributaries as 3.5 mlrd m^3/yr, and the sum of precipitation and the lake and the run-off from the catchment to it as 8 mlrd m^3/yr. The annual out-flow from Lake Kyoga to the Lower Victoria Nile was considered as 19.7 mlrd m^3/yr. These figures leave 12.4 mlrd m^3/yr for evaporation from the lake (1760 km^2) and evapotranspiration from the swamps (4500 km^2). The average rates of eva-poration and evapotranspiration have been estimated at 3.9 mm/day and 6.1 mm/day, respectively.

The mean monthly and annual Piche evaporation as observed in the Victoria Nile Basin are as follows:

Evaporation, in mm/day, for

Jan.	Feb.	Mar.	Apr.	May	June	July	Aug.	Sep.	Oct.	Nov.	Dec.	Year
7.4	7.7	6.0	4.3	3.5	4.1	3.9	3.7	4.2	4.5	5.0	4.9	4.9

The reduction factor to be applied to the above values so as to convert them into open water evaporation was described by Hurst as "unfortunately very uncertain". As a rough approximation he adopted a factor of 0.67 to the Piche evaporation. This figure is the mean of 0.83, the factor proposed for Lake Victoria and 0.50, the factor found for Egypt and the Sudan. If this factor is applied to the mean annual evaporation, 3.3 mm/day evaporation is obtained from the free water surface of Lake Kyoga.

The method of Olivier (1961) gives a mean annual evaporation and a mean annual evapotranspiration from the swamp, both around Soroti, close to the northern shore of Lake Kyoga, of 7.0 and 4.7 mm/day respectively.

Assuming the ratio open water to standard tank evaporation to be 0.75, the evaporation from a free water surface in the Victoria Nile Basin becomes equal to 5.25 mm/day. This figure is slightly less than the average value of 5.75 mm/day estimated for the surroundings of Soroti by the method of Penman.

Once more consider the water-balance drawn by Hurst for Lake Kyoga and surroundings. It is neither understandable nor justifiably why the rate of evapotranspiration from the swamps is taken as 1.61 times (6.1:3.9) the evaporation rate from the lake itself. If it is arbitrarily assumed that both rates are equal, a general figure of 5.25 mm is obtained.

From the foregoing analysis, one may draw the following conclusions:

i)   A mean annual evaporation of 3.9 mm/day seems to be fairly small. This figure needs to be increased to about 5.25 mm/day.

ii) The factor to convert Piche evaporation to free water surface evaporation is not 0.67 as originally suggested by Hurst. This factor needs to be changed to 1.05 or 1.10.

iii) The evaporative demand of the Victoria Nile Basin ranges from 5.5 mm/day to 6.0 mm/day.

5.1.3   Lakes George and Edward, and Lake Albert

In the volumes of the Nile Basin up to and including Vol IV there is no information at all about evaporation from Lakes George and Edward. In Vol. V it is mentioned that no direct information about evaporation from these two lakes

is available. Their size and geographical position suggest, however, that the
evaporation from Lakes George and Edward should be similar to that from Lake
Albert (Hurst, H.E., and Philips, P., 1938).

Evaporation measurement at Butiaba near Lake Albert started in 1932 using the
Piche atmometer. The mean monthly and yearly Piche readings over the period
1932-1934 are as follows:

Evaporation, in mm/day, for

Jan.	Feb.	Mar.	Apr.	May	June	July	Aug.	Sep.	Oct.	Nov.	Dec.	Year
6.0	6.2	5.5	5.0	4.3	4.5	3.9	3.9	4.1	4.3	4.4	4.6	4.7

Hurst assumed a reduction factor of 0.7 so as to deduce open water evaporation from the Piche evaporation. Based on this value for the reduction factor,
the evaporation from Lake Albert would be 3.3 mm/day.

Later, in his book "The Nile" (1952), Hurst gave the balance of the Lakes
George and Edward. The total inflow and run-off to the lakes was assumed to be
2.2 mlrd m^3/yr, the direct precipitation 3.4 mlrd m^3/yr and the outflow from
Edward to the River Semleeki 2.0 mlrd m^3/yr. These figures leave a yearly volume
of 3.6 mlrd m^3 for evaporation. The two lakes have been considered as one lake
2500 km^2 in size, though Lake George is at a slightly higher level than Lake
Edward. The mean annual depth of evaporation is 1440 mm, or almost 3.9 mm/day.

In the same reference, i.e. "The Nile" (1952), the balance for Lake Albert
was prepared on the assumption that the annual volume of direct precipitation on
the lake is 4.6 x 10^9 m^3, the annual volume brought by the Lower Victoria Nile,
River Semleeki and the other tributaries is 25.0 x 10^9 and the outflow from the
lake is 22.0 x 10^9 m^3/yr. These figures are balanced by a mean annual evaporation of 3.9 mm/day. The same value for the evaporation rate from Lake Albert
appears in Vol. X of the Nile Basin (Hurst, H.E., Black, R.P., and Simaika,
Y.M., 1966). In the latter, the balance was drawn between the inflow to and the
outflow from, the lake and the change in storage averaged over the period 1940-
1957.

According to the method of Olivier, eq. 5.3, the wet-bulb depression measured
at Butiaba gives a mean annual evaporation from a standard tank of about 4.2
mm/day. This value corresponds to probably a figure of about 3.5 mm/day free
water surface evaporation. The annual evaporation measured by the Piche tube and
the Kenya pan at Gulu, almost 120 km north-east of Butiaba, have been reported
as 76 and 91 inches respectively (WMO, 1966). These figures correspond to nearly
5.3 mm/day Piche evaporation and 5.6 mm/day open water evaporation. The author
has estimated the evaporative demand at some places between the Lakes Albert and
Edward using Penman's method. The estimates obtained range from 3.5 to 3.9 mm/day.

The above review may lead us to the following conclusions:

i) The water balance of Lake Albert yields a mean annual evaporation of 3.9
    mm/day from the lake. The similarity and the nearness of this lake to Lakes
    George and Edward suggest that the evaporation from these two lakes can
    also be considered as 3.9 mm/day. According to this figure, the Piche read-
    ing at Butiaba needs to be multiplied by a reduction factor of 0.82 to
    deduce the open water evaporation.

ii) The evaporative demand of the catchment areas of Lakes George, Edward and
    Albert varies considerably from one location to the other. The range of
    values is from about 3.5 mm/day, somewhere between Lake Edward and Lake
    Albert, to more than 5.5 mm/day, in the neighbourhood of the northern edge
    of Lake Albert (see Fig. 5.3.).

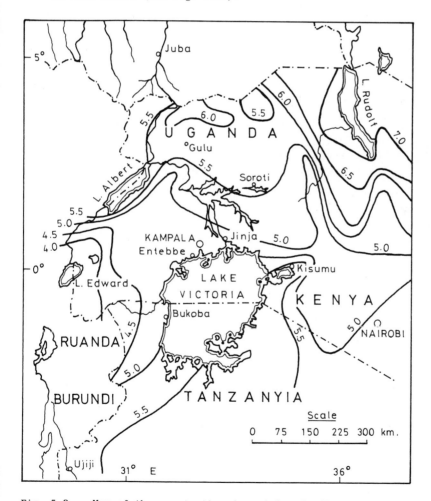

Fig. 5.3.   Map of the evaporative demand for the Equatorial Lakes Plateau

180

iii) The Piche evaporation of 4.7 mm/day at Butiaba and of 5.3 mm/day measured
at Gulu needs to be multiplied by a factor of from 1.1 to 1.3 to convert it
to the evaporative demand of the catchment at these two locations.

5.2   BAHR EL JEBEL BASIN
5.2.1   Bahr el Jebel Basin (outside the swamps)
    In Vol. V of the Nile Basin it is mentioned that a small tank 0.3 m square
was floated in the river near the discharge site at Nimule. Evaporation from
that tank was observed for some time. The difficulty in maintaining the tank to
function properly has led to neglecting all the observations collected from it
(Hurst, H.E., and Philips, P., 1938).
    Both Volumes I and V of the Nile Basin give the mean annual evaporation from
the Piche tube at a number of stations as follows:

Station	Period of observation	Piche evaporation, mm/day
Malakal	1915-1934	9.0
Masindi Port	1934	4.2
Lira	1933-1934	5.0
Lerua	1928-1934	4.9
Torit	1922-1934	7.1
Juba	1925-1928 and 1931-1934	7.0
Mongalla	1906-1930	6.1
Wau	1906-1929	7.4

    Hurst took the mean of these stations after excluding Malakal and Wau and
multiplied the mean value by 0.57 to deduce the open water evaporation from the
southern part of the Bahr el Jebel Basin. The results he got were

Evaporation, in mm/day, for

Jan.	Feb.	Mar.	Apr.	May	June	July	Aug.	Sep.	Oct.	Nov.	Dec.	Year
6.3	6.3	4.5	3.2	2.2	2.1	1.7	1.6	2.2	2.4	3.2	4.2	3.3

    He commented on these results saying "the whole question of evaporation from
open water in these southern districts needs investigation, but is not easy on
account of the difficulty of maintaining floating tanks free from wave action.
The distribution throughout the year is similar at all the stations and so the
monthly mean values taken from all the stations hve been reduced to give a total
of 1200 mm for the year".
    In order to estimate the free water surface evaporation from the Bahr el
Jebel and its Basin outside the swamps, the author has worked out the formulas
of Penman and of Hargreaves using the available meteorological data at Malakal,

Wau and Mongalla/Juba. These stations are situated close to the latitudes of 10°, $7^{\circ}30'$ and $5^{\circ}N$, respectively.

The formulas used read as follows:

Penman's formula (1948)

$$E_o = \frac{\Delta \frac{H}{60} + \gamma E_a}{\Delta + \gamma} \qquad (5.6)$$

where

E_o = free water surface evaporation, in mm/day,

H = net heat budget at the surface, in cals/cm^2.day,

Δ = slope of the tangent of the saturated vapour pressure - temperature curve at the mean air temperature, in mm Hg/$^{\circ}C$,

γ = psychrometer constant = 0.49 mm Hg/$^{\circ}C$, and

E_a = evaporation corresponding to the hypotetical case of equal air and water temperatures.

The quantities H and E_a can be calculated from the equations

$$H = 0.95 R_A (0.18 + 0.55 \frac{n}{D}) - 117.4 \times 10^{-9} . T_a^4 (0.56-0.092\sqrt{e_a})(0.10+0.90 \frac{n}{D})$$

$$(5.7)$$

and

$$E_a = 0.35 (e_s - e_a) (1 + 0.01 u_2) \qquad (5.8)$$

where

R_A = amount of energy reaching the outer limit of the atmosphere, in cals/cm^2 (horizontal) . day,

$\frac{n}{D}$ = relative duration of bright sunshine,

T_a = mean air temperature, in degrees absolute,

e_a = actual vapour pressure in the air, in mm Hg,

e_s = saturation vapour pressure in the air, in mm Hg, and

u_2 = mean wind speed in miles/day at a height of 2 m above the water surface.

Hargreaves' formula (1956)

$$E_v = 0.38 d (1 - h_n) (T - 32) \qquad (5.9)$$

where

E_v = monthly evaporation from a class A pan, in inches,

d = monthly day-time coefficient,

h_n = mean monthly relative humidity at noon, in decimals, and

T = mean monthly temperature, in degrees Fahrenheit

Eq. 5.9 is based on an average wind speed of 100 km/day. Evaporation increases or decreases about 9% with each 50 km/day increase or decrease in the wind speed. Furthermore, where the sunshine is materially less than 90% one ought to correct the computed evaporation as follows:

sunshine in percent	30	40	50	60	70	80	90
correction in percent	-34	-28	-24	-20	-16	- 9	0

When observations on sunshine were not available, we used the equation developed by Palayasoot (1965) to compute the sunshine, S, from the cloud cover, C, scale 0 - 8, as

$$S = 74.5 + 9.5\ C - 2.0\ C^2 \tag{5.10}$$

Eq. 5.9 is based on data collected from locations with an average elevation of 500 ft, say 150 m. Since evaporation increases with elevation, eq. 5.9 can be corrected by increasing the calculated values by 1% for each 100 m increase in elevation.

The figures obtained from Hargreaves' equation have been reduced by a factor of 0.7 so as to convert the class A pan evaporation to evaporation from open water. The mean monthly evaporation computed from eq. 5.6 and from eq. 5.9 and observed by the Piche instrument is listed in Table 5.4. In this table, the station at Malakal is considered as representing the northern part of the Bahr el Jebel and Mongalla/Juba as representing the southern part, both outside the swamps. The station at Wau can be regarded as a transient point between the outside and the inside of the swamps.

The comparison between the evaporation figures presented in Table 5.4 can be made somewhat easier when they are plotted versus the months of the year, as shown in Fig. 5.4. The remarkable feature in this figure is that the three methods yield, for all stations considered, evaporation curves showing an identical pattern of variation with time during the year. The evaporation curve, starting from January, exhibits a slight rise till it reaches the peak evaporation in February then falls steeply, especially in the beginning, till the minimum evaporation is reached between July and August. From this time onwards, the evaporation increases gradually till January and February in the next year.

TABLE 5.4   Mean monthly open water evaporation and Piche evaporation at some
stations in the Bahr el Jebel Basin, outside the swamps

Method	Evaporation, in mm/day, for												
	Jan.	Feb.	Mar.	Apr.	May	June	July	Aug.	Sep.	Oct.	Nov.	Dec.	Year
1. Malakal													
Penman	5.5	6.6	6.8	6.7	5.8	4.5	4.0	4.1	4.8	5.0	5.2	5.8	5.4
Hargreaves x 0.7	10.6	10.8	9.4	9.3	6.3	4.2	2.6	1.9	2.5	2.9	5.4	9.2	6.3
Piche	16.5	18.2	15.7	10.9	7.0	4.4	2.9	2.5	2.9	3.7	8.3	13.6	9.0
2. Wau													
Penman	6.4	7.0	7.0	6.7	6.0	5.2	4.5	4.8	5.0	5.1	5.4	6.1	5.8
Hargreaves x 0.7	9.5	9.1	8.4	7.5	6.4	4.9	3.4	3.5	3.6	3.8	5.8	8.1	6.2
Piche	11.9	12.6	12.0	9.4	6.4	4.7	3.7	3.5	3.9	4.7	7.7	10.2	7.6
3. Mongalla/Juba													
Penman	6.4	6.9	6.6	5.8	5.2	4.7	4.1	4.5	4.9	5.2	5.3	5.7	5.5
Hargreaves x 0.7	8.3	8.2	7.7	5.3	4.1	3.2	2.3	2.8	3.2	3.8	4.6	6.6	5.0
Piche	12.3	12.0	9.7	7.4	4.9	4.3	3.3	3.4	4.4	5.4	7.1	9.2	7.0

The figures obtained from the method of Penman differ from those obtained
from the Hargreaves formula and from the Piche evaporimeter in two distinct ways.
The first is that the mean annual Penman evaporation increases from Mongalla/
Juba to Wau and decreases at Malakal, whereas the other two methods show
increasing evaporation from the south to the north. The second is that the
variation in the mean monthly evaporation estimated from the method of Penman is
small compared to the variation of evaporation estimated from the other two
methods. The ratio of the minimum to the maximum evaporation in the Penman's
method of about 0.59 for both Malakal and Mongalla/Juba and 0.64 for Wau. The
same ratios but obtained from the Hargreaves formula and from the Piche tube are
0.18 and 0.14 for Malakal, 0.37 and 0.28 for Wau, and 0.28 and 0.27 for
Mongalla/Juba, respectively.

The class A pan evaporation estimated by the Hargreaves method has been
multiplied by a factor of 0.7 to deduce the open water evaporation. This value
should be modified to 0.77 for Mongalla/Juba, 0.66 for Wau, and 0.60 for Malakal
so as to yield the same mean annual evaporation as obtained from Penman's
method. The modified values lie within 0.60-0.81, the range found for the class
A pan factor (Linsley, R.K., Kohler, M.A., and Paulhus, J.L., 1958).

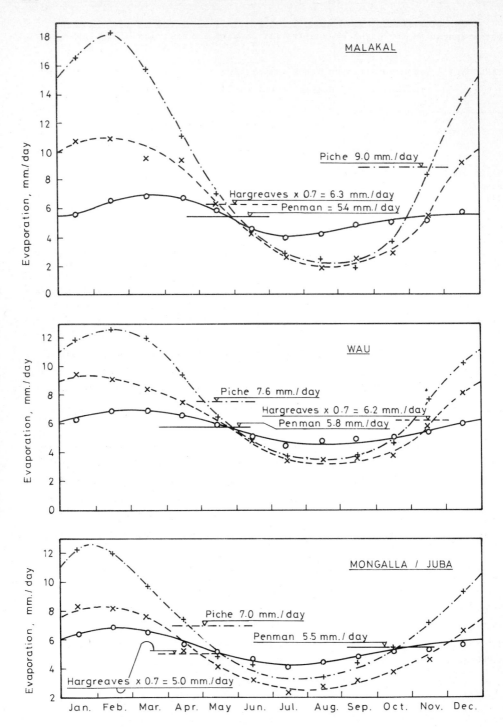

Fig. 5.4.   Mean monthly evaporation at Malakal, Wau and Mongalla/Juba

The old data given by Keeling and the more recent ones given by Olivier for
the mean annual evaporation from the so-called standard tank can be summarized
in the following:

Station	Keeling, mm/day	Olivier, mm/day
Malakal	6.4	6.0
Wau	5.4	9.3
Juba	5.0	7.4

These values, except the one obtained from Olivier's formula for Wau, seems
to be in reasonable agreement with the figures obtained from the methods we have
already described.

The ratio of the Piche evaporation to the Penman evaporation shows an
increase from the south to the north. This ratio is 1.27 for Mongalla/Juba, 1.31
for Wau and 1.67 for Malakal.

5.2.2 Bahr el Jebel and Bahr el Ghazal Basins (inside the swamps)

Evaporation observations from a 10 m square open water tank floating in
Shambe Lagoon are mentioned in Vol. X of the Nile Basin (Hurst, H.E., Black,
R.P., and Simaika, Y.M., 1966). The reported readings for the years 1948 and
1949 are 4.94 mm/day and 6.02 mm/day. The difficulty in measuring that part of
the rainfall that reached the tank is probably one of the factors responsible
for this difference.

Unfortunately, the available meteoeological data for the area inside the
swamps (channels, lagoons, etc.) are neither complete nor quite reliable. It is
therefore not worthwhile employing these data for estimating the evaporation
inside the swamps, as this was done for that part of the basin outside the
swamps.

One should expect, however, less evaporation from the free water inside the
swamps than from outside. This argument is supported by the figures given in
Vol. V of the Nile Basin (Hurst, H.E., and Philips, P., 1938). The average
temperature inside the swamps is $2^{o}C$ less than outside the swamps, the relative
humidity is about 37% more, the cloudiness is 10% more and the wind force is
only 20% of the wind force outside the swamps. The question that then arises is
by how much the evaporation inside the swamps is less than the evaporation outside. In an attempt to answer this question, Olivier (1961) referred to the old
findings of Ebermayer about the evaporation inside and outside the forests in
Germany. Ebermayer suggested a reduction factor of 0.36 to be applied to the
free water evaporation outside the forests to deduce the evaporation from inside
the forests. If this reduction factor holds in the case of the swamps, one
should expect the free water evaporation from the swamps of Bahr el Jebel and

Bahr el Ghazal to be about 2.1 mm/day. The Piche readings suggest the ratio of
0.41 for the evaporation from the swamps in proportion to the free water surface evaporation. Substituting the difference in the meteorological data between the inside and the outside of the swamps in Hargreaves' formula, eq. 5.9,
one comes to a reduction factor of about 0.5. So one has to expect a mean annual
evaporation from inside the swamps in the range of from 2.5 to 3.0 mm/day. This
range constitutes about 50% only of the observed floating-tank evaporation.

The above discussion can lead us to the following general conclusions:

i) The open water evaporation outside the swamps shows a gradual increase in
a northerly direction from slightly more than 5 mm/day at Mongalla to
slightly less than 6 mm/day at Malakal.

ii) The mean monthly evaporation at all stations shows a well-defined pattern
with time in the year.

iii) The range of variation of the mean monthly evaporation estimated from the
Penman method is much smaller compared to the same range in the observed
Piche readings or the class A pan evaporation estimated from the Hargreaves
formula.

iv) The reduction factor of the Piche readings to deduce the free water surface
evaporation increases in a northerly direction from 1.27 at Mongalla to
1.67 at Malakal.

v) A reasonably accurate measurement of the open water evaporation inside the
swamps is not available. Indirect estimates yield a mean annual evaporation in the range of 2.5 to 3.0 mm/day.

5.3 THE WHITE NILE BASIN (from Malakal to Khartoum)

Evaporation has been measured by means of floating tanks at Malakal and
Khartoum, as well as by the Piche evaporimeter at a number of stations. The
floating tanks have been reported as being not entirely satisfactory, though
they give more direct and useful information. The reason behind that state of
affairs was the difficulty in protecting the tanks completely from the effects
of strong winds.

The evaporation from the White Nile Basin has become important after the construction of the Jebel Aulia dam, some forty kilometres south of Khartoum. The
storage reservoir, when full, has a relatively large surface compared to its
volume. The mean monthly open water evaporation at Khartoum, Jebel Aulia, Dueim,
Rabak/Kosti, Renk and Malakal has been deduced from the Piche readings using a
reduction factor of 0.5. The mean annual figures are presented both in Table
5.1 and Fig. 5.1. Hurst, in Vol. VIII of the Nile Basin (1950), commented on
the changing climate as one travels south along the White Nile starting from
Khartoum. The range of the Piche evaporation at the northern end is from 10 to

20 mm/day. At Dueim, 200 km south, the maximum is about the same but the minimum decreases to 7 mm/day. At Malakal, 800 km south, the maximum remains almost
the same and the minimum drops further to 3 mm/day and the evaporation is less
than the minimum at Khartoum for seven months, from May up to and including
November. The mean monthly evaporative demand for Khartoum estimated from the
methods of Penman and Hargreaves together with the Piche evaporation is included in Table 5.5.

TABLE 5.5   Mean monthly open water evaporation at Khartoum and Piche evapora-
             tion from the White Nile Basin

Method	Evaporation, in mm/day, for												
	Jan.	Feb.	Mar.	Apr.	May	June	July	Aug.	Sep.	Oct.	Nov.	Dec.	Year
Penman	7.2	7.8	9.3	10.5	10.4	10.1	8.4	6.8	7.1	8.9	8.2	7.1	8.5
Hargreaves x 0.7	10.1	11.6	13.6	16.6	17.0	15.2	10.8	7.6	8.6	12.6	12.6	11.0	12.1
Piche	12.0	14.0	17.0	18.0	18.0	16.0	13.0	10.0	12.0	14.0	13.0	12.0	14.1
Piche x 0.5[+]	8.0	9.0	10.0	10.0	8.0	7.0	5.0	3.0	4.0	5.0	7.0	7.0	7.0

[+]Average of Piche readings at Khartoum, Jebel Aulia, Dueim, Rabak/Kosti, Renk
and Malakal. Other data belong to Khartoum only. The data about Malakal is
included in Table 5.4.

The mean monthly evaporation given in Table 5.5 shows the same pattern for
all methods used. There are two seasons for the peak evaporation, though the
second peak has an average value equal to about 75% of the value of the first
peak. The first peak occurs in April-May and the second in October-November.
The proportion of the minimum monthly evaporation to the maximum monthly evaporation varies from 0.65 in the method of Penman to 0.45 in the method of
Hargreaves to 0.55 for the Piche evaporation. The results obtained from the
Hargreaves formula when multiplied by 0.7 are about 43% higher than those
obtained from the Penman method whereas the Piche readings need to be reduced
by about 40% to become equal to the Penman evaporation. Additional estimates of
evaporation from a standard tank are 11.3 and 10.5 mm/day as obtained from the
Olivier method and the data of Keeling, respectively.

At Khartoum observatory, comparisons were made among evaporation data from
(1) a 12-ft diameter tank, 4-ft deep; (2) a class A pan; (3) a Piche atmometer;
(4) lake evaporation computed by Penman equation, and (5) lake evaporation by
Kohler's method, which is based on correcting class A pan evaporation for heat
transfer through its sides and bottom. The results obtained for the years 1961
and 1962 are given in Table 5.6 (WMO, 1966).

TABLE 5.6   Comparison between evaporation obtained from different methods and
             evaporimeters at Khartoum (WMO, 1966)

Year	Month	12-ft pan / Class A pan	12-ft pan / Piche	12-ft pan / Penman	12-ft pan / (Kohler)
	Jan.	0.68	0.50	1.23	1.07
	Feb.	0.60	0.43	1.13	0.99
	Mar.	0.65	0.48	1.18	1.02
	Apr.	0.62	0.46	1.25	0.97
	May	0.64	0.53	1.37	0.99
	June	0.67	0.53	1.20	1.03
1960	July	0.62	0.54	1.06	0.94
	Aug.	0.65	0.60	1.14	1.00
	Sep.	0.69	0.61	1.17	1.01
	Oct.	0.69	0.55	1.31	1.05
	Nov.	0.64	0.45	1.25	0.99
	Dec.	0.61	0.47	1.16	0.98
	Annual	0.64	0.51	1.20	1.00
	Jan.	0.67	0.51	1.20	1.06
	Feb.	0.66	0.50	1.24	1.01
	Mar	0.64	0.50	1.21	1.00
	Apr.	0.62	0.49	1.28	0.98
	May	0.59	0.48	1.36	0.97
	June	0.61	0.58	1.24	0.95
1961	July	0.68	0.78	0.97	0.98
	Aug.	0.74	0.63	0.96	1.03
	Sep.	0.69	0.63	1.20	1.04
	Oct.	0.67	0.53	1.25	1.01
	Nov.	0.66	0.50	1.26	1.00
	Dec.	0.69	0.50	1.22	1.03
	Annual	0.65	0.55	1.20	1.00

The figures listed in Table 5.6 can be converted to other figures giving the
ratio between evaporation from a free water surface and evaporation as observed
by an evaporimeter or estimated from a formula. This has been done using the
data in Table 5.2 where the ratio of 12-ft pan diameter to an extended water
surface is 1.099.

The monthly average ratios found for 1960 and 1961 of the 12-ft pan: Piche
and the 12-ft pan: Penman were each divided by 1.099 to deduce the corresponding free water surface: Piche and free water surface: Penman. By this procedure
two sets of mean monthly evaporation values from the free water surface at
Khartoum have been developed. The new set of data originating from Piche readings is somewhat on the low side, and that originating from Penman's method is
somewhat on the high side. A fair compromise is to take the arithmetic mean of
these two sets. Accordingly, one can list the free water surface evaporation
at Khartoum as follows:

Evaporation, in mm/day, for

Jan.	Feb.	Mar.	Apr.	May	Jun.	Jul.	Aug.	Sep.	Oct.	Nov.	Dec.	Year
6.74	7.16	8.85	9.94	10.60	9.64	7.78	6.05	7.21	8.62	7.49	6.49	8.04

The foregoing analysis leads us to the conclusion that the free water surface evaporation from the White Nile Basin ranges from slightly less than 6 mm/
day at Malakal to 8 mm/day at Khartoum in a northerly direction over a distance
of about 800 km. The mean monthly evaporation at Khartoum has two peaks; the
first peak of 10.6 mm/day occurring in October. The minimum takes place in
August and in December where the mean monthly evaporation falls to 6.0 and 6.5
mm/day respectively. Evaporation from open water at Malakal shows a single peak
only. The experiments conducted at Khartoum when combined with other experiments show that the Penman evaporation is about 10% less than the evaporation
from a free water surface. The reduction factor of the Piche readings needs to
be increased to 0.55-0.60 instead of 0.5 to deduce the open water evaporation.

5.4   THE ETHIOPIAN PLATEAU

5.4.1   The Sobat Basin

Estimation of free water evaporation from the Sobat Basin has been confined
to three stations, namely: Gambeila, Akobo and Malakal. As a matter of fact the
station at Malakal belongs to the White Nile Basin. However, since it is situated close to the mouth of the Sobat on the White Nile, its data can be used to
complete the estimates of evaporation from the catchment of the Sobat.

The average Piche readings over different periods for Gambeila and Akobo are
listed in Table 5.7 whereas the data belonging to Malakal have already been
included in Table 5.4.

TABLE 5.7   Piche evaporation at Gambeila and Akobo in the Sobat Basin (Hurst,
H.E., 1950)

Station	Period *	Evaporation, in mm/day, for												
		Jan.	Feb.	Mar.	Apr.	May	June	July	Aug.	Sep.	Oct.	Nov.	Dec.	Year
Gambeila	a	9.0	10.0	11.0	8.0	5.0	4.0	3.0	3.0	3.0	4.0	5.0	7.0	5.9
	b	8.0	9.8	9.6	7.5	4.0	2.5	2.2	2.1	2.6	3.3	4.3	5.9	5.1
Akobo	c	10.3	11.8	10.9	8.8	5.3	3.9	2.9	2.4	3.1	3.6	4.6	7.8	6.3

*a = 1909-34; b = 1950-57; c = 1950-57 (in 1950 from January to March only).

The mean annual evaporation as given in Table 5.7 is not really consistent with the regression relation between the annual rainfall, R in mm, and the Piche evaporation, in mm/day, which was developed by Hurst in Vol. X of the Nile Basin (1966). The regression equation

$$E_{piche} = 10.3 - 0.0076 \ (R - 714) \tag{5.11}$$

gives E_{piche} for Gambeila and Akobo as 6.6 and 8.9 mm/day respectively. The discrepancy between these figures and the observed Piche evaporation may throw some doubt on the validity of eq. 5.11, or the data in Table 5.7, or both.

The available meteorological data of the Sobat Basin, except for Malakal, are inadequate to allow for estimating the free water evaporation from any formula. We therefore need to approach the problem from a different angle. The weighted average Piche evaporation (total 32 years of record) is 5.7 mm/day for Gambeila. Assuming a linear proportion between the evaporation at the two stations, an average Piche evaporation of 7.04 mm/day, over the same period, can be obtained for Akobo. The two stations are situated outside the Machar swamps and apparently have similar conditions affecting the free water evaporation to the stations outside the swamps in the Bahr el Jebel and Bahr el Ghazal Basins. From the data presented in Table 5.4 it is obvious that the reduction factor to the Piche readings at Mongalla and Wau has an average of 0.75.

Applying this factor to the estimated Piche evaporation for Gambeila and Akobo one obtains a mean annual open water evaporation of 4.3 and 5.3 mm/day respectively. Furthermore, if we assume that these figures are distributed throughout the year similar to the Piche readings in Table 5.7, one can obtain the mean monthly evaporation as follows:

Station	Jan.	Feb.	Mar.	Apr.	May	June	July	Aug.	Sep.	Oct.	Nov.	Dec.	Year
Gambeila	6.6	7.7	8.0	6.0	3.5	2.5	2.1	2.1	2.2	2.8	3.6	5.0	4.3
Akobo	8.7	9.9	9.2	7.4	4.4	4.5	3.3	2.4	2.0	2.6	3.9	6.5	5.3

Evaporation, in mm/day, for

The Sobat Basin includes the Machar swamps. The surface area of these swamps is exceedingly variable. The Jonglei investigation team reported that when the rains were below the normal, they became dry, and when they were above the normal, the swamps covered about 20.000 km^2. This team used what they reported as the maximum surface area for estimating the evaporation from the Machar swamps. From the water-balance of the swamps the annual evaporation was estimated at 0.95 m or about 2.6 mm/day. Hurst condemned this figure as being absurdly low. Instead, in Vol. X of the Nile Basin (1966), he estimated the annual volume of water disappearing by evaporation from the Machar swamps at 9.9 mlrd m^3. He further assumed that this volume corresponded to 6.700 km^2 surface of the swamps. If this is really so, the mean annual free water surface evaporation becomes 4.1 mm/day and not 5.0 mm/day as computed by Hurst. The figure of 4.1 mm/day seems to be fairly consistent with those found by us as 4.3 and 5.3 mm/day evaporation at Gambeila and Akobo respectively.

5.4.2   The Blue Nile Basin

The open water evaporation deduced from the Piche readings at a number of places in the Blue Nile Basin is already included in Table 5.1.   The mean annual evaporation as given in this table varies from 3.0 mm/day for Lake Tana to 6.5 mm/day at Wad Medani and eventually to 7.8 mm/day at Khartoum.

Rzóska, upon compiling some hydrological data from Hurst's book , "The Nile" (1976) concluded that the mean annual losses of open water measured in tanks were 1.09 (0.54-1.29) m for Lake Tana and 2.30 m for the reach from Roseires up to Wad Medani.

Lake Tana has a surface area of about 3150 km^2 and a catchment of about 13400 km^2. The isohyetal maps show a direct rainfall of 1.2 m in a normal year over the lake and about 1.3 m over the catchment.

The run-off coefficient for this catchment as obtained from "The Water Resources of The Earth" (1974) is 0.15. Since the outflow from Lake Tana in a normal year is known to be about 3.5 mlrd m^3, one can easily arrive at an annual depth of evaporation of 0.9 m, from the water balance, eq. 5.2. This depth corresponds to a mean annual evaporation of 2.5 mm/day, which is 20% less than the figure presented in Table 5.1. The inaccuracies involved in the rain depths, in the estimate of the outflow from the lake, and probably the run-off coefficient from the catchment obviously affect the accuracy of the estimate of evaporation.

Hurst, in Vol. VIII of the Nile Basin (1950), mentions that the next point (next to Lake Tana) where there is any hydrological information is at Roseires in the Sudan. We have tried to estimate the free water surface evaporation from the available meteorological data for three stations, namely: Addis Abbaba, Roseires and Wad Medani. The computed evaporation, together with the available Piche readings are listed in Table 5.8.

The evaporation from the Blue Nile Basin, whether computed from the formula of Penman or that of Hargreaves, or measured by the Piche instrument, shows an increase in a northerly direction from Addis Abbaba to Wad Medani and eventually to Khartoum.

TABLE 5.8   Computed free surface water evaporation and measured Piche evaporation at some stations in the Blue Nile Basin

Station and Method*	Evaporation, in mm/day, for												
	Jan.	Feb.	Mar.	Apr.	May	June	July	Aug.	Sep.	Oct.	Nov.	Dec.	Year
Addis Abbaba													
a	5.2	5.3	5.4	5.7	5.6	4.9	4.2	4.1	4.7	5.0	5.1	5.2	5.0
b	7.3	7.5	7.7	8.5	8.3	7.0	5.6	5.3	6.5	6.9	7.1	7.3	7.1
Roseires													
a	6.3	6.5	7.8	8.3	7.6	6.4	5.0	5.1	5.3	5.5	5.7	5.8	6.3
b	10.0	10.7	11.9	12.3	9.2	7.2	3.9	4.0	4.6	4.8	6.3	8.4	7.9
c	13.6	14.7	15.8	16.2	13.2	8.4	5.0	5.1	5.8	6.8	8.8	11.4	10.6
d	6.8	7.4	7.9	8.1	6.6	4.2	2.5	2.6	2.9	3.4	4.4	5.7	5.3
Wad Medani													
a	5.8	6.3	7.8	9.2	8.9	8.3	6.4	5.3	6.3	7.5	6.6	5.6	7.0
b	7.8	8.7	10.2	12.9	12.5	11.3	8.1	5.7	6.5	9.5	9.2	8.1	9.2
c	14.3	16.5	20.4	21.5	20.4	18.7	11.3	6.6	7.1	11.4	14.6	13.4	14.7
d	7.1	8.2	10.2	10.8	10.2	9.4	5.6	3.3	3.6	5.7	7.3	6.7	7.3

*a = Penman; b = Hargreaves x 0.7; c = Piche; d = Piche x 0.5

The free water evaporation computed by Penman's formula has been plotted
versus the time in months for the stations at Addis Abbaba, Roseires, Wad
Medani and Khartoum. The four evaporation curves illustrated in Fig. 5.5. present some marked features. The ratio of the minimum evaporation to the maximum
evaporation for Addis abbaba, being 0.72, is certainly high compared with the
average of the corresponding ratios for the other stations, which is about 0.60
(0.57-0.64). The evaporation curves for Addis Abbaba and Roseires are similar
to those plotted for the stations in the Bahr el Jebel Basin (Fig. 5.4.) in that
each curve has a single peak. The other two curves for Wad Medani and Khartoum
show two peak evaporation rates, the first occurs in April-May and the second,
the smaller of the two, in October.

The figures listed in Tables 5.6 and 5.8 indicate that the ratio of the mean
annual evaporation computed by the Penman to the mean annual evaporation read
from the Piche tube is nearly the same for both Khartoum and Roseires, 0.60 and
0.59 respectively. The same ratio can be found from Table 5.4 as 0.60 for
Malakal also. It might be that for those parts of the White Nile and the Blue
Nile Basins outside the swamps the factor 0.6 can reasonably be used to convert
the mean annual Piche evaporation to the corresponding Penman evaporation. Note
that this is slightly different from the factor of 0.5 used by Hurst for converting the Piche readings to free water surface evaporation.

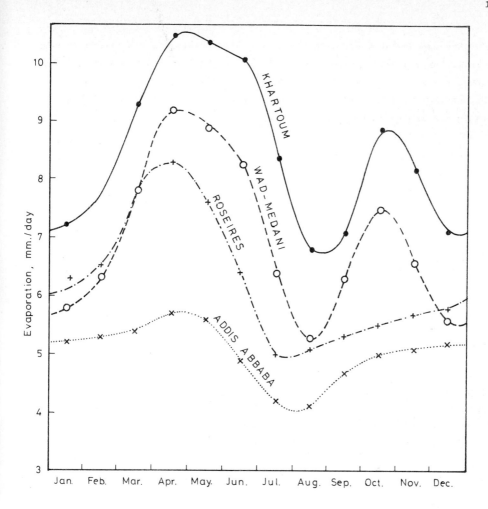

Fig. 5.5. Mean monthly evaporation from open water at some stations in the Blue Nile Basin estimated from the Penman method

The Penman: the Piche ratio for the mean monthly evaporation has been found to vary from as small as 0.44 to as large as 1.0 for Roseires and Khartoum stations. The range of variation for Malakal is even much wider since the minimum and the maximum values for this ratio are 0.33 and 1.66 respectively.

In addition to the stations at Addis Abbaba, Roseires, Wad Medani and Khartoum, there are three other stations in the Blue Nile Basin for which the Piche readings over the period 1921-1950 are available. These are: Singa and Sennar, nearly halfway between Roseires and Khartoum, and Kurmuk, to which the nearest station is Roseires.

The average of the monthly coefficients for Khartoum and Roseires have been multiplied by the Piche evaporation at Singa and at Sennar whereas the monthly coefficients for Roseires alone have been multiplied by the Piche evaporation at Kurmuk (Hurst, H.E., Black, R.P. and Simaika, Y.M., 1959). The results obtained are listed in Table 5.9.

TABLE 5.9   Free water surface evaporation for Singa, Sennar and Kurmuk in the Blue Nile Basin

Station	Evaporation, in mm/day, for												
	Jan.	Feb.	Mar.	Apr.	May	June	July	Aug.	Sep.	Oct.	Nov.	Dec.	Year
Singa	7.2	8.3	8.8	9.7	8.3	8.2	5.8	3.7	3.8	5.8	8.6	7.5	7.0
Sennar	7.8	9.5	11.4	12.9	9.2	9.5	6.6	4.2	5.0	7.2	9.6	7.9	7.4
Kurmuk	7.2	7.9	8.2	6.7	4.6	3.7	3.1	2.6	2.8	2.9	4.8	6.9	5.3

5.4.3   The Atbara Basin

While discussing the hydrology of the Atbara Basin in Vol. IX of the Nile Basin, Hurst mentions that there is very little information about the evapora- tion from this basin. Though Lake Tana is just outside the Atbara Basin, the surrounding country is similar to that of the Atbara catchment. Hurst used this characteristic together with the average of the Piche readings at Kassala and Gallabat, reduced by a factor of 0.5, in order to sketch an approximate picture of the open water evaporation in or near the Atbara Basin (Hurst, H.E., Black, R.P., and Simaika, Y.M., 1959). The figures obtained by Hurst and the figures we have estimated by the Penman method directly for Atbara and indirectly as an average for Kassala and Gallabat are given in Table 5.10.

The foregoing discussion of the free water surface evaporation from the Ethiopian Plateau may lead to the following conclusions:

i)   The evaporation from Lake Tana is not known to a fair degree of accuracy. The inedaquate information about the hydrology of this lake points to a mean annual evaporation of 3.0-3.5 mm/day.

ii)   The mean annual evaporation from the Machar swamps in the Sobat Basin may be taken as about 4.1 mm/day.

iii) The open water evaporation for the Blue Nile Basin increases gradually in a north-westerly direction. The mean annual evaporation at Kurmuk is 5.3 mm/day, increasing to about 6.3 mm/day at Roseires, 7.0-7.5 for Singa, Sennar and Wad Medani and becomes 8.0 mm/day at Khartoum. The average for Kassala and Gallabat in the Atbara basin is nearly 6.7 mm/day. This figure increases to 8.0 mm/day at Atbara.

iv) The evaporative demand for the White Nile Basin and the Ethiopian Plateau
     is presented in Fig. 5.6.

v)   The reduction coefficient for the Piche readings, on an annual basis,
     decreases from 0.8 for the Sobat Basin to about 0.6 for the basins of the
     Blue Nile and the Atbara. The monthly coefficient varies in a wide range
     with an indication to a heavy seasonal component.

TABLE 5.10   Open water evaporation in the Atbara Basin

Station + Method *	Evaporation, in mm/day, for												
	Jan.	Feb.	Mar.	Apr.	May	June	July	Aug.	Sep.	Oct.	Nov.	Dec.	Year
Piche x 0.5													
a	3.9	4.8	5.1	5.7	4.1	2.4	1.1	1.1	1.4	2.2	2.4	3.2	3.1
b	5.6	6.5	7.8	8.3	7.1	5.2	3.3	2.4	2.8	4.3	5.2	5.3	5.3
Penman (direct)													
c	5.5	6.2	7.5	10.6	9.3	9.0	8.8	8.6	8.6	9.9	6.5	5.6	8.0
(indirect)													
b	6.7	7.3	8.6	9.6	8.2	6.6	4.3	3.3	3.3	5.5	6.6	6.3	6.7

*a = Lake Tana; b = Kassala and Gallabat; c = Atbara

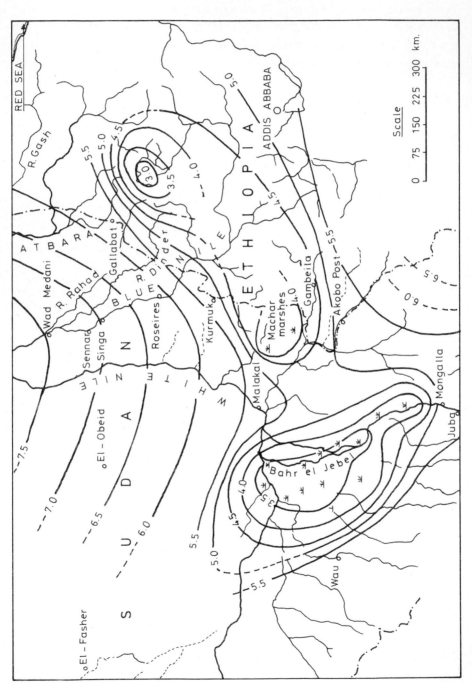

Fig. 5.6. Evaporative demand for the basins of the Bahr el Jebel and the White Nile, and for the Ethiopian Plateau

5.5   THE MAIN NILE

5.5.1   The Main Nile from Khartoum to Aswan

The mean monthly as well as the mean annual evaporation as observed by the Piche instrument at a number of locations from Khartoum to Aswan can be found in Vol. IX of the Nile Basin (Hurst, H.E., Black, R.P., and Simaika, Y.M., 1959). A summary of these data is given in Table 5.11.

TABLE 5.11   Piche evaporation for a number of stations on the Main Nile from Khartoum to Aswan

Station	Evaporation, in mm/day, for												
	Jan.	Feb.	Mar.	Apr.	May	June	July	Aug.	Sep.	Oct.	Nov.	Dec.	Year
Khartoum[*]	13.0	16.0	19.0	20.0	19.0	18.0	14.0	11.0	12.0	15.0	14.0	13.0	15.0
Zeidab	10.0	12.0	15.0	16.0	16.0	17.0	15.0	13.0	13.0	13.0	12.0	10.0	14.0
Atbara	13.8	16.2	18.9	20.8	20.8	20.2	18.5	16.1	16.6	16.4	14.8	13.5	17.2
Merowe	11.6	13.7	16.8	20.0	21.2	21.0	18.6	16.7	19.0	17.6	14.0	11.5	16.8
Dongola	8.4	10.9	13.5	15.6	17.9	17.0	17.8	16.6	15.8	14.1	10.9	9.1	14.0
Wadi Halfa	8.8	10.9	14.4	18.1	19.4	21.5	19.4	17.5	18.2	15.9	11.7	8.6	15.4
Aswan	7.6	9.1	12.9	16.8	18.5	21.6	19.5	19.2	18.1	15.5	10.8	7.3	14.7

[*]Mean of three stations at Khartoum

Hurst mentions that the evaporation values for Dongola are definetly lower than would have been expected from the observations at the other stations. This station was closed and the observations covered 7 years only. Furthermore, it is not possible now to say whether the difference is real or due only to some circumstance connected with the instrument and its exposure.

The free water surface evaporation at Khartoum and Atbara has already been given in the last two sections. The long-term meteorological data available for Merowe, Wadi Halfa and Aswan have been worked out for estimating the open water evaporation for these stations using the Penman and Hargreaves methods. The results obtained are listed in Table 5.12.

In addition to the computation results presented in Table 5.12, it may be worthwhile giving here the mean annual evaporation from a standard tank as obtained from the method of Olivier. This is 10.5 for Khartoum, 11.0 for Wadi Halfa and 10.8 mm/day for Aswan. The mean monthly values for Aswan, as an example, are as follows:

Month	Jan.	Feb.	Mar.	Apr.	May	June	July	Aug.	Sep.	Oct.	Nov.	Dec.
Evaporation, mm/day	4.9	6.4	9.5	12.6	14.1	16.0	15.1	14.7	13.2	10.9	7.3	5.1

TABLE 5.12   Estimated free water surface evaporation for Merowe, Wadi Halfa and
Aswan on the Main Nile

Station + Method *	Jan.	Feb.	Mar.	Apr.	May	June	July	Aug.	Sep.	Oct.	Nov.	Dec.	Year
Evaporation, mm/day, for													
Merowe													
a	5.7	6.5	7.6	8.9	9.6	9.8	9.4	9.3	9.1	8.2	6.7	5.6	8.0
b	7.6	8.8	11.7	15.5	20.3	19.2	17.1	17.5	15.7	14.1	11.2	8.6	13.9
Wadi Halfa													
a	4.5	5.8	7.1	8.7	9.3	9.8	9.6	9.3	8.6	7.7	6.1	4.0	7.5
b	4.3	5.8	8.3	12.0	14.6	15.8	14.7	14.3	12.8	10.4	7.7	4.8	10.5
Aswan													
a	3.5	4.6	6.2	7.7	8.5	9.4	9.3	8.8	8.2	6.5	4.5	3.7	6.8
b	5.3	6.1	9.4	12.5	15.4	16.6	16.6	15.6	13.7	11.5	8.2	5.7	11.4

*a = Penman; b = Hargreaves x 0.7

The River Atbara is the last tributary of the Main Nile. The annual flow downstream of the confluence is the sum of the flow above it, at Hassanab, and the supply of the Atbara. The annual flow at Mongalla, some 810 km downstream of the confluence, at Hassanab and at Atbara, has been averaged over a period of 30 years, 1943-72. The mean annual loss in the reach from just downstream of the confluence to Dongola for the period considered is about 1060 million m^3. This figure is most definitely only an approximate one, as it varies in a very wide range from -6236 million m^3/yr in 1958 to +5623 million m^3/yr in 1964. Our figure is also high compared to that obtained by Hurst - 800 million m^3/yr - as an average for the period 1912-1952, for the reach from just downstream of the confluence down to Wadi Halfa, 1210 km downstream of Atbara. Anyhow, using our figure and assuming an average width of 500 m for the river between the rising and the falling stages, one gets an evaporation depth of 2.62 m/yr. This figure corresponds to a mean annual evaporation of 7.2 mm/day, which is about 10% less than what we have obtained from Penman's method for Atbara and Merowe.

The reach of the Main Nile from Atbara to Aswan is characterized by the almost constant ratio between the Penman and the Piche evaporation. This is 0.465 for Atbara, 0.476 for Merowe, 0.487 for Wadi Halfa and 0.463 for Aswan, with a mean of 0.475 for the entire reach. The pan evaporation figures obtained from the Hargreaves method need to be multiplied by a coefficient of 0.44 instead of 0.7 to reduce them to Penman's evaporation. On the other hand, the evaporation from a standard tank computed by the Olivier formula needs to be corrected by a factor of 0.68 for Wadi Halfa and 0.63 for Aswan. One should not forget that all these ratios are based on mean annual evaporation. For the comparison between the mean monthly evaporation as obtained from the different methods, reference is made to Fig. 5.7.

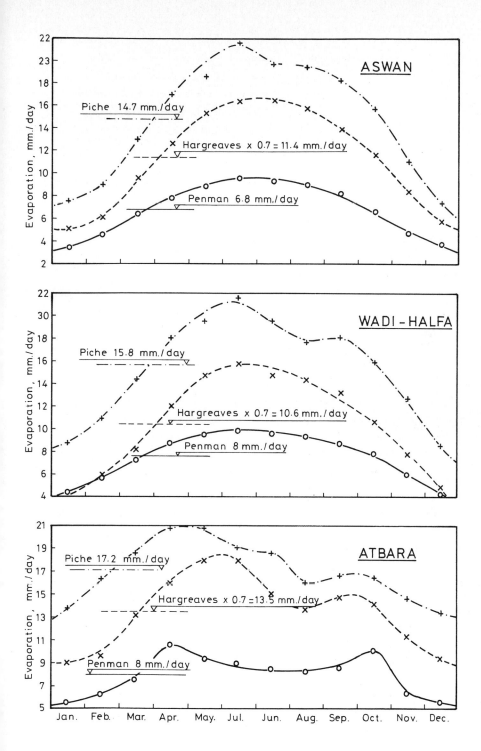

Fig. 5.7. Mean monthly evaporation at Atbara, Wadi Halfa, and Aswan

5.5.2 The Nile from Aswan to Giza, the Oases, and the Red Sea Coast

In this stretch of the river, in the oases of the Libyan desert, and along the coast of the Red Sea, there is a reasonable number of meteorological screens. Till the time of writing this book, the published data of these screens are up to 1945 only (Ministry of War and Marine, 1950). For most of the screens, the data up to then have therefore been used. This covers a period of, say, 20 to 40 years for each station.

Other than the station at Aswan, we have selected four stations on, or very near to, the river: Qena, Assiut, Minya and Giza; two along the Red Sea Coast: Qusseir and Hurghada; a semi-desertic oasis: El Fayum, and El-Kharga Oasis. Since the climatological normals for El Dakhla Oasis are nearly the same as those for El Kharga, the estimated evaporation for the latter is assumed to represent the free water surface evaporation for the two oases.

Each of the stations considered here is equipped with a Piche instrument. Additionally, two class A pans have been installed at Giza and El Kharga stations since 1957 and 1964 respectively. The evaporation data obtained from these two pans have, among others, been reported by the author in an earlier work (Shahin, M.M., 1970). Some of these data, together with the observed Piche evaporation and the free water surface evaporation estimated by the methods of Penman and Hargreaves, are given in Table 5.13. Furthermore, the lines of equal evaporative demand for the area considered are plotted on the map, Fig. 5.8. From this map, the gradual decline in the evaporative demand from Wadi Halfa in a northerly direction to Giza can easily be detected. This is also true for the coast of the Red Sea, except at Hurghada, nearly halfway between Qusseir and Suez, where a positive jump in the evaporative demand can be seen.

With the exception of the stations at Qusseir and Hurghada, which are situated at the coast, the ratio of the Hargreaves evaporation to the evaporation estimated by the Penman method is very nearly constant for all the stations. In fact, it varies from 2.01 to 2.21 with an overall average of 2.08. The same ratio but for Qusseir and Hurghada is 1.54 and 1.66 respectively. This marked reduction is obvious as the humidity at noon, which is a parameter in the Hargreaves equation, does not differ considerably from the mean of the day as in the case of the stations in the interior.

The pan evaporation computed by Hargreaves' equation is seen to be less than the observed class A pan evaporation for El-Kharga by about 16%, and more than the observed pan evaporation at Giza by 27%. This emphasizes the very strong effect of the humidity at noon in Hargreaves' method compared to the mean humidity which influences the pan evaporation.

TABLE 5.13   Observed and estimated evaporation for the Nile from Aswan to
              Giza, the Oases, and the Red Sea Coast

Station and Method[*]	Evaporation, in mm/day, for												
	Jan.	Feb.	Mar.	Apr.	May	June	July	Aug.	Sep.	Oct.	Nov.	Dec.	Year
Qena													
a	2.1	2.9	5.0	6.6	8.4	9.2	8.9	8.0	7.2	5.2	3.7	2.2	5.8
b	3.2	4.3	7.0	10.4	13.3	14.0	13.4	13.0	10.2	7.4	5.2	3.5	8.7
c	3.3	4.2	6.5	9.2	10.9	12.4	11.8	11.4	8.6	6.1	4.4	3.4	7.7
Kharga (Oasis)													
a	2.4	3.5	5.2	6.9	8.6	9.4	9.2	8.5	7.7	5.7	3.6	2.4	6.1
b	3.5	4.5	6.8	9.6	11.5	13.1	13.2	12.7	10.2	8.3	5.8	4.0	8.6
c	9.5	11.6	16.3	20.3	24.2	25.1	24.5	23.9	22.6	19.8	13.7	9.9	18.4
d	6.4	8.5	12.0	16.1	20.0	23.3	21.7	19.5	18.2	14.4	9.9	6.8	14.7
Assiut													
a	1.8	3.1	4.3	6.6	8.3	8.7	8.5	8.2	6.4	4.9	3.4	2.0	5.5
b	2.9	3.8	6.1	9.2	11.8	12.9	12.6	11.5	9.3	6.6	4.7	3.1	7.9
c	3.9	5.1	7.6	10.8	13.7	15.6	13.9	12.8	10.6	7.1	4.9	3.7	9.1
Qusseir (sea coast)													
a	3.1	3.9	5.1	6.2	7.3	8.2	8.5	8.0	7.3	5.4	4.4	3.2	5.9
b	4.2	4.6	5.5	6.2	7.6	8.9	8.4	8.3	7.3	6.0	5.0	4.5	6.4
c	8.0	5.6	9.6	10.0	11.3	12.9	11.5	11.9	11.2	9.4	8.8	7.9	9.8
Minya													
a	1.9	2.6	3.8	5.3	7.0	7.7	7.7	6.9	5.6	4.1	3.3	2.0	4.8
b	2.5	3.2	4.7	7.4	11.0	11.7	11.5	10.5	7.9	6.4	4.6	2.9	7.0
c	4.2	5.7	7.1	9.8	14.2	15.2	13.7	11.8	8.7	7.3	5.5	3.7	8.8
Hurghada (sea coast)													
a	3.7	4.4	5.6	7.0	8.3	9.5	9.6	9.0	7.7	6.2	5.2	3.8	6.7
b	4.8	5.2	6.4	8.0	10.3	10.8	11.0	10.5	8.7	7.0	5.8	5.1	7.8
c	10.7	11.5	13.6	15.5	17.6	21.5	19.5	19.8	17.6	13.7	11.5	10.4	15.2
Fayum (semi-oasis)													
a	1.5	2.6	3.9	5.4	7.6	8.2	8.2	7.3	5.5	4.0	3.0	1.6	4.9
b	2.7	3.5	5.3	8.2	11.8	13.5	12.3	10.8	9.1	6.6	4.6	2.8	7.6
c	3.4	4.4	6.0	8.6	11.8	13.3	11.8	10.1	8.0	6.5	4.5	3.2	7.6
Giza													
a	1.5	2.6	3.3	5.2	6.7	7.9	7.4	6.2	5.0	3.5	2.3	1.5	4.4
b	2.7	3.2	4.6	7.0	9.0	10.2	9.7	8.5	6.9	5.7	4.1	3.0	6.2
c	3.3	4.2	5.3	7.2	9.1	9.2	8.4	7.0	5.5	4.8	3.5	2.9	5.9
d	2.8	3.9	5.8	8.1	10.3	11.4	10.4	9.5	8.1	6.1	4.1	3.0	7.0

[*] a = Penman; b = Hargreaves x 0.7; c = Piche; d = class A pan

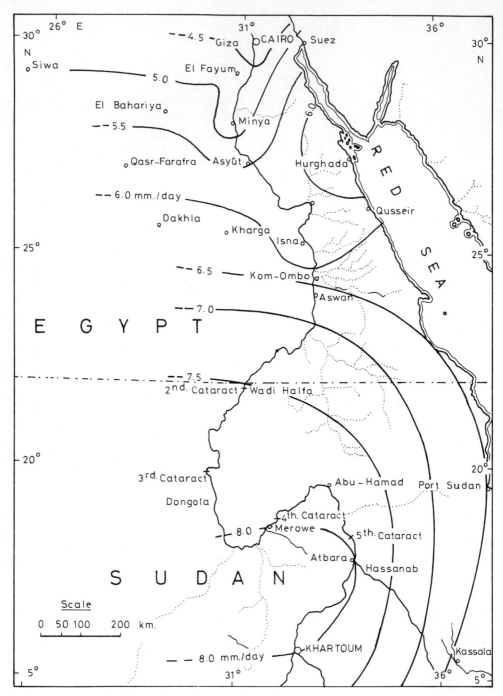

Fig. 5.8.   Evaporative demand for the Main Nile from Atbara to Giza, the
coastal line of the Red Sea, and the Libyan desert oases

Another example of the role played by humidity in evaporation estimates is
the results obtained from Olivier's method

Evaporation, in mm/day, for
*

	Jan.	Feb.	Mar.	Apr.	May	Jun.	Jul.	Aug.	Sep.	Oct.	Nov.	Dec.	Year
a	5.02	6.71	10.75	14.03	14.66	17.43	16.71	15.62	12.97	10.42	6.35	5.11	11.31
b	1.76	2.48	4.02	5.59	7.68	8.83	6.71	5.92	4.78	3.56	2.48	1.81	4.64

*a = El Kharga Oasis station; b = Giza station

where the ratio of the mean annual evaporation for El Kharga to Giza is 2.44.
The ratio of the Penman evaporation at these two stations is 1.39 only.

Further analysis and discussion of the relationships between the evaporation
figures obtained for the different parts of the Nile Basin can be found in section 5.6.

5.5.3   The Nile Delta Area from the Apex to the Mediterranean Sea Coast

For the evaporation study in the area extending from the apex of the delta
to the Mediterranean Sea Coast, eleven stations have been selected. These are:
Cairo, near the apex of the delta; Suez, at the farthest end of the Red Sea;
Zagazig, El Sirw, Tanta, Edfina and El Tahrir, inside and outside the Damietta
and Rosetta branches of the Nile and Port Said, Alexandria, El Kasr and Sallum,
along the Mediterranean Coast. The evaporation figures obtained from Penman's
method, from Hargreaves' method, corrected by a factor of 0.7, and from the
Piche instrument, are listed in Table 5.14. The stations at El Tahrir and El
Kasr have each been provided with a class A pan since 1964. The average pan evaporation from the period 1964-1968 is also included in Table 5.14. In this table
the evaporation data representing Cairo are for Almaza Airport. Actually, there
are four stations near, or in, Cairo itself; Almaza, Heliopolis, Ezbekiya and
Abbassiya. No Piche evaporation data are available for the last station. The
average of the first three stations is 70% of Almaza alone. Accordingly, the
Penman figure of 4.9 mm/day obtained for Almaza Airport has been reduced to 4
mm/day in order to represent the Cairo area. This last figure has been used,
among others, for preparing the map of the evaporative demand for the Nile delta
area shown in Fig. 5.9.

Fig. 5.9 shows that the evaporative demand decreases from the east to the
centre of the delta area and then increases in a westerly direction. The evaporation increases gradually from 4.4 mm/day for Alexandria to 4.9 mm/day for
Sallum on the Mediterranean Coast. The evaporation decreases from the apex of
the delta to the centre then increases in a northerly direction to the coast.

TABLE 5.14   Evaporation data for the Nile delta area from the apex to the
              Mediterranean Sea Coast

Station and Method[*]	Evaporation, in mm/day, for												
	Jan.	Feb.	Mar.	Apr.	May	June	July	Aug.	Sep.	Oct.	Nov.	Dec.	Year
Almaza (airport)													
a	2.3	3.3	4.1	6.2	7.1	7.7	7.0	6.7	4.8	4.3	3.3	1.9	4.9
b	2.8	3.7	5.3	7.9	10.7	12.0	10.8	9.8	7.0	6.5	4.1	2.7	7.0
c	7.1	7.7	8.8	11.2	14.2	14.1	12.5	10.9	9.2	8.9	7.2	5.2	9.8
Suez (Canal)													
a	2.3	3.1	4.6	5.7	7.0	7.9	7.9	7.4	6.3	5.0	3.6	2.2	5.2
b	2.8	3.5	5.3	6.4	8.5	11.0	10.8	10.2	8.4	6.3	4.1	3.2	6.7
d	5.1	5.8	7.6	10.1	12.5	13.7	13.2	12.4	10.8	8.8	6.5	5.2	9.3
Zagazig													
a	1.6	2.6	3.5	4.9	6.3	7.0	6.9	6.4	5.2	4.0	3.0	1.7	4.4
b	2.4	3.2	4.6	6.9	9.2	10.2	9.6	8.5	7.2	5.8	4.1	2.6	6.2
c	1.9	2.5	3.2	4.5	5.6	5.9	5.2	4.4	3.5	3.1	2.3	1.8	3.7
Tanta													
a	1.5	2.1	2.8	4.2	5.3	6.4	6.1	5.8	4.6	3.6	2.6	1.5	3.8
b	1.8	2.2	3.4	5.5	7.9	10.3	9.0	7.5	5.7	5.3	3.2	2.1	5.3
c	2.1	2.8	3.8	5.6	7.2	7.8	6.7	5.7	4.5	3.6	2.7	2.0	4.5
El Tahrir													
a	1.5	2.8	3.6	5.3	6.0	7.3	6.9	5.8	4.9	3.2	2.5	1.6	4.3
b	2.1	2.5	3.5	5.6	7.2	9.2	8.6	7.7	6.4	4.5	3.1	2.3	5.2
e	3.8	4.8	6.8	9.1	10.8	12.5	11.4	10.3	8.7	6.3	4.2	3.3	7.7
El Sirw													
a	1.9	2.6	3.7	4.8	6.3	7.4	7.4	6.6	5.6	4.5	3.1	1.7	4.5
b	1.9	2.3	2.9	4.0	5.7	7.0	7.4	6.2	5.0	3.8	2.5	2.0	4.2
c	2.9	3.5	4.4	5.3	6.0	6.3	6.1	5.2	4.7	4.2	3.3	2.7	4.6
Edfina													
a	1.6	2.1	3.1	4.3	5.2	6.7	6.9	6.2	4.6	3.3	2.4	1.5	4.0
b	1.8	2.2	3.0	4.3	5.5	7.1	7.0	5.9	4.7	3.9	2.7	1.8	4.2
c	3.1	3.5	4.6	5.9	6.9	7.2	7.0	6.5	5.6	4.6	3.8	3.0	5.1
Port Said (airport)													
a	2.1	2.8	3.7	5.2	6.2	7.0	7.1	6.6	5.5	4.1	3.2	2.1	4.6
b	2.1	2.7	3.1	3.7	4.5	4.9	5.7	5.8	5.4	5.0	2.9	2.1	4.0
c	5.9	6.2	7.7	7.4	8.5	8.8	9.2	9.1	10.0	9.3	7.8	5.5	8.0
Alexandria													
a	2.3	3.0	3.6	4.8	5.3	6.8	6.6	5.6	5.5	4.0	3.2	2.1	4.4
b	2.4	2.8	3.4	4.2	4.6	5.3	5.0	5.4	5.1	4.3	3.1	2.5	4.0
c	4.8	5.2	5.0	5.3	5.0	4.7	4.7	5.0	5.5	5.2	4.9	4.9	5.0
El Kasr													
a	1.9	2.6	3.6	4.8	5.9	6.9	8.3	6.6	5.4	3.6	2.6	1.7	4.7
b	1.7	2.1	3.3	3.7	4.4	5.5	7.4	5.5	4.6	3.3	2.2	1.7	3.8
e	4.8	5.8	7.0	8.4	8.2	10.4	11.0	10.6	9.6	7.8	5.4	5.1	7.8

[*]a = Penman; b = Hargreaves x 0.7; c = Piche; d = Piche (Port Tewfik);
e = Class A pan

TABLE 5.14   (continued)

Station and Method[*]	Evaporation, in mm/day, for												
	Jan.	Feb.	Mar.	Apr.	May	June	July	Aug.	Sep.	Oct.	Nov.	Dec.	Year
Sallum (Observatory)													
a	2.4	3.0	4.1	5.4	5.6	7.6	8.5	7.6	5.7	3.9	3.1	2.3	4.9
b	2.3	2.8	3.6	4.8	4.5	7.5	8.6	6.6	5.3	4.2	3.1	2.6	4.7
c	6.6	7.3	7.9	8.9	7.9	9.4	9.7	8.2	7.9	7.2	6.9	7.1	7.9

[*] a = Penman; b = Hargreaves x 0.7; c = Piche

In other words, the evaporation from the delta has a bowel shape with decreasing evaporation towards the centre.

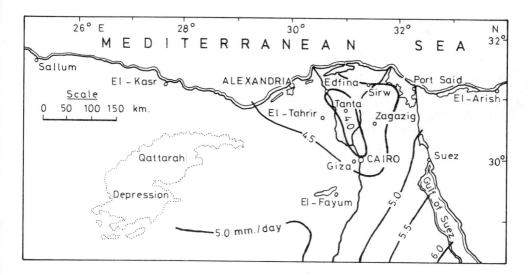

Fig. 5.9.   Map of the evaporative demand for the Nile delta area and the coast of the Mediterranean Sea

5.6   ANALYSIS OF THE OBSERVED AND THE COMPUTED EVAPORATION FOR THE NILE BASIN

General

In many instances in the previous sections, the observed Piche evaporation at a number of stations in the Nile Basin is given. This evaporation has been multiplied by the reduction factor proposed by Hurst to deduce the open water evaporation for the whole basin (see Fig. 5.1).

The evaporation measured from the USWB class A pan at four locations in Egypt is given in sections 5.5.2 and 5.5.3. The supposed class A pan evaporation has been computed for nearly all the stations in the basin, using the Hargreaves formula. The results have been reduced by a factor 0.7 in order to convert the pan evaporation into free water surface evaporation.

The evaporation from the so-called 'standard tank' has been computed, but for a fewer number of stations using Olivier's method.

The free water surface evaporation has been computed directly using the Penman formula for most of the stations where the relevant climatological standards are available. The results are presented not only in tabular form, but also in maps for quicker and easier use. The maps shown in Figs. 5.3., 5.6., 5.8., and 5.9. have been combined together to compose the general map, Fig. 5.10.

5.6.1   The free water surface evaporation computed by Penman's formula

For the purpose of estimating the open water evaporation for the Nile Basin, the direct method that has been used is that of Penman. The necessary equations have been given earlier. The new or modified version of the Penman equation employs a wind function other than that given in eq. 5.8 which we used in our calculation.

When combined with the findings of Hickox, the experiments conducted at the Khartoum Observatory in 1961 and 1962 showed that the mean annual open water evaporation was about 10% larger than that obtained from Panman's method. If we trust the results of those experiments, the computed evaporation from Penman's formula needs then to be adjusted. The monthly and yearly adjustment factors are as follows:

	Adjustment Factor
January	1.106
February	1.078
March	1.087
April	1.051
May	1.242
June	1.110
July	0.924
August	0.955
September	1.078
October	1.165
November	1.142
December	1.083
Year	1.092

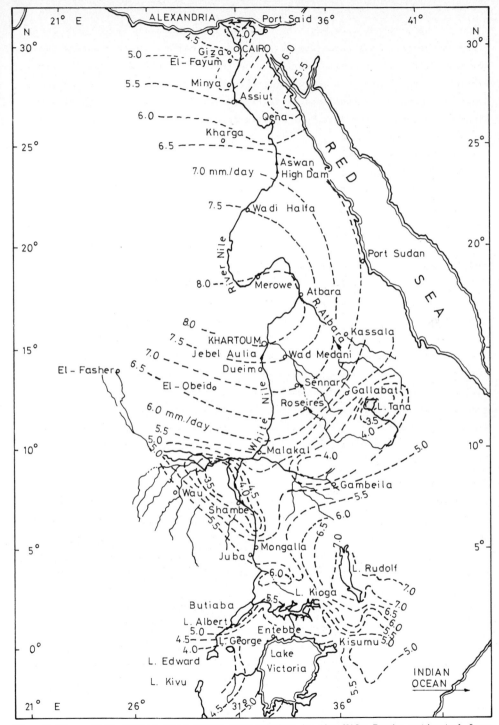

Fig. 5.10. Mean annual free water evaporation in the Nile Basin estimated from Penman's method

To the best of our knowledge the results of the Khartoum experiments are probably the only ones which may help to judge the accuracy of Penman's formula for estimating the free water surface evaporation from the Nile Basin. Accordingly, the results presented in the previous section either in tabular or graphical form may be described as fairly accurate. More accurate evaporation figures may be obtained using our results together with the appropriate adjustment.

5.6.2   Relationship between estimated evaporation from a class A pan and estimated free water surface evaporation

In order to estimate evaporation from a class A pan, the method of Hargreaves has been used. The estimated pan evaporation has been reduced by a constant factor, 0.7, so as to deduce the open water evaporation. Most of the evaporation figures thus obtained have proved to be larger than the free water surface evaporation estimated directly from Penman's method. This discrepancy leads us to the conclusion that either the reduction factor, 0.7, is, for most of the stations considered, larger than it should be, or the Hargreaves evaporation is not equal to the pan evaporation.

The simple graphical plot of the Hargreaves evaporation (without any reduction), Y, versus the Penman evaporation, X, for 30 different stations, suggests the possibility of having the pairs X and Y divided into four groups. The first group comprises 18 interior stations. The second group is for the three stations along the Red Sea Coast and the third is for the six stations at, or near to, the coast of the Mediterranean Sea. The fourth group comprises the three stations located outside the swamps of the Bahr el Jebel and Bahr el Ghazal Basins. This grouping has the advantage that it reduces the scatter of the points about the respective regression line, though does not eliminate it. Further grouping may still reduce further the scatter.

In view of the rather approximate nature of the climatological data used for estimating the evaporation, at least for some of the stations, it has been found unnecessary to describe the regression of Y on X by a polynomial of a degree higher than 2. Under this condition the regression equations of the four groups are:

1st group $Y = 1.839 + 0.924\ X + 0.125\ X^2$ $(r_{XY} = 0.951)$ (5.12a)

2nd group $Y = 1.239\ \mathrm{x}\ 1.331\ X + 0.020\ X^2$ $(r_{XY} = 0.950)$ (5.12b)

3rd group $Y = 1.222 + 0.822\ X + 0.041\ X^2$ $(r_{XY} = 0.953)$ (5.12c)

4th group $Y = -6.271 + 1.489\ X + 0.187\ X^2$ $(r_{XY} = 0.947)$ (5.12d)

where X and Y are in mm/day and r_{XY} is the correlation coefficient between X and Y.

The fit of the regression lines given by eq. 5.12 to the respective group of stations is shown in Fig. 5.11a. thru' c.

Estimation of the free water surface evaporation by the method of Penman has already been described by the equations 5.6, 5.7, and 5.8. It requires a knowledge of the temperature, t, the mean relative humidity, h, the short-wave radiation, R_A, the wind speed, u, and the relative duration of the bright sunshine, $\frac{n}{D}$. Estimation of evaporation from a class A pan using the Hargreaves method, eq. 5.9 requires a knowledge of the temperature, t, the relative humidity at noon, h_n, and a coefficient, d, which depends on the duration of the day-time. Correction of the figures obtained from eq. 5.9 requires, however, a knowledge of the sunshine, $\frac{n}{D}$, the wind speed, u, and the elevation of the station above the mean level of the sea. Since R_A, in the case of Penman's method, and d, in the case of Hargreaves' method, are available for a given month and the geographical latitude of any point, the basic difference in the climatological parameters needed for these two methods is confined to h and h_n. Both humidities are recorded at every meteorologic screen, though h_n at a slightly fewer number of screens.

It is quite apparent that the value of having the pan evaporation first computed by the Hargreaves method then converted into a free water surface evaporation, using either Fig. 5.11., or Table 5.13, is questionable!

5.6.3   <u>Relationship between observed evaporation from a class A pan and</u>
        <u>estimated free water surface evaporation</u>

From the stations for which the free water surface evaporation has been estimated by the Penman method, there are five stations only for which class A pan evaporation is available. These are: El Kharga oasis, El-Tahrir, El Kasr and Giza in Egypt and Khartoum Observatory in the Sudan.

The monthly and yearly ratios of the Penman evaporation to the pan evaporation for these ratios are listed in Table 5.15, from which is evident that the ratio, E penman : E pan A (pan coefficient) varies from one month to the other. The largest value of the pan coefficient occurrs in July for the three northern stations and in August for the two southern stations. The ratio of the maximum pan coefficient to the minimum pan coefficient varies considerably from one station to the other. For the five stations in their tabulated order, this ratio is 2.27, 1.53, 1.42, 1.16 and 1.37 respectively. The mean coefficient of the pan also varies from one location to the other. The mean annual relative humidity for the stations in their tabulated order is 72.3, 66.8, 71.7, 31.4 and 32.1, respectively. The mean annual value of the pan coefficient and the amplitude of the mean monthly coefficient both seem to some extent, to be dependent on the mean annual humidity.

210

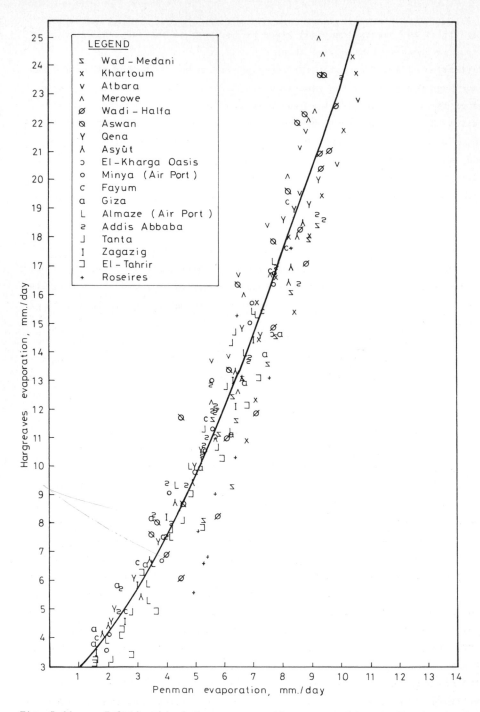

Fig. 5.11a.　Relationship between evaporation computed by the Hargreaves formula and evaporation computed by the Penman formula for the interior stations in
the Nile Basin

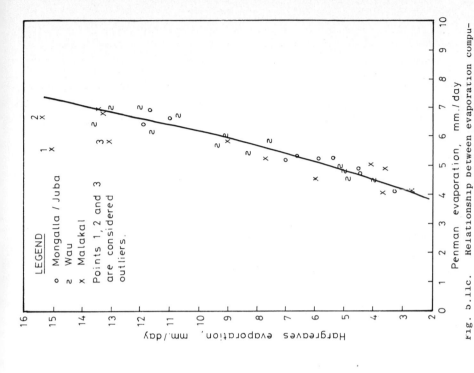

Fig. 5.11c. Relationship between evaporation computed by Hargreaves' formula and that by Penman's formula for stations just outside the swamps of the Bahr
el Jebel Basin

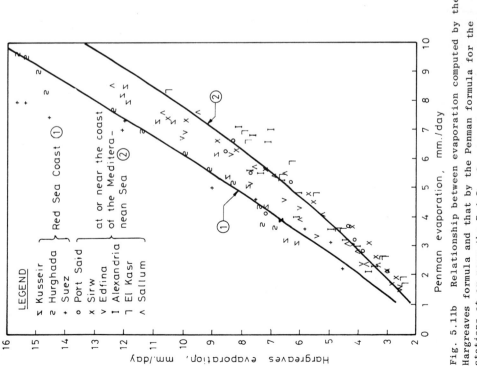

Fig. 5.11b Relationship between evaporation computed by the
Hargreaves formula and that by the Penman formula for the
stations at or near the Red Sea Coast and the Mediterranean
Sea Coast

TABLE 5.15   Monthly and yearly ratios of Penman evaporation to Class A pan
              evaporation

Station	E penman : E pan A, for												
	Jan.	Feb.	Mar.	Apr.	May	June	July	Aug.	Sep.	Oct.	Nov.	Dec.	Year
El-Kasr	.396	.448	.414	.571	.720	.663	.755	.623	.563	.462	.481	.333	.603
El-Tahrir	.395	.583	.529	.582	.555	.584	.605	.563	.563	.508	.595	.485	.558
Giza	.536	.667	.569	.642	.650	.693	.712	.653	.617	.574	.561	.500	.629
Kharga Oasis	.375	.412	.433	.429	.430	.404	.424	.436	.423	.396	.364	.353	.415
Khartoum	.556	.532	.540	.490	.450	.525	.643	.670	.583	.532	.518	.546	.538

The observed pan evaporation, E_p, has been plotted versus the free water surface evaporation estimated from the method of Penman, X. The polynomial regression equations fitted to the plotted points are:

El-Kasr
$$E_p = 2.082 + 1.652 \, X - 0.068 \, X^2 \qquad (r_{Ep, \, X} = .963)$$

El-Tahrir
$$E_p = 0.856 + 1.592 \, X - 0.005 \, X^2 \qquad (r_{Ep, \, X} = .993)$$

Giza
$$E_p = 0.520 + 1.687 \, X - 0.040 \, X^2 \qquad (r_{Ep, \, X} = .993) \quad (5.13)$$

Kharga Oasis
$$E_p = 2.407 + 1.696 \, X + 0.046 \, X^2 \qquad (r_{Ep, \, X} = .995)$$

Khartoum
$$E_p = 15.613 - 2.667 \, X + 0.316 \, X^2 \qquad (r_{Ep, \, X} = .945)$$

where E_p and X are in mm/day.

These regression relations when fitted to the respective pairs of E_p and X
give the lines shown in Fig. 5.12. Though eqs. 5.13 are somewhat different from
those developed earlier by the author while analyzing the evaporation pan data
in Egypt (Shahin, M.M., 1970), both sets of equations, for the usual range of X,
still yield almost the same values of E_p.

5.6.4   Relationship between class A pan evaporation and Piche evaporation

The ratio class A pan evaporation to Piche evaporation for Giza, Kharga and
Khartoum stations is given in Table 5.16. The same ratio, averaged over the
period from June 1954 up to and including May 1960 for the Lod Airport, Israel
(WMO, 1966) is also included in Table 5.16 for comparison. The conclusion one
can draw from the figures listed in this table is that the mean annual ratio of
class A pan to Piche evaporimeter in screen varies from about 0.8 for Kharga and
Aswan (mean relative humidity about 32%) to 1.03 for Giza (mean relative humidity about 72%). Even for the Lod, which is situated outside the Nile Basin,

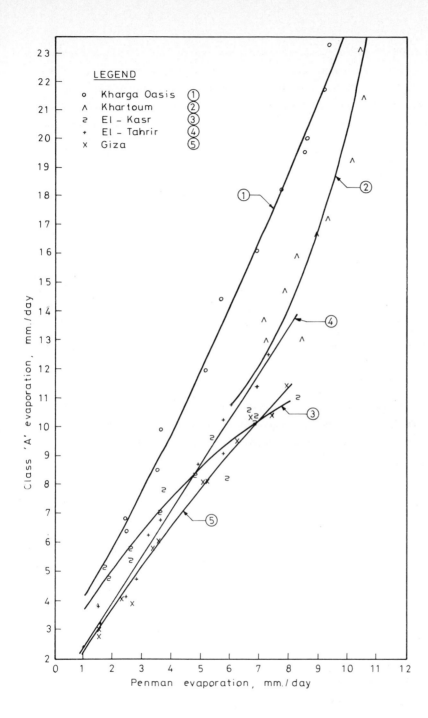

Fig. 5.12.   Relationship between observed class A pan evaporation and free water surface evaporation estimated from the Penman method

the ratio is 1.06. This ratio as obtained, but indirectly, from the 2-year evaporation studies in Kenya and Uganda is 3.3 for Entebbe, 1.6 for both Equator
and Gulu, and 1.47 for Kisumu. The Piche evaporation at Entebbe has been
reported as being abnormal. If we neglect this station, we will still be left
with the rather high ratio for the other stations. This supports our opinion
that the Piche readings in the Equatorial Lakes area are generally small and
much more variable than are tanks.

TABLE 5.16   Monthly and annual ratios of class A pan evaporation to Piche
evaporation

Station	E pan A : E piche, for												
	Jan.	Feb.	Mar.	Apr.	May	June	July	Aug.	Sep.	Oct.	Nov.	Dec.	Year
Giza a[*]	.54	.57	.60	.59	.61	.67	.73	.74	.69	.62	.58	.53	.64
b[*]	.92	.95	1.01	.98	.99	1.07	1.20	1.24	1.12	.98	.92	.86	1.03
Kharga	.67	.74	.74	.79	.83	.93	.89	.82	.81	.73	.72	.69	.80
Khartoum	.75	.74	.76	.77	.82	.87	1.01	1.09	.90	.79	.73	.75	.82
Lod (Airport)	.73	.88	1.00	1.05	1.13	1.29	1.35	1.34	1.15	.94	.74	.68	1.06

[*]a = Piche unscreened; b = Piche screened

The results obtained from the Namulonge Cotton Research Station, Uganda, are
quite impressive and illuminating. A graphical plot of these results has been
presented by Olivier ((1961) and reproduced here in Fig. 5.13.

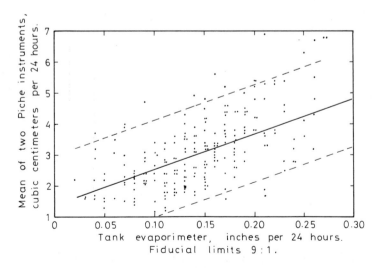

Fig. 5.13. Comparison of readings of Piche and tank evaporimeters at
Namulonge Cotton Research Station, Uganda

5.6.5   Relationship between Piche evaporation and estimated free water surface
         evaporation

We have already mentioned at the beginning of this chapter that the Piche
evaporimeter does not, strictly, present "free water surface" conditions. The
reasons behind this are, as given by Olivier (1961), that the Piche evaporation
takes place from a porous blotting paper and therefore the surface from which
water evaporates is not as a uniform "free" water surface. Furthermore, radiation absorption is probably different for this type of evaporimeter compared
with a plain water surface. Additionally, the area of the evaporation surface
is infinitely small compared with any surface area of a reservoir or a lake.

However, in areas where only Piche readings are available, these have to be
used, with suitable corrections. In other words, the Piche evaporation has to
be multiplied by a coefficient in order to bring it to its equivalent of a free
water surface. For this purpose, Table 5.17 has been prepared to give the
monthly and the annual Piche coefficients for 30 stations in the Nile Basin.
For other stations, especially in Uganda and Kenya, where the available data
allow for an indirect computation of the open water evaporation, only the
annual coefficient of each station has been estimated as follows:

Station	Piche Coefficient	Station	Piche Coefficient
Gambeila	0.78	Masindi Port	1.20
Akobo	0.84	Jinja	1.30
Gulu	1.07	Entebbe	1.85
Eldoret	1.10	Kampala	1.35
Soroti	1.12	Kisumu	1.03
Butiaba	1.17	Narok	1.00

The areal distribution of the mean annual Piche coefficient over the basin
of the Nile can be seen from the map in Fig. 5.14. Examination of this map
shows that the mean annual value of the Piche coefficient changes in two principal directions. The first direction is south-north where the coefficient
decreases from a probable maximum of, say, 1.3 around the northern shore of
Lake Victoria (neglecting the readings at Entebbe) to a minimum of about 0.35
for El Kharga Oasis in Egypt. This is followed by an increase in the Piche
coefficient up to about 0.9 which is reached at, or near, the coastal line of
the Mediterranean Sea between Damietta and Alexandria.

The second direction is east-west where the Piche coefficient decreases from
about 0.65 to about 0.50 for Khartoum, Merowe, Halfa and Aswan on the Main Nile
and to 0.35 further to the west for El Kharga Oasis. This pattern is interrupted
by the rather high coefficient in the strip between Qusseir on the Red Sea
Coast and Qena on the Nile River. This interruption is probably due to the
several torrents and Wadi's which intersect this strip. The upheaval in the

TABLE 5.17 Monthly and annual Piche coefficients for the Nile Basin

$E_{Penman} : E_{Piche}$, for

Station	Jan.	Feb.	Mar.	Apr.	May	June	July	Aug.	Sep.	Oct.	Nov.	Dec.	Year
Sallum	.364	.411	.519	.607	.709	.809	.876	.927	.722	.542	.449	.324	.620
Alexandria	.479	.577	.720	.906	1.060	1.447	1.404	1.120	1.000	.769	.653	.429	.880
Port Said	.356	.452	.481	.703	.729	.795	.772	.725	.550	.441	.410	.382	.575
Edfina	.516	.600	.674	.729	.754	.931	.986	.954	.821	.717	.632	.500	.784
Sirw	.655	.743	.841	.906	1.050	1.175	1.213	1.270	1.191	1.071	.940	.630	.978
Tanta	.714	.750	.737	.750	.736	.821	.910	1.018	1.022	1.000	.963	.750	.844
Zagazig	.842	1.040	1.094	1.089	1.125	1.186	1.327	1.455	1.486	1.290	1.304	.944	1.189
Giza	.455	.619	.623	.722	.736	.859	.881	.857	.909	.729	.657	.517	.746
Almaza (Airport)	.324	.429	.466	.554	.500	.546	.560	.615	.522	.483	.458	.365	.500
Suez	.451	.534	.605	.564	.560	.577	.598	.597	.583	.568	.554	.423	.559
Fayum	.441	.591	.650	.628	.644	.617	.695	.723	.688	.615	.667	.500	.645
Minya (Airport)	.452	.456	.535	.541	.493	.507	.562	.585	.644	.562	.600	.541	.600
Hurghada	.346	.383	.412	.452	.472	.442	.492	.455	.438	.453	.452	.365	.441
Asyût	.462	.608	.566	.611	.606	.558	.612	.641	.604	.690	.694	.541	.604
El Kharga (Oasis)	.253	.302	.319	.340	.355	.375	.376	.356	.347	.288	.263	.242	.332
Qena	.636	.690	.769	.717	.771	.742	.754	.702	.837	.852	.841	.647	.753
Qusseir	.388	.696	.531	.620	.646	.636	.739	.672	.652	.574	.500	.405	.602
Aswan	.461	.505	.481	.458	.459	.435	.477	.458	.453	.419	.417	.507	.463
Wadi Halfa	.511	.532	.493	.481	.479	.455	.495	.531	.473	.484	.521	.465	.487
Merowe	.491	.474	.452	.445	.453	.467	.505	.557	.479	.466	.479	.487	.476
Atbara	.399	.383	.397	.510	.447	.497	.497	.534	.518	.604	.439	.415	.465
Khartoum	.518	.446	.466	.497	.558	.536	.556	.550	.601	.575	.535	.499	.536
Kassala/Gallabat	.598	.562	.551	.578	.577	.635	.651	.688	.590	.640	.635	.594	.632
Wad Medani	.406	.382	.382	.428	.436	.444	.566	.803	.887	.658	.452	.418	.476
Sennar	.530	.573	.531	.595	.579	.699	.832	.819	.758	.729	.640	.551	.573
Singa	.535	.494	.514	.545	.583	.695	.844	.831	.760	.731	.601	.546	.591
Roseires	.463	.442	.494	.513	.576	.762	1.000	1.000	.914	.809	.648	.509	.594
Kurmuk	.459	.434	.490	.508	.573	.771	1.010	.995	.903	.810	.649	.503	.581
Malakal	.333	.363	.433	.606	.829	1.023	1.380	1.640	1.656	1.351	.627	.426	.600
Wau	.538	.555	.583	.713	.937	1.106	1.216	1.371	1.282	1.095	.701	.598	.763
Mongalla/Juba	.520	.575	.680	.784	1.061	1.093	1.243	1.324	1.113	.963	.746	.620	.786

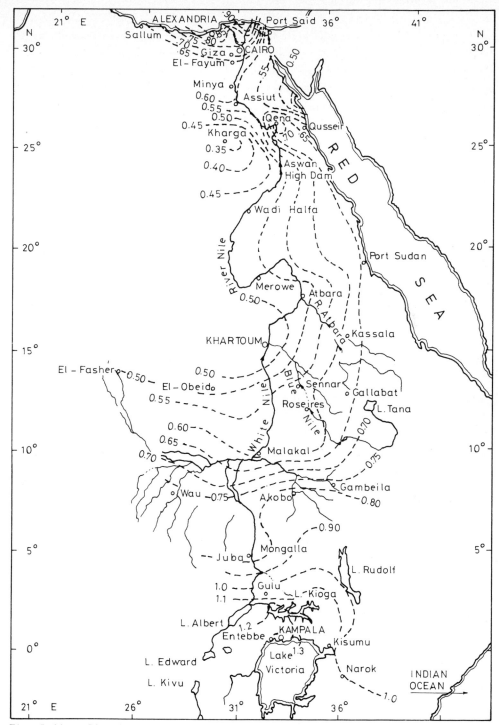

Fig. 5.14. Distribution of the mean annual coefficient for converting the Piche evaporation to free water evaporation using Penman's method

value of the Piche coefficient around Qena to the west is counter-balanced by
the drop around Hurghada on the coast of the Red Sea between Qusseir and Suez.
There is a rapid increase in the value of the coefficient along the Mediterranean Coast from Port Said, in the east, to Alexandria, in the west. This
increase may be due to the high relative humidity brought up locally by the
continuous irrigation and the very many canals in the Nile Delta. West of
Alexandria in the direction of Sallum, the Piche coefficient decreases from
about 0.88 for the former to about 0.62 for the latter. In general, the mean
monthly coefficient of the Piche evaporimeter undergoes a seasonal cycle. The
amplitude of this cycle varies from one station to the other. Based on the figures listed in Table 5.17 the maximum value of the coefficient occurs mostly in
July and August. In 84% of the stations investigated the peak has been found to
take place in the three-month period July-September. The minimum value of the
monthly coefficient occurs for 90% of the stations in the three-month period
December-February. The frequency histograms of the occurrence of the maximum
and the minimum values of the Piche coefficient for all stations are shown in
Figs. 5.15a and 5.15b, respectively.

The ratio of the maximum to the minimum, excluding the very high value at
Malakal, varies from 1 to 3.5. The frequency distribution of this ratio for
the stations investigated is shown in Fig. 5.15c. Most of the stations that
enjoy a fairly stable value for the Piche coefficient, i.e. the ratio of the
maximum monthly to the minimum monthly coefficient is in the range of from 1.0
to 1.5, are those on the Nile River in the arid zone. The less stable ratio,
from 1.5 to 2.0, takes place in those stations in the Nile Delta and Valley
areas where intensive irrigated agriculture is practised. The least stable
ratio, i.e. larger than 2.0 is a characteristic of the stations at, or close to,
the Mediterranean Coast and just outside the swamps of the Bahr el Jebel, Bahr
el Ghazal and the Sobat Basins.

In order to illustrate the pattern of variation of the Piche monthly coefficient, the figures listed in Table 5.17 have first been converted to modular
values. This is done by dividing the coefficient for each month by the mean
annual coefficient for the station considered.

Fig. 5.16 shows smooth curves fitted to the monthly coefficients versus the
months of the year for six stations only as an example. These curves present
other characteristics in addition to the seasonal cycle, the months in which
the maximum and the minimum values occur, and the ratio of the maximum to the
minimum.

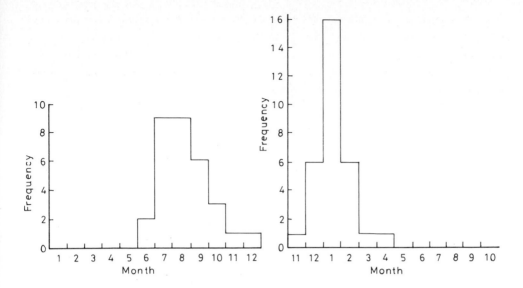

Fig. 5.15a. Frequency histogram of the month of occurrence
of the maximum value of the
Piche coefficient

Fig. 5.15b. Frequency histogram of the
month of occurrence of the minimum value
of the Piche coefficient

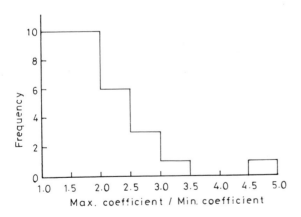

Fig. 5.15c. Frequency histogram of the ratio of the maximum monthly Piche
coefficient to the minimum monthly Piche coefficient

Curves similar to those for Khartoum and Minya Airport indicate the presence
of two cycles instead of one. The rising limbs of the curves approach, or equal,
the modular value 1.0 between the end of March and mid-April. The same modular
value is equalled or approached by the falling limbs of the curves in the period
between mid-September and mid-November.

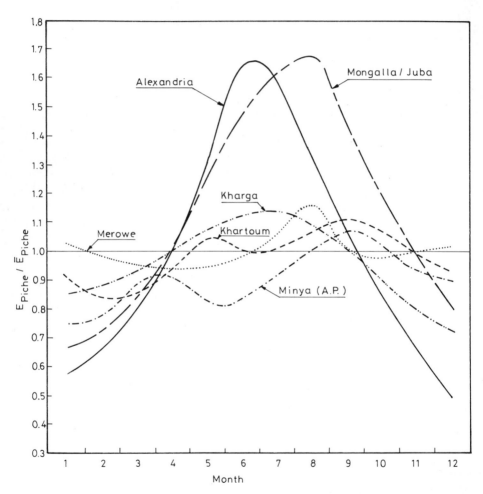

Fig. 5.16. Variation of the modular value of the Piche coefficient with the
month of the year for some evaporation stations in the Nile Basin

REFERENCES

Baumgartner, A., and Reichel, E., 1975. The world water balance, R. Oldenburg
    Verlag, Munich.
Dagg, M., 1972. East Africa: its peoples and resources (edited by W.T.W. Morgan).
    Chapter 10: Water requirements of crops, 119-125. Oxford University Press,
    London, New York.
Hargreaves, G.H., 1956. Irrigation requirements based on climatic data. Paper
    No. 1105 Journ. Ir. & Dr. Div., ASCE, Vol. 82, No. IR-3: 1-10.
Hickox, G.H., 1946. Evaporation from a free water surface. Trans. Amer. Soc.
    Civ. Engrs., 111, Paper No. 2266, 1-33.
Hurst, H.E., 1950. The Nile Basin, Vol. VIII: The hydrology of the Sobat and
    White Nile and the topography of the Blue Nile and Atbara. Physical Depart-
    ment paper No. 55, The Government Press, Cairo, Egypt, 125 pp.
Hurst, H.E., 1952. The Nile, a general account of the river and the utilization
    of its waters. Constable, London, 326 pp.
Hurst, H.E., Black, R.P., and Simaika, Y.M., 1959. The Nile Basin, Vol. IX: The
    hydrology of the Blue Nile and Atbara and of the main Nile at Aswan, with
    some reference to projects. Nile Control Department Paper No. 12, The General
    organization for Government Printing Offices, Cairo, Egypt, 206 pp.
Hurst, H.E., Black, R.P., and Simaika, Y.M., 1966. The Nile Basin, Vol. X, The
    major Nile projects. Nile Control Department Paper No. 23, General Organiza-
    tion for Government Printing Offices, Cairo, 217 pp.
Hurst, H.E., and Philips, P., 1931. The Nile Basin, Vol. I, General description
    of the basin, meteorology and topography of the White Nile Basin. Physical
    Department Paper No. 26, Government Press. Cairo, 128 pp.
Hurst, H.E., and Philips, P., 1938. The Nile Basin, Vol. V, The hydrology of the
    Lake Plateau and Bahr el Jebel. Physical Department Paper No. 35, Schnidlers'
    Press. Cairo, 235 pp.
Krishnamurthy, K.V., and Ibrahim, A.H., 1973. Hydrometeorological studies of
    Lakes Victoria, Kyoga and Albert, Geophysical nomograph 17: Man-Made Lakes
    (Ackermann, W.C., et al: editors), A.G.U., Washington D.C., 272-277.
Linsley, R.K., Kohler, M.A., and Paulhus, J.L., 1958. Hydrology for engineers.
    McGraw-Hill Book Company Inc., New York, London, 340 pp.
Ministry of War and Marine, Egypt, 1950. Climatological normals for Egypt.
    Meteorological Department of Egypt. C, Tsoumas & Co. Press, Cairo.
Olivier, H., 1961. Irrigation and Climate, Edward Arnold (publishers) Limited,
    London, 250 pp.
Palayasoot, P., 1965. Estimation of pan evaporation and potential evapotranspi-
    ration of rice in the central plain of Thailand by using various formulas
    based on climatological data. M.Sc. thesis presented to the Utah State
    University, Utah, U.S.A.
Penman, H.L., 1948. Natural evaporation from open water, bare soil and grass.
    Proceeding Roy. Soc. Agric., 193: 120-145.
Rijks, D.A., 1969. Evaporation from a papyrus swamp. Quart. Journ. Roy. Meteo.
    Soc., 95: 643-649.
Rzóska, J. (editor), 1976. The Nile, Biology of an ancient river. Dr W. Junk
    B.V. Publishers, The Hague, The Netherlands, 417 pp.
Shahin, M.M., 1970. Analysis of evaporation pan data in U.A.R. (Egypt). Annual
    Bulletin of ICID, New Delhi, India, 53-69.
Sleight, R.B., 1927. Discussion of Houk's paper: Evaporation on United States
    reclamation projects. Trans. Amer. Soc. Civ. Engrs., 90, 303-316.
US Geological Survey, 1954. Water loss investigations: Vol. I - Lake Hefner
    studies. Geological Survey Professional Paper No. 269.
USSR National Committee for IHD, 1974. World water balance and water resources
    of the earth (translated from Russian), UNESCO, Paris.

222

WMO, 1966. Measurement and estimation of evaporation and evapotranspiration. Technical note No. 83, Geneva, 121 pp.

WMO, 1974. Hydrometeorological survey of the catchments of Lakes Victoria, Kyoga and Albert, Vol. I: Meteorology and hydrology of the basin, Parts 1 & 2, Geneva.

Chapter 6

EVAPOTRANSPIRATION

Evapotranspiration is the total quantity of water consumed by evaporation
and transpiration. Evaporation has already been defined in Chapter 5. Transpiration is simply evaporation from the plant. It is a process by which water vapour
is released to the atmosphere through surface pores in plant foliage, mainly
through stomatal openings. A small portion of the emitted moisture, generally
less than 10%, comes from the younger plant stems. Release of moisture by transpiration ocurs principally during the day-time hours of the growing season.
Probably not more than 5 to 10% of the daily transpiration takes place during
the night. The rate of transpiration usually reaches a maximum value shortly
after noon, and a minimum just before sunrise.

Evapotranspiration can be defined as the sum of the volumes of water used
per unit area by the vegetative growth in transpiration and that evaporated
from the soil, snow or intercepted precipitation on a given area in any specified time. Evapotranspiration is usually expressed in units of depth per unit
of time. The term consumptive use, commonly used in irrigation hydrology, is
essentially synonymous with evapotranspiration. The definition of evapotranspiration implies that the factors affecting it are essentially those affecting
evaporation and transpiration. So the rate of evapotranspiration depends on the
climate, crop raised, stage of plant development, density of vegetative cover,
soil moisture supply, salinity and length of growing season. Factors included
in climate are: solar radiation, temperature, day-time hours, duration of sunshine, humidity and wind speed.

When evapotranspiration takes place from a soil surface completely covered by
actively growing vegetation and there is no limitation in soil moisture, it is
referred to as potential evapotranspiration. In other words, potential evapotranspiration can be considered as the upper limit of the actual evapotranspiration. According to Thornthwaite, potential evapotranspiration depends only on
the amount of solar energy received by the earth's surface, consequently the
resulting temperature, and not on the kind of plant (Thornthwaite, C.W., 1948).

The most common methods used for determining the evapotranspiration are the
water-balance, the tank and lysimeter experiments, soil moisture depletion
studies in field plots, correlation with evaporation from a pan or from an open
water body, and methods based on the physics of the vapour transfer and/or the
heat energy-balance.

A general idea about the actual and potential evapotranspiration from the
Nile Basin can be drawn from at least two sources. Reichel and Baumgartner
(1975) gave the actual evapotranspiration for the different 5-degree latitude
zones of the globe, which was based on the simple equation

$$ET = P - R \tag{6.1}$$

where
ET = actual evapotranspiration,
P = precipitation, and
R = run-off

The three variables, ET, P, and R, are all expressed in a depth unit per
year. For the 5-degree latitude zones covered by the Nile Basin the values of
these variables are as follows:

	Latitude, degree	Precipitation, P, mm/yr	Run-off, R, mm/yr	Actual evapotranspiration, ET, mm/yr
north	35-30	151	-7	158
	30-25	29	-5	34
	25-20	31	1	30
	20-15	151	5	146
	15-10	741	55	686
	10- 5	1203	233	970
	5- 0	1329	342	987
south	0- 5	1488	343	1145

The second source of information is the two maps, Figs. 6.1. and 6.2., which
give the actual evapotranspiration and the potential evapotranspiration redrawn
from the Water Resources of the Earth (Korzun, V., et al, 1978). Evidently
these two maps give more detailed information than the very general figures
given by Reichel and Baumgartner. As an example, the 5-degree latitude zone
extending from the equator up north shows an actual evapotranspiration increasing from, say, 300 mm/yr in the east to about 1250 mm/yr in the west at the
shore of Lake Edward. For this zone, the above tabulated figures give an actual
evapotranspiration of 987 mm/yr.

The countries sharing the Nile water have been using several but also different practices to determine or to estimate evapotranspiration. It may, therefore,
be wiser to give an account of the practices used and the results obtained
therefrom on a country-wise basis instead of a river basin-wise basis.

225

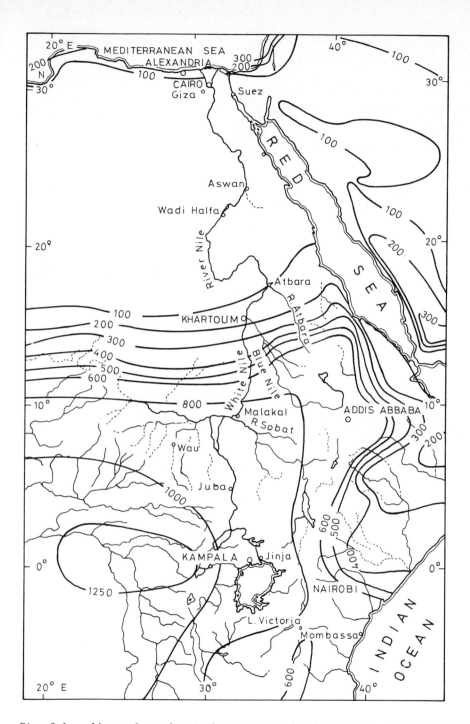

Fig. 6.1.   Lines of equal actual evapotranspiration, mm/yr, from the Nile Basin
and surroundings (redrawn from the World Water Balance and Water Resources of
the Earth (Korzun, V., et al, 1978)

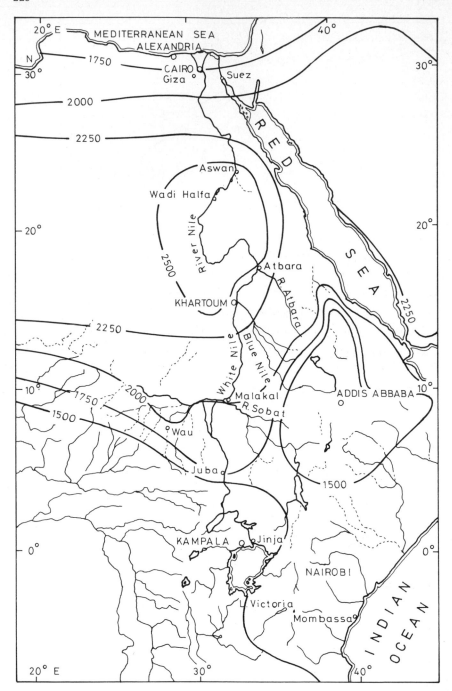

Fig. 6.2.   Lines of equal potential evapotranspiration, mm/day, from the Nile Basin and surroundings (redrawn from the World Water Balance and Water Resources of the Earth (Korzun, V., et al, 1978)

6.1   EVAPOTRANSPIRATION STUDIES IN THE ARAB REPUBLIC OF EGYPT

6.1.1   Tank and lysimeter experiments

The results of tank experiments at the agrometeorological station at Giza have been reported by Omar, M.H. (1960). During the 3-year period, 1957-1959, potential evapotranspiration was measured from three evapotranspirometers of the type known as "modified evapotranspirometer" developed by Mather. This was a tank with an opening at the bottom where a tube was fixed in order to enable the measurement of percolation water. Two tanks were cylinderical, about 60 cm in diameter each, and the third tank had a rectangular cross-section 130 x 90 cm. The tanks were planted with libya grass and installed in the station's grass field. Care was taken to maintain the same vegetation level inside and outside the tanks by cutting the grass when necessary. The difference in area of the evaporating surface of the tanks used was reported to be of no significant influence. Results generally bore this out.

Other measurements were made by Popoff evapotranspirometers which have a smaller area (500 cm^2) compared to Mather's evapotranspirometers. The former were reported to give results almost 10% larger than those obtained from the bigger tanks. A comparison between the evapotranspiration from the small tanks and the bigger tank, all planted with cotton, can be seen from Fig. 6.3.

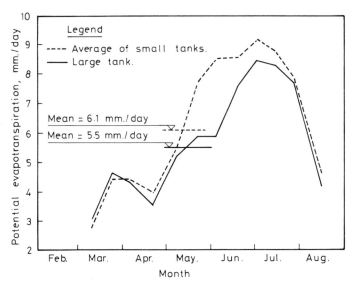

Fig. 6.3.   Comparison between potential evapotranspiration from small and large tanks raising cotton at Giza (Omar, M.H., 1960)

Zein el Abedin et al (1967) compared the evapotranspiration from field plots to that from weighable lysimeters, all raising late maize, under two different irrigation treatments at the experimental farm of Cairo University. The soil moisture changes were detected by a neutron moisture metre. A summary of some of the results obtained from this investigation is presented in Table 6.1. From this summary it is evident that the evapotranspiration from the lysimeters as well as the field plots responds to the frequency of irrigation, consequently the available soil moisture. The evapotranspiration from the lysimeter was 12.7% larger than that from the field plot in the treatment that received more frequent irrigation, and 22.9% in the other treatment. These differences in the amount of evapotranspiration resulted in an increase in the crop yield in the lysimeter than in the field plot by 10.5 and 5.2% for treatments 1 and 2 respectively.

TABLE 6.1 Evapotranspiration from lysimeters and field plots raising maize under two different irrigation treatments, Faculty of Agriculture Experimental Farm, Cairo University, Egypt (Z. el Abedine, A., Abdallah, M., and Abdel-Samie, 1967)

	Treatment 1				Treatment 2		
Year 1966	depth of root zone cm	evapo-transpiration, mm		Year 1966	depth of root zone cm	transpiration, mm	
		lysimeter	plot			lysimeter	plot
24.07-01.08	40	34.4	31.8	24.07-07.08	40	52.6	48.8
02.08-09.08	60	40.8	39.2	08.08-20.08	80	50.6	64.2
10.08-20.08	80	57.6	59.2	21.08-07.09	100	75.2	73.8
21.08-04.09	100	76.6	62.4	08.09-25.09	130	119.6	67.2
05.09-17.09	120	80.2	67.0	26.09-11.10	140	119.2	88.2
18.09-01.10	140	72.8	77.0	12.10-05.11	140	141.4	112.0
02.10-11.10	140	82.8	72.8				
12.10-05.11	140	147.3	116.2				
Total		592.5	525.6	Total		558.6	454.2
Yield, ton/ha		4.30	3.89			3.63	3.45

6.1.2 Soil moisture depletion studies in controlled field plots

Israelsen, O.W. (1956) mentioned that the early measurements of consumptive use were made on selected field plots of irrigated crops where the water table was at a considerable depth below the surface. The procedure was to measure the volume of water applied to the plot at each irrigation and to measure any surface run-off that might occur. In order to avoid percolation of water below the

plant root zone, it was necessary to apply the water in small depths, not
exceeding five inches in a single irrigation on ordinary soils. Precise measurements of the change in soil moisture were not undertaken in most of the early
studies.

The field plot method was adopted for the first time in Egypt in 1948 by the
then Ministry of Public Works, nowadays the Ministry of Irrigation. A few years
later, El-Shal, M.I. (1954) investigated the water use by cotton and maize at
Shebin el Kom, El-Menufiya Governrate, by measuring the changes in soil moisture
using the soil tube. In view of the proximity of the water table to the ground
surface, the crop water use was partly supplied through capillary movement of
the subsoil to the root zone. Since that part could not be determined, it was
not possible at that time to determine the actual water use by either crop.

Consumptive use of water by various crops has been determined by intensive
soil moisture studies. Soil samples have been taken at 10 cm intervals before
and after each irrigation, with some samples in between the successive irrigation cycles. The depth from the samples taken varied between 60 and 90 cm.
Standard laboratory practices have been applied for determining the moisture
content in the soil samples. The quantity of water removed from each layer of
soil was computed by the equation

$$d = \frac{(p_{fc} - p_r) \cdot \rho_s \cdot D}{100} \tag{6.2}$$

where

d   = depth of water removed, in cm,
p_{fc} = soil moisture content at field capacity, in percent,
p_r = remaining soil moisture content, in percent,
ρ_s = apparent specific gravity of soil, and
D   = depth of layer considered, in cm.

The total depth of water removed from the root zone could thus be obtained
by summing up the moisture extracted from the successive layers during a certain time period. That depth was corrected for the extraction of moisture from
the layers below the sampling depth and for the consumption of water in the
interval of time from the moment of irrigation application till the field capacity of the soil moisture had been reached.

The main findings from those extensive experiments which lasted more than
ten years can be summarized as follows:
i)   The soil moisture in the surface layer reaches the saturation capacity
     upon the completion of land irrigation. This moisture level drops rapidly
     in the first few days after irrigation followed by a less rapid decrease

in the subsequent days. A study of the cyclic change in soil moisture has
shown that the moisture content in the top 10 cm of soil fell below the
wilting point 18 days after irrigation application, whereas the underlying
layers behaved differently, even with elongated cycles of 30 days or more.
Moreover, the moisture depleted from the surface 10 cm of soil was negatively correlated with the atmospheric relative humidity. The total
moisture removed from the upper 30 cm of soil was strongly positively correlated with the average air temperature (Z. el Abedine, A., and Abdallah,
M., 1949).

ii) Further studies have shown that the significant cyclic changes in soil
    moisture are confined to the top 50 cm of soil, and the rate of moisture
    depletion decreases with depth below the surface. This can readily be seen
    from Fig. 6.4. The amount of moisture depleted from the root zone varies
    from one irrigation cycle to another, depending on the stage of plant
    growth as well as the climatological conditions. Fig. 6.5. illustrates the
    cyclic variation in moisture content in the top 50 cm of soil, as observed
    by El-Warith (Khafagi, A., Z. el Abedine, A., Shahin, M., and El-Warith,
    M., 1964), in a field plot raising cotton at Sids.

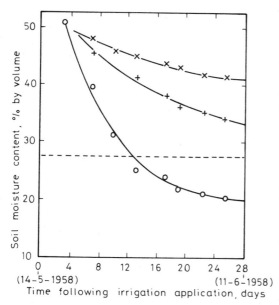 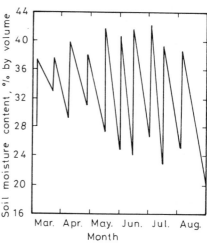

Fig. 6.4. Moisture depletion from soil
layers of a field plot raising cotton at
Sids under an elongated irrigation treatment (Khafagi, A., Z. el-Abedine, A.,
Shahin, M., and El-Warith, M., 1964)

Fig. 6.5. Cyclic variation of
the moisture content in the top
50 cm. of soil in a field plot
raising cotton at Sids (Khafagi,
A., Z. el-Abedine, A., Shahin,
M., and El-Warith, M., 1964)

iii) The moisture depletion characteristic curves of the soil layers from which
the plant extracts its moisture use can be obtained by plotting the daily
loss in moisture from each layer versus the times during the growing season
of the plant. Four sets of characteristic curves are shown in Fig. 6.6.
These sets represent the daily moisture removal from the individual layers
of field plots raising cotton, wheat, maize and berseem (Egyptian clover)
at Giza and Sids. Each characteristic curve assumes the shape of a forced
oscillation, the amplitude of which damps with depth. The crown point of
the curve belonging to any layer lags behind the crown point of the curve
belonging to the overlying layer.

iv) The contribution of each layer to the seasonal consumptive use is simply
the integration of the corresponding characteristic curve, or simply the
the area under it. This contribution evidently decreases with depth below
the surface. Nevertheless, layers deeper than 50 cm still contribute to
the total consumptive use, though by smaller amounts. The contribution of
the successive layers, u_z, in proportion to the total extraction, u, when
plotted versus the depth below surface, z, yields the set of curves shown
in Fig. 6.7. (El-Shal, M.I., 1966). These curves can approximately be described by the general equation

$$\frac{u_z}{u} = A \cdot e^{-\alpha z} \qquad\qquad\qquad (6.3)$$

where A and α are constant for each crop, and almost the same for the four
crops presented in Fig. 6.7. For A = 0.35, α = 3.5, and 0.8 m as depth of
moisture extraction, the contribution in percent of the total extraction
is 55 for the first quarter, 25.3 for the second, 12.4 for the third and
7.3 for the fourth quarter. These results are in agreement with the basic
finding of Schockley, D.R., that most irrigated crops have similar extraction patterns. His conclusion that the four successive quarters of the root
zone, starting from the surface, contribute by 40, 30, 20 and 10% to the
total extraction, is not in agreement with ours. The figures we obtained
are more in line with values obtained by Stanberry, C.O. (United States
Department of Agriculture, 1955) from the analysis of the data from 28
alfalfa tests conducted in 10 states in USA. Those were 47, 26, 17 and 10%
for the first, second, third and fourth quarter of the extraction depth,
respectively.

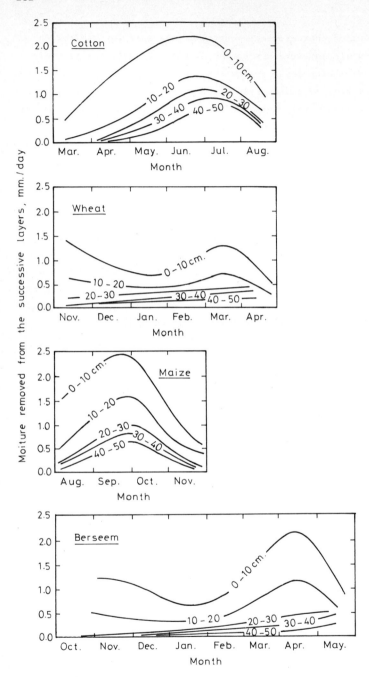

Fig. 6.6.   Moisture depletion characteristic curves for the top 50 cm layers
of soil under four different crops, average for Giza and Sids stations (El-Shal,
M.I., 1966)

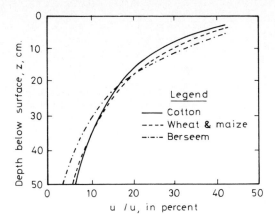

Fig. 6.7.   Pattern of seasonal moisture extraction by some crops in Egypt

A further support for our idea that the moisture extraction pattern is
more or less exponential rather than linear, as was originally proposed by
Schockley, came from the analysis of the data collected from experiments
at Scotts Bluff Station, Nebraska, USA (United States Department of Agriculture, 1955). A summary of some of the results is as follows:

Depth, ft	Moisture extraction in percent of total extraction for			
	Oats	Beets	Potatoes	Average
0-1	64	62	57	61.0
1-2	16	19	23	19.3
2-3	12	12	13	12.3
3-4	8	7	7	7.3

v)   The corrected cyclic loss of moisture when divided by the duration of each
cycle gives the daily loss or the crop water use per day. Daily, monthly
and seasonal consumptive use of water by a large number of crops in Egypt
are given in Tables 6.2 up to and including 6.8. Specifications of these
tables are:

Table	Number of experiments	Number of experiment stations	Crop investigated
6.2	13	4	Cotton
6.3	7	4	Wheat
6.4	6	3	Late maize
6.5	1	1	Early maize
6.6	1	1	Fenugreek
6.6	1	1	Chick pea
6.6	1	1	Egyptian Lupine
6.6	1	1	Lentil
6.7	2	2	Berseem (Egyptian clover)
6.8	4	2	Citrus trees

TABLE 6.2   Consumptive use of water for cotton in Egypt

Sr. No.	Expt. Stat. & Investigator	From	To	Interval days	Consumptive use, mm daily	Consumptive use, mm monthly	From	To	Interval days	Consumptive use, mm daily	Consumptive use, mm monthly
1	Sakha	22.03.1957	31.03.1957	9	1.19	10.71	21.06.1957	30.06.1957	10	7.11	186.10
		01.04.1957	10.04.1957	10	1.19		01.07.1957	10.07.1957	10	7.54	
		11.04.1957	20.04.1957	10	1.63		11.07.1957	10.07.1957	10	7.67	
	El-Shal[a]	21.04.1957	30.04.1957	10	2.29	51.10	21.07.1957	30.07.1957	11	6.74	226.24
		01.05.1957	10.05.1957	10	2.29		01.08.1957	10.08.1957	10	4.18	
		11.05.1957	20.05.1957	10	3.68		11.08.1957	20.08.1957	10	3.35	
		21.05.1957	31.05.1957	11	4.27	106.67	20.08.1957	31.08.1957	11	2.10	98.40
		01.06.1957	10.06.1957	10	5.38		01.09.1957	10.09.1957	10	2.10	
		11.06.1957	20.06.1957	10	6.12		11.09.1957	14.09.1957	4	2.10	29.40
	Season			176 days						708.62 mm	
2	Sakha	18.03.1958	10.03.1958	2	1.34		11.06.1958	20.06.1958	10	6.73	
		21.03.1958	31.03.1958	11	1.34	17.42	21.06.1958	30.06.1958	10	7.09	201.80
		01.04.1958	10.04.1958	10	1.34		01.07.1958	10.07.1958	10	7.20	
		11.04.1958	20.04.1958	10	1.72		11.07.1958	20.07.1958	10	7.30	
	El-Shal[a]	21.04.1958	30.04.1958	10	2.60	56.60	21.07.1958	31.07.1958	11	4.90	198.90
		01.05.1958	10.05.1958	10	2.60		01.08.1958	10.08.1958	10	4.00	
		11.05.1958	20.05.1958	10	3.69		11.08.1958	20.08.1958	10	3.00	
		21.05.1958	31.05.1958	11	4.42	111.52	21.08.1958	31.08.1958	11	3.00	103.00
		01.06.1958	10.06.1958	10	6.36		01.09.1958	06.09.1958	6	3.00	18.00
	Season			172 days						707.24 mm	
3	Sakha	17.03.1958	20.03.1958	3	1.60		12.06.1958	30.06.1958	19	5.20	146.10
		21.03.1958	31.03.1958	11	1.60	22.40	01.07.1958	04.07.1958	4	5.25	
		01.04.1958	16.04.1958	16	1.60		05.07.1958	22.07.1958	18	7.55	
		17.04.1958	30.04.1958	14	2.10	55.00	23.07.1958	31.07.1958	9	5.80	209.10
	A. El-Warith[b]	01.05.1958	13.05.1958	13	2.10		01.08.1958	09.08.1958	9	5.80	
		14.05.1958	31.05.1958	18	3.50	90.30	10.08.1958	31.08.1958	22	3.40	127.00
		01.06.1958	11.06.1958	11	4.30		01.09.1958	14.09.1958	13	3.40	44.20
	Season			180 days						694.10 mm	

a = El-Shal, M.I., 1966;   b = Abdel-Warith, M., 1965

TABLE 6.2   (continued)

Sr. No.	Expt. Stat. & Investigator	From	To	Interval days	Consumptive use, mm daily	Consumptive use, mm monthly	From	To	Interval days	Consumptive use, mm daily	Consumptive use, mm monthly
	Sakha	12.03.1960	20.03.1960	9	1.10		20.06.1960	30.06.1960	11	6.80	177.40
		21.03.1960	31.03.1960	11	1.10	22.00	01.07.1960	04.07.1960	4	6.80	
4		01.04.1960	11.04.1960	11	1.10		05.07.1960	19.07.1960	15	7.60	
	A. El-Warith[b]	12.04.1960	30.04.1960	19	2.20	53.90	20.07.1960	31.07.1960	12	5.60	208.40
		01.05.1960	17.05.1960	17	2.80		01.08.1960	07.08.1960	7	5.60	
		18.05.1960	31.05.1960	14	4.90	116.20	08.08.1960	31.08.1960	24	3.70	128.00
		01.06.1960	19.06.1960	19	5.40		01.09.1960	03.09.1960	2	3.70	7.40
	Season	175 days								713.30 mm	
	Giza	05.03.1958	10.03.1958	5	0.85		01.06.1958	10.06.1958	10	6.26	
		11.03.1958	20.03.1958	10	0.92		11.06.1958	20.06.1958	10	6.87	
		21.03.1958	31.03.1958	11	1.59	30.94	21.06.1958	30.06.1958	10	7.30	204.30
5	El-Shal[a]	01.04.1958	10.04.1958	10	2.14		01.07.1958	10.07.1958	10	7.32	
		11.04.1958	20.04.1958	10	2.42		11.07.1958	20.07.1958	10	6.55	
		21.04.1958	30.04.1958	10	2.48	70.40	21.07.1958	31.07.1958	11	5.78	202.28
		01.05.1958	10.05.1958	10	2.97		01.08.1958	10.08.1958	10	4.21	
		11.05.1958	20.05.1958	10	3.35		11.08.1958	20.08.1958	10	3.57	
		21.05.1958	31.05.1958	11	4.65	114.35	21.08.1958	31.08.1958	11	2.09	100.79
	Season	179 days								723.06 mm	
	Giza	05.03.1958	18.03.1958	14	1.80		18.05.1958	31.05.1958	14	4.50	129.30
		19.03.1958	31.03.1958	13	2.00	51.20	01.06.1958	14.06.1958	14	5.30	
6		01.04.1958	02.04.1958	2	2.00		15.06.1958	30.06.1958	16	6.40	176.80
	A. El-Warith[b]	03.04.1958	15.04.1958	13	2.70		01.07.1958	14.07.1958	14	7.30	
		16.04.1958	30.04.1958	15	3.00	84.10	15.07.1958	31.07.1958	17	5.90	202.50
		01.05.1958	17.05.1958	17	3.90		01.08.1958	30.08.1958	30	3.47	104.10
	Season	179 days								748.00 mm	

a = El-Shal, M.I., 1966;   b = Abdel-Warith, M., 1965

TABLE 6.2 (continued)

Sr. No.	Expt. Stat. & Investigator	From	To	Interval days	Consumptive use, mm daily	monthly	From	To	Interval days	Consumptive use, mm daily	monthly
	Giza	01.03.1958	23.03.1958	23	1.0		23.06.1958	30.06.1958	8	4.7	158.30
		23.03.1958	31.03.1958	8	1.0	31.00	01.07.1958	06.07.1958	6	4.7	
		01.04.1958	15.04.1958	15	1.0		07.07.1958	15.07.1958	9	7.3	
		16.04.1958	30.04.1958	15	1.8	42.00	16.07.1958	27.07.1958	12	10.4	261.90
7	A. El Samie & Y. Barrada[c]	01.05.1958	02.05.1958	2	1.8		28.07.1958	31.07.1958	4	10.8	
		03.05.1958	18.05.1958	16	5.5		01.08.1958	05.08.1958	5	10.8	
		19.05.1958	31.05.1958	13	5.4	161.80	06.08.1958	20.08.1958	15	6.4	
		01.06.1958	03.06.1958	3	5.4		21.08.1958	29.08.1958	9	3.2	185.20
		04.06.1958	22.06.1958	19	5.5		30.08.1958	31.08.1958	2	3.2	
							01.09.1958	03.09.1958	3	3.2	9.60
	Season		187 days							849.80 mm	
	Sids	06.03.1957	10.03.1957	4	0.74		01.06.1957	10.06.1957	10	6.25	206.10
		11.03.1957	20.03.1957	10	0.74		11.06.1957	20.06.1957	10	7.10	
		21.03.1957	31.03.1957	11	1.11	22.57	21.06.1957	30.06.1957	10	7.26	
		01.04.1957	10.04.1957	10	1.75		01.07.1957	10.07.1957	10	7.30	
8	El-Shal[a]	11.04.1957	20.04.1957	10	1.99		11.07.1957	20.07.1957	10	6.52	
		21.04.1957	30.04.1957	10	2.35	60.90	21.07.1957	31.07.1957	11	5.93	203.43
		01.05.1957	10.05.1957	10	2.78		01.08.1957	10.08.1957	10	5.22	
		11.05.1957	20.05.1957	10	3.42		11.08.1957	20.08.1957	10	3.63	
		21.05.1957	31.05.1957	11	5.33	120.63	21.08.1957	31.08.1957	11	2.04	110.94
	Season		178 days							724.57 mm	
	Sids	03.03.1958	10.03.1958	7	0.79		01.06.1958	10.06.1958	10	6.12	
9		11.03.1958	20.03.1958	10	0.79		11.06.1958	20.06.1958	10	6.90	
		21.03.1958	31.03.1958	11	1.37	28.50	21.06.1958	30.06.1958	10	7.63	206.50
		01.04.1958	10.04.1958	10	1.63		01.07.1958	10.07.1958	10	7.33	

a = El-Shal, M.I., 1966; c = Abd El-Samie, A.G., and Barrada, Y., 1960

TABLE 6.2   (continued)

Sr. No.	Expt. Stat. & Investigator	From	To	Interval days	Consumptive use, mm		From	To	Interval days	Consumptive use, mm	
					daily	monthly				daily	monthly
9	El-Shal[a]	11.04.1958	20.04.1958	10	1.98		11.07.1958	20.07.1958	10	1.76	
		21.04.1958	30.04.1958	10	2.33	59.40	21.07.1958	31.07.1958	11	5.62	202.70
		01.05.1958	10.05.1958	10	3.16		01.08.1958	10.08.1958	10	4.88	
		11.05.1958	20.05.1958	10	3.69		11.08.1958	20.08.1958	10	4.14	
		21.05.1958	31.05.1958	11	4.94	122.84	21.08.1958	31.08.1958	11	2.46	117.26
		Season	181 days				Season				737.20 mm
10	Sids	03.03.1958	22.03.1958	20	1.30		01.06.1958	15.06.1958	15	6.60	
		23.03.1958	31.03.1958	9	2.00	44.00	16.06.1958	30.06.1958	15	5.80	186.00
		01.04.1958	08.04.1958	8	2.00		01.07.1958	16.07.1958	16	7.00	
	A. El-Warith[b]	09.04.1958	26.04.1958	18	2.20		17.07.1958	31.07.1958	15	5.60	196.00
		27.04.1958	30.04.1958	4	3.20	68.40	01.08.1958	02.08.1958	2	5.60	
		01.05.1958	16.05.1958	16	3.20		03.08.1958	31.08.1958	29	3.50	112.70
		17.05.1958	31.05.1958	15	5.00	126.20	01.09.1958	02.09.1958		3.50	3.50
		Season	183 days				Season				736.80 mm
11	Sids	22.02.1959	28.02.1959	7	0.80		01.06.1959	06.06.1959	6	6.50	
		01.03.1959	16.03.1959	16	0.80	5.60	07.06.1959	18.06.1959	12	7.40	
		17.03.1959	31.03.1959	15	1.40		19.06.1959	27.06.1959	9	8.40	
		01.04.1959	13.04.1959	13	1.40	33.80	28.06.1959	30.06.1959	3	8.20	228.00
	A. El-Warith[b]	14.04.1959	30.04.1959	17	2.60	62.40	01.07.1959	05.07.1959	5	8.20	
		01.05.1959	16.05.1959	16	3.40		06.07.1959	14.07.1959	9	8.10	
		17.05.1959	27.05.1959	11	5.40		15.07.1959	27.07.1959	13	6.30	
		28.05.1959	31.05.1959	4	6.50	139.80	28.07.1959	31.07.1959	4	4.20	212.60
							01.08.1959	23.08.1959	23	4.20	92.00
		Season	183 days				Season				774.20 mm

a = El-Shal, M.I., 1966;   b = Abdel-Warith, M., 1965

TABLE 6.2 (continued)

Sr. No.	Expt. Stat. & Investigator	From	To	Interval days	Consumptive use, mm daily	monthly	From	To	Interval days	Consumptive use, mm daily	monthly
	Sids	09.03.1960	10.03.1960	1	1.06		01.06.1960	10.06.1960	10	5.61	
		11.03.1960	20.03.1960	10	1.06		11.06.1960	20.06.1060	10	7.36	
		21.03.1960	31.03.1960	11	1.06	23.32	21.06.1960	30.06.1960	10	7.92	208.90
		01.04.1960	10.04.1960	10	1.51		01.07.1960	10.07.1960	10	7.62	
12	El-Shal [a]	11.04.1960	20.04.1960	10	1.63		11.07.1960	21.07.1960	11	6.10	
		20.04.1960	30.04.1960	10	2.70	58.40	21.07.1960	31.07.1960	10	5.75	200.45
		01.05.1960	10.05.1960	10	3.29		01.08.1960	10.08.1960	10	4.02	
		11.05.1960	20.05.1960	10	4.37		11.08.1960	20.08.1960	10	3.29	
		21.05.1960	31.05.1960	11	5.05	132.15	21.08.1960	31.08.1960	11	3.29	109.29
	Season			175 days							732.51 mm
	Mallawi	22.02.1960	29.02.1960	7	1.20	8.40	01.06.1960	10.06.1960	10	5.84	
		01.03.1960	10.03.1960	10	1.20		11.06.1960	20.06.1960	10	7.72	
		11.03.1960	20.03.1960	10	1.38		21.06.1960	30.06.1960	10	8.26	218.20
		21.03.1960	31.03.1960	11	1.66	44.06	01.07.1960	10.07.1960	10	7.72	
13	El-Shal [a]	01.04.1960	10.04.1960	10	1.82		11.07.1960	20.07.1960	10	6.48	
		11.04.1960	20.04.1960	10	1.97		21.07.1960	31.07.1960	11	5.73	205.03
		21.04.1960	30.04.1960	10	2.96	67.50	01.08.1960	10.08.1960	10	4.35	
		01.05.1960	10.05.1960	10	3.51		11.08.1960	20.08.1960	10	3.43	
		11.05.1960	20.05.1960	10	4.57		21.08.1960	31.08.1960	11	3.33	114.43
		21.05.1960	31.05.1960	11	5.03	136.13	01.09.1960	04.09.1960	4	3.28	13.12
	Season			195 days							806.87 mm

a = El-Shal, M.I., 1966

TABLE 6.3   Consumptive use of water for wheat in Egypt

Sr. No.	Expt. Stat & Investigator	From	To	Interval days	Consumptive use, mm daily	Consumptive use, mm monthly	From	To	Interval days	Consumptive use, mm daily	Consumptive use, mm monthly
1	Sakha / El-Shal [a]	15.11.1957	20.11.1957	5	1.81	27.15	21.02.1958	28.02.1958	8	2.64	54.22
		21.11.1957	30.11.1957	10	1.81		01.03.1958	10.03.1958	10	2.68	
		01.12.1957	10.12.1957	10	1.81		11.03.1958	20.03.1958	10	3.15	100.87
		11.12.1957	20.12.1957	10	1.73	53.00	21.03.1958	31.03.1958	11	3.87	
		21.12.1957	31.12.1957	11	1.60		01.04.1958	10.04.1958	10	2.57	
		01.01.1958	10.01.1958	10	1.60		11.04.1058	20.04.1958	10	2.24	
		11.01.1958	20.01.1958	10	1.60		21.04.1958	30.04.1958	10	2.24	70.50
		21.01.1958	31.01.1958	11	1.60	49.60	01.05.1958	10.05.1958	10	1.89	
		01.02.1958	10.02.1958	10	1.60		11.05.1958	16.05.1958	6	1.80	29.10
		11.02.1958	20.02.1958	10	1.71						
	Season	182 days									385.04 mm
2	Sakha / El-Shal [a]	22.11.1958	30.11.1958	9	1.79	16.11	21.02.1959	28.02.1959	8	2.27	57.69
		01.12.1958	10.12.1958	10	1.79		01.03.1959	10.03.1959	10	2.27	
		11.12.1958	20.12.1958	10	1.79		11.03.1959	20.03.1959	10	2.34	
		21.12.1958	31.12.1958	11	1.70	54.50	21.03.1959	31.03.1959	11	2.95	78.61
		01.01.1959	10.01.1959	10	1.63		01.04.1959	10.04.1959	10	2.95	
		11.01.1959	20.01.1959	10	1.63		11.04.1959	20.04.1959	10	2.09	
		21.01.1959	31.01.1959	11	1.63	50.53	21.04.1959	30.04.1959	10	1.72	67.66
		01.02.1959	10.02.1959	10	1.68		01.05.1959	10.05.1959	10	1.72	
		11.02.1959	20.02.1959	10	2.27		11.05.1959	12.05.1959	2	1.72	20.69
	Season	172 days									345.79 mm
3	Giza / Shahin [d]	17.11.1957	30.11.1957	14	1.56	21.84	01.03.1958	04.03.1958	4	2.10	84.00
		01.12.1957	18.12.1957	18	1.56		05.03.1958	31.03.1958	27	2.80	
		19.12.1957	31.12.1957	13	1.92	53.04	01.04.1958	03.04.1958	3	2.80	
		01.01.1958	23.01.1958	23	1.92		04.04.1958	30.04.1958	27	1.70	54.30
		24.01.1958	31.01.1958	8	2.10	60.76	01.05.1958	04.05.1958	4	1.70	6.80
		01.02.1958	28.02.1958	28	2.10	58.80					
	Season	169 days									339.54 mm

a = El-Shal, M.I., 1966; d = Shahin, M., 1959

TABLE 6.3 (continued)

Sr. No.	Expt. Stat. & Investigator	From	To	Interval days	Consumptive use, mm daily	Consumptive use, mm monthly	From	To	Interval days	Consumptive use, mm daily	Consumptive use, mm monthly
	Giza	10.11.1958	20.11.1958	10	1.91		11.02.1959	20.02.1959	10	1.85	
		20.11.1958	30.11.1958	10	1.91	38.20	21.02.1959	28.02.1959	8	2.46	55.18
		01.12.1958	10.12.1958	10	1.88		01.03.1959	10.03.1959	10	2.46	
		11.12.1958	20.12.1958	10	1.58		11.03.1959	20.03.1959	10	2.48	
4	El-Shal[a]	21.12.1958	31.12.1958	11	1.58	51.98	21.03.1959	31.03.1959	11	2.63	78.33
		01.01.1959	10.01.1959	10	1.58		01.04.1959	10.04.1959	10	2.63	
		11.01.1959	20.01.1959	10	1.66		11.04.1959	20.04.1959	10	2.63	
		21.01.1959	31.01.1959	11	1.70	51.10	21.04.1959	30.04.1959	10	2.63	70.90
		01.02.1959	10.02.1959	10	1.70		01.05.1959	10.05.1959	10	1.68	16.80
	Season		182 days								370.49 mm
	Sids	09.11.1957	10.11.1957	1	1.79		11.02.1958	20.02.1958	10	2.36	
				10	1.79	37.59	21.02.1958	28.02.1958	8	2.36	62.28
		01.12.1957	10.12.1957	10	1.79		01.03.1958	10.03.1958	10	2.49	
		11.12.1957	20.12.1957	10	1.79		11.03.1958	20.03.1958	10	3.01	
		21.12.1957	31.12.1957	11	1.60	53.40	21.03.1958	31.03.1958	11	3.01	88.11
5	El-Shal[a]	01.01.1958	10.01.1958	10	1.60		01.04.1958	10.04.1958	10	2.83	
		11.01.1958	20.01.1958	10	1.60		11.04.1958	20.04.1958	10	2.56	
		21.01.1958	31.01.1958	11	1.60		21.04.1958	30.04.1958	10	2.30	76.90
		01.02.1958	10.02.1958	10	1.98		01.05.1958	09.05.1958	9	1.90	17.10
	Season		181 days								384.98 mm
	Sids	09.11.1958	10.11.1958	1	2.00		11.02.1959	20.02.1959	10	2.24	
		11.11.1958	20.11.1958	10	2.00	42.00	21.02.1959	28.02.1959	4	2.23	57.44
		21.11.1958	30.11.1958	10	2.00		01.03.1959	10.03.1959	10	2.24	
		01.12.1958	10.12.1958	10	1.68		11.03.1959	20.03.1959	10	2.37	
6	El-Shal[a]	11.12.1958	20.12.1958	10	1.61	50.70	21.03.1959	31.03.1959	11	3.16	80.90
		21.12.1958	31.12.1958	11	1.62		01.04.1959	10.04.1959	10	3.16	

a = El-Shal, M.I., 1966

241

TABLE 6.3 (continued)

Sr. No.	Expt. Stat. & Investigator	From	To	Interval days	Consumptive use, mm daily	Consumptive use, mm monthly	From	To	Interval days	Consumptive use, mm daily	Consumptive use, mm monthly
6	El-Shal[a]	01.01.1959	10.01.1959	10	1.63		11.04.1959	20.04.1959	10	2.52	
		11.01.1959	20.01.1959	10	1.63	50.53	21.04.1959	30.04.1959	10	1.74	74.20
		21.01.1959	31.01.1959	11	1.63		01.05.1959	08.05.1959	8	1.74	13.92
		01.02.1959	10.02.1959	10	1.72						
	Season	180 days									369.69 mm
	Mallawi	15.11.1957	20.11.1957	5	1.61		20.02.1958	28.02.1958	8	2.66	61.18
		21.11.1957	30.11.1957	10	1.61	24.15	01.03.1958	10.03.1958	10	3.25	
		01.12.1957	10.12.1957	10	1.61		11.03.1958	20.03.1958	10	3.30	
		11.12.1957	20.12.1957	10	1.56		21.03.1958	31.03.1958	11	3.40	102.90
7	El-Shal[a]	21.12.1957	31.12.1957	11	1.56	48.86	01.04.1958	10.04.1958	10	2.78	
		01.01.1958	10.01.1958	10	1.614		11.04.1958	20.04.1958	10	2.52	
		11.01.1958	20.01.1958	10	1.614		21.04.1958	30.04.1958	10	2.52	78.20
		21.01.1958	31.01.1958	11	1.614	50.03	01.05.1958	10.05.1958	10	1.50	
		01.02.1958	10.02.1958	10	1.93		11.05.1958	12.05.1958	4	1.50	18.00
		11.02.1958	20.02.1958	10	2.06						
	Season	178 days									383.22 mm

a = El-Shal, M.I., 1966

TABLE 6.4 Consumptive use of water for late maize in Egypt

Sr. No.	Expt. Stat. & Investigator	From	To	Interval days	Consumptive use, mm		From	To	Interval days	Consumptive use, mm	
					daily	monthly				daily	monthly
	Sakha	30.07.1959	31.07.1959	1	3.06	3.06	01.10.1959	10.10.1959	10	6.67	
		01.08.1959	10.08.1959	10	3.06		11.10.1959	20.10.1959	10	5.006	167.25
		11.08.1959	20.08.1959	10	3.06		21.10.1959	31.10.1959	11	4.59	
1	El-Shal[a]	21.08.1959	31.08.1959	11	4.61	111.87	01.11.1959	10.11.1959	10	1.936	
		01.09.1959	10.09.1959	10	5.06		11.11.1959	20.11.1959	10	1.936	53.67
		11.09.1959	20.09.1959	10	6.086		21.11.1959	30.11.1959	10	1.495	
		21.09.1959	30.09.1959	10	6.36	175.86	01.12.1959	08.12.1959	8	1.05	8.40
	Season			131 days						520.11 mm	
	Giza	02.08.1959	10.08.1959	8	2.93		01.10.1959	10.10.1959	10	6.99	
		11.08.1959	20.08.1959	10	2.93		11.10.1959	20.10.1959	10	6.76	
2	El-Shal[a]	21.08.1959	31.08.1959	11	4.49	102.13	21.10.1959	31.10.1959	11	4.07	182.27
		01.09.1959	10.09.1959	10	5.43		01.11.1959	10.11.1959	10	2.26	
		11.09.1959	20.09.1959	10	6.25		11.11.1959	20.11.1959	10	1.48	
		21.09.1959	30.09.1959	10	6.886	185.66	21.11.1959	30.11.1959	10	1.48	52.20
	Season			120 days						522.26 mm	
	Giza	09.07.1962	27.07.1962	18	3.155		01.09.1962	11.09.1962	11	6.64	
		28.07.1962	31.07.1962	4	5.59	79.15	12.09.1962	30.09.1962	19	3.19	147.33
3	El-Gibali[e]	01.08.1962	09.08.1962	9	5.59		01.10.1962	03.10.1962	3	3.91	
		10.08.1962	23.08.1962	14	5.905		04.10.1962	21.10.1962	18	3.225	58.05
		24.08.1962	31.08.1962	8	6.64	186.10					
	Season			104 days						470.63 mm	
	Giza	07.07.1963	31.07.1963	24	3.445	82.68	14.09.1963	30.09.1963	17	5.20	152.30
		01.08.1963	10.08.1963	10	3.445		01.10.1963	04.10.1963	4	5.20	
4	El-Gibali[e]	11.08.1963	24.08.1963	14	4.995		05.10.1963	17.10.1963	13	5.04	
		25.08.1963	31.08.1963	7	6.39	149.11	18.10.1963	31.10.1963	14	3.315	132.73
		01.09.1963	13.09.1963	13	6.39		01.11.1963	06.11.1963	6	3.315	19.89
	Season			122 days						536.72 mm	

a = El-Shal, M.I., 1966; e = El-Gibali, A., 1966.

TABLE 6.4 (continued)

Sr. No.	Expt. Stat. & Investigator	From	To	Interval days	Consumptive use, mm daily	Consumptive use, mm monthly	From	To	Interval days	Consumptive use, mm daily	Consumptive use, mm monthly
5	Sids	28.07.1959	31-07.1959	3	2.95	8.85	01.10.1959	10.10.1959	10	6.52	
		01.08.1959	10.08.1959	10	2.95		11.10.1959	20.10.1959	10	6.26	
		11.08.1959	20.08.1959	10	3.12		21.10.1959	31.10.1959	11	4.73	179.83
	El-Shal a	21.08.1959	31.08.1959	11	4.68	112.18	01.11.1959	10.11.1959	10	3.36	
		01.09.1959	10.09.1959	10	4.68		11.11.1959	20.11.1959	10	2.19	
		11.09.1959	20.09.1959	10	6.25		21.11.1959	30.11.1959	10	1.15	67.00
		21.09.1959	30.09.1959	10	6.45	173.80	01.12.1959	08.12.1959	8	1.04	8.32
	Season	133 days								549.98 mm	
6	Sids	15.07.1964	26.07.1964	12	3.62		13.09.1964	24.09.1964	12	5.56	
		27.07.1964	31.07.1964	5	2.66	56.74	25.09.1964	30.09.1964	6	5.31	143.58
	Shenouda	01.08.1964	07.08.1964	7	2.66		01.10.1964	06.10.1964	6	5.31	
	El-Gibali	08.08.1964	19.08.1964	12	5.74		07.10.1964	18.10.1964	12	5.52	
	Tawdros & Gamal f	20.08.1964	31.08.1964	12	6.21	162.02	19.10.1964	31.10.1964	19	3.31	163.99
		01.09.1964	12.09.1964	12	3.75						
	Season	109 days								526.33 mm	

TABLE 6.5 Consumptive use of water for early maize in Egypt

Sr. No.	Expt. Stat. & Investigator	From	To	Interval days	Consumptive use, mm daily	Consumptive use, mm monthly	From	To	Interval days	Consumptive use, mm daily	Consumptive use, mm monthly
1	Sids	16.04.1964	30.04.1964	15	1.85		12.06.1964	23.06.1964	12	6.91	
		01.05.1964	06.05.1964	6	1.85	27.75	24.06.1964	30.06.1964	7	7.47	200.44
	Shenouda	07.05.1964	18.05.1964	12	3.06		01.07.1964	05.07.1964	5	7.47	
	El-Gibali	19.05.1964	30.05.1964	12	6.25		06.07.1964	17.07.1964	12	5.44	
	Tawdros & Gamal f		31.05.1964	1	5.93		18.07.1964	29.07.1964	12	5.60	
		01.06.1964	11.06.1964	11	5.93	128.75	30.07.1964	31.07.1964	2	3.20	176.23
							01.08.1964	18.08.1964	8	3.20	25.60
	Season	115 days								558.77 mm	

a = El-Shal, M.I., 1966; f = Shenouda, El-Gibali, Tawdros and Gamal, 1966

TABLE 6.6 Consumptive use of water for some winter crops in Egypt

Sr. No.	Expt. Stat. & Investigator	Month	Fenugreek[+]		Chicken pea[+]		Egyptian Lupin[++]		Lentil[++]	
			daily	monthly	daily	monthly	daily	monthly	daily	monthly
1	Assiut El-Gibali g	Nov. 1965	1.541	30.82	1.414	28.28	1.532	30.65	1.405	28.11
		Dec. 1965	2.087	64.70	1.721	53.34	2.065	64.00	1.778	55.12
		Jan. 1966	1.880	58.24	1.360	42.16	1.852	57.40	1.417	43.94
		Feb. 1966	2.394	67.03	1.923	53.85	2.325	65.10	1.729	53.59
		Mar. 1966	1.709	12.88	0.735	8.09	1.221	18.31	0.819	12.29
		Season		233.67		185.72		235.46		193.05

+ = Growth season: Nov. 10-Mar. 11; ++ = Growth season: Nov. 10-Mar. 15

TABLE 6.7 Consumptive use of water for Berseem (Egyptian clover) in Egypt

Sr. No.	Expt. Stat. & Investigator	From	To	Interval days	Consumptive use, mm daily	Consumptive use, mm monthly
1	Giza Shahin d	10.11.1957	27.11.1957	17	1.82	
		28.11.1957	30.11.1957	3	2.15	37.39
		01.12.1957	21.12.1957	21	2.50	
		22.12.1957	31.12.1957	10	1.45	67.00
		01.01.1958	31.01.1958	31	0.80	
		01.02.1958	06.02.1958	6	0.45	
		07.02.1958	14.02.1958	8	1.65	24.80
		17.02.1958	26.02.1958	10	3.15	
		27.02.1958	28.02.1958	2	3.50	57.70
		01.03.1958	27.03.1958	27	3.95	
		28.03.1958	31.03.1958	4	2.80	117.85
		01.04.1958	06.04.1958	6	2.50	
		07.04.1958	16.04.1958	10	4.10	
		15.04.1958	30.04.1958	16	4.35	117.40
		01.05.1958	07.05.1958	7	1.85	12.95
		Season	214 days (no records were available between Oct. 15 and Nov. 10)			435.09 mm

g = El-Gibali, M.H., 1969; d = Shahin, M., 1959

TABLE 6.7 (continued)

Sr. No.	Expt. Stat. & Investigator	From	To	Interval days	Consumptive use, mm daily	Consumptive use, mm monthly
	Sids	20.10.1957	31.10.1957	11	1.564	17.204
		01.11.1957	10.11.1957	10	1.564	
		11.11.1957	20.11.1957	10	1.530	
		21.11.1957	30.11.1957	10	1.526	46.200
		01.12.1957	10.12.1957	10	2.047	
		11.12.1957	20.12.1957	10	2.068	
2	El-Shal[a]	21.12.1957	31.12.1957	11	2.261	66.021
		01.01.1958	10.01.1958	10	1.659	
		11.01.1958	20.01.1958	10	0.757	
		21.01.1958	31.01.1958	11	0.757	32.487
		01.02.1958	10.02.1958	10	1.229	
		11.02.1958	20.02.1958		1.700	
		21.02.1958	28.02.1958	8	1.821	43.858
		01.03.1958	10.03.1958	10	2.671	
		11.03.1958	20.03.1958	10	2.671	
		21.03.1958	31.03.1958	11	4.953	107.903
		01.04.1958	10.04.1958	10	4.433	
		11.04.1958	20.04.1958	10	3.859	
		21.04.1958	30.04.1958	10	4.337	126.290
		01.05.1958	10.05.1958	10	4.165	
		11.05.1958	20.05.1958	10	4.091	
		21.05.1958	31.05.1958	11	2.940	114.900
		01.06.1958	07.06.1958	7	1.558	10.906
	Season			230 days		565.769 mm

TABLE 6.8 Consumptive use of water for citrus trees in Egypt

Sr. No.	Expt. Stat. & Investigator	From	To	Interval days	Consumptive use, mm daily	Consumptive use, mm monthly
	Giza	01.01.1957	31.01.1957	31	0.72	22.32
		01.02.1957	28.02.1957	28	1.47	41.16
		01.03.1957	16.03.1957	16	2.25	71.25
		17.03.1957	31.03.1957	15	2.35	
1	Shahin[d]	01.04.1957	28.04.1957	28	2.81	84.42
		29.04.1957	30.04.1957	2	2.87	
		01.05.1957	28.05.1957	28	2.87	90.65
		29.05.1957	31.05.1957	3	3.43	
		01.06.1957	12.06.1957	12	3.53	
		03.07.1957	21.07.1957	19	3.93	121.25
		22.07.1957	31.07.1957	10	3.91	
		01.08.1957	06.08.1957	6	3.91	118.96
		07.08.1957	31.08.1957	25	3.82	
		01.09.1957	22.09.1957	22	3.50	104.28
		23.09.1957	30.09.1957	8	3.41	
		01.10.1957	08.10.1957	8	3.41	91.68
		09.10.1957	31.10.1957	23	2.80	
		01.11.1957	26.11.1957	26	2.33	

a = El-Shal, M.I., 1966; d = Shahin, M., 1959

TABLE 6.8 (continued)

Sr. No.	Expt. Stat. & Investigator	From	To	Interval days	Consumptive use, mm		From	To	Interval days	Consumptive use, mm	
					daily	monthly				daily	monthly
1	Shahin[d]	13.06.1957	30.06.1957	18	3.74	109.68	27.11.1957	30.11.1957	4	2.06	68.82
		01.07.1957	02.07.1957	2	3.74		01.12.1957	31.12.1957	31	1.72	53.32
	Annual										977.77 mm
	Giza	01.01.1958	31.01.1958	31	0.91	28.21	01.07.1958	06.07.1958	6	3.68	
		01.02.1958	28.02.1958	28	1.38	38.64	07.07.1958	31.07.1958	25	3.86	118.58
		01.03.1958	03.03.1958	3	1.38	64.90	01.08.1958	26.08.1958	26	3.77	115.92
		04.03.1958	31.03.1958	28	2.17		27.08.1958	31.08.1958	5	3.58	
		01.04.1958	30.04.1958	30	2.40	72.00	01.09.1958	09.09.1958	9	3.50	102.06
2	Shahin[d]	01.05.1958	08.05.1958	8	2.72		10.09.1958	30.09.1958	21	3.36	
		09.05.1958	27.05.1958	19	3.48	101.88	01.10.1958	08.10.1958	8	3.36	
		28.05.1958	31.05.1958	4	3.50		09.10.1958	31.10.1958	23	2.93	94.27
		01.06.1958	15.06.1958	15	3.58	108.00	01.11.1958	30.11.1958	30	2.15	64.50
		16.06.1958	30.06.1958	15	3.62		01.12.1958	31.12.1958	31	1.79	55.49
	Annual										964.45 mm
	Delta Barrage	01.01.1959	31.01.1959	31	1.59	49.29	11.07.1959	26.07.1959	16	4.22	
		01.02.1959	14.02.1959	14	1.59		27.07.1959	31.07.1959	5	4.57	142.35
		15.02.1959	28.02.1959	14	2.12	51.64	01.08.1959	09.08.1959	9	4.57	
		01.03.1959	06.03.1959	6	2.12		10.08.1959	22.08.1959	13	3.65	
		07.03.1959	27.03.1959	21	2.53		23.08.1959	31.08.1959	9	3.34	118.59
3	El-Nokrashy[h]	28.03.1959	31.03.1959	4	2.62	76.33	01.09.1959	05.09.1959	5	3.34	
		01.04.1959	18.04.1959	18	2.62		06.09.1959	20.09.1959	15	2.86	
		19.04.1959	30.04.1959	12	5.30	110.76	21.09.1959	30.09.1959	10	3.02	89.80
		01.05.1959	18.05.1959	18	3.74		01.10.1959	05.10.1959	5	3.02	
		19.05.1959	29.05.1959	11	6.07		06.10.1959	20.10.1959	15	2.28	
		30.05.1959	31.05.1959	2	4.48	143.08	21.10.1959	31.10.1959	11	1.53	74.83
		01.06.1959	13.06.1959	13	4.48		01.11.1959	09.11.1959	9	1.53	

d = Shahin, M., 1959 ; h = El-Nokrashy, M.A., 1963

TABLE 6.8 (continued)

Sr. No.	Expt. Stat. & Investigator	From	To	Interval days	Consumptive use, mm daily	Consumptive use, mm monthly	From	To	Interval days	Consumptive use, mm daily	Consumptive use, mm monthly
3	El-Nokrashy[h]	14.06.1959	26.06.1959	13	5.10		10.11.1959	30.11.1959	21	1.29	40.86
		27.06.1959	30.06.1959	4	5.20	197.44	01.12.1959	24.12.1959	24	0.625	
		01.07.1959	10.07.1959	10	5.20		25.12.1959	31.12.1959	7	1.59	26.13
	Annual										1121.10 mm
	Delta Barrage	01.01.1961	31.01.1961	31	1.42	44.02	19.07.1961	31.07.1961	13	4.14	141.10
		01.02.1961	06.02.1962	6	1.42		01.08.1961	12.08.1961	12	3.84	
		07.02.1961	28.02.1961	22	2.02	52.56	13.08.1961	24.08.1961	12	4.22	120.03
		01.03.1961	06.03.1961	6	2.02		25.08.1961	31.08.1961	7	3.33	
		07.03.1961	31.03.1961	25	2.77	81.37	01.09.1961	06.09.1961	6	3.33	
		01.04.1961	21.04.1961	21	2.81		07.09.1961	18.09.1961	12	3.79	
		22.04.1961	30.04.1961	9	4.07	95.64	19.09.1961	30.09.1961	12	2.53	95.82
4	El-Nokrashy[h]	01.05.1961	07.05.1961	7	4.07		01.10.1961	20.10.1961	20	1.60	
		08.05.1961	18.05.1961	11	6.70		21.10.1961	31.10.1961	11	1.88	52.68
		19.05.1961	31.05.1961	13	6.07	181.10	01.11.1961	05.11.1961	5	1.88	
		01.06.1961	12.06.1961	12	4.35		06.11.1961	25.11.1961	20	1.50	
		13.06.1961	24.06.1961	12	5.31		26.11.1961	30.11.1961	5	0.80	43.40
		25.06.1961	30.06.1961	6	5.27	149.54	01.12.1961	25.12.1961	25	0.80	
		01.07.1961	06.07.1961	6	5.27		26.12.1961	31.12.1961	6	1.42	28.52
		07.07.1961	18.07.1961	12	4.64						
	Annual										1085.80 mm

h = El-Nokrashy, M.A., 1963

The consumptive use data listed in Tables 6.2 thru' 6.8, together with the
same type of data for other crops not included in this survey, should, no doubt,
form the basis of the irrigation requirements for agriculture in Egypt. Before
undertaking this step, it was felt necessary to compare these data with the use
requirements for the same kind of crops raised outside Egypt under more or less
similar climatic and soil conditions. Comparison may, however, be extended to
consumptive use requirements for the same crops in Egypt as obtained by other
methods whenever needed.

The data in Table 6.2 have been used for developing the consumptive use
curves for cotton in Sakha, northern central part of the Delta, Giza, apex of
the Delta, and Sids and Mallawi, Middle Egypt (Shahin, M., and El-Shal, M.I.,
1969). These curves are shown in Fig. 6.8a. By neglecting the differences in
water use produced mainly by the differences in temperature, length of growing
season, depth to water table, fertilizers, and frequency and quantity of irrigation applications, one can immediately see that Fig. 6.8a. is quite comparable
to Fig. 6.8b. for Mesa and Tempe Arizona, as given by Erie (1963), and to
Fig. 6.8c. for the southern part of Bulgaria (Shahin, M., et al, 1973).

Olivier estimated the consumptive use of water for cotton as follows (1961):

Location	Growing season		Station	Seasonal consumptive use, mm
	from	to		
Lower Egypt (Nile Delta and Giza)	20.2.-10.4	20.8.-10.11	Giza	710
Middle Egypt (from Giza to Assiut)	20.2.-05.4	5.8.-20.10	Assiut	1085
Upper Egypt (from Assiut to southern border of Egypt)	20.2.-05.4	5.8.-15.10	Aswan	1465

These values were obtained from his formula

$$Cu \phi = c \times \frac{L_0}{L^2} \tag{6.2}$$

where

c = average depression of wet-bulb in $^\circ$C for a particular month, and

L_0/L^2 = cyclic (radiation/latitude) factor for the particular month (tabulated
by Olivier (1961)).

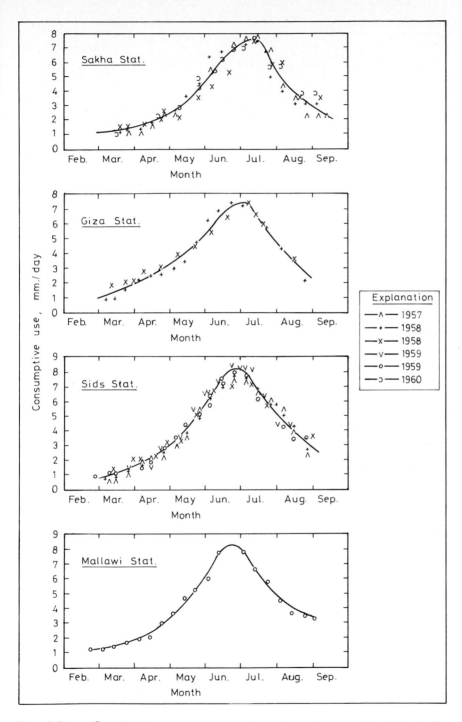

Fig. 6.8a.   Consumptive use of water for cotton at some locations in Egypt
(Shahin, M., and El-Shal, M.I., 1969)

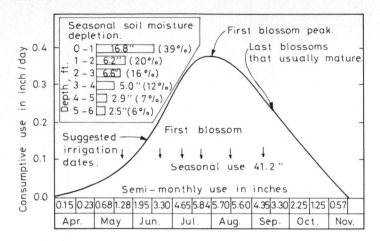

Fig. 6.8b.   Mean consumptive use for cotton at Mesa and Tempe, Arizona, 1954-1962 (Erie, L.J., 1963)

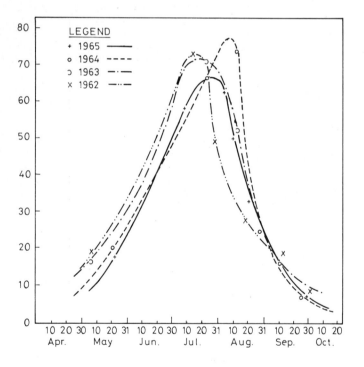

Fig. 6.8c.   Consumptive use for cotton in the southern part of Bulgaria (Shahin, M., et al, 1973)

The figure obtained by Olivier for Lower Egypt and Giza is exactly like that given in Table 6.2. The figure for Mallawi can be compared to the average of the consumptive use at Giza and Assiut. This gives a figure almost 10% larger than the measured water use by cotton at Mallawi.

The figures obtained from Olivier's method for wheat are as follows:

Location	Growing season		Station	Seasonal consumptive use, mm
	From	To		
Lower Egypt	25.10	15.5	Giza	400
Middle Egypt	10.10	20.4	Assiut	472
Upper Egypt	5.10-30.11	10.4-10.5	Aswan	800

The figures obtained by Olivier are about 8% and 10% larger than the figures in Table 6.3 for wheat at Giza and Mallawi respectively.

The consumptive use for wheat at Dujailah experimental station, Iraq, was determined by means of the water-balance method. The seasonal water use is 485 mm with a probable error of from 10 to 20% (Boumans, J.H., et al, 1963). The daily consumptive use from this experiment station is plotted versus the time during growth, so as to compare it with the characteristic use curves derived from the data in Table 6.3. Figs 6.9a and 6.9b show the characteristic curves for Egypt and Iraq, respectively.

In order to illustrate the effects of the moisture level in the soil, the quantity of fertilizers on the evapotranspiration for wheat and the water use efficiency, Haise, R., and Viets, F., have used unpublished data from Marvin E. Jensen, Amarillo experiment station, USDA, Bushland, Texas, for the period 1955-1956 (1957). These data are presented in Table 6.9. The interesting feature about the figures in this table is that experiment M-1 was run under irrigation conditions very similar to those in Egypt before the construction of the High Aswan Dam. The irrigation practice then was to stop irrigating the winter crops for 6 weeks, from 25 December to 5 February each year. This period was known as the winter closure of canals. In the post-dam condition, i.e. from 1965 and onwards, the duration of the canal closure was reduced to about 3 weeks. From that experiment, we see that the average consumptive use for wheat under the three different nitrogen treatments is about 500 mm for the season. This figure is very close to the one obtained by Olivier for Assiut in the southern part of Middle Egypt. It also agrees fairly well with the result obtained from the Dujailah experiment station in Iraq. The estimate of the seasonal water use by the indian corn and Dura maize raised in Lower Egypt (from 5.7-30.8 to 15.10-30.11) using the Olivier method, is 350 mm. The same method gives a seasonal consumption of 950 mm by flood sorghum raised at Aswan (from 20.8 to 15.12).

252

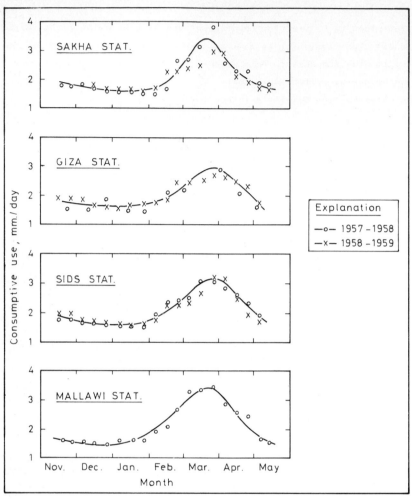

Fig. 6.9a.   Consumptive use of water for wheat at some locations in Egypt (Shahin, M., and El-Shal, M.I., 1969)

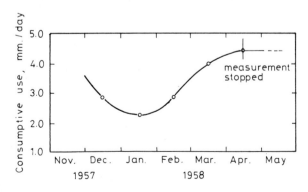

Fig. 6.9b.   Consumptive use of water for wheat at Dujailah, Iraq (Boumans, J.H., et al, 1963)

TABLE 6.9   Evapotranspiration (ET) and water use efficiency (W.U.E.) of winter
             wheat for various nitrogen and moisture levels (Haise, H.R., and
             Viets, F.G., 1957)

Nitrogen applied lbs/acre	Moisture level[*]							
	M-1		M-2		M-3		M-4	
	ET, in	W.U.E. bu/in	ET, in	W.U.E. bu/in	ET, in	W.U.E. bu/in	ET, in	W.U.E. bu/in
0	19.4	0.87	21.6	1.03	22.9	1.28	23.6	1.42
80	19.7	0.92	24.2	1.08	24.8	1.67	30.4	1.51
120	20.3	0.86	23.9	1.18	28.3	1.51	30.2	1.74

[*]M-1: No spring application
M-2: One 4-inch application at jointing stage, March 28
M-3: One 4-inch application prior to boot stage, April 16 and a 4-inch
     application at the flowering stage, May 15
M-4: One 4-inch application at jointing stage, March, 28; one 4-inch
     application at early boot stage, April 30, and a 4-inch application
     just after flowering, May 15

The data listed in Tables 6.4 and 6.5 for late corn and early corn are
presented graphically in Figs. 6.10a and 6.11a, respectively. Fig. 6.10a can be
compared to Fig. 6.10b for zea maize grown in Central California (FAO, 1971).
The latter consumes seasonally 500 mm which is about 5% less than the corresponding figure for Sakha and Giza stations, where the climatic conditions are
almost identical. Moreover, the consumptive use for sorghum at Mesa, Arizona
(from July to October) as found by Harris and compiled by Blaney (1957) is 20.4
inches or 520 mm for the season. This figure is in full agreement with ours for
the stretch from Sakha to Sids. The results obtained by Harrold, L.L., and
Dreibelbis, F.R., (1959) about evapotranspiration from a lysimeter 8.4 m^2 in
surface area naar Coshocton, Ohio, raising corn in the period from May to
October are shown graphically in Fig. 6.11b. This figure can be compared to
Fig. 6.11a showing the use for early corn at Sids, Middle Egypt. The average
evapotranspiration from the lysimeter for the two years 1949 and 1953 is 570 mm
per season. The corresponding seasonal use at Sids in 1964 was 558 mm. The consumptive use for corn has been reported by Blaney, H.F., and Criddle, W.D.,
(1966) as 440, 525 and 740 mm/season for Davis, California, Manden, N. Dakota,
and Redfield, S. Dakota.

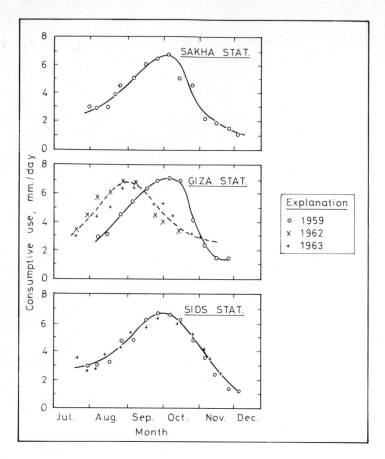

Fig. 6.10a.   Consumptive use of water for late corn at some locations in Egypt (Shahin, M., and El-Shal, M.I., 1969)

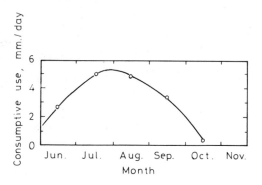

Fig. 6.10b.   Consumptive use of water for zea maize at Central California (FAO, 1971)

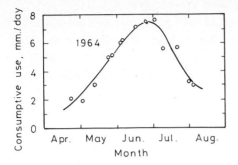

Fig. 6.11a.   Consumptive use of water for early corn at Sids, Egypt (Shenouda, El-Gibali, Tawdros and Gamal, 1966).

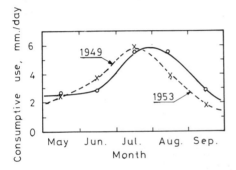

Fig. 6.11b.   Consumptive use of water for corn from a lysimeter near Coshocton, Ohio (Harrold, L.L., and Dreibelbis, F.R., 1959)

The strangest figures for water use by corn (maize) for Egypt one can meet are those obtained by Doorenbos and Pruitt (1977). The estimated figures are based on the climatological data of Cairo, which is very close to Giza station. Using their own method, they came up with the following monthly and seasonal (mid-May to mid-September) figures:

May	June	July	August	September	Season, mm
50	150	310	255	70	855
50	170	340	280	70	910

These estimates are certainly too high and need to be reduced by about 35%. Furthermore, they contradict the range of values given in the same reference. The seasonal consumptive use ranges from 400 to 750 and for sorghum from 300 to 650 mm. The author does not intend to condemn the method Doorenbos and Pruitt have developed for estimating the evapotranspiration, but the crop coefficients, at least for some of the crops, undoubtedly need drastic changes.

The consumptive use curve of berseem as shown in Fig. 6.12a, has an undulating shape which consists of a number of connected curves, each having a base
width equal to the time interval between two consecutive cuttings. The number of
cuttings during the growing season is usually three to four, after which the
land raising berseem is left for seed developing. The marked decline in the
water use by this crop during January and February is caused, in addition to the
low temperature and the first cutting of the crop, by the winter closure already
mentioned in connection with the consumptive use for wheat. As berseem cultivation is confined to a rather limited number of countries, one can hardly find
any information about its use of water in the literature of agricultural hydrology. The results obtained from experiments on berseem irrigation using saline
water in Tunisia (Combremont, R., 1972) are presented in Fig. 6.12b. There, the
growing season of berseem is about 30 days shorter than that in Egypt. The crop
seasonal use of water is 530 mm in Tunisia and 570 mm in the central part of
Middle Egypt.

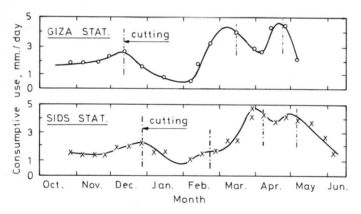

Fig. 6.12a.   Consumptive use of water for berseem at two locations in Middle
Egypt (Shahin, M., and El-Shal, M.I., 1969)

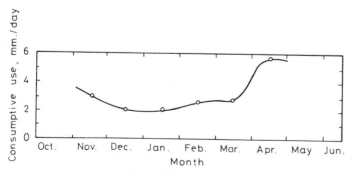

Fig. 6.12b.   Consumptive use of water for berseem at Tunisia (Cambremont, R.,
1972)

The data listed in Table 6.8 are presented graphically, as shown in Fig. 6.13a. Using the climatological data of Cairo and taking the crop coefficient by Doorenbos and Pruitt for citrus trees, which are clean cultivated and provide almost 50% ground cover, one can obtain curve (1) shown in Fig. 6.13b. The annual consumption of the specified orange trees using this method amounts to 1215 mm. This figure is nearly 17% larger than the average water use by the same crop raised in the Delta barrage area and Giza station. The consumptive use of water for oranges raised in the Salt River Valley in USA is presented by curve (2) in Fig. 6.13b. This curve compares fairly well with the two curves shown in Fig. 6.13a.

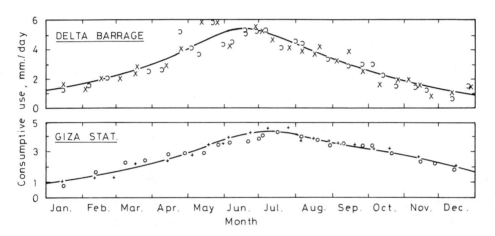

Fig. 6.13a.   Consumptive use of water for citrus trees at the Delta barrage area (El-Nokrashy, M.A., 1963) and at Giza station, Egypt (Shahin, M., 1959)

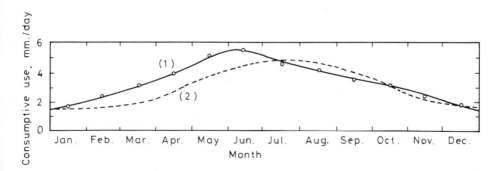

Fig. 6.13b.   Consumptive use of water for citrus trees, (1) estimated from the climatological data of Cairo, Egypt, using the method of Doorenbos and Pruitt (1977) and (2) measured at the Salt River Valley, USA (Criddle, W.D., Harris, K., and Williardson, L.S., 1962)

The data presented in this section, though not representing the consumptive use of water for all crops raised in Egypt, do represent the water use by the major crops which cover the largest areas of the agricultural land there. These crops, together with rice and sugar cane, are certainly the largest consumers of the irrigation water.

The question that arises here is whether the data listed in Tables 6.2 thru' 6.8 represent the actual evapotranspiration or whether they represent the potential one. Before answering this question, one has to remember that these data have been derived from the soil moisture changes in the extraction zone where the moisture content used to reach the field capacity 2 to 3 days after irrigation application and not to fall below the lower limit of the readily available moisture just before the next irrigation application. In other words, two-thirds of the available moisture (field capacity minus permanent wilting point) have been consumed in every irrigation cycle. The proposed concepts about the relation between the actual evapotranspiration, ET_a, and the potential evapotranspiration, ET_p, against the soil moisture content have been reviewed by Tanner (1967) and summarized as shown in Fig. 6.14.

The available measurements of the soil moisture change during an irrigation cycle under the different irrigation, crop and climatic conditions show such a wide scatter that the derivation of a single and consistent relationship between ET_a/ET_p and the moisture content in the range of the readily available moisture is hardly possible. Nevertheless there is strong evidence that the ratio ET_a/ET_p is equal to unity in the upper third of the range of the available moisture. The scatter of the points representing ET_a/ET_p versus moisture in the middle-third can be fitted more or less by a straight line. This line connects the upper limit of the middle-third at $ET_a/ET_p = 1$ with the lower limit of the same one-third portion of the available moisture at $ET_a/ET_p \simeq 0.5$. This line is indicated as b-c in Fig. 6.14. In other words, the change in ET_a/ET_p within the irrigation cycle can be very approximately represented by the broken line abc. This means that the cyclic consumptive use and consequently the monthly and seasonal crop water use is about 10 to 12 percent less than the potential evapotranspiration.

6.1.3   Empirical methods

The last three or four decades have witnessed the development of a large number of empirical formulas which relate the consumptive use or evapotranspiration to climatological measurements. These formulas include one or more of the climatological standards such as mean air temperature, humidity, radiation, wind speed, day-time length, bright sunshine, and tank and pan evaporation.

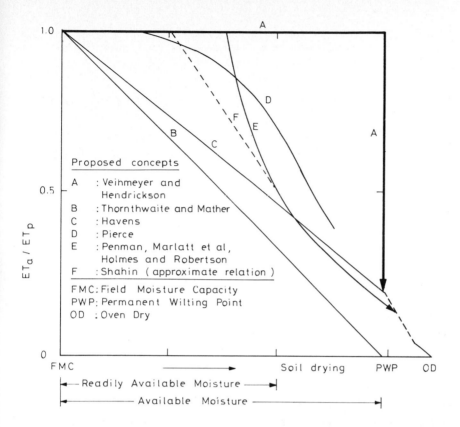

Fig. 6.14.   Proposed relations of actual evapotranspiration ET_a to potential ET_p as affected by soil water content (Tanner, C.B., 1967)

Examples of excellent reviews of the empirical methods can be found in the work of Rijtema (1959) and Slatyer and McIlroy (1961). A comprehensive summary has been given, also by Tanner (1967).

In this section we shall try to give a brief summary of some of the empirical methods. Emphasis will be placed on those methods which can be used for completing the picture of the evapotranspiration from the Nile Basin.

a) Temperature methods - The general form of the formula covering these methods is:

$$ET_p = a + b \cdot d \left(\frac{C_0 T - C_1}{C_2}\right)^n (C_3 - C_4 h) \tag{6.3}$$

where

a, b, C_0, C_1, C_2, C_4 and n are coefficients or constants,

d = measure of day length,

T = temperature in degrees Fahrenheit, and

h = relative humidity

The Lowry-Johnson formula (Lowry, R.L., and Johnson, F.R., 1942) can be written as:

$$ET_p = 0.8 + \frac{0.156}{1000} \Sigma \ (T - 32) \qquad\qquad (6.3.1)$$

where T is the maximum daily temperature and the summation is done over the length of the growing season. ET_p is the seasonal consumptive use in feet of water. The monthly water use is ET_p multiplied by the ratio of the cumulative day degrees (T - 32) for the month considered, to the seasonal day degrees.

The Blaney-Morin formula (Blaney, H.F., and Morin, K.V., 1942) is:

$$ET_p = b \ . \ d \ . \ T \ (114 - h) \qquad\qquad (6.3.2)$$

where

ET_p = monthly evapotranspiration in inches,
b   = monthly crop coefficient,
d   = mean monthly percent of day-time hours of the year,
T   = mean monthly air temperature, and
h   = mean monthly relative humidity in percent

The monthly values of d for the different latitudes can be found in several sources among which is the original paper of Blaney and Criddle (1950, slightly revised 1952).

The Blaney-Criddle method is nothing but a modified form of eq. 6.3.2, in which the term (114 - h) is put equal to unity. It reads

$$ET_p = b \ . \ d \ . \ T \qquad\qquad (6.3.3)$$

Pruitt's formula (Pruitt, W.O., 1960) can be expressed as

$$ET_p = -0.115 + 1.618 \ dT \qquad\qquad (6.3.4)$$

where

ET_p = potential evapotranspiration in inches/day,
T   = mean daily temperature in degrees Fahrenheit, and
d   = daily percentage of daylight hours, expressed decimally.

Quackenbush and Phelan (1965) suggested that the coefficient b in the Blaney-Criddle method, eq. 6.3.3 be split into two coefficients b_1 and b_2. The coefficient b_1 is a crop coefficient that varies during the season, whereas b_2 is a linear function of the mean temperature expressed by the equation

$$b_2 = 0.173 \ T - 0.314 \tag{6.3.5}$$

Al-Barrak (1964) considered that the b coefficient in eq. 6.3.3 of Blaney and Criddle was a linear function of the mean monthly temperature. For irrigated crops in central Iraq the expression he developed for b was

$$b = 0.43 + 0.0074 \ T \tag{6.3.6}$$

where T is the mean monthly temperature in degrees Fahrenheit.

The empirical formula developed by Thornthwaite (1948) is

$$ET_p = \frac{Nd}{30 \times 12} \cdot 1.6 \ (\frac{10 \ T}{C_2})^n \tag{6.3.7}$$

where

N   = actual number of days in the month considered,

d   = mean monthly day length in hours for the month considered,

C_2 = seasonal or annual heat index = $\Sigma \ (T/5)^{1.514}$ where the summation is done for the months of the season or of the year, and

n   = $675 \times 10^{-9} \ C_2^{3} - 771 \times 10^{-7} \ C_2^{2} + 17921 \times 10^{-6} \ C_2 + 0.49239$

Values of the heat index C_2 and the power n for a wide range of temperatures T are available in a number of references as the paper of Thornthwaite and Mather (1955).

Hargreaves developed the formula (1956):

$$ET_p = 0.38 \ b \cdot d \ (T - 32)(1 - h) \tag{6.3.8}$$

where

d = day-time measure = 0.12 the monthly percent of the day-time hours of the year,

h = mean monthly relative humidity at noon, expressed decimally

Monthly values of d for the different latitudes are tabulated in the original paper of Hargreaves. Later he proposed some corrections for the deviation of the actual sunshine percentage, wind speed, and elevation from the standard values on which eq. 6.3.8 had been based (Hargreaves, G.H., 1966). These corrections are given in Chapter 5. In 1968 Hargreaves developed the consumptive use coefficients for a large number of crops considering 10% successive increments of the crop growing season (1968).

The formula developed recently by Doorenbos and Pruitt (1977) suggests the replacement of the coefficient b in the formula of Blaney and Criddle, eq. 6.3.3, by two sub-coefficients b_1 and b_2. The sub-coefficient b_1 depends on the crop and its rate of development during the growing season. The sub-coefficient b_2 is an adjustment factor which depends on the minimum relative humidity, sunshine hours and day-time wind estimates. Values of the adjustment factor b_2 can be read from the graphs prepared by Doorenbos and Pruitt.

For ease of reference, the constants and coefficients included in the above temperature formulas have been summed up and listed in Table 6.10.

Temperature methods are often criticized on the grounds that the effect of the climate on the crop water use cannot be defined adequately by the temperature and a measure of the length of day only. In the humid tropics where temperature remains fairly constant, the crop water use changes as a consequence of change in other meteorological parameters. At high altitudes the high level radiation may cause reasonably high consumption of water by plants, despite the fairly low temperature. Moreover, the temperature lag corrections are rarely introduced explicitly in the temperature formulas. On the other hand, the availability of the temperature records everwhere in quantity and quality compared to other climatological parameters encourages many professionals to use the temperature methods extensively. One should not forget that they are easier to apply than most of the remaining methods. In any case, estimates of consumptive use by the temperature methods have, as a rule, to be calibrated before being used for any practical purpose.

The current practice in Egypt is to use the soil moisture depletion studies for calibrating such widely used formulas as Blaney-Criddle and Thornthwaite, given by eqs. 6.3.3 and 6.3.7 respectively. The monthly and seasonal consumptive use coefficients for a large number of crops in Egypt obtained from this calibration are given in Tables 6.11a and 6.11b. These coefficients are not increased by the 10 to 12% that we proposed in connection with the relation between ET_a and ET_p during an irrigation cycle.

TABLE 6.10 Summary of the constants and coefficients included in some of the temperature formulas

Formula	a	b	d	c_0	c_1	c_2	n	c_3	c_4	Remarks
Lowry-Johnson[i]	0.8	156×10^{-6}	1.0	1.0	32	1.0	1.0	1.0	0	ET_p, ft/season T=maximum temperature
Blaney-Morin[ii]	0.0	coefficient	coeff.	1.0	0.0	1.0	1.0	114	1.0	ET_p, in/month in percent
Blaney-Criddle[iii]	0.0	coefficient	coeff.	1.0	0.0	1.0	1.0	1.0	0.0	ET_p, in/month
Pruitt[iv]	-.115	coefficient	coeff.	1.0	0.0	1.0	1.0	1.0	0.0	ET_p, in/month
Quackenbush-Phelan[v]	0.0	coefficient	coeff.	1.0	0.0	1.0	1.0	1.0	0.0	ET_p, in/month. b is a product of crop coefficient and function of temperature
Al-Barrak[vi]		linear func. of temp.	coeff.	1.0	0.0	1.0	1.0	1.0	0.0	ET_p, in/month
Thornthwaite[vii]		const. = 1.6	adjusted coeff.	10.0	0.0	func. of temp.	func. of c_2	1.0	0.0	ET_p, cm/month
Hargreaves[viii]	0.0	coefficient	coeff.	1.0	32	1.0	1.0	1.0	1.0	ET_p, in/month. h=relative humidity at noon
Doorenbos-Pruitt[ix]	0.0	coefficient	coeff.	.46	-8	1.0	1.0	1.0	0.0	ET_p, mm/day. T in °C. b is a product of crop coeff. and func. of min. hum. wind speed sunshine

i = Lowry, R.L., and Johnson, F.R., 1942; ii = Blaney, H.F., and Morin, K.V., 1942; iii = Blaney, H.F., and Criddle, W.D., 1950, slightly revised 1952; iv = Pruitt, W.O., 1960; v = Quackenbush, T., and Phelan, J.T., 1965; vi = Al-Barrak, A.H., 1964; vii = Thornthwaite, C.W., 1948; viii = Hargreaves, G.H., 1966; ix = Doorenbos, J., and Pruitt, W.O., 1977

TABLE 6.11a Consumptive use coefficients for some crops in Egypt to be used with the Blaney-Criddle formula

Crop	Consumptive use coefficient for												Season or year
	Jan.	Feb.	Mar.	Apr.	May	June	July	Aug.	Sep.	Oct.	Nov.	Dec.	
Cotton	-	-	.26	.39	.66	1.07	1.03	.56	.51	-	-	-	0.71
Wheat	.49	.59	.66	.49	.31	-	-	-	-	-	.46	.47	0.51
Early Corn	-	-	-	.36	.71	1.06	.87	.60	-	-	-	-	0.79
Late Corn	-	-	-	-	-	-	.51	.91	1.13	.80	-	-	0.88
Berseem	.31	.42	.72	.82	.64	.25	-	-	-	.31	.37	.62	0.56
Citrus Orchards	.35	.51	.55	.57	.60	.64	.70	.64	.59	.56	.42	.32	0.54
Fenugreek & Lupin	.56	.57	.26	-	-	-	-	-	-	-	.38	.59	0.46
Chickpea & Lentil	.42	.46	.18	-	-	-	-	-	-	-	.34	.50	0.37
Sugar Cane	.29	.72	.87	.84	.90	1.00	1.36	1.37	1.32	1.00	.79	.41	0.91
Field Beans	.54	.73	.67	.48	-	-	-	-	-	-	-	-	0.58
Potatoes (summer)	-	.63	.77	1.01	.69	-	-	-	-	-	-	-	0.77
Potatoes (fall)	.86	-	-	-	-	-	-	-	-	.19	.49	.98	0.59
Snap Beans	-	-	-	-	.35	.80	-	-	.37	.80	1.14	.78	0.77
Cow Peas	-	-	-	.30	-	-	.99	.60	-	-	-	-	0.64
Squash	-	-	-	-	-	-	.44	1.02	1.01	-	-	-	0.87
Cucumber	-	-	-	-	-	-	.30	.63	.87	.79	-	-	0.69

TABLE 6.11b Consumptive use coefficients for some crops in Egypt to be used with Thornthwaits's formula

Crop	Consumptive use coefficient for												Season or year
	Jan.	Feb.	Mar.	Apr.	May	June	July	Aug.	Sep.	Oct.	Nov.	Dec.	
Cotton	–	–	0.75	0.72	0.98	1.33	1.23	0.66	0.73	–	–	–	0.97
Wheat	2.15	2.78	1.89	.89	.42	–	–	–	–	–	.97	1.67	1.54
Early Corn	–	–	–	.37	.96	1.28	1.01	.56	–	–	–	–	0.93
Late Corn	–	–	–	–	–	–	.59	1.00	1.54	1.40	–	–	1.08
Berseem	1.49	1.65	2.18	1.59	.98	.50	–	–	–	.28	.82	2.43	1.29
Citrus Orchards	1.47	2.46	1.51	1.14	.88	.82	.78	.76	1.25	1.00	.86	1.14	0.94
Fenugreek & Lupin	2.86	2.32	0.40	–	–	–	–	–	–	–	.44	1.43	1.96
Chickpea & Lentil	2.13	1.88	0.26	–	–	–	–	–	–	–	.48	1.21	1.56
Sugar Cane	1.22	3.44	2.39	1.68	1.32	1.28	1.52	1.63	2.80	1.79	1.62	1.46	1.85
Field Beans	2.44	3.44	1.84	.96	–	–	–	–	–	–	–	–	2.08
Potatoes (summer)	–	2.97	2.06	2.02	1.01	–	–	–	–	–	–	–	1.51
Potatoes (fall)	3.77	–	–	–	–	–	–	–	0.78	0.34	1.00	3.49	2.01
Snap Beans	–	–	–	–	–	–	–	–	–	1.43	2.33	2.78	1.82
Cow Peas	–	–	–	0.55	0.52	0.99	1.18	0.71	–	–	–	–	0.83
Squash	–	–	–	–	–	–	.52	1.21	2.14	–	–	–	1.36
Cucumber	–	–	–	–	–	–	.36	0.75	1.85	1.41	–	–	1.16

b) Radiation methods - The radiation methods developed for estimating potential evapotranspiration are either based on the heat energy budget, together with some empirical approximations to utilize the available climatic data, or they are completely empirical. In the first group is the formula of Penman, which we already presented in Chapter 5 as a means for estimating evaporation from a free water surface, E_o. The same formula when used for estimating the potential evapotranspiration reads:

$$ET_p = \alpha \cdot E_o \tag{6.4}$$

where α is a coefficient verying with the month.

A large number of empirical formulas, all based on one form or another of radiation, have been developed for estimating potential evapotranspiration. The general form of the equation embracing these formulas can be written as

$$ET_p = K \cdot C \cdot (R + a) \tag{6.5}$$

In 1961, Turc, L. (1961), presented his formula:

$$ET_p = 0.40 \left(\frac{t}{t + 15}\right)(R + 50) \qquad mm/month \tag{6.5.1}$$

where

t = mean monthly temperature in degrees centigrade, and

R = mean incoming radiation in gm cals/cm^2.day.

R can be computed from the theoretical radiation, R_A, reaching the earth's atmosphere using the relation $R = R_A(0.18 + 0.0062 S)$, where S is the percentage of possible sunshine.

The formula developed by Jensen and Haise (1963) is:

$$ET_p = (0.014 T - 0.37) R_{so} (0.35 + 0.61 S) \quad inches/month \tag{6.5.2}$$

where

T   = mean monthly temperature in degrees Fahrenheit,

R_{so} = solar radiation on cloudless days, and

S   = possible sunshine expressed decimally.

Christiansen, J.E., (1969) considered a = 0 in eq. 6.5 in order to estimate evaporation or evapotranspiration. He and his co-workers have developed an extensive number of equations, in which k is kept as a dimensionless constant and the product kc represents the ratio of the energy utilized in the evaporation process to the energy available at the outer surface of the atmosphere. The coefficient C is the product of a large number of sub-coefficients, each related to a climatic or other parameter that is likely to affect evaporation or evapotranspiration. The formula developed by Christiansen which relates ET_p to extraterrestrial radiation as a base is

$$ET_p = 0.324 \ R_A \cdot C_{TT} \cdot C_{WT} \cdot C_{HT} \cdot C_{ST} \cdot C_{El} \quad \text{inches/month} \tag{6.5.3}$$

where

R_A = extraterrestrial radiation in equivalent evaporation in inches/month,

C_{TT} = $0.174 + 0.428 \ (T/T_o) + 0.398 \ (T/T_o)^2$

T = the mean air temperature in degrees Fahrenheit and $T_o = 68^o F$,

C_{WT} = $0.672 + 0.406 \ (W/W_o) - 0.078 \ (W/W_o)^2$

W = the wind speed in miles per day 2.0 m above ground surface and W_o = miles per day,

C_{HT} = $1.035 + 0.240 \ (H_m/H_{mo})^2 - 0.275 \ (H_m/H_{mo})^3$

H_m = the mean relative humidity in decimals and $H_{mo} = 0.60$,

C_{ST} = $0.340 + 0.856 \ (S/S_o) - 0.196 \ (S/S_o)^2$

S = mean sunshine percentage in decimals and $S_o = 0.80$, and

C_{El} = $0.970 + 0.030 \ (El/El_o)$

El = elevation in ft above mean sea level and $El_o = 300$ ft

The sub-coefficients C_{TT}, C_{WT}, C_{HT}, C_{ST} and C_{El} can be read directly from the tables prepared by Christiansen for any given value for T, W, H_m, S and El respectively.

The formula of Olivier (1961) has already been given as eq. 6.2.

Doorenbos recommends two relationships (Doorenbos, J., and Pruitt, W.O., 1977). The one that suits the empirical type of radiation formulas is given by

$$ET_p = k \cdot C \cdot W \cdot R_s \qquad , \ \text{mm/day} \tag{6.5.4}$$

where

k = crop coefficient,

C = adjustment factor which depends on mean humidity and day-time wind conditions,

W = weighting factor which depends on temperature and altitude, and

R_s = solar radiation in equivalent evaporation in mm/day.

R_s is related to R_A as $R_s = R_A$ $(0.25 + 0.50 \frac{n}{N})$ where $\frac{n}{N}$ is the ratio between the actual measured bright sunshine hours and the maximum possible sunshine hours.

Tables and graphs needed to give C and W have already been prepared by Doorenbos and Pruitt. Eq. 6.5.4 is very similar to eq. 6.5.3 originally developed by Christiansen.

The second relation recommended by Doorenbos and Pruitt (1977) is an adjustment to Penman's formula. They gave it as

$$ET_p = k \cdot c \left\{ W.R_n + (1 - W) \cdot f(u) \cdot (e_s - e_a) \right\} \qquad mm/day \qquad (6.6)$$

where

k   = crop coefficient,

c   = adjustment factor to compensate for the effect of day and night weather conditions,

W   = temperature-related weighting factor,

R_n = net radiation in equivalent evaporation in mm/day

            R_n is the sum of the net short wave radiation, R_{ns}, and net long wave radiation, R_{nl},

f (u)   = wind function = 0.27 (1 + 0.01 u), u is the 24-hr wind run in km/day at 2 m height, and

$(e_s - e_a)$ = saturation vapour pressure deficit = $e_s(1 - h)$ in millibar.

Values of c, W, (1 - W), e_s, e_d, R_A in equivalent evaporation units, N, R_{ns} and R_{nl} are given in tables in the Doorenbos and Pruitt paper (1977). Tanner, C.B. (1967) mentions that because radiation methods are tied more closely to energy supply, they show greatest promise for short-term, as well as long-term, estimates. Doorenbos and Pruitt also reported on this matter saying that "the radiation method should be more reliable than the Blaney-Criddle approach". In fact, in equatorial zones, on small islands or at high altitudes, the radiation method may be more reliable even if measured sunshine or cloudiness data are not available; in this case, solar radiation prepared for most locations in the world should provide the necessary solar radiation data (Doorenbos, J., and Pruitt, W.O., 1977).

The author does not want to argue here about either the superiority of one method or one formula to another. The point which is necessary to bear in mind is that every formula needs to be calibrated. When the author had to calibrate a large number of formulas, much larger than those included in this text, using measurements from various countries, the conclusion was that each formula needed to be adjusted. There is not, nor very likely will be, a single formula of

universal applicability without correction or adjustment. The importance of
local factors, biological factors, and time factors have been realized and
reported by several investigators, including Penman, H., in his study of evaporation over the British Isles (1950). The stress should then be laid on determining or knowing the adjustment factor precisely and not on the method or the
formula only. The monthly and seasonal figures of the adjustment factor needed
for the formulas of Penman and Olivier are given in Tables 6.12a and 6.12b respectively.

TABLE 6.12a   Consumptive use coefficients for some major crops in Egypt to be
used with Penman's formula

| Crop | Consumptive use coefficient for | | | | | | | | | | | | Season or year |
	Jan.	Feb.	Mar.	Apr.	May	June	July	Aug.	Sep.	Oct.	Nov.	Dec.	
Cotton	–	–	0.27	0.38	0.57	0.88	0.91	0.50	0.45	–	–	–	0.61
Wheat	0.82	0.73	0.84	0.48	0.28	–	–	–	–	–	0.74	0.93	0.73
Early Corn	–	–	–	0.34	0.62	0.89	0.80	0.52	–	–	–	–	0.70
Late Corn	–	–	–	–	–	–	0.45	0.90	1.15	1.12	0.89	–	1.00
Berseem	0.48	0.68	0.79	0.80	0.60	0.22	–	–	–	0.43	0.62	1.30	0.72
Citrus Orchards	0.54	0.56	0.60	0.55	0.56	0.57	0.62	0.62	0.60	0.77	0.71	0.67	0.62
Sugar Cane	0.45	0.79	0.95	0.81	0.84	0.89	1.20	1.33	1.34	1.37	1.33	0.86	1.03

TABLE 6.12b   Consumptive use coefficients for some major crops in Egypt to be
used with Olivier's formula

| Crop | Consumptive use coefficient for | | | | | | | | | | | | Season or year |
	Jan.	Feb.	Mar.	Apr.	May	June	July	Aug.	Sep.	Oct.	Nov.	Dec.	
Cotton	–	–	0.50	0.67	0.72	1.33	1.41	0.87	–	–	–	–	0.92
Wheat	2.00	1.57	1.60	0.75	0.33	–	–	–	–	–	1.54	2.22	1.45
Early Corn	–	–	–	0.53	0.80	1.30	1.23	0.86	–	–	–	–	1.02
Late Corn	–	–	–	–	–	–	0.76	1.28	1.65	2.25	2.66	–	1.39
Berseem	1.28	1.25	1.46	1.20	0.72	0.31	–	–	–	0.82	1.24	2.86	1.09
Citrus Orchards	1.42	1.31	0.99	0.86	0.81	0.92	0.92	1.02	1.06	1.32	1.46	1.74	1.01
Sugar Cane	1.18	1.85	1.57	1.27	1.22	1.44	1.78	2.19	2.37	2.35	2.73	2.23	1.68

c) Evaporation pans - Pan evaporation and plant evapotranspiration are similar,
since each of them can be considered as a measure of the integrated effect of
the climatological factors on the loss of water by evaporation or evapotranspiration. The differences in both reflection of the solar radiation and the
exchange of heat energy from a vegetated surface compared to those from water in

a pan, added to the influences of pan size, colour, exposure and placement are among the causes that can make pan evaporation differ considerably from plant evapotranspiration. Adjusting the pan reading is, therefore, inevitable, in order to convert it to its equivalent of evapotranspiration. The general form of the equation needed for adjusting the pan evaporation, E pan, is

$$ET_p = a \ (E_{pan})^n + b \qquad\qquad (6.7)$$

The formula given by Penman, and by Doorenbos and Pruitt is

$$ET_p = a \ . \ E_{pan} \qquad\qquad (6.7.1)$$

where a is a monthly or a seasonal coefficient as given by Penman, and a product of two sub-coefficients for the pan, k_p, and the crop, k_c, as given by Doorenbos and Pruitt (1977).

Stanhill's regression relation between the daily evapotranspiration, in mm, from Alfalfa and the daily evaporation, in mm, from a class A pan (1961) is:

$$ET_p = 0.70 \ E_{pan} + 0.47 \qquad\qquad (6.7.2)$$

Butler and Prescott (1955) observed that the power n in eq. 6.7 is in the neighbourhood of 0.75 instead of 1.0, so their formula can be written as

$$ET_p = a \ . \ (E_{pan})^{0.75} \qquad\qquad (6.7.3)$$

The overall adjustment coefficient a, eq. 6.7.1 will just be called the crop coefficient. Values of a for the major crops in Egypt are included in Table 6.13.

TABLE 6.13   Monthly and seasonal crop coefficients to be used with USWB class A evaporation pan for the major crops in Egypt

Crop	Consumptive use coefficient for												Season or year
	Jan.	Feb.	Mar.	Apr.	May	June	July	Aug.	Sep.	Oct.	Nov.	Dec.	
Cotton	–	–	0.24	0.27	0.42	0.53	0.69	0.44	0.39	–	–	–	0.43
Wheat	0.64	0.52	0.45	0.26	0.14	–	–	–	–	–	0.42	0.56	0.49
Early Corn	–	–	–	0.22	0.37	0.52	0.52	0.33	–	–	–	–	0.45
Late Corn	–	–	–	–	–	–	0.34	0.57	0.62	0.62	–	–	0.55
Berseem	0.29	0.53	0.66	0.48	0.18	–	–	–	–	–	0.46	0.72	0.51
Citrus Orchards	0.39	0.42	0.42	0.37	0.40	0.41	0.41	0.40	0.37	0.39	0.44	0.44	0.41
Sugar Cane	0.32	0.49	0.67	0.55	0.60	0.64	0.79	0.86	0.83	0.69	0.82	0.56	0.66

Investigation of the figures in Table 6.13 shows that the maximum values of
the monthly crop coefficients to be used with pan evaporation are distributed
throughout the year more uniformly than with any other method. The same conclusion was reached by Tanner while using ET_p data of irrigated rye grass which
were obtained by Pruitt. The distribution of the monthly coefficient is very
nearly rectangular, with a value of about 0.8 (Tanner, C.B., 1967).

6.1.4   The integration method

We should have liked to use the data from the USWB class A pan, after converting them, preparing a map showing the potential evapotranspiration for Egypt.
Unfortunately, only a few stations are equipped with class A pan. This state of
affairs compels us to abandon the pan method, though it seems promising. The
method that serves as a reasonable alternative is that of Blaney-Criddle since
it has been calibrated frequently using a wide variety of irrigated crops. The
only major crop for which the Blaney-Criddle formula has not yet been calibrated
is rice. To cover this deficiency we shall employ the consumptive use coefficients for rice grown in the dry areas in California, USA. These are approximately 1.1, 1.2, 1.3, 1.3, and 1.0 for May through September respectively. These
figures, together with use coefficients of other crops which are given in Table
6.11a, have been plotted against the months of the year, and the curve envelopping the monthly maximum values drawn as shown in Fig. 6.15. This curve represents the potential evapotranspiration requirement averaged for Egypt. The
ET_p coefficient averaged over the year is about 1.1. It may therefore be
reasonable to apply an annual ET_p coefficient of 1.00 for the Delta area, 1.05
for the area north of Minya up to Giza, 1.10 for the area from Assiut up to
Minya, 1.15 to the area from Qena up to Assiut and 1.20 for the area south of Qena.
These figures have been used in preparing the map of the potential evapotranspiration requirement for Egypt, Fig. 6.16. This map shows an evapotranspiration
requirement slightly different, $\pm$ 5%, from that for the part of Egypt shown in
the map in Fig. 6.2. Moreover, the contour interval in Fig. 6.16 is 100 mm/yr,
whereas the contour interval in the other map is 250 mm/yr.

Both maps, Figs. 6.2 and 6.16 give the requirements for potential evapotranspiration. A survey of the water use by crops, even at the potential evapotranspiration level, requires a knowledge of the crop pattern, area and development
both in time and space. This can be done using the so-called integration method
(Israelsen, O.W., 1956). In this method the agricultural area is divided into a
number of blocks where the crop pattern is fairly homogeneous and the year is
divided into a number of intervals. The water use by a crop is computed for each
interval during the growing season, and the total water use for a given interval
is the sum of water use by the different crops in that interval.

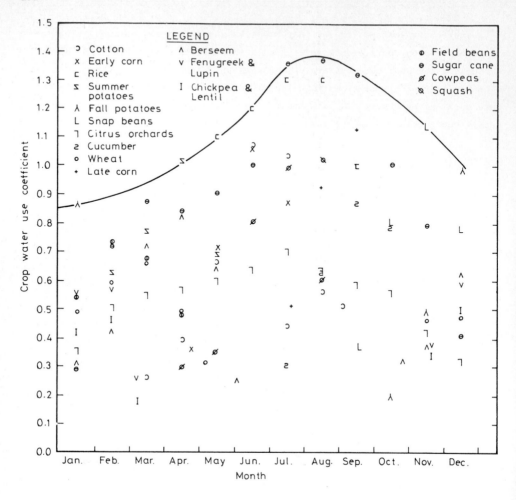

Fig. 6.15.   Monthly coefficients of water use by crops in Egypt for the method of Blaney-Criddle. The envelopping curve represents the average potential evapotranspiration requirement for Egypt.

We have used the integration method to determine the water use of the cropped area in Egypt twice. One time is for the 10.4 million feddans[*] cropped area in 1962, i.e. shortly before the High Dam at Aswan, and the other time for the 11.2 million feddans cropped area in 1975, i.e. after the operation of the dam. In each case the month is chosen as a time interval and the Governrate as a

---

[*] 1 feddan = 4200.6 m^2

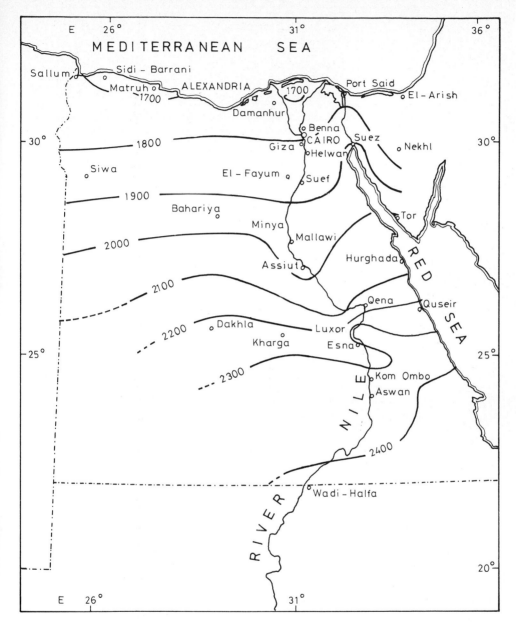

Fig. 6.16.   Map showing lines of equal potential evapotranspiration requirement mm/yr, for Egypt

geographic unit. The monthly consumptive use is the sum of the water uses by the different crops raised in the month considered for all the Governrates of the country. The computation results are presented in Table 6.14.

TABLE 6.14   Quantities of water used by the crops in 1962 and 1975 in Egypt, 10^6 m^3

Year	Jan.	Feb.	Mar.	Apr.	May	June	July	Aug.	Sep.	Oct.	Nov.	Dec.	Year
1962	1047	856	1610	1833	2968	3304	3532	3174	2467	2248	1666	1904	26609
1975	1118	919	1740	2245	4303	5147	5065	2997	1591	1195	1414	2034	29769

The remarkable feature about these results is the dispropirtionate increase in the total use of water by crops, 12%, compared to the increase in the cropped area, 7.7%. This is quite understandable since the greatest bulk of the increase in the cropped area is occupied by summer crops which consume much water, especially rice. Also, most of the area previously irrigated during and following the flood (August-November) in the pre-dam time is now also irrigated during the summer season (April-August) in the post-dam time. Last but not least is the shortening of the duration of the canals' winter closure from 40-45 days before building the dam to 18-21 days after the dam has been built.

6.2   EVAPOTRANSPIRATION IN THE SUDAN

Agriculture in the Sudan does not depend entirely on water supplied by irrigation as in Egypt. The Sudan can be divided into three zones: the southern rain belt which extends from the southern frontier of the country up to an isohyetal line of about 350 mm/yr, and the third zone which covers the remaining part of the country. It is in this last zone that irrigation is practised extensively. Supplementary irrigation is used to some extent in the northern rain belt and to a much less extent in the southern one. The development of irrigation in the Sudan is very much connected with the construction of the Sennar and later the Roseires Dams on the Blue Nile and Khashm el Girba Dam on the River Atbara.

Various crops grow in the Sudan, of which cotton and corn (Dura) can be considered as the principal ones. The introduction and thereafter the expansion of irrigation in the Gezirah area (the triangle confined between the White and the Blue Niles) has resulted in a considerable increase in the area growing cotton from 250 feddans in 1912 to 194000 feddans in 1932 and to about 217000 feddans in 1946. The total area raising cotton in the Sudan has expanded from about 450000 feddans in 1946 to about 1180000 feddans in 1972. Whereas cotton is the chief source of the Sudan's earning from the foreign currency, the Dura (corn, millet, sorghum, bulrush millet, ...), which grows in an area of more than five million feddans, is the main element entering people's food and drink.

The weighted, drained, and floating types of lysimeters have been used for measuring evapotranspiration rates from a number of crops such as wheat, cotton, lucerne and broad beans. Moisture changes have been studied in controlled field

plots and evapotranspiration estimated for some crops from Olivier, Penman, and
the heat-balance methods.

6.2.1   Estimates of evapotranspiration using empirical methods

Olivier, using his method, eq. 6.2, estimated the evapotranspiration for some
crops at Wad-Medani, for grass at Malakal and for the marshes in the Sudd region
at Shambe (inside the swamps of the Bahr el Jebel). The results obtained from
his method are given in Table 6.14.

TABLE 6.14   Evapotranspiration for some crops and vegetation in the Sudan as
estimated by Olivier (1961)

Month	Evapotranspiration, mm/day, for					
	Wad-Medani				Malakal	Swamps (Shambe)
	Cotton	Wheat	Dura	Lubia	Grass	Papyrus
January	4.65	4.65	–	4.65	6.10	6.62
February	6.16	6.16	–	–	7.60	7.43
March	8.69	8.69	–	–	8.69	7.22
April	10.08	–	–	–	5.99	6.01
May	–	–	–	–	3.61	4.46
June	–	–	–	–	2.22	3.17
July	–	–	3.46	–	1.40	2.86
August	2.23	–	2.23	–	1.35	2.70
September	2.95	–	2.95	2.95	1.72	2.98
October	4.75	–	4.75	4.75	1.98	3.42
November	5.06	5.06		5.06	4.12	4.50
December	3.92	3.92		3.92	5.28	5.61
Season or year	1265	689	261	562	1517	1729

It has been reported by El-Nadi (1969) that wheat grown in the experimental
farm of the Faculty of Agriculture, Khartoum University, can consume water up
to 675 mm during its growing season. He also reported that increased yields of
broad beans were obtained from the sixth to the ninth irrigation, the depth per
irrigation being 75 mm (1970). We have estimated the consumptive use requirement
for the same four crops at Wad-Medani, using two different empirical formulas.
The first one is that given by Hargreaves, which estimates the evapotranspiration as the product of the measured or the calculated pan evaporation, eq. 5.9,
times a crop coefficient which depends on the percentage of the growing season
(1966). The second formula is the Blaney-Criddle one. The results obtained from
the calculations are given in Table 6.15.

TABLE 6.15   Estimates of water use by some crops at Wad-Medani using the
              Hargreaves and the Blaney-Criddle formulas

Month	Crop water use, mm/day, for							
	Hargreaves formula				Blaney-Criddle formula			
	Cotton	Wheat	Dura	Lubia	Cotton	Wheat	Dura	Lubia
January	9.47	6.69	–	4.46	4.71	5.41	–	2.65
February	9.07	10.44	–	–	3.75	4.44	–	–
March	7.58	10.35	–	–	1.93	2.23	–	–
April	6.45	–	–	–	1.27	–	–	–
May	–	–	–	–	–	–	–	–
June	–	–	–	–	–	–	–	–
July	–	–	3.82	–	–	–	1.83	–
August	1.43	–	5.86	–	1.00	–	5.14	–
September	3.25	–	7.43	2.32	3.45	–	4.80	0.86
October	7.36	–	7.46	6.11	5.40	–	1.34	3.43
November	10.19	2.50	–	11.57	5.75	1.88	–	5.36
December	10.24	4.41	–	8.91	5.32	3.61	–	3.92
Season or year	1842	885	544	943	964	480	347	461

The discrepancy between the results obtained from the three methods, i.e.
Olivier, Hargreaves and Blaney-Criddle, is wide indeed. All the figures obtained
from the Hargreaves method are on the high side, whereas the figures found from
the Blaney-Criddle method are, except for Dura, on the low side.

Since the available evapotranspiration measurements for the Sudan are rather
limited, one has to search for the most reasonable estimates from the empirical
and/or other methods that may prove relevant. We have investigated the validity
of the estimates of the class A pan evaporation from the Hargreaves equation,
using the direct measurements available at Khartoum and the indirect figures for
Wad-Medani. The data listed in Table 6.16 show that the mean annual estimated
pan evaporation for Khartoum is almost 10% bigger than the measured one. The
largest differences were 20% and 14% for December and January respectively. For
the remaining months of the year the differences are at, or below, 10%. The
estimated pan evaporation from the equation of Hargreaves on an annual basis
agrees perfectly with the pan evaporation found indirectly. The largest difference in a month is in the order of 10%. This short discussion leads us to the
conclusion that the wide discrepancy between the evapotranspiration computed
from Hargreaves' method and the other two methods is not mainly caused by the
basic element in the method, i.e. the estimate of the class A pan evaporation.

It is quite probable that Hargreaves developed his equation on the grounds
that the evaporation from a free water surface is about 0.75 times the evaporation from a class A pan. This is true in many cases but not in every case. The
measured pan evaporation at Khartoum is 5771 mm/yr, whereas the free water

TABLE 6.16   Comparison between measured and estimated class A pan evaporation
from Hargreaves' method for Khartoum and Wad-Medani

| Month | Class A pan evaporation, mm/day, for | | | |
| | Khartoum | | Wad-Medani | |
	measured	estimated	measured[*]	estimated
January	12.60	14.43	10.44	11.14
February	14.70	16.57	11.88	12.43
March	17.23	19.43	14.45	14.57
April	21.42	23.71	18.77	18.43
May	23.13	24.28	19.78	17.86
June	19.31	21.71	15.87	16.14
July	13.17	15.43	10.03	11.57
August	10.38	10.86	8.09	8.14
September	12.19	12.29	10.82	9.29
October	16.75	18.00	14.12	13.57
November	15.83	18.00	12.74	13.14
December	13.03	15.71	10.28	11.57
mean annual	15.81	17.53	13.11	13.15

[*]Penman evaporation times the ratio E_{pan}:E_{Penman} available for
Khartoum

surface evaporation is about 8 mm/day or 2920 mm/yr (see Chapter 5). The
indirectly found pan evaporation at Wad-Medani is 4785 mm/yr whereas the evaporation from a free surface of water is 7.5 mm/day or 2738 mm/yr (see Chapter 3).
These pairs of evaporation data show that the pan coefficient is about 0.51 for
Khartoum and about 0.56 for Wad-Medani. Such relatively small values of the pan
coefficient can be caused by low humidity, strong wind, or distance to a green
cropped area or dry fallow land or any combination of the two. This state of
affairs inclines us to agree with Doorenbos and Pruitt (1977) that the crop
water use can be expressed as the product of a crop coefficient, k_c, times the
so-called reference evapotranspiration, ET_o. The latter is also the product of
the pan evaporation times the pan coefficient, k_{pan}. In other words, the crop
evapotranspiration can be related to the pan evaporation by the equation

$$ET_p = k_c \cdot ET_o = k_c \cdot k_p \cdot E_{pan} \qquad (6.8)$$

The remarkable thing here is that the estimated seasonal evapotranspiration
by the Hargreaves method when calibrated by the corresponding estimates from the
Blaney-Criddle method shows that k_p has the values of 0.523, 0.542, 0.638 and
0.489 for cotton, wheat, dura and lubia respectively. The average value, being
0.55, is very close to the 0.56 for the annual evaporation from open water com-

pared to the annual evaporation from a class A pan at Wad-Medani. It is becoming clear that the empirical methods may provide us with two possibilities to solve the problem. The first is to observe or calculate the evaporation from a class A pan, using Hargreaves' equation, reduce it by the pan coefficient and reduce it further by the Hargreaves crop coefficient. The second possibility, which is faster and easier than the first one, is to apply the Blaney-Criddle formula with the appropriate consumptive use coefficient. These two possibilities seem to be adequate at least for the arid and the semi-arid zones in the Sudan. However, before any measure can be undertaken to work out this statement, one needs to check whether the figures obtained from the Blaney-Criddle method are valid or not.

6.2.2   The energy-balance method

One of the methods used for estimating the water use by irrigated cotton in the Gezirah area is a combination of micrometeorological techniques. These techniques, as described by Rijks, D.A., (1971), consist of four components: the vertical and horizontal fluxes during the day and night. Vertical fluxes were estimated either from the net radiation and Bouwen ratios or from the Thornthwaite-Holzman formula; horizontal fluxes from wind speeds and wet- and dry-bulb temperatures measured at various heights above the crop and at different distances from the leading edge of the experimental field.

The measurements were taken on two fields adjoining the Gezirah research station (14° 24'N, 33° 29'E, 407 m altitude). The two fields together are 280 m in the north-south direction and 150 m in the east-west direction, the total area is 4.2 hectares. The fields were bordered by other cotton fields to the east and west, by a field of sorghum to the south and by bare fallow to the north. Three masts were placed at variable distances from the northern edge of the first field. Each mast carried four pairs of wet- and dry-bult thermometers. As the crop grew, the thermometers were gradually raised, so that the lowest was 20-25 cm above the top of the plants.

Referring to the definition sketch (Fig. 6.17.), the day-time values of E_v were calculated from the equations

$$E_v = \frac{R_n + G}{1 + \beta} \tag{6.9}$$

and

$$\beta = 0.64 \frac{\delta T}{\delta e} \tag{6.10}$$

where

R_n = net radiation, cal/cm^2 . unit time,

G = soil heat flux, cal/cm^2 . unit time,

β = Bowen's ratio, and

$\frac{\delta T}{\delta z}$ = slope of the temperature, T, versus the vapour pressure, e, at elevation z.

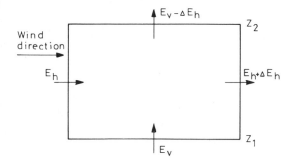

Fig. 6.17.   Diagram of horizontal and vertical fluxes of water vapour and the symbols used; z is the height of the evaporating surface; z_2 the height at which the Bouwen ratio was measured (Rijks, D.A., 1971)

The vertical flux during the night-time was calculated from the formula of of Thornthwaite-Holzman

$$E_v = \frac{0.62 \; L \; \rho \; k^2 \; (e_1-e_2)(u_2-u_1)}{\ln \; (\frac{z_2-d}{z_1-d})^2}$$

(6.11)

where

L = latent heat of vaporization, cals/gm,

ρ = density of air, gm/cm^3

k = Von Karman's constant,

u = wind speed, cm/sec,

z = height, cm, and

d = zero plane displacement, cm

Since the values obtained for E_v for the night-time were very small, they have been omitted.

The divergence of the horizontal flux of the latent heat, ΔE_h, between elevations z_1 and z_2 over a given period is given by the equation

$$\Delta E_h \simeq \int_{z_1}^{z_2} \frac{L.\rho.\varepsilon}{p} \cdot \bar{u}_z \cdot \frac{\delta e_z}{\delta x} \, \delta z \qquad (6.12)$$

where

ε = ratio of mole weight of water vapour to mole weight of air,

p = air pressure, m bars, and

x = distance from leading edge, cm

The daily evaporation totals E_1 and E_2 (E_2 being the more downwind of the two estimates) were taken to refer to the respective locations of ΔE_h. If the upwind mast had been at the leading edge of the field, they would have been the sum of ΔE_h and the downwind estimate of E_v. If both masts were inside the field, the estimate of E is the sum of ΔE_h and the mean of the upwind and downwind values of E_v.

The accuracy of the combination of methods employed in that work and of the various assumptions made was tested by the energy-balance. The test showed that the income and expenditure of energy for a given period (mostly an irrigation cycle) were always different, and the difference decreased with increase in fetch. When the whole area was taken, the error dropped to 8%. Calculations for periods earlier in the growing season have indicated larger differences in the energy balance, as have calculations for some individual days.

Despite those differences, the results obtained from that investigation can be viewed as the best that we have. We shall therefore consider the mean of E_1 and E_2 as a fair representation of the evapotranspiration from an irrigated cotton field in the Gezirah area. The daily and the mean monthly evapotranspiration figures are listed in Table 6.17. The mean daily evapotranspiration in October, November and December given in this table deviates from the corresponding figures estimated by the Blaney-Criddle method by only 6.7, 5.7 and 7.3% respectively. These small differences are quite acceptable, as the energy-balances for October and December did not cover the whole month, but 15 and 20 days respectively. Moreover, the results obtained from the balance method have been reported to include some error as the expenditure was not equal to the incoming energy (Rijks, D.A., 1971). In any case, the satisfactory similarity between the figures obtained from the two methods encourages us to apply the Blaney-Criddle method. It is only for the purpose of estimating the annual potential evapotranspiration in the arid zone in the Sudan that we shall increase the product of the day-time hours and temperature by about 10%.

TABLE 6.17   Daily evapotranspiration from irrigated cotton in the Gezirah area
estimated from the energy-balance method

Date	Evapotranspiration,		Date	Evapotranspiration,		Date	Evapotranspiration,	
	day, mm	mean monthly, mm/day		day, mm	mean monthly, mm/day		day, mm	mean monthly, mm/day
18.10.1965	3.29		11.11.1965	6.40		05.12.1965	5.85	
19.10.1965	3.77		12.11.1965	5.66		06.12.1965	6.70	
20.10.1965	4.07		13.11.1965	6.12		07.12.1965	6.02	
21.10.1965	3.82		14.11.1965	–		08.12.1965	6.36	
22.10.1965	4.13		15.11.1965	5.85		09.12.1965	6.08	
23.10.1965	5.56		16.11.1965	6.35		10.12.1965	5.02	
24.10.1965	7.21	5.76	17.11.1965	6.72	6.08	11.12.1965	5.16	5.71
25.10.1965*	7.72		18.11.1965*	6.35		12.12.1965	5.93	
26.10.1965	7.75		19.11.1965	6.92		13.12.1965	5.58	
27.10.1965	7.78		20.11.1965	5.36		14.12.1965	5.26	
28.10.1965	7.79		21.11.1965	7.08		no	–	
29.10.1965	6.35		22.11.1965	6.10		records	–	
30.10.1965	–		23.11.1965	6.61		24.12.1965	6.29	
31.10.1965	5.60		24.11.1965	6.42		25.12.1965	5.57	
			25.11.1965*	5.27		26.12.1965	5.11	
01.11.1965	–		26.11.1965	6.31		27.12.1965	5.53	
02.11.1965	4.64		27.11.1965	5.20		28.12.1965	5.47	
03.11.1965	5.20		28.11.1965	4.90		29.12.1965	5.31	
04.11.1965	5.85		29.11.1965	5.92		30.12.1965	4.89	
05.11.1965	6.91		30.11.1965	5.52		31.12.1965*	6.21	
06.11.1965*	7.80							
07.11.1965	5.20		01.12.1965	5.55		01.01.1966	6.95	
08.11.1965	6.12		02.12.1965	5.24		02.01.1966	7.42	
09.11.1965	6.97		03.12.1965*	5.99		03.01.1966	6.78	
10.11.1965	6.49		04.12.1965	6.53		04.01.1966	5.65	6.70

* field is irrigated

6.2.3   Penman's method

In the semi-humid and humid areas south of isohyet of 700 mm/yr rainfall, the
use of the Blaney-Criddle formula may be unjustified on the grounds that the
mean daily temperature is nearly constant, whereas other factors influencing
evapotranspiration could be widely variable.

Dupriez, G.L., (1959) raised serious objections to the use of Thornthwaite's
formula, eq. 6.3.7, in tropical areas or in areas where the air temperature
varies but only a little during the course of the year. The comparisons he made
between measured and calculated evaporation and potential evapotranspiration at
a number of stations in the Congo and in Rwanda/Burundi have shown that the
Penman formula provides the best estimate for the loss of water. The absence of
any expression for the saturation deficit in the Thornthwaite formula is likely

to be the reason why this formula overestimates the ET_p in the humid season and underestimates it in the dry season.

The monthly potential evapotranspiration is related to Penman's evaporation by the relation

$$ET_p = 0.91 \, E_{penman} + 2.5 \qquad \text{(in mm)} \tag{6.13}$$

The ratio of the annual ET_p to the annual $E_{penman} = 0.97$, and to the annual measured evaporation is 0.93.

Rijks investigated the evaporation from a papyrus swamp at Namulonge in Uganda in the period from 8 March up to and including 11 April. He used the Bowen's ratio method and the Penman method to estimate the evaporation from a free water surface as well as the evapotranspiration from the swamp. In that period where the average daily temperature fluctuated between 20.9°C and 24.2°C the average evaporation from open water estimated by Penman's method was 5.5 mm/day, potential evapotranspiration from the swamp using Penman's evaporation reduced by 0.8 was 4.4 mm/day and from Bowen's ratio method using measured meteorological parameters was 3.81 mm/day (Rijks, D.A., 1959). Those results suggest the use of a reduction coefficient of 0.69 to be multiplied by Penman's evaporation so as to deduce the evapotranspiration from a swamp. Though the experiment took place during the rainy season, it is not clear whether the estimated evapotranspiration was the actual or the potential one.

The evaporation from a free water surface estimated from the method of Penman for Malakal, Mongalla/Juba, and Wau is 5.4, 5.5 and 5.8 mm/day for the whole year. The mean monthly values for these stations are given in Table 5.4 and in Fig. 5.4. We shall assume the Penman evaporation for Shambé in the Sudd region to be the arithmetic mean of the evaporation for the said stations, i.e. 2033 mm/yr. The potential evapotranspiration at Shambé estimated from the method of Olivier and from the tank measurements is 1729 mm/yr (see Table 6.14). These figures give a reduction factor of about 0.85 for the swamps and not 0.69 as proposed by Rijks. Accordingly, the Penman open water evaporation will be considered as the basis for estimating the potential evapotranspiration from the semi-humid and humid areas in the Sudan. The reduction factors we shall use are from 0.95 to 0.90 for the areas outside the swamps and from 0.85 to 0.80 for those areas inside the swamps of the Bahr el Jebel and Bahr el-Ghazal Basins, and the Machar swamps. These values together with the map, Fig. 5.10, showing the Penman free water surface evaporation, have been incorporated in preparing the potential evapotranspiration requirement for the Sudan. This requirement on an annual basis can be read directly from the map in Fig. 6.18.

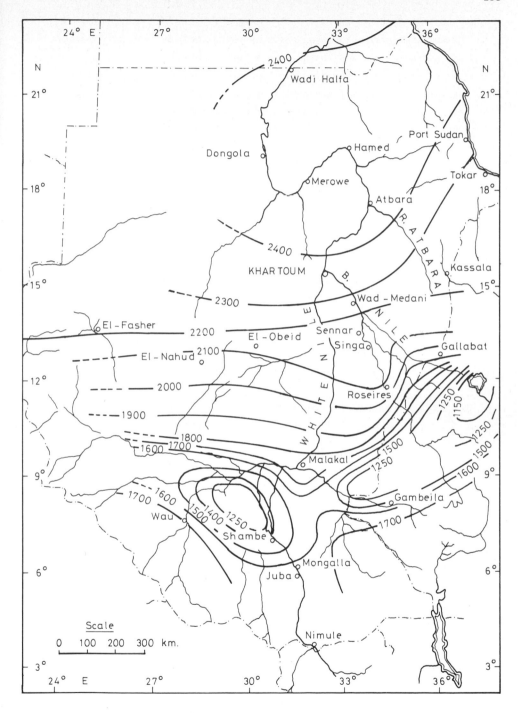

Fig. 6.18.   Map showing lines of equal potential evapotranspiration requirement, mm/yr, for the Sudan

6.3 EVAPOTRANSPIRATION IN THOSE PARTS OF RWANDA-BURUNDI, UGANDA, TANZANIA
     AND KENYA WHICH ARE SITUATED WITHIN THE NILE BASIN

   In these areas the evaporation from open water and the evapotranspiration
from vegetated surfaces are measured from pans and lysimeters respectively.
Estimates are obtained from the method of Penman and/or some microclimatologic
techniques. Dupriez, G.L., (1959) criticized severely the application of
Thornthwaite's method in equatorial areas. He measured the ET_p from lysimeters
at the stations of Kisozi and Musas which are located within the territory of
Rwanda-Burundi very close to the divide of the Nile Catchment (see Fig. 6.19.).

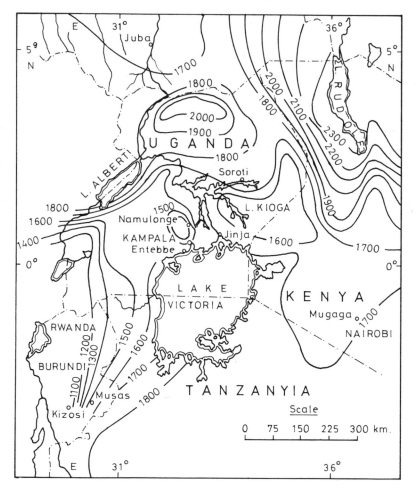

Fig. 6.19.   Map showing lines of equal potential evapotranspiration requirement,
mm/yr, for some parts of Rwanda-Burundi, Uganda, Tanzania, and Kenya

The monthly ET_p at these two stations as observed by Dupriez are tabulated below. The ET_p at Kisozi is smaller than that at Musas mainly because of its high altitude and thereupon its lower temperature. The most important thing here in this area is that the ET_p, in mm/day, is nearly 95% of the Penman free water surface evaporation, also in mm/day.

Month	Measured ET_p, mm/month, average of 1957-58	
	Kisozi	Musas
	(S 3°33′, E 29°41′, h=2155 m)	(S 3°39′, E 30°21′, h=1260 m)
January	76.0	137.9
February	67.9	109.7
March	86.1	142.7
April	83.7	136.9
May	79.7	105.9
June	78.6	100.4
July	93.7	121.7
August	99.1	111.3
September	119.3	138.3
October	103.7	148.5
November	87.0	151.3
December	75.5	123.5
Year	1050.3 mm	1527.9 mm

Hanna, L.W., (1971), while studying the effects of water availability on tea yields in Uganda, concluded that ET_p from a full cover of tea is nearly 0.85 times Penman evaporation. In East Africa the best estimate of potential evapotranspiration from crops is derived from Penman's formula. The network of Entebbe, Kabanyolo, Jinja and Kituza stations for which the ten-day means of Penman evaporation (see Fig. 6.20.) was reported by Hanna to permit a reasonably accurate estimate of the potential evapotranspiration within a narrow zone adjacent to Lake Victoria. In that zone, many of the estates of sugar cane and tea are concentrated, which makes the zone densely populated.

A short distance north of Kabanyolo is situated the Cotton Research Station at Namulonge (0°32′N, 32°37′E, and h=1100 m). At the northern boundary of the station there is a swamp about 500 m wide and running from NW to SE. Meteorological measurements were taken in the swamp and discussed by Rijks (1969). The evaporation from an old stand of papyrus was estimated by the Bowen ratio method. It was found that about 30% of the net radiation reaching the canopy was converted into sensible heat. The layer formed by the heads of the papyrus seemed to act as a major resistance to turbulent exchange of water vapour from the undergrowth. The evapotranspiration from the old stand of papyrus was estimated to be 60 ± 15 percent of Penman estimates of evaporation from open water, E_o. The mean evapotranspiration from the swamp for the period from 8 March up to and

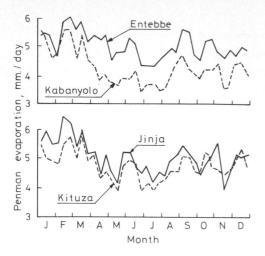

Fig. 6.20.   Ten-day means of Penman evaporation at four stations in Southern Uganda (Hanna, L.W., 1971).

including 11 April is 3.81 mm/day. This figure is small compared to the mean annual evapotranspiration from the Lake Kyoga swamps given as 4.72 and 5.1 mm/ day by Olivier and Hurst, respectively. If we trust Olivier's estimates (1961), the mean monthly evapotranspiration figures are then 7.0, 8.1, 5.9, 4.2, 3.3, 3.8, 3.7, 3.4, 4.1, 4.1, 4.9 and 4.6 mm/day for the months from January to December, respectively.

According to Dagg, M., (1972) the water use will rise from about 0.4 E_o when the plants are young to 1.0 when full cover is achieved. The maximum ET_p/E_o = 1.0 for cotton is at the beginning of the fourth month after planting, Namulonge, Uganda, and 1.1 for sugar cane after four months from planting, Moshi, Tanzania. The evergreen forests have a ratio ET_p/E_o = 0.9 as an average for the whole year. At Muguga, Kenya, the seasonal potential evapotranspiration of maize is 560 mm, whereas the open water evaporation for the same season is about 840 mm, i.e. ET_p/E_o = 0.67. The evapotranspiration from a lysimeter raising peren- nial grass at the same location was observed by Glover, J., and Forsgate, J., for a period of 126 days in 1962 (1964). From this investigation they concluded that $ET = E_o - 1.17$ for 5-day periods ($r^2 = 0.94$) and $ET = 1.03 E_o - 1.35$ for 10-day periods ($r^2 = 0.97$), both E_o and ET are expressed in mm/day. In these relations ET = evapotranspiration from the lysimeter and E_o = Penman evaporation using the ratio $\frac{n}{N}$ for the relative duration of the bright sunshine. For the range of E_o from 5 to 6 mm/day the corresponding ratio $ET/E_o \simeq 0.75$. The obser- vations from this experiment supports the concept of Veihmeyer and Hendrickson that the actual evapotranspiration remains at the potential level as long as the soil moisture in the top 4 ft is in the range from field capacity to wilting point.

Last but not least, we have the map of evaporative demand for East Africa prepared by Rijks, Owen, and Woodhead, and presented by Dagg (1972). As a basis for converting that map to another giving the potential evapotranspiration requirement of East Africa, we considered a conversion factor of 0.9 except for the swamps where a factor of 0.8 was used. The open water evaporation thus converted to potential evapotranspiration has been compared to the measured ET_p at the locations mentioned in this section and at other locations and the necessary corrections made. The map thus obtained is as shown in Fig. 6.19.

The partial maps of the Nile Basin shown in Figs. 6.16., 6.18., and 6.19., have been combined in one map representing the potential evapotranspiration requirement for the whole basin, as shown in Fig. 6.21. Comparing this figure to Fig. 6.2., one can easily notice the following:

i)   Both maps in Figs. 6.2., and Fig. 6.21., show almost the same pattern of ET_p requirement from the Mediterranean Sea coast and southward up to Malakal. The map in Fig. 6.21., shows an ET_p requirement almost 5% smaller than that given by the map in Fig. 6.2. The maximum difference between the two does not, however, exceed 10%.

ii) The map in Fig. 6.21., shows two depressions: one to the west, inside the swamps of Bahr el Ghazal and Bahr el Jebel Basins, and the second to the east, inside the Machar swamps, in the basin of the Sobat. In these two depressions the potential evapotranspiration requirement drops to less than 1250 mm/yr, i.e. a mean annual of 3.4 mm/day.

iii) South and east of Juba there is a sharp rise in the ET_p requirement. In warm Kenya it may reach 2300 mm/yr and in northern Uganda it reaches 2000 mm/yr, though in a limited area only. Between Lake Victoria and each of the adjacent lakes, Kyoga, Albert, Edward and Tanganiyka, there is a sharp decline or rise in the potential evapotranspiration requirement. ET_p may reach 2000 mm/yr or fall to 1100 mm/yr, depending on the fall or the rise in the ground surface level.

iv) The details in ii) and iii) do not appear in Fig. 6.2. All that can be seen from it is that the ET_p requirement of the area east and south-east of Juba exceeds 1500 mm/yr. The area west, south and south-west of Juba has an ET_p requirement of less than 1500 mm/yr, but still above 1250 mm/yr.

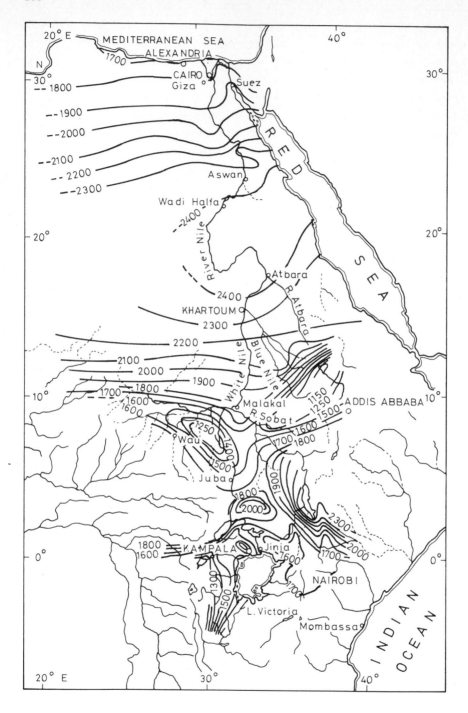

Fig. 6.21.   Map showing lines of equal potential evapotranspiration requirement mm/yr, for the Nile Basin

REFERENCES

Abd el-Samie, A.G., and Barrada, Y., 1960. Consumptive use and irrigation water
    requirements of Ashmoony cotton in Giza under regular irrigation schedule.
    Bull. No. 227: 3-39, Faculty of Agriculture, Cairo University, Egypt.
Abd el-Warith, M., 1965. Consumptive use and irrigation requirements for cotton
    plant in Egypt. Thesis submitted to the Faculty of Engineering, Cairo Univer-
    sity for the M.Sc. degree, Egypt.
Al-Barrak, A.H., 1964. Evaporation and potential evapotranspiration in Central
    Iraq. Thesis submitted for the degree of M.Sc., Utah State University,
    Logan, Utah, USA.
Blaney, H.F., 1957. Monthly consumptive use of water by irrigated crops and
    natural vegetation. IASH General Assembly of Toronto, Vol. II, Publ. 44:
    431-439.
Blaney, H.F., and Criddle, W.D., 1950 (slightly revised in 1952). Determining
    water requirements in irrigated areas from climatological and irrigation
    data. USDA, SCS Tech. Paper 96, 48 pp.
Blaney, H.F., and Morin, K.V., 1942. Evaporation and consumptive use of water
    empirical formula, Part I, Trans. American Geophysical Union, Washington,
    DC.
Blaney, H.F., and Criddle, W.D., 1966. Determining consumptive use for planning
    water developments. Methods for estimating evapotranspiration: irrigation
    and drainage speciality conference, Las Vegas (published by ASCE), 1-34.
Boumans, J.H., et al, 1963. Reclamation of salt affected soils in Iraq, Publ.
    No. 11 ILRI, Wageningen, The Netherlands.
Butler, P.F., and Prescott, J.A., 1955. Evapotranspiration from wheat and pas-
    ture in relation to available moisture. Austr. Journ. Agr. Res. No. 6: 52-61.
Christiansen, J.E., and Hargreaves, G.H., 1969. Irrigation requirements from
    evaporation, Trans. 7th ICID Congress, New Mexico, 23.569-23.596.
Combremont, R., 1972. Water use seminar, Damascus, FAO Irrigation and Drainage
    Paper No. 13: 138-141, Rome.
Criddle, W.D., Harris, K., and Willardson, L.S., 1962. Consumptive use and water
    requirements for Utah, Tech. Bull. No. 8, Office of the Utah State Engineer,
    47 pp, Utah, USA.
Dagg, M., 1972. East Africa: its peoples and resources (edited by W.T.W. Morgan).
    Chapter 10: Water requirements of crops, 119-125. Oxford University Press,
    London, New York.
Doorenbos, J., and Pruitt, W.O., 1977. Guidelines for predicting crop water
    requirements, FAO Irrigation and Drainage Paper No. 24 (Revised), 144 pp,
    Rome.
Dupriez, G.L., 1959. La curve Lysimetrique de Thornthwaite, comme instrument de
    mesure de l'evapotranspiration en regions equatoriales. IASH Symposium of
    Hannoversch-Münden, Vol. II (Lysimeters) Publ. No. 49: 84-98.
El-Gibali, A. et al, 1966. Irrigation requirements and frequency of late corn,
    Agr. Res. Rev. Vol. 44, No. 1: 139-157, Cairo, Egypt.
El-Gibali, M.H., 1969. The effect of soil moisture regime, phosphorous and
    nitrogen application on the consumptive use of some crops in Assiut (UAR),
    Trans. 7th ICID Congress, New Mexico, 23.53-23.71.
El-Nadi, A.H., 1969. Efficiency of water use by irrigated wheat in the Sudan.
    Journ. Agr. Sci., Vol. 73, Part 2: 261-266, Cambridge University Press.
El-Nadi, A.H., 1970. Water relations of beans: II - Effects of differential
    irrigation on yield and seed size of broad beans, Exper. Agr. No. 6: 107-111.
El-Nokrashy, M.A., 1963. Effect of frequency of irrigation on quality and
    quantity of bearing-orange varieties budded on three root stocks in loamy
    clay soil. Thesis submitted to the Faculty of Agriculture, Cairo University
    for the M.Sc. degree, Cairo, Egypt.
El-Shal, M.I., 1954. Water requirements of crops for the Menoufia Province.
    Thesis submitted to the Faculty of Engineering, Cairo University for the
    degree of M.S., Cairo, Egypt.

El-Shal, M.I., 1966. Consumptive use and water requirements for some major crops in Egypt. Thesis submitted to the Faculty of Engineering, Cairo University for the Ph.D. degree, Cairo, Egypt.

Erie, L.J., 1963. Irrigation management for optimum cotton production. Cotton gin. and oil mill press No. 64: 30-32.

FAO, 1971. Irrigation practice and water management. Irrigation and Drainage Paper No. 1, Rome.

Glover, J., and Forsgate, J., 1964. Transpiration from short grass. Quart. Jou. Roy. Met. Soc., Vol. 90: 320-324.

Haise, H.R., and Viets, F.G., 1957. Water requirements as influenced by ferti- lizer use. Trans. 3rd ICID Congress, San Francisco, Vol. III, 8.497-8.508..

Hanna, L.W., 1971. The effects of water availability on tea yields in Uganda, Jou. App. Eco., Vol. VIII: 791-813.

Hargreaves, G.H., 1956. Irrigation requirements based on climatic data. Paper No. 1150, Jou. Irr. & Dr. Div. ASCE, Vol. 82, No. IR 3.

Hargreaves, G.H., 1966. Consumptive use computation from evaporation. Paper presented to the Symposium on methods for estimating evapotranspiration, irrigation and drainage speciality conference, Las Vegas (published by ASCE), 35-42.

Hargreaves, G.H., 1968. Consumptive use derived from evaporation pan data. Paper No. 5863, Jou. Irr. & Dr. Div., ASCE Vol. 94, No. IR 1.

Harrold, L.L., and Dreibelbis, F.R., 1959. Weighing monolith lysimeters and evaluation of agricultural hydrology. IASH Symposium of Hannoversch-Münden, Vol. II, Publ. No. 49: 105-115.

Israelsen, O.W., 1956. Irrigation principles and practices, 2nd edition, John Wiley & Sons Inc., New York.

Jensen, M.E., and Haise, H.R., 1963. Estimating evapotranspiration from solar radiation. Paper No. 3737, Proc. ASCE, Jou. Irr. & Dr. Div., Vol. 89, No. IR 4: 15-41.

Khafagi, A., Z-el-Abedine, A., Shahin, M., and El-Warith, M., 1964. Consumptive use and irrigation wqter requirements for cotton plant in Egypt. Ann. Bul. Fac. Engrg., Cairo University, 3-29.

Korzun, Y., et al, 1978. World water balance and water resources of the earth, UNESCO Publ. (translated from the Russian to English by Nace, R.L.).

Lowry, R.L., and Johnson, F.R., 1942. Consumptive use of water for agriculture. Trans. ASCE, Vol. 107: 1243.

Olivier, H., 1961. Irrigation and climate. Edward Arnold (publishers) Limited, London, 250 pp.

Omar, M.H., 1960. Evapotranspiration at Giza. Publ. UNESCO regional training course in microclimatology for ecology and soil science, Cairo, Egypt.

Penman, H.L., 1950. Evaporation over the British Isles. Quart. Jou. Roy. Met. Soc. No. 76: 372-383.

Pruitt, W.O., 1960. Relation of consumptive use of water to climates. Trans. ASCE, 3(1): 9-13, 17.

Quackenbush, T.H., and Phelan, J.T., 1965. Irrigation water requirements of lawns. Jou. Irr. & Dr. Div., ASCE, Vol. 91, No. IR 2: 11-19.

Reichel, E., and Baumgartner, A., 1975. The world water balance. R. Oldenburg Verlag, Munich.

Rijks, D.A., 1969. Evaporation from a papyrus swamp. Quart. Jou. Roy. Met. Soc., No. 95: 643-649.

Rijks, D.A., 1971. Water use by irrigation cotton in Sudan, III: Bowen ratios and advective energy. Jou. App. Eco., Vol. VIII: 643-663.

Rijtema, P.E., 1959. Calculation methods of potential evapotranspiration. ILRI Tech. Bull. No. 7, Wageningen, The Netherlands.

Shahin, M.M., 1959. Tile drainage of irrigated lands in Egypt. Thesis submitted to the Faculty of Engineering, Cairo University in partial fulfillment of the requirement for Ph.D. degree, Cairo, Egypt.

Shahin M.M., and El-Shal, M.I., 1969. An investigation of the consumptive use of water for crops and the frequency of irrigation in the United Arab Republic. Trans. 7th Congress of ICID, New Mexico: 23.27-23.51.

Shahin, M.M., et al, 1973. Irrigated cotton: A world-wide survey (edited by Framji, K.K., and Mahajan, I.K.), ICID Publ., New Delhi, 321 pp.

Shenouda, El-Gibali, Tawdros and Gamal, 1966. Irrigation requirement of early corn and the best irrigation frequency for the crop with a comparative study for the plant requirements of early corn and late corn in Middle Egypt. Agr. Res. Rev., Vol. 44, No. 1: 159-170, Cairo, Egypt.

Slatyer, R.O., and McIlroy, I.C., 1961. Practical microclimatology. CSIRO, Plant Ind. Div., Canberra.

Stanhill, G., 1961. A comparison of methods of calculating potential evapotrans-piration from climatic data. Isr. Jou. Agr. Res., No. 11: 159-171, Jerusalem.

Tanner, C.B., 1967. Irrigation of agriculture land (edited by Hagan, R., Haise, H., Edminster, T., and Dinauer, R.). Chapter 29: Measurement of evaporation: 534-574, (published by the American Society of Agronomy). Wisconsin, USA.

Thornthwaite, C.W., 1948. An approach toward a rational classification of climate. The Geogr. Rev., Vol. 38, No. 1: 55-94.

Thornthwaite, C.W., and Mather, J.R., 1955. The water balance, Publication in climatology No. 8: 1-104.

Turc, L., 1961. Evaluation des besoins en ear d'irrigation, evapotranspiration potentielle. Ann. Agron. 12(1): 13-50.

United States Department of Agriculture, 1955. Yearbook of agriculture: Water. US Government Printing Office.

Zein el-Abedine, A., and Abdallah, M., 1949. Cyclic and seasonal moisture variations on Cairo (Fouad I) University farm, Giza district, Egypt. Soil Science, Vol. 68, No. 3: 213-227.

Zein el-Abedine, A., Abdallah, M., and Abd el-Samie, 1967. Evapotranspiration studies on maize in Giza, UAR. Paper presented at the Symposium on the use of radio-isotopes in evapotranspiration studies, Istanbul, Turkey.

Chapter 7

GEOLOGY AND GEOHYDROLOGY OF THE NILE BASIN

7.1   GEOLOGY

7.1.1   The Equatorial Lakes Plateau

The old Precambrian formations of East Africa have been dated between, say, 400, and more than 3000 million years. Of these formations the Nyanza Shield is the one located in those parts of Uganda and Tanzania within the Nile Basin. This shield comprises rocks of the Gneiss Complex of Northern Uganda and of other systems of North-Western Uganda and Central Tanzania and has a predominantly east-west trend. The Nyanzian system is thus made up of basic, intermediate and acid volcanic rocks with interbedded sediments of coarse material. The depth of this system extends some thousands of metres and is developed around Lake Victoria (see map, Fig. 7.1.) where its rocks are associated with intrusive granites. In some other parts of Uganda, Kenya and Tanzania the system consists mainly of boulder conglomerates, mudstones and quartzites. Where this is the case the system is called Kavirondian, and in some locations it simply rests on the Nyanzian formations. The rocks of both systems, the Nyanzian and the Kavirondian, form part of the Nyanza Shield (Saggerson, E., 1972).

In Uganda the Buganda-Toro rocks, predominantly argillaceous, occupy a broad arc extending from the northern shores of Lake Victoria to the Ruwenzori in the western, as well as the northern and eastern shores of Lake Kyoga. The series known as the Karagwe-Ankolean, which is thought to be 1400 million years old, consists of metamorphosed rock and is about 10000 metres thick. This series occupies the major part of the area west of Lake Victoria and the borders with Rwanda and Burundi. The Karagwe-Ankolean series is separated from the southwestern shore of Lake Victoria by the Bukoban system, as shown on the map, Fig. 7.1. The latter system comprises sandstones, shales, quartizites and conglomerates overlain by marles and limestone or sandstone.

There is hardly any evidence of Palaezoic rocks in those parts of Uganda, Tanzania and Kenya which lie within the Basin of the Nile River. The same remark holds for the Jurassic and Cretaceous periods of the Middle and Upper Mesozoic ages.

The movement of the rigid block of Africa led, among others, to great fractures on the eastern side of the continent and locally elsewhere. The gigantic troughs which extend from north to south and contain the Eastern Rift Valley, are occupied in part by the East-African Lake system (Stamp, L.D., and Morgan, W.T., 1972). The Equatorial Nile system subsequently drains the tectonically and yet volcanically active Ugandan Plateau area where many drainage changes

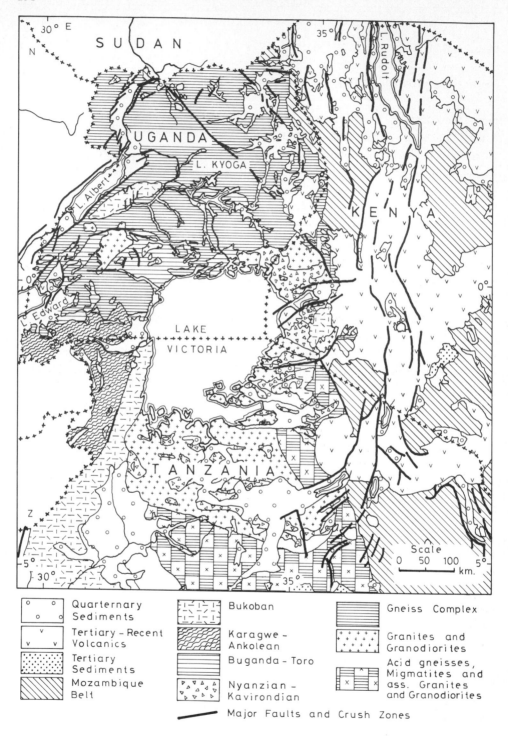

Fig. 7.1.　Geological map of the catchments of the Equatorial Lakes

have been caused by rift-associated geophysical activity. It is thus supposed
that the depression occupied by Lake Victoria was formed in the Miocene period,
whereas Lake Kyoga seems to have existed since the Pliocene period (Saggerson,
E., 1962). These two lakes have become linked together as a result of the subsidence of the Albert depression. This event has led to the conclusion that the
Upper Victoria Nile is probably a very young river formed in the Pleistocene
period, as is the Albert Nile (Hepworth, J., 1964).

The deposits of the Pleistocene period of the Quarternary age cover a great
part of the Albert and Edward troughs and could well exceed 1200 m in thickness.
They are argillaceous rocks with which mud volcanoes, salt and gypsum deposits
are associated. In some places they are overlain by the Semliki Series derived
from the outwash obtained from the Western-Rift margin.

The Tertiary to Recent period is significant for its volcanic activity. The
volcanic rocks are mildly to strongly alkaline, containing, among others,
basalts. They have given rise to extensive lava plains or are associated with
the major central volcanoes such as Mounts Elgon, Kenya and Kiliminjaro.

7.1.2   The Nile in the Sudan

The Basement Complex in the Sudan forms over two-thirds of the rock-exposures
there. It consists of igneous and sedimentary rocks out of which the platform
was carved. In the northern part of the Sudan the rock types that prevail are
crystalline schists, gneiss, limestone, graphite-slate and quartzite of various
degrees of metamorphism. These are intruded by igneous rocks. In the north-
eastern part of the Sudan there are non-metaporphic sediments associated with
volcanic rocks. The Basement Complex of the southern part of the Sudan consists
of granoblastic-foliated gneiss with felspars intruded into foliated paraschists
and paragneisses.

The platform, on which the Nubian Series of quasi-horizontal sandstones and
mudstones was deposited, is of folded metamorphic sediments with intruded volcanic rocks (Andrew, G., 1948).

The Sudan was inundated by the sea during the Upper Palaeozoic and Mesozoic
ages, at which time the deposition of part of the Nubian Series took place. The
Palaeozoic succession is represented by sandstones with a black shale inter-
calation, and is closed by limestone (Sandford, K.S., 1935). These Series are
overlain by the Nubian shales, mudstones and sandstones. The latter are regarded
to have occurred during the Mesozoic age. They can be found in the north-western
area of the Sudan (continuation of the Libyan region), the eastern area (continuation of the Abyssinian-Arabian-Somalian area) and west of the Bahr el Jebel in
the well-known Yirol beds (see the geological map, Fig. 7.2.).

296

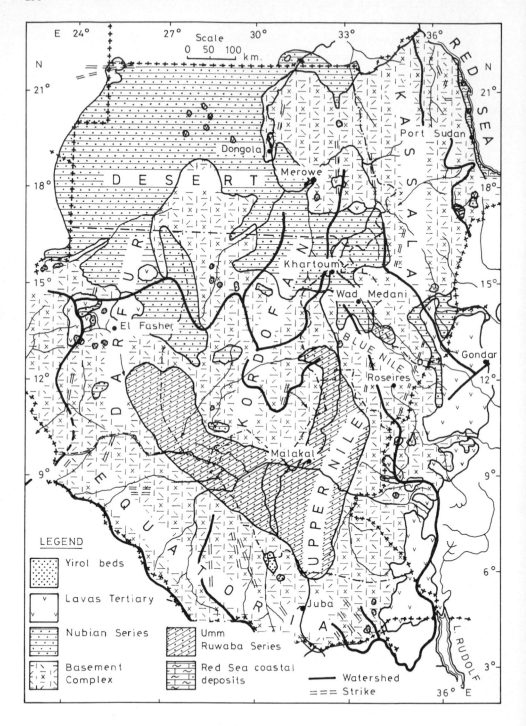

Fig. 7.2.   Geological map of the Sudan showing the distribution of rocks
(Andrew, G., 1948)

The sea coast retreated steadily through the Eocene period. This was followed by a very long period of erosion which resulted in the removal of the Nubian Series from several places.

The Upper Eocene period is characterized by the rise of the eastern plateau and the Red Sea hills and the formation of the Red Sea. The rise of the plateau was supported by the extensive volcanic activity which took place in the Upper Tertiary (Miocene) and which produced lavas capped by volcanoes (see Fig. 7.2.). The hollow running from south to north as a result of the elevation of the eastern plateau is now occupied by the plain through which the Nile River runs in Egypt and the northern part of the Sudan. The bouldery masses known as the Hudi Chert Series belong to the Lower Tertiary. It is thought that the Chert masses have travelled from some distance in the upstream to their present location. These masses can be found east of the Atbara mouth and near Zeidab. The depressions penetrated by the White Nile Valley, the Sudd region and some parts of the Blue Nile Basin have been filled in by deposits similar to the fill of the present desert area in the north. These deposits are referred to as the Umm Ruwaba Series. They consist of unconsolidated sands with or without clay, with some gravel deposits. The Bahr el Jebel north of Juba flows over a formation of fluviatile and lacustrine sediments belonging to this series. They are laid down in a succession of land deltas with extremely low surface gradients. This river has a reasonably defined valley as far as the north of Mongalla. From Juba to Mongalla there are several channels each with marked levees. As the proportion of the coarse material carried by the river decreases, the river levees become weaker and less effective in confining the high-flow within the river section. In the Sudd region the levees are mainly formed of fine sand, whereas the back swamp deposits are composed of silt and clay with occasional sandy lenses.

The Sobat joins the White Nile near Malakal. It has been suggested that the geological history of its basin must have been similar to that of the Blue Nile. It may have originated in post-Oligocene times on the surface of the Ethiopian Traps (Berry, L., and Whiteman, A., 1968). The White Nile from the Sobat junction down to Khartoum has a very small gradient, which is responsible for the smallness of flow velocity in it. The valley in this reach is rather broad and for most of the distance the river itself has a well-defined channel.

The White Nile is joined by the Blue Nile at Khartoum to form one river. The geological and geomorphological evidence indicates that the Blue Nile is an ancient river system. It rises on the great volcanic plateau of Ethiopia, which is formed by the lavas extruded during the Oligocene period. Lake Tana is supposed to have been formed in the Pliocene and the Abbai canyon was excavated during the Pliocene and Pleistocene. It is probable that an overall tectonic process took place during the Late Cretaceous-Late Eocene interval, which led to roughly parallel courses of the Blue Nile, Dinder, Rahad and Atbara.

It is quite possible that the Atbara and/or the Blue Nile could not have joined the White Nile or the Main Nile during the Pleistocene dry periods. As a result, the Atbara could not have flushed its suspended load into the Main Nile. Instead, it was laid down in its own plain to form thick alluvial deposits. The geology of the Blue Nile at the Roseires area was investigated in connection with the construction of the Roseires Dam. The bed-rock there is composed of a varied assemblage of metamorphic and granite rocks, forming part of the Basement Complex of the Sudan. Rock is exposed for nearly 8 km along the Damazin Rapids, but, further away from the river, outcrops are limited to local inselberg or jebels. The original formation of the gneisses was followed by the emplacement of early granites and pegmatites. This formation was later intruded with granite sills and pegmatites. The extensive erosion which took place in a long interval of time led to the exposure of the bed-rock at a number of locations (Fitt, R.L. et al, 1967).

The Blue Nile, from Roseires northwards, flows across its own sediments. Its valley widens northwards to merge with the Gezira Plain north of Sennar, though the river is still incised below the general country level. The surface of the Gezira is covered with clays. Nevertheless, there are some patches occupied by dunes and sand-spreads as a result of overbank floods. These sandy patches are associated with shallow discontinuous channel systems. Another feature of the Gezira Plain is the comparative uniformity of the surface clays.

The Gezira Plain has been formed by the deposits brought mainly by the heavily laden Blue Nile water. During its flood, the Blue Nile may carry in suspension up to 3000 parts per million of clay and fine sand, mostly from Ethiopian soils, plus a significant amount of dissolved load. These sediments, together with the product of aggradation by rivers carrying coarse materials, have filled the then-existing depressions and eventually produced a heterogeneous alluvial fan. The whole structure is mounted on a thick layer of Nubian sandstone (100-150 m thick), which was formed in the Late or Upper Mesozoic.

In the Khartoum area the thickness of the alluvium is variable, and in some spots there are infilled scour pools where the thickness of the sediments may exceed 20 m. Terraces made up of clay and gravel flank the river in the same area, however, and these terraces narrow rapidly northwards towards the 6th Cataract.

West of Khartoum the formation known as the qōz is spread over the low ground of the west-central part of the Sudan (Darfur and northern Kordufan). This formation is an accumulation of dune-sand consisting almost entirely of quartz grains, and, possibly, derived from the Nubian Series.

The Main Nile north of Khartoum flows north in the tectonic low behind the uplifted rim of the continent - the Red Sea hills. The interplay of long-continued tectonic activity, river processes of erosion and deposition, and the effects of climate fluctuations, over the last one million years or so, combine to produce a complex geological structure of the area.

From Khartoum to the 6th Cataract the Nile flows largely over the Basement Complex and Nubian formations (see Fig. 7.2.). North of the 6th Cataract (known as Sabaloka) the valley widens again. In the Shendi area the surface is developed mainly on the Nubian formation. The terraces flanking the river there could be up to 5 km wide. The river runs in a northerly direction on Nubian sandstone as far as Ed-Damer, where the Basement Complex appears again.

In the distance from the 5th to the 4th Cataract the river runs across the Basement Complex. Between these two cataracts the river falls some 90 m. The river bed throughout the major part of this distance is rocky. The Basement consists of gneisses, slates and marble.

The river valley in the section from the 4th to the 3rd Cataract has been cut since the Early Pleistocene times. Towards Dongola there is an extensive gravel and silt terrace, west of which the Nubian formation is covered with coarse gravel for a distance of not less than 10 km wide. The 3rd Cataract itself is composed of gneiss with marble veins orientated to the west. North of this cataract the river passes between two mounts or hills, each about 380 m high. The other feature of this area is that the Nubian formation and the Basement Complex are cut by dykes and volcanic necks at some places.

The Nile runs through the Basement Complex in the section from the 3rd to the 2nd Cataract. Halfway between these two cataracts, terraces have been distinguished at 15 and 30 m above the present flood plain level. Both terraces are covered with coarse gravel. The Semna Cataract, famous for its Nilometer (see Chapter 1), is formed of hard gneiss and crushed granite which confines the Nile, from about 400 m wide, to just a 40 m wide channel.

In general, very little silt is deposited by the Nile in the Sudan except during occasional overbank floods. On the contrary, floods tend to scour rather than deposit in the Sudan and much of the scour product in the form of suspended matter used to be deposited in the Nile Delta, at least in the pre-High Aswan Dam era. The river valley from the 2nd Cataract to the southern frontier of Egypt is cut in Nubian sandstone with outcropping of igneous and metamorphic rocks at some places, such as Aswan.

7.1.3   The Nile in Egypt

During the Archeozoic and Proterozoic eras (some 500 million years ago) the surface of Egypt was covered by igneous, metamorphic and crystallized rocks such as granite, gneiss and schist. Following that time the Mediterranean Sea covered all the surface of Egypt and a considerable area of the Western Desert and the Sudan for a long period estimated at between 300 and 325 million years. In that period, the Palaezoic era, biological sea sediments were deposited. Most of them, however, were washed out by weathering factors, while some other deposits were buried down under the sedimentary rocks which came later. The fossils of that era, some of which are still present in Sainai, belong to the Carboniferous period. At the end of that period the sea receded for a long time ranging between 50 and 75 million years. The recession of the sea was followed by a subsidence of a considerable area of North Africa, including Egypt and the Sudan. Consequently, the sea water flooded the subsided parts for a period of 50 to 75 million years during the Mesozoic era, thereby depositing a layer of about 1500 metres in thickness of sands and pebbles, but free from fossils. The lowest one-third of this layer was deposited in the Jurassic period and the remaining two-thirds in the Cretaceous period. The lower half of the latter is known as Nubian sandstone.

By the end of the Mesozoic era the sea receded gradually and a number of changes took place in the Cainozoic era (50 million years ago). Most of the surface in Egypt subsided during the Eocene period, thus permitting the sea once more to flood the country and to deposit layers of calcareous stones. These deposits have formed the chain of hills east and west of the Nile Valley. Again, the sea receded to about Cairo in the Olegocene period. A number of minor changes followed during the Miocene, Pliocene and Pleistocene periods.

A simplified geological map of Egypt is shown in Fig. 7.3. The geological divisions shown in this map occupy the following areas (Said, R., 1962):

Geologic division	Area, km^2
Pleistocene and Recent	165 000
Pleiocene, Miocene and Oligocene	136 000
Paleocene and Eocene	204 000
Cretaceous	130 000
Nubian sandstone (mainly Cretaceous)	290 000
Igneous and metamorphic rocks	95 000
Total:	1 020 000

The plateaus which make up the greater part of the surface of Egypt consist, in the extreme south-west, of metamorphic rocks and granites. These are covered in a northerly direction by gently inclined sediments, giving rise as a result of weathering, to great flat-topped, table-like hills. North of the southern outcrops of ancient rocks are found the wide stretches of Nubian sandstone, then northward, wide expanses of limestone.

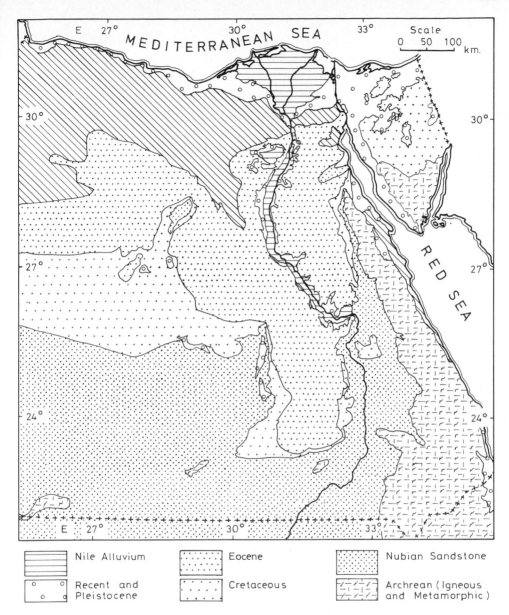

Fig. 7.3. Geological map of Egypt (with minor approximations)

The principal oases in the Western Desert are the Kharga and Dakhla Oases.
Each of them rests on Nubian sediments consisting of alternations of clays,
shales, sands and sandstones. These sediments have been gradually exposed as a
result of erosion. The plateau bordering the Kharga Oasis on the north and east
is of Cretaceous and Eocene rocks, which consist essentially of limestone. Erosion is claimed to have removed some 200 to 300 m of these limestones to expose
the underlying Nubian formations. These are the formations which contain the
important aquifer from which the water of the oases is drawn (Paver, G.L., and
Pretorius, D.A., 1954).

The geology of the coastal desert of Egypt, the desert area west and east of
the Nile Delta, and the geology of the Red Sea hills was investigated and
reported in connection with water supply in the Middle East campaigns during
World War II. For information about the results of those investigations the
reader is referred to the six articles which were published by Shotton, F.W.,
and by Paver, G.L., 1946.

The Nile judged by its present form appears to be made up of several distinct
systems which became joined at a much too late stage in geological history. Each
of these systems is related to a different structural setting and/or a different
geological period. The Nile in Egypt began to form its valley in the Upper
Miocene age. The then Nile, Eonile, cut its gorge at a much lower level than the
present sea level. The bottom of the canyon formed by the Eonile reached depths
from 170 m in Aswan to more than 2500 m north of Cairo and to even greater
depths in the northern Delta embayment (Said, R., 1982). As a result of faults
and shifts the Eonile changed its course from north-west to a more northerly
course.

In the Lower Pliocene time the Eonile Valley became covered with the sediments brought by the sea as it advanced along the valley as far as Aswan. Later,
sediments were brought by the Paleonile, during the Upper Pliocene some 3.2
million years ago. The sedimentation was very much augmented by the contribution
of the Red Sea hills via the then existing Wadis, which carried the sediment-
laden torrents of water. By the end of the Paleonile sedimentation, the Eonile
canyon was completely filled up and the Delta surface became more or less even
with a gentle northward slope.

The interval from about 1.85 million to 0.7 million years ago is claimed to
be a period of great cooling and dryness in Egypt. It is quite possible that the
Paleo-Proto-Nile (the then Nile in Egypt) ceased to flow and Egypt itself became
a desert. The abraded material was transported by the blowing wind and began to
form the large depressions in the Western Desert. Deposition of gravels and
sands was quite active in the Protonile period, that is, from 0.7 to 0.5 million

years ago. The deposited material took the form of terraces parallel to the
modern Nile Valley.

The tectonic movements which took place from the end of the Protonile period
up to about 125 000 years before present helped to form some hydraulic connection between the Nile in Egypt (the Prenile) and the Ethiopian Plateau. Accordingly, much larger floods and sediments were carried by the Prenile than by the
Proto or by the Palaeo-Proto Niles. As a result, a large delta was developed,
with sediments extending into the sea. One can thus conclude that the Nile Delta
occupies a great tectonic depression, and is bounded on both sides by gravelly
plains which rise up to 100 m above mean sea level. On the eastern side, the
Delta region is bounded by a major upwarp zone which occupies most of North-
Central Sinai. This is followed in a northward direction by a downwarp zone
which occupies most of the Delta region and the north-western part of Sinai.
This downwarp zone is affected by a number of faults. The plains on the eastern
and western sides of the Delta merge into the elevated table lands (higher than
200 m above mean sea level), which act as watershed areas. The eastern table
lands are dissected by a number of Wadis which act during rainy seasons as
drainage arteries. The sedimentary section in the Delta has an expected thickness of more than 10 000 m (Shata, A., and El-Fayoumy, I., 1969).
The Prenile was so vigorous that it shifted its course to the east of what used
to be the course of the Protonile, though yet west of the present course of the
Nile. The last section of the Prenile period, which terminated some 125 000
years before present had been characterized by uplifts which led to reduced
river flow in Egypt. This was followed by the Pre-Neonile, which lasted from
125 000 to 30 000 years before present. In the wet episodes of this period heavy
rainfalls resulted in coarse sand and gravel deposits. The Late Acheulean
pluvial was an important one and it ended some 35 000 years ago. The Prenile
sediments were laid down by the river and the final shape of the present Nile
Valley was formed (Said, R., 1982). The Neolithic pluvial lasted from about
10 000 years to, say, 5 000 years before present. The climate that prevailed
then was so wet that the southern part of Egypt had an annual rainfall of 100 to
300 mm (Said, R., 1981). At present the annual rainfall at the same place is
practically nil.

The above discussion about the regression and transgression of the sea during
the Cainozoic era, and the development of the course of the river during the
successive geological periods can be seen in Figs. 7.4 a thru' f.

304

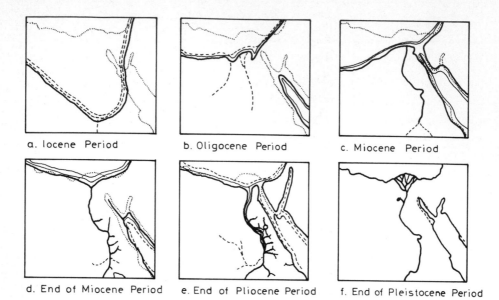

a. locene Period   b. Oligocene Period   c. Miocene Period

d. End of Miocene Period   e. End of Pliocene Period   f. End of Pleistocene Period

Fig. 7.4.   Regression and transgression of the sea during the Cainozoic era

7.2   GROUNDWATER POTENTIAL

7.2.1   Groundwater in Kenya, Uganda and Tanzania

In the previous section it has been mentioned that some areas in those parts of Kenya, Uganda and Tanzania which are situated within the Nile Basin are covered with Precambrian rocks. In these areas it is only the shallow water bearing layers close to the surface that yield water unless wells or holes are drilled to depths usually of more than 100 m.

In the volcanic areas between the eastern and western Rift Valleys, higher yields of water can be expected from the highly elevated grounds. Groundwater, on the contrary, is quite scarce in many of the sedimentary areas, partly due to the low rainfall and partly because of excessive evaporation. Accordingly, most of the groundwater - if available - is of relatively poor quality.

Drilling of shallow wells and water holes is practised for water supply to villages and small communities. Probably 90% of the dry season demands in the arid parts of Tanzania are supplied by water holes and wells. The water-bearing formations are often found on the slopes of the bedrock hills and they consist of gravel, sand, laterites, granite (crevices) and calcrete. The valleys them- selves are covered with mbuga clays as shown in Figs. 7.5a and 7.5b.

Some of the leading figures of the shallow water wells as practised in Tanzania are as follows (D.H.V., 1976 and 1978):

Maximum depth of well	$\simeq$ 10 m
Spacing between wells	5-10 m
Internal diameter	1.25 m
Operation	hand pump
Duration	10 hrs/day
Volume of pumped water	6-10 m^3/day
Number of consumers/well	250 inhabitants

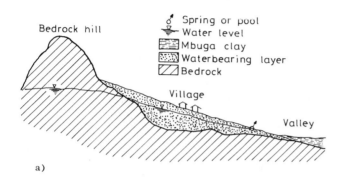

a)

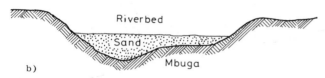

b)

Fig. 7.5.   Shallow water-bearing formations in Tanzania (D.H.V., 1976)

7.2.2   Groundwater in the Sudan

In general, water is not evenly distributed underneath the surface of the
Sudan. This is quite understandable since the groundwater supplies depend on
the rainfall and on the local geology of the area from which water has to be
extracted. When the soil has a low absorption capacity a high proportion of the
rainfall remains on the surface and is lost by evaporation. It is only in a few
places that groundwater supplies depend on the rainfall at some distance from
the local underground reservoir.

In the western part of Equatoria Province rivers may run dry, but if it
happens this will be for a short time. In such a case supplies can be obtained
from pools or shallow wells in stream beds. In the eastern part of the same
province the hills are traversed by perennial, though small, streams. The plains
remain, however, dry in the non-rainy season. In the plain east of the Bahr el
Jebel, groundwater can be found in some localities at depths of 40-50 m. In some

other localities where the shortage could be acute, the water table can be found
at depths of 100 m or more.

The underground supplies in the central part of the Sudan are often quite
poor because of the extremely low permeability of the clay cap covering the surface of the plains. The isolated hills (inselberg) which protrude through this
aquiclude are usually surrounded by cones of coarser material, which are highly
permeable. In such hills, local supplies can be found in pools and in wells
located near the foot.

Water supply by wells is widely practised for domestic purposes and for
raising livestock in the Gash Delta in Kassala Province. Water is drawn from the
wells throughout the year and other wells function only seasonally. Replenish-
ment of groundwater is accomplished annually by flooding through channels taking
from the canals. The basalt country of Gedaref has water in joints in the lava,
but the deeper lavas have no open joints and are dry.

The thick accumulations of consolidated sands, gravels and clays in the
depressions in the Kordofan Province carry water. Some of these depressions,
however, have too saline water for consumption. The marginal areas of these
depressions have no water to a depth of 100 m or more, except close to the hills
where the Wadis spill out on to the outwash fan at the margin of the plain. The
areas which have been geologically referred to as qōz frequently have small
shallow supplies at the foot of the sands. The distribution of these supplies is
mainly dependent on the buried topography. In the southern half of the undulating country the number of wells supplying good water has been steadily increasing.

The volcanic areas of the Darfur Province are similar to the basalt country
of Gedaref. The sandstones belonging to the Nubian Series are considered as a
source of fairly deep groundwater (e.g. the area between El-Nahoud and El-
Fasher). The groundwater supplies can be described as fairly good.

The tube wells which are located throughout the Gezira region give the piezometric levels. These have been interpolated to give the contours of the phreatic
surface as shown in Fig. 7.6. (Salaam, A., 1966). The consistency of this map
has been questioned by Eagleson, P., and Miller, S. (1983).

In the northern desert region of the Sudan there are two groundwater resources. One is a permanent groundwater reservoir in the sandstones of the Nubian
Series (see the map, Fig. 7.2.). It usually exists at a fairly considerable
depth from the surface though it could be locally bared by erosion to form some
oases resting on a mudstone layer. The other source is formed by local concentrations of subsoil water derived from local rains and mainly supported by
seepage from floods. In the western part of the sandstone area the groundwater
reserves are reliable and plentiful when compared to the existing demand on

them. These reserves receive, however, modest amounts of recharge, characterizing such waters as mainly non-renewable. Only in the area along the Nile, between Dongola and Wadi Halfa, due to infiltration of the Nile and Lake Nasser into the banks, a considerable area of renewable groundwater will become available (Gischler, C.E., 1979). In the areas covered with crystalline rocks, which are mainly to the east, resources are much more restricted, except along the line of major Wadis. As such they are dependent on the occasional floods. Thus, the yield of a resource of this type is roughly proportional to the size of the catchment area upstream of the well-point and to the rainfall. Obviously, the salinity of water increases with distance from the head due to evaporation.

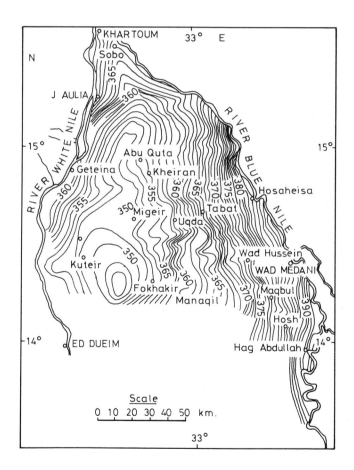

Fig. 7.6.   Phreatic surface; El Gezira, Sudan (Salaam, A., 1966)

7.2.3   Groundwater in Egypt

7.2.3.1   Groundwater in the Western Desert

The Nubian Basin, part of which has already been discussed in connection with the groundwater in the Sudan, covers a surface area of about 1.8 million km^2. This basin extends to Egypt west of the Nile, to the extreme north-east of Chad, and to south and eastern Lybia.

The geology of the Western Desert in Egypt has been explained earlier. The Nubian Basin is characterized by its extreme aridity. The inhabited part of it is a series of oases. The most important oases in Egypt are the Kharga, Dakhla, Farafra, Bahariya and Siwa.

A number of theories regarding the formation and flow of groundwater in this desert have been published. According to Ball, J., (1927) the origin and source of the artesian water supply of the oases in Egypt is a vast subterranean stream originating in the Erdi and Ennedi region on the border between the Chad Basin and the Nile Basin in the Sudan. This underground water flows in the direction of the north-west, north and north-east.

Hellström, B., (1940) developed the flow net shown in the map, Fig. 7.7a. This map also includes the north-western part of the Sudan. The works of Ball and Hellström were discussed in a paper which was prepared by the Survey Department of Egypt (Murray, G.W., 1952). The fact that was stressed in this paper is that the groundwater flowing freely into the oases has nothing to do with the Nile. It is merely fossil water imprisoned in the Nubian rocks for several thousands of years; the lowest layer of all groundwater may date back to the Pliocene. The age of the groundwater has recently been determined at some locations using ^{14}C and was found to originate from 40 000 to 20 000 years BP (before present), i.e. from the late Acheulean pluvial. Not much groundwater formation took place in the Sahara in the interval between 20 000 and 14 000 years BP.

The period from 14 000 years BP to present times is interpreted as a post-pluvial humid phase with the humid peaks alternating with arid phases. All the younger waters originate from shall groundwaters (Sonntag, C., et al, 1976).

Groundwater moving downhill, readily flows into Cretaceous, Eocene and Miocene strata; doing so, it usually becomes contaminated with salt. The analysis of the water has shown that it is relatively highly mineralized, particularly in iron, carbonates and sodium chloride, while the gasses contain at least some hydrogen sulphide and possibly carbon dioxide. This quality of water calls for having the well casing adequately covered by protective coatings (Paver, G.L., and Pretorious, D.A., 1954). The same reference estimates the groundwater extraction from the Kharga and Dakhla oases only as follows:

	Kharga	Dakhla
Approximate yield of shallow wells, million m^3/yr	38.7	92.7
Approximate yield of deep wells, million m^3/yr	3.8	16.8
Total annual withdrawal, million m^3	42.5	109.5

Hellström, B., (1940) estimated the groundwater supply to the Quattara depression at about 3.1 million m^3/day or 1130 million m^3/yr and the permeability of the Nubian sandstone to lie between 2.4 x 10^{-5} and 2.4 x 10^{-4} m/day.

The Government of Egypt has undertaken a programme for exploiting the groundwater reservoir, especially in the area occupied by the Kharga and Dakhla oases. Since 1959 till recently the number of wells drilled there has exceeded 350 wells with depths ranging between 400 and 1200 m. As a result of the extensive withdrawal of the groundwater, the hydrostatic pressure in the wells has fallen by more than 30 m, with the obvious consequence that many of the wells have ceased to flow freely (artesian). This alarming situation has drawn the attention of the Egyptian General Desert Development Organization, as well as other organs and individuals.

The future of the groundwater discharge in the Kharga and Dakhla Oases was forecasted till the year 2000. Using the flow net originally developed by Hellström and more recent data, the steady guaranteed recharge was estimated at about 204 million m^3/yr, of which 64 million to El-Kharga and 140 million to El-Dakhla. Installation of pumps was reported to increase the discharge of wells in these two oases by 15% and 7%, respectively (Hammad, Y.H., 1969). Also, the hydrological aspects of the groundwater reservoir underlying the Dakhla Oasis were extensively studied and reported to the Ministry of Land Reclamation, Egypt (Shahin, M., et al, 1970). A more recent flow net has been prepared (see Fig. 7.7b.). From the recent data it was found that the groundwater discharges through natural springs, drilled wells and losses due to evapotranspiration from areas covered by natural vegetation and due to leakage have been estimated as included in Table 7.1.

In the course of the last few years many a model study has been carried out with the aim of arriving at the quantities of groundwater which can possibly be withdrawn under the constraint of maximum lift of 100 m below the ground surface. It has been reported that the total annual exploitable water is about 1350 million m^3 for the next 50 years. This amount is distributed as 140, 182, 363, 509 and 156 million m^3/yr for the Siwa, Bahariya, Farafra, Dakhla and Kharga Oases, respectively.

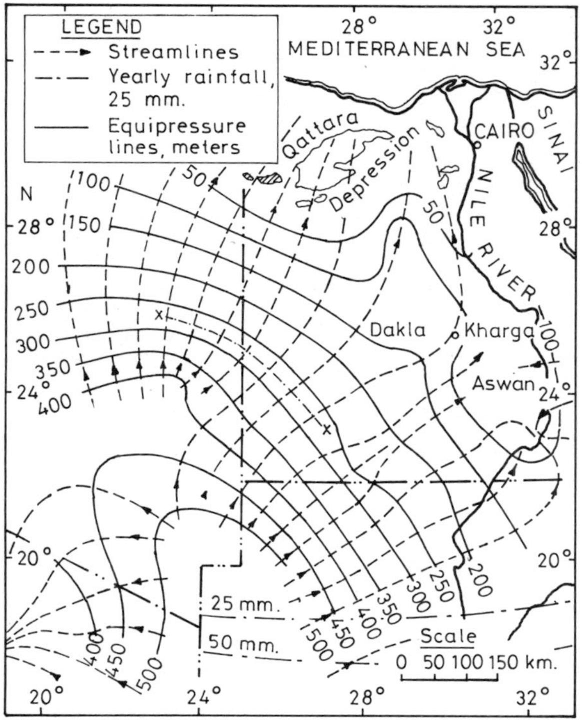

310

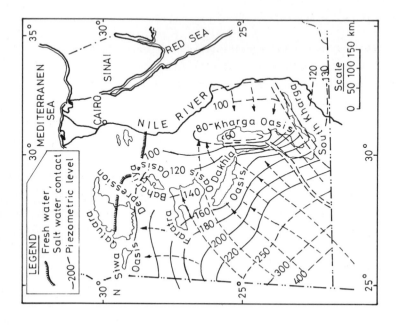

Fig. 7.7b.   Flow net in the Western Desert of Egypt
(Ezzat, M., 1976)

Fig. 7.7a.   Flow net in Sahara aquifer (Hellström,
B., 1940)

TABLE 7.1   Total discharge of wells and natural losses in Western Desert (Ezzat, M.A., 1976)

Area	Discharge of wells, m³/day	Natural losses, m³/day	Total m³/day
Siwa Oasis	120.000	300.000	420.000
Quattara Depression		1.400.000	1.400.000
Farafra & Bahariya Oases	145.000	400.000	545.000
Dakhla Oasis	557.000	141.000	698.000
Kharga Oasis	225.000	190.740	416.000
South Kharga area		100.000	100.000
Total	1.047.000	2.531.740	3.579.000

7.2.3.2   Groundwater in the Eastern Desert

There is too little known about groundwater extraction from the aquifer underlying the Eastern Desert. Nearly all of the available information belongs to the desert between the Nile Delta and the Suez Canal, some parts of the Sainai Peninsula and the south-eastern desert of Upper Egypt. These areas used to be, and probably are still, occupied by military establishments.

The wells and boreholes existing between the Delta and Suez Canal were investigated and reported by Shotton, F.W., (1946). According to this source they were 73 in total and the depth to water table varied between 2 m and 75 m, and the yield was in the range of between a few hundred to more than 40 000 gals/hr. Some water holes were described as yielding very sweet water and others were abandoned because the water was brackish or saline.

Forty boreholes in the south-eastern desert of Upper Egypt were also investigated. It was concluded that supplies from 1000 to 5000 gals/hr of good quality water are obtainable by boreholes along the Nile Valley. Contrarily, supplies in the consolidated sedimentary formations between the Nile and the Red Sea are of low yield and of bad quality. In some locations the depth to the water table in the Nubian sandstone could reach 80 m below the ground surface. Supplies of 500 to 1000 gals/hr drinkable water are obtainable by bores of approximately 30 m in depth in the Wadis traversing the Pre-Cambrian area, west of the watershed, almost midway the line between Qena, on the Nile, and Safaga, on the coast of the Red Sea. Similar supplies are obtainable east of the watershed, but the quality of water tends to be bad. The sodium sulphate in particular reaches a high level. The supplies obtained from the coastal sedimentary strip are extremely poor both in quantity and quality (Paver, G.L., 1946).

7.2.3.3   Groundwater in Upper Egypt

The cliffs which bound the course of the Nile River in Upper Egypt are made up of either limestone or sandstone. The contact surface between them and the

312

valley, in which the river has cut its course, forms the boundary of the aquifer system. Such an aquifer has been formed by the deposits brought up by the Nile during the wet periods. The aquifer can be divided into two distinct types: unconfined, occupying almost one-quarter of the surface area of Upper Egypt and semi-confined, occupying the rest of the area. The latter is overlain by a relatively thin layer of silt, loam and clay. Both types of aquifers are built up, however, of the same materials: graded sands and gravels.

Lithological sections across the Valley at some locations are shown in Fig. 7.8., (Attia, F. et al, 1983). The study of the lithology, and the analysis of the pumping-test data have led us to the results included in Table 7.2 (Shahin, M., 1983).

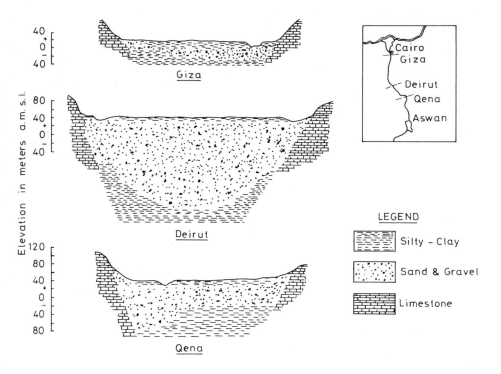

Fig. 7.8.   Lithological sections across the Nile Valley in Upper Egypt

The water balance of the river stretch between Aswan and Cairo, the water balance of the cultivated area traversed by the same stretch and the weighted average recharge method have been worked out. The estimated recharge to the river was found to be in the range of between 1.33 mlrd m^3/yr to 2.67 mlrd m^3/yr for the period from 1972 to 1980.

The downward percolation of the excess of irrigation water and the seepage from high-level canals are the two principal sources of recharge of the aquifer system, especially in the post-High Dam period.

TABLE 7.2   Geohydrological constants of the aquifer system underlying Upper Egypt

Type of aquifer	Aquifer				Semi-confining layer		
	k, m/day	D, m	kD, m^2/day	S	k´, mm/day	d´, m	C, day
Unconfined	40- 80	15- 60	1000- 5000	0.10-0.12	–	–	–
Semi-confined	40-120	15-240	1000-20000	$(5-50)10^{-4}$	1-10	3-20	200-4000

7.2.3.4   Groundwater in the Nile Delta area

The Nile Delta aquifer is one of the most important groundwater reservoirs in Egypt. Systematic hydrogeological investigation in the Delta area began in 1954 and is still underway. This investigation so far has led to some under-standing of the groundwater properties, parameters and movement in the aquifer system there.

The aquifer has recently been reported as consisting of three different types (Shahin, M., 1983) and not only of the unconfined type (Shata, A., et al, 1969) or of the open-leaky flow system (Farid, M.S., 1980). The approximate boundaries separating the three types of aquifer can be seen from the map, Fig. 7.9. An impression about the stratigraphy and lithology can be seen from the longitudi-nal and cross-sections which were prepared by Solait, M.L. (1964).

The geohydrologic constants of the three types of aquifer underlying the Nile Delta area together with those belonging to the reservoir connecting the Delta and Upper Egypt are included in Table 7.3.

Each year the Nile Delta area receives something like 35×10^9 m^3 of surface water from the Nile for irrigation, industry and water supply. This recharges the aquifer through infiltration of excess irrigation water and through seepage from an extensive network of canals and drains. The water budget of the Nile Delta area has been studied by a number of authorities resulting in a range of aquifer recharge volumes. This range coincides with the range of 5 to 10% of the input value, i.e. from about 2 to 4 mlrd m^3/yr. Kashef, A., (1983) suggests 3.98×10^9 m^3/yr to represent the potential water surplus for future development, unless unaccounted losses or errors in the estimated values of the items com-prising the water balance appear in the future. Shahin, M., (1983) has obtained a figure of 2.2×10^9 m^3/yr. Which figure is more accurate is not the main ques-tion. What is important is that the aquifer underlying the Nile Delta area receives a certain recharge at the end of each year. That part which does not escape to the sea or flow here and there to any of the existing depressions

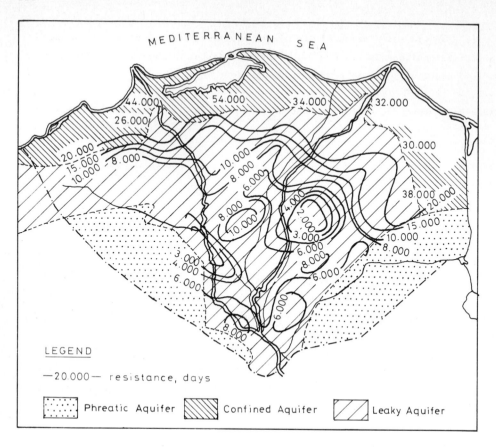

Fig. 7.9. The three types of aquifer underlying the Nile Delta area and the
resistance of the overlying cap.

TABLE 7.3 Summary of the geohydrologic constants of the reservoirs underlying
the Nile Delta and its connection with Upper Egypt

Type of aquifer	Aquifer				Semi-confining layer		
	k, m/day	D, m	kD, m^2/day	S	k', mm/day	d', m	C, day
Unconfined	50- 80	50-150	2500-12000	0.15-0.25	-	-	-
Semi-confined	25-100	100-250	2500-25000	$(1-10)10^{-4}$	1-3	8-20	3000-20000
Confined		U N K N O W N			<1	20-50	20000-50000
Greater Cairo			1000-10000	$(1-28)10^{-3}$	1-5	8-10	1500-10000

produces a rise in the water table level. It is quite possible to prevent this
rise by pumping the groundwater. Would the pumping wells be considered then as a
means of drainage or as a means of extracting water for, say, irrigation or for
the two purposes combined? The answer to these questions depends primarily on
the quality of the pumped water and its variability both in time and in space.

The results which have been obtained from a very recent study on the salinity of the groundwater in the Delta aquifer (Kashef, A., 1983) are rather alarming. They show a deeper penetration of the sea water into the aquifer than it was ever thought. The inland extent of the zone of saline groundwater (unsuitable for most purposes), instead of being about 50 km, has been suggested to be not less than 120 km from the sea coast.

REFERENCES

Andrew, G., 1948. The agriculture in the Sudan (edited by J.D. Tothill). Chapter VI: Geology of the Sudan: 84-128. Oxford University Press, London.
Attia, F., Amer, A., and Hefny, K., 1983. Effect of High-Aswan Dam on groundwater conditions in Upper Egypt. Proceedings of the International Conference on water resources development in Egypt, Cairo: 99-119.
Ball, J., 1927. Problems of the Libyan Desert. Geographical Journal, Vol. 70, No. 1, 2 and 3, London.
Berry, L., and Whiteman, A.J., 1968. The Nile in the Sudan (with discussion). The Geographical Journal, Vol. 134, Part I: 1-38.
Eagleson, P.S., and Miller, S.A., 1983. Water table depression in the Gezira region. Proceedings of the International Conference on Water Resources Development in Egypt, Cairo, 201-212.
Ezzat, M.A., 1976. Regional groundwater models, El-Wadi El-Gadid project. Working Document No. 2, UNDP/FAO report, AGON: EGY 71/561, Cairo.
Farid, M.S., Hefny, K., and Amer, A., 1979. Hydrological aspects of the Nile Delta reservoir. Proceedings of the International Conference on Water resources planning in Egypt, Cairo: 299-320.
Fitt, R.L., Marwick, R., and Whitaker, F.W., 1967. The Roseires Dam, Sudan: planning and design. Proceedings Institution of Civil Engineers, Paper 7047, Vol. 38: 21-51.
Gischler, C.E., 1979. Water resources in the Arab Middle East and North Africa. Middle East and North African Studies Press Ltd., Cambridge, England, 132 pp.
Hammad, Y.H., 1969. Future of groundwater in African Sahara Desert. Journal of the irrigation and drainage division, ASCE, Vol. 95 No. IR 4: 563-580.
Hellström, B., 1940. The subterranean water in the Libyan Desert. Geografiska Annaler, Stockholm.
Hepworth, J.V., 1964. Explanation of the geology of sheets 19, 20, 28 and 29. Southern West Nile, Geological Survey of Uganda.
Kashef, A.I., 1983. Salt water intrusion in the Nile Delta. Groundwater Vol. 21, No. 2: 160-167.
Murray, G.W., 1952. The artesian water of Egypt. Survey Department of Egypt. Paper No. 52, Cairo.
Paver, G.L., 1946. Water supply in the Middle East campaigns, VI - The southeastern desert of Upper Egypt (Red Sea Hills). Water and water Engineering, London, 10 pp.
Paver, G.L., and Pretorious, D.A., 1954. Report on hydrological investigations in Kharga and Dakhla Oases. Publications of the Desert Institute of Egypt, No. 4, 108 pp, Cairo, Egypt.
Saggerson, E.P., 1962. The physiography and geology of East Africa. Natural Resources of East Africa, Nairobi, Kenya.
Saggerson, E.P., 1972. East Africa: its peoples and resources (edited by W.T.W. Morgan). Chapter 7: Geology, 67-94. Oxford University Press, Nairobi, Kenya.
Said, R., 1962. The geology of Egypt. Elsevier, Amsterdam and New York, 377 pp.
Said, R., 1981. The geological evolution of the River Nile. Springer-Verlag, New York, 151 pp.
Salaam, A., 1966. The groundwater geology of the Gezira, M.Sc. thesis, University of Khartoum.

Sandford, K.S., 1935. Sources of water in the North-Western Sudan. The Geographical Journal: lXXXV, 412-431.

Shahin, M.A., et al, 1970. Hydrological aspects of the groundwater reservoir underlying the Dakhla Oasis. Progress Report No. 1 submitted to the Ministry of Land Reclamation, Egypt, 183 pp.

Shahin, M.A., 1983. Report on a TOKTEN-UNDP Assignment at the Groundwater Research Institute. Academy of Science and Technology and Office of UNDP, Cairo.

Shata, A., and El-Fayoumy, I., 1969. Remarks on the hydrogeology of the Nile Delta, UAR. Proceedings of the International Symposium on the hydrology of deltas, Bucharest, IASH/UNESCO, Vol. II: 385-396.

Shotton, F.W., 1946. Water supply in the Middle East campaigns V. The desert between the Nile Delta and the Suez Canal. Water and Water Engineering, London, 12 pp.

Solait, M.L., 1964. Groundwater at several places in the Nile Delta and Valley, The United Arab Republic. Report submitted to the Seventh Arab Engineering Conference (in Arabic), Baghdad, 79 pp.

Sonntag, C., Klitzsch, E., and El-Shazly, E.M., 1976. ^{14}C-dating of Sahara groundwaters and palaeoclimatic information by deiterium and oxygen 18.

Stamp, L.D., and Morgan, W.T., 1972. Africa: a study in tropical development, John Wiley and Sons Inc., New York, 520 pp.

Chapter 8

THE BASIN SURFACE RUN-OFF AND THE RIVER LEVELS AND DISCHARGES

Most of the studies of ground water in the Nile Basin point to the relative
small quantities of ground water discharge, recharge and in storage. This statement should neither be understood nor interpreted as an attempt to belittle the
importance of ground water investigation and/or exploitation. Both are certainly
important in some special situations. The conclusion as given here is based,
however, on the proportion of the quantity of ground water to the quantity of
surface water. For example, the volume of the ground water contained within the
fissure system in the Equatorial Lakes Plateau represents only about one half of
the mean annual precipitation on the catchments of these lakes. The ground water
discharge to the lakes is probably less than 0.2% of the total mean annual run-
off to them. For this reason our discussion on the basin run-off shall be confined to the surface run-off only.

8.1   CATCHMENT OF LAKE VICTORIA
8.1.1   Run-off coefficient
The catchment area of Lake Victoria has been given in Chapter 2 as 193 000
km^2, and the area covered by the lake water as 67 600 km^2. The surface area of
the gauged basins is 153 640 km^2, or about 78.5% of the total catchment area.
The rest, which comprises 21.5% of the catchment area, is ungauged. It seems
that the contribution of the ungauged streams to the total run-off to Lake
Victoria is quite inferior compared to the contribution of the gauged basins.
This can be attributed to the fact that much of the ungauged surface is covered
by swamps and relatively low-lying areas. Consequently, any error in the estimation of the contribution of the ungauged part will not affect the total inflow
to the lake to any considerable extent. The rainfall and run-off data for the
period 1906-1932 were reported by Hurst in Vol. V of the Nile Basin (Hurst,
H.E., and Philips, P., 1938). A summary of the results are given in Table 8.1.
Additionally, the 1969 and 1970 rainfall and run-off data collected by the
hydrometeorological project of the Equatorial Lakes (WMO, 1974) have been
worked out so as to evaluate the monthly and the annual run-off coefficients.
This coefficient is the percentage of the rainfall on the catchment that reaches
the lake. The results obtained are presented in Tables 8.2 and 8.3.
The run-off coefficients for the ungauged sub-basins have been taken arbitrarily at 2% for the northern and the southern shores of the lake, and at 3%
for the eastern and the western shores, similar to the hydrometeorological

survey project of the lakes. These values of the run-off coefficient are somewhat different from those assumed by Hurst.

From Table 8.3 it is clear that the run-off coefficients for the year 1969
are much closer to those averaged over the period 1946-1970 than the 1970 coefficients. It is quite natural that the run-off coefficient increases with the
precipitation depth. In 1969 the rainfall was 1054 mm and in 1970, 1213 mm. This
increase in rainfall resulted in the rise of the coefficient from 8.95% in 1969
to 10.09% in 1970. It is therefore quite reasonable to take 9% as the yearly
coefficient, averaged over a long sequence of years. For the 27-year period,
from 1906 up to and including 1932, the value of the yearly coefficient suggested by Hurst was 8% (Table 8.1).

The Kagera River has always been described as the principal feeder to Lake
Victoria. The basin of the Kagera has a total surface of 58370 km^3 and it comprises six sub-basins. The area, mean annual rainfall and annual run-off coefficient of each of these sub-basins over the period 1958 up to 1962, except for
the last two sub-basins where the data apply to 1970 only, are as follows:

River sub-basin	Area, km^2	Mean annual rainfall, mm	Run-off coefficient, %
Akyanaru	5285	1180	11.8
Nyvarongo	13315	1182	16.7
Ruvuvu	12300	1158	14.6
Middle Kagera	22835	1100	2.3
Mwisa	2035	1000	2.5
Ngono	2600	1200	18.0
Total and means	58370	1139	$\simeq$ 10.0

TABLE 8.1   The run-off of Lake Victoria for the period 1906-1932 (Hurst, H.E.,
and Philips, P., 1938)

District	Mean annual rainfall, in mm	Area, in km^2	Estimated percentage of rainfall reaching L. Victoria	Total water reaching Lake Victoria, in 10^6 m^3/yr
Kagera Basin	1.242	60.000	10.6	7.900
North-western and northern portion	1.302	30 000	1	400
North-eastern	1.248	50.000	10	6.200
South-eastern and southern portion	995	50.000	7	3.500
Islands in L. Victoria	1.150	3.000	10	300
Means & Totals	1.190	193.000	8	18.300

TABLE 8.2   The run-off coefficients of the sub-basins contained in the catchment of Lake Victoria for 1969 and 1970, computed from the data collected by the Hydrometeorological Survey Project (WMO, 1974)

| Catchment area | | Monthly and yearly values of the run-off coefficient | | | | | | | | | | | | |
Name	Area, km²	Jan.	Feb.	Mar.	Apr.	May	June	July	Aug.	Sep.	Oct.	Nov.	Dec.	Year
Sio	1080	8.57	4.48	8.72	7.89	21.74	53.32	18.73	14.82	22.27	29.38	67.91	21.17	17.96
		10.23	4.80	11.54	10.14	27.53	67.10	23.84	18.87	19.48	36.62	98.21	31.48	24.74
		9.40	4.64	10.13	9.01	24.64	60.21	21.29	16.85	20.88	33.00	83.06	26.33	21.35
Nsioa	11980	8.19	26.47	6.71	4.37	15.20	15.71	11.48	9.42	14.61	8.35	17.36	19.22	11.24
		9.25	23.28	8.16	10.95	19.63	27.02	23.44	25.23	30.80	34.89	42.60	21.70	20.81
		8.72	24.88	7.44	7.66	17.42	21.36	17.46	18.32	22.71	21.62	29.98	20.46	16.03
Yala	2650	11.59	25.64	11.85	20.86	32.40	33.08	32.48	38.70	42.99	45.69	48.67	66.44	30.01
		8.37	17.50	5.01	12.77	23.23	23.05	23.71	26.04	31.24	32.70	45.58	33.04	20.61
		9.98	21.57	8.43	16.82	27.81	28.06	28.10	32.37	37.12	39.20	47.12	49.74	25.31
Kibos	490	3.88	8.96	4.14	7.97	18.90	20.90	26.84	10.80	13.06	11.95	9.60	5.49	10.20
		5.07	7.33	4.71	10.05	16.07	16.30	19.26	5.74	9.72	10.20	8.70	6.44	9.11
		4.47	8.14	4.42	9.01	17.48	18.60	23.05	8.27	11.39	11.07	9.15	5.97	9.65
Nyando	2650	3.54	39.14	8.74	4.43	12.23	7.09	7.68	6.50	10.09	8.58	9.81	7.94	8.44
		5.05	17.47	8.26	15.57	25.65	15.51	11.62	19.75	24.18	16.24	11.41	10.76	14.83
		4.30	28.30	8.50	10.00	19.94	11.30	9.65	13.12	17.13	12.41	10.61	9.35	11.64
Cheronoit	560	2.95	14.49	4.05	2.70	7.87	0.56	9.53	1.70	2.42	2.55	1.09	0.00	3.73
		3.82	5.98	5.10	8.11	8.54	12.32	12.93	4.67	8.00	6.87	3.57	2.56	6.36
		3.38	10.24	4.58	5.40	8.20	6.44	11.23	3.18	5.21	4.71	2.33	1.28	5.04
Sondu	3230	13.50	72.64	21.67	15.76	28.29	24.55	16.85	10.53	34.13	25.60	15.48	7.83	21.09
		6.02	25.52	24.44	40.84	55.88	53.99	32.88	30.96	52.58	58.40	28.38	21.57	34.48
		9.76	49.08	23.06	28.30	42.09	39.27	24.86	20.74	43.36	42.00	21.93	14.70	27.78

For each catchment area, the top line gives the run-off coefficient for 1969, the middle line for 1970, and the bottom line gives the average run-off coefficient for 1969 and 1970

TABLE 8.2 (continued)

| Catchment area | | Monthly and yearly values of the run-off coefficient | | | | | | | | | | | | |
Name	Area, km²	Jan.	Feb.	Mar.	Apr.	May	June	July	Aug.	Sep.	Oct.	Nov.	Dec.	Year
Awach-Kaboun	610	23.68	124.50	33.02	15.19	15.46	48.92	56.28	14.20	16.68	10.02	14.01	9.36	29.51
		10.93	22.72	10.87	19.45	47.44	67.95	95.63	30.17	74.47	47.90	29.25	20.42	31.65
		17.30	73.61	21.95	17.32	31.45	58.43	75.95	22.18	45.57	28.96	21.63	14.88	30.58
Gucha-Migoei	6840	3.62	11.70	11.15	11.32	23.00	16.39	11.30	3.13	5.95	2.99	5.98	4.90	9.40
		5.00	18.77	14.79	17.22	34.48	44.12	15.78	8.75	14.34	8.96	9.39	6.32	16.05
		4.31	15.23	12.97	14.27	28.74	30.25	18.54	5.94	10.15	5.98	7.69	5.61	12.72
Mori	590	2.71	3.99	3.08	23.87	34.90	38.52	16.95	6.16	4.35	6.05	1.71	4.89	10.17
		15.09	10.97	13.86	74.94	85.95	88.14	33.15	10.80	4.25	4.60	2.13	4.49	28.74
		8.90	7.48	8.47	49.40	60.43	63.33	20.05	8.48	4.30	5.32	1.92	4.69	19.45
Mara (Mines)	10830	2.13	4.62	1.80	6.25	16.24	17.82	18.06	18.76	21.06	12.93	9.79	6.10	9.78
		3.14	8.54	5.29	11.17	18.05	23.14	14.82	11.40	14.44	12.41	9.04	4.58	10.22
		2.63	6.58	3.55	8.71	17.15	20.48	16.44	15.08	17.75	12.67	9.42	5.34	10.00
Suguti	1020	1.84	2.97	1.74	9.80	18.59	39.22	21.01	56.02	14.71	6.54	4.67	2.13	7.06
		0.00	1.79	4.07	21.19	0.00	201.96	0.00	0.00	0.00	0.00	0.00	0.34	7.33
		0.92	2.38	2.91	15.50	9.30	120.59	10.50	28.01	7.36	3.27	2.33	1.24	7.20
Ruana Gruneti	11430	3.44	5.47	1.21	7.29	1.75	7.88	2.63	5.26	2.69	1.62	0.66	0.93	2.98
		4.59	7.08	1.62	9.77	2.40	6.79	5.15	45.50	4.06	2.45	0.88	1.33	4.44
		4.02	6.28	1.42	8.53	2.07	7.34	3.89	25.38	3.38	2.03	0.77	1.13	3.71
Mbalageti	3730	0.16	3.38	0.72	1.56	3.30	22.34	22.12	53.62	0.00	0.00	10.23	1.37	2.38
		3.33	3.41	6.79	9.14	2.92	29.10	24.78	131.64	13.38	6.22	3.46	6.72	6.33
		1.75	3.40	3.76	5.35	3.11	25.72	23.45	92.63	6.69	3.11	6.85	4.05	4.36

For each catchment area, the top line gives the run-off coefficient for 1969, the middle line for 1970, and the bottom line gives the average run-off coefficient for 1969 and 1970

TABLE 8.2 (continued)

Monthly and yearly values of the run-off coefficient

Name	Area, km²	Jan.	Feb.	Mar.	Apr.	May	June	July	Aug.	Sep.	Oct.	Nov.	Dec.	Year
Siniyu-Duma	10790	5.40	20.93	2.49	13.73	14.80	103.20	47.27	44.49	14.30	1.08	2.22	2.17	5.51
		3.34	5.64	8.14	10.10	10.84	153.94	39.10	91.11	21.62	2.89	11.09	9.90	7.71
		4.37	13.28	5.31	11.91	12.82	128.57	43.18	67.80	17.96	1.98	6.60	6.04	6.61
Ngogo	1200	1.64	2.60	4.50	1.67	4.10	0.00	0.00	0.00	0.00	0.00	10.51	5.28	3.85
		2.00	2.21	5.18	1.71	2.65	0.00	0.00	0.00	0.00	0.00	0.00	0.00	4.58
		1.82	2.40	4.84	1.69	3.38	0.00	0.00	0.00	0.00	0.00	5.26	2.64	4.22
Moame	2090	0.81	1.28	1.11	13.33	18.40	39.88	95.70	47.85	20.51	5.04	1.14	0.00	4.37
		0.75	0.84	1.83	5.61	1.06	0.00	0.00	0.00	0.00	0.00	0.00	0.00	2.86
		0.78	1.06	1.47	9.47	9.73	19.94	47.85	23.92	10.26	2.52	0.57	0.00	3.62
Isanga	4780	2.62	0.77	1.29	3.04	44.83	83.68	62.76	31.38	16.74	0.44	0.64	0.38	2.62
		5.93	1.17	3.97	7.78	8.37	0.00	0.00	0.00	0.00	0.00	0.00	0.00	2.86
		4.28	0.97	2.63	5.41	26.60	41.84	31.38	15.69	8.37	0.22	0.32	0.19	2.74
Kagera	58370	13.58	16.05	9.19	7.36	17.52	179.04	61.56	32.40	18.01	11.89	8.61	8.52	14.73
		10.04	10.85	5.97	5.36	14.38	196.65	88.92	34.20	21.18	12.97	8.50	8.87	13.56
		11.81	13.45	7.58	6.36	15.95	187.85	75.24	33.30	19.60	12.43	8.55	8.70	14.15
Ruizi	5670	1.42	1.21	0.81	0.85	1.64	10.88	3.85	1.76	0.57	0.52	0.48	1.01	1.35
		2.13	2.05	1.18	1.18	3.28	21.71	5.67	2.29	1.70	0.74	0.68	1.73	1.18
		1.78	1.63	1.00	1.02	4.61	16.30	4.61	2.02	1.13	0.63	0.58	1.37	1.26
Katonga	13020	0.47	2.28	0.50	0.34	1.28	9.45	1.26	0.58	0.51	0.29	0.34	0.91	0.70
		0.68	3.07	0.75	0.68	2.14	17.92	0.30	0.94	0.79	0.33	0.61	1.66	0.79
		0.58	2.68	0.63	0.51	1.71	13.69	0.78	0.76	0.65	0.31	0.48	1.29	0.75

For each catchment area, the top line gives the run-off coefficient for 1969, the middle line for 1970, and the bottom line gives the average run-off coefficient for 1969 and 1970

TABLE 8.2 (continued)

Catchment area		Monthly and yearly values of the run-off coefficient												
Name	Area, km^2	Jan.	Feb.	Mar.	Apr.	May	June	July	Aug.	Sep.	Oct.	Nov.	Dec.	Year
Northern shores	5060	2.0	2.0	2.0	2.0	2.0	2.0	2.0	2.0	2.0	2.0	2.0	2.0	2.0
Southern shores	16140	2.0	2.0	2.0	2.0	2.0	2.0	2.0	2.0	2.0	2.0	2.0	2.0	2.0
Eastern shores	12100	3.0	3.0	3.0	3.0	3.0	3.0	3.0	3.0	3.0	3.0	3.0	3.0	3.0
Western shores	6990	3.0	3.0	3.0	3.0	3.0	3.0	3.0	3.0	3.0	3.0	3.0	3.0	3.0

For each catchment area, the top line gives the run-off coefficient for 1969, the middle line for 1970, and the bottom line gives the average run-off coefficient for 1969 and 1970

TABLE 8.3   Average overall monthly and yearly run-off coefficients for the
years 1969 and 1970, and for the period 1946-70

Month and Year	Run-off coefficient, in percent for			
	1969	1970	Average of 1969 and 1970	Average of 1946-1970
January	6.00	5.33	5.67	5.0
February	11.15	7.62	9.38	10.0
March	5.91	5.50	5.71	7.0
April	5.47	6.88	6.18	7.0
May	12.69	13.44	13.06	8.5
June	30.16	37.10	33.63	25.0
July	17.46	24.91	21.18	17.0
August	12.11	16.49	14.30	12.0
September	12.96	17.07	15.01	10.0
October	7.51	10.49	9.00	7.5
November	6.26	7.56	6.91	6.5
December	5.82	6.31	6.06	6.5
Year	8.95	10.09	9.52	8.94

The figures given by Hurst for the run-off from the Kagera Basin are 10.6% (see Table 8.1) and 11% (Hurst, H.E., and Philips, P., 1938). Any of these two values is, in view of the rainfall causing the run-off, small compared to the more recent findings. For estimating the run-off coefficient from a given rainfall on the Kagera Basin we recommend the expression

$$r = 6 + 0.05 (P - 1000), \text{ in percent} \tag{8.1}$$

where r is the run-off coefficient and P is the annual precipitation in mm. The graph presented in Vol. IV of the report on the hydrometeorological project of the Equatorial Lakes can be translated to the expression

$$r = 4.6 + 0.17 (P - 1100), \text{ in percent} \tag{8.2}$$

The range of P is from 1000 to 1200 mm for eq. 8.1 and from 1100 t0 1200 mm for eq. 8.2. These two expressions give for P = 1200 mm/yr an annual coefficient of 16% and 21.6%, respectively, instead of the 10.6 or 11.0% proposed by Hurst.

The monthly and the annual discharges of the Kagera at Kyaka Ferry for the period from 1940 up to and including 1971 are listed in Table 1, Appendix D. The mean hydrograph of the river for the same period is shown in Fig. 8.1., whereas the basic statistical descriptors of these data are included in Table 8.4.

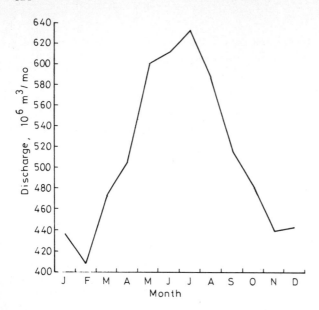

Fig. 8.1. The average hydrograph of the River Kagera at Kayaka Ferry for the period 1940-71

The results presented in Table 8.4 show clearly that the mean monthly and annual discharges are all serially correlated. Since the twelve monthly and the annual series have almost the same pattern of serial correlation, the model that can be used to describe any of the series is the same for all the 13 series. A first-order autoregressive scheme has proved to be adequate for this purpose. This model can be written as

$$(X_t - \bar{X}) = \alpha(X_{t-1} - \bar{X}) + \varepsilon_t \tag{8.3}$$

where X_t is the variate value at any time t, X_{t-1} is the value at time t-1, $\bar{X}$ is the mean, α is the first serial correlation coefficient, r_1, and ε_t is the residual at time t.

8.1.2   Water level and storage

The total surface run-off to Lake Victoria in a normal year is estimated at $(1.050 \times 0.895 \times 193 \times 10^9) + (1.25 \times 0.10 \times 1400 \times 10^6) = 18.3 \times 10^9$ m^3, and in a high year at $(1.25 \times 0.10 \times 193 \times 10^9) + 1.7 \times 0.12 \times 1400 \times 10^6) = 24.4 \times 10^9$ m^3.

The run-off to the lake, R, plus the direct precipitation on the lake surface, P_1, less the evaporation from it, E_1, plus the change in the storage content, ΔS, equals the outflow from the lake.

TABLE 8.4   The basic statistical descriptors of the Kagera monthly and annual discharges at Kyaka Ferry for the period 1940-1971

Month and Year	Basic statistical descriptors					Serial correlation coefficients							
	$\bar{X}, 10^6$ m^3	$s, 10^6$ m^3	C_v	C_s	C_k	r_1	r_2	r_3	r_4	r_5	r_6	r_7	r_8
January	436.1	120.9	0.2772	1.3597	5.4370	0.7023	0.4174	0.3005	0.3714	0.4760	0.4357	0.1460	$\overline{0}.0832$
February	408.3	133.2	0.3262	1.5377	5.7681	0.7055	0.4953	0.3294	0.3787	0.4609	0.3568	0.1309	$\overline{0}.0695$
March	472.3	174.5	0.3694	0.3170	4.1510	0.6735	0.3690	0.1672	0.2828	0.4386	0.3815	0.1690	$\overline{0}.0878$
April	505.4	203.2	0.4020	1.1882	3.6699	0.6911	0.5769	0.3497	0.4513	0.4365	0.4116	0.1327	0.0085
May	601.0	270.4	0.4498	1.1135	3.3572	0.6766	0.5576	0.3494	0.4713	0.5074	0.5020	0.1546	0.0590
June	612.0	286.1	0.4674	1.0813	3.3385	0.5911	0.5864	0.3390	0.4680	0.4514	0.5198	0.1652	0.1161
July	634.1	289.2	0.4561	1.1078	3.7140	0.5712	0.5171	0.3236	0.3824	0.4423	0.4450	0.1565	0.0781
August	589.5	246.4	0.4180	1.3136	4.8183	0.5793	0.3592	0.2239	0.2458	0.4755	0.4130	0.1628	0.0627
September	518.0	187.8	0.3625	1.3541	5.5765	0.5612	0.2966	0.2274	0.2021	0.5004	0.4104	0.1557	0.0615
October	485.9	145.6	0.2997	0.9973	4.5614	0.5816	0.3055	0.2594	0.2414	0.5303	0.4328	0.1767	0.0662
November	441.4	117.4	0.2660	0.7227	3.5304	0.6821	0.4234	0.3651	0.3372	0.5553	0.4622	0.2270	0.1179
December	450.8	116.1	0.2575	0.7421	3.5192	0.7190	0.4593	0.3938	0.4199	0.5265	0.4788	0.2300	0.0681
Year	6154.7	2168.6	0.3523	1.0744	3.5682	0.7270	0.5122	0.3413	0.3949	0.5193	0.4829	0.2036	0.0580

The water balance of Lake Victoria has already been discussed in connection
with evaporation from the lake surface. Also, the balance for each of the years
1969 and 1970, and for the so-called normal year, is given in Part II, Vol. I of
the report on the hydrometeorological survey of the catchments of Lakes Victoria,
Kyoga and Albert (WMO, 1974). The balance given for the so-called normal year
has been based on the average of the data for the period 1946-1970. This period
can be split into two sections, the first section lasting for 16 years, from
1946 up to 1961, and the second for 9 years, from 1962 to 1970. The fact that
the latter was excessively wet has caused the average over the whole period to
be rather high. The mean annual outflow from the lake for the 1946-61 period is
20.12×10^9 m^3, for the 1962-70 period, 43.92×10^9 m^3 and for the 1946-1970
period, 28.61×10^9 m^3/yr, all measured at Jinja. The average outflow for the
70-year period, 1900-1970, is about 23.5×10^9 m^3/yr. For the same period, the
change in storage, ΔS, showed an average value of 23 mm/yr. Expressing all terms
in the balance equation in mmillimeters over the surface of the lake: R = 270 mm,
O = 347 mm, ΔS = 23 mm; the difference $P_1 - E_1$ = 100 mm.
A summary of some of the water balances for Lake Victoria is as follows:

Year	P_1, mm/yr	R, mm/yr	O mm/yr	ΔS, mm/yr	E_1 mm/yr	mm/day
1969	1765	269	680	-248	1602	4.37[a]
1970	1827	352	654	+128	1399	3.83[a]
1906-1932	1151	276	311	–	1116	3.06[b]
1946-1970	1691	278	426	+ 70	1473	4.03[a]
1900-1970	1560	270	347	+ 23	1460	4.00

a = WMO, 1974;   b = Hurst, H.E., and Phillips, P., 1938

The continuous variation of the inflow to, and the outflow from, the lake has
resulted in the change in the volume of water storage in the lake. This is
reflected in the continuous variation of the lake water surface level. The
change in both lake water level and storage versus time, from 1900 up to and
including 1970 is shown in Fig. 8.2. The first thirty years of record show a
distinct maximum lake surface level and a distinct minimum lake level once every
10 to 11 years. The next thirty years show instead a cycle of about 5 years
length, and in the last ten years no rythmic cyclicity in the water level can be
seen.
In Vol. V of the Nile Basin (1938), Hurst wrote: "Some years ago, Dr C.E.P.
Brooks published a paper on 'Variations in the Levels of the Central African
Lakes' in which he pointed out a connection between the levels of Lakes
Victoria and Albert and the frequency of sunspots".

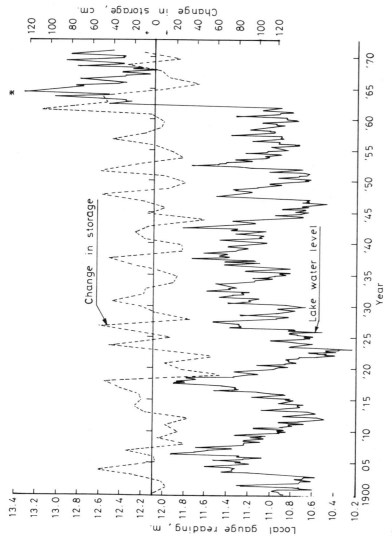

*For a further explanation of the sudden and considerable rise in the lake level from 1962 and onwards the reader can be referred to Kite, G.W. (1981) and Shahin, M., (1983)

Fig. 8.2. Lake Victoria level at Jinja deduced from 10-day means from gauge readings, 1900-52, and from daily means from charts 1953-70. The dashed line represents the annual change in lake storage

In his discussion on the paper on "Flood-Stage Records of the River Nile"
published by C.S. Jarvis in 1935, J.W. Shuman mentioned the different cycles
and showed with some graphical techniques and data smoothening procedures that
both annual rainfall on the catchment of Lake Victoria and the level of water
surface in the lake follow the solar cycle (11 years) and that the lake levels
lag one year with the annual rainfall curve. In his discussion on the same paper,
Hurst mentioned the periodic analyses made by Turner, Craig and Brooks. He added
that periods varying from 2 to 240 years have been found (Jarvis, C.S., 1935).

Again, Hurst, in Vol. V of the Nile Basin, concludes that the high correlation which existed between lake level and sunspot numbers of the years 1896-1922,
which were used by Brooks, has practically disappeared for the subsequent years
1923-1934. Neither is there any correlation between sunspot numbers and change
of lake level or rainfall. He adds: "In fact the only correlation which remains
is the straightforward one between rainfall and change of lake level" (Hurst,
H.E., and Philips, P., 1938).

The relationship between the hydrologic series of monthly precipitation,
annual precipitation and annual run-off to sunspot numbers was investigated by
I. Rodriguez and V. Yevjevich using cross-correlation analysis for various time
lags. No significant correlation was found between these hydrologic series and
sunspot numbers (1967).

The relationship between the cyclicity in the climatic and the hydrologic
time series and the 11.1-year sunspot cycle or the 22.2-year Brückner cycle has
been strongly emphasized by King, J.W. (1975), and equally denied by Ghani and
others. The whole question of the cyclicity in the hydrologic time series would
therefore seem far from being settled.

The level of the water surface in Lake Victoria undergoes a seasonal variation. The maximum occurs between April and May and the minimum occurs once in
January-February and another time between July and October. The pattern is so
regular that the mean seasonal variation in the level of Victoria has been
represented as the sum of six harmonic components. This can be written as

$$X_t = X_o + \sum_{j=1}^{6} A_j \sin \left(\frac{2 \pi j t}{12} + \theta_j \right) \tag{8.4}$$

where A_j and θ_j are the amplitude and phase angle of the jth harmonic, X_o is
the mean level and t is the number of the month. It has been found that the
12-monthly and the 6-monthly cycles are responsible for about 96% of the total
seasonal variation, and the 4-monthly oscillation is responsible for just 3% of
the total variation.

The curve representing eq. 8.4 is shown in Fig. 8.3. The means of the observed monthly departures from the annual mean lake level are supposed to coincide with this curve.

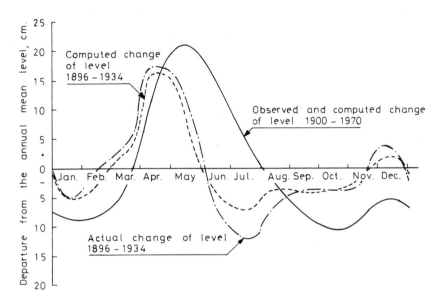

Fig. 8.3.   Seasonal oscillation of water level in Lake Victoria

Hurst reports that the level of Lake Victoria has a yearly oscillation and the normal range of the lake is only 30 cm. The extreme range in the period 1896-1934 was 1.74 m (Hurst, H.E., and Philips, P., 1938). The observed variation in lake level and the computed one from the lake balance for that period are also shown in Fig. 8.3. The increase in the amplitude of oscillation seems to be rather small, but the location of the maximum and minimum levels seems to have undergone a one-month retardation.

The seasonal range of variation in the lake storage based on the mean monthly depths for the period from 1900 up to and including 1970 has been analyzed statistically and the values of the basic descriptors are: mean = 44 cm/yr, standard deviation = 13.15 cm/yr and skewness = 0.296. The data have also shown to be serially uncorrelated (see Fig. 8.4.). The Pearson type III has proven to be the distribution function of best fit to the data. The graphical plot of the observed data and of the distribution function is shown in Fig. 8.5. The theoretically obtained ranges of variation for a number of non-exceedance probabilities are given in the same figure.

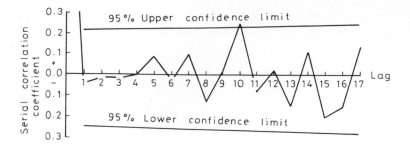

Fig. 8.4.   Serial correlogram of the seasonal variation of the storage depth
in Lake Victoria

8.1.3   Lake outflow

The outflow leaves Lake Victoria over Ripon Falls and flows in a ravine
obstructed by rocks and rapids for a distance of about 65 km. From just upstream
of Namasagali down to Lake Kyoga, a distance of about 50 km, the Victoria
Nile has a mild sloping channel. The total length of the river from the exit of
Lake Victoria at Jinja to its escape into Lake Kyoga is 125 km.

The site at Namasagali had been used for measuring the discharge of the
Victoria Nile. A detailed analysis of the discharge measurement at this site
was given by C.L. Berg (1953). The conclusion drawn from this analysis is that
the measurement practice based on considering the mean velocity for a vertical
as 0.96 times the velocity at half the depth of the vertical gives a discharge
5.3% in excess of the actual discharge. He proposed that the constant 0.96
should be reduced to 0.91. After the construction and the putting into operation of the Owen Falls Dam as from 1954, the release through the dam constitutes
the most reliable estimate of the discharge of the Victoria Nile.

The relationship between the measured discharges at Namasagali and the gauge
reading at Jinja was set up, and used to be termed "the agreed rating curve".
This curve was later extended by means of hydraulic model tests to give discharges corresponding to levels above 12.0 m Jinja local gauge reading. Unfortunately, the modified curve gave discharges which were, on average, 15% less
than the measured ones.

The gauging site at Mbulamuti situated some 50 km downstream of Jinja was
recently chosen for discharge measurement and for constructing the rating curve
that represents the natural regime of the Victoria Nile between Lakes Victoria
and Kyoga. The agreed curve and its extension, the Mbulamuti rating curve and
the discharges reported by Hurst are summarized and presented in Fig. 1,
Appendix E.

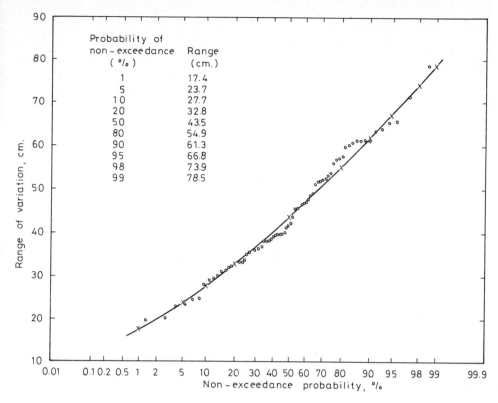

Probability of
non - exceedance   Range
(%)   (cm.)

1	17.4
5	23.7
10	27.7
20	32.8
50	43.5
80	54.9
90	61.3
95	66.8
98	73.9
99	78.5

Fig. 8.5.   Fit of Pearson Type III function to the distribution of the seasonal
range of variation of the storage depth in Lake Victoria

   The monthly and annual discharges of the Victoria Nile at Jinja in the 25-yr
period 1946–1970 are given in Table 2, Appendix D. The average discharge hydrograph for this period is shown in Fig. 8.6., whereas Table 8.5 gives the basic
statistical descriptors and serial correlations for the monthly and annual data.

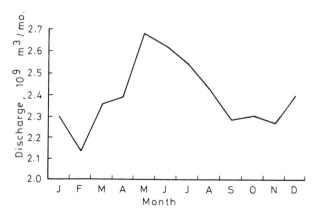

Fig. 8.6.   The average hydrograph of the Victoria Nile at Jinja for the period
1946–70

TABLE 8.5 The basic statistical descriptors of the outflows from Lake Victoria at Jinja for the period 1946-1970

Month and Year	Basic statistical descriptors					Serial correlation coefficients					
	$\bar{X}$, 10^9 m^3	s, 10^9 m^3	C_v	C_s	C_k	r_1	r_2	r_3	r_4	r_5	r_6
January	2.2964	1.0298	0.4484	0.8070	2.5252	0.8803	0.7080	0.5412	0.4242	0.2627	0.0509
February	2.1376	0.9375	0.4386	0.7498	2.4221	0.8541	0.6761	0.5295	0.4159	0.2796	0.1213
March	2.3548	1.0774	0.4575	0.7820	2.3322	0.8730	0.6891	0.5763	0.4985	0.3965	0.2098
April	2.3872	1.0482	0.4391	0.7184	2.1296	0.8653	0.7071	0.5816	0.5158	0.4156	0.2429
May	2.6772	1.1203	0.4185	0.6142	1.9226	0.8457	0.7252	0.5829	0.5220	0.3927	0.2356
June	2.6160	1.1135	0.4256	0.5993	1.8727	0.8421	0.7185	0.5735	0.4988	0.3793	0.2371
July	2.5404	1.0738	0.4227	0.6352	2.0348	0.8267	0.6716	0.4907	0.4668	0.4167	0.3057
August	2.4240	1.0167	0.4194	0.6024	1.9469	0.8273	0.6668	0.4866	0.4584	0.4136	0.2970
September	2.2808	0.9532	0.4179	0.6200	1.9993	0.8164	0.6802	0.4836	0.4637	0.3943	0.2936
October	2.3024	0.9607	0.4173	0.6250	2.0063	0.8116	0.6999	0.5021	0.4679	0.3751	0.2756
November	2.2676	0.9228	0.4070	0.4244	1.6900	0.7787	0.6872	0.5347	0.4609	0.3929	0.2213
December	2.4040	1.0135	0.4216	0.5188	1.7861	0.9093	0.7542	0.6185	0.5230	0.4364	0.3039
Year	28.6152	12.1148	0.4234	0.6301	1.9134	0.8755	0.7257	0.5739	0.4994	0.3974	0.2435

The mean annual outflow from Lake Victoria varies a great deal from one period of time to another and from one duration of the period to another. This can be seen from the following figures:

Period		Duration	Mean
from	to	years	10^9 m^3/yr
1896	1945	50	21.4
1906	1932	27	20.6
1946	1961	16	20.1
1962	1970	9	43.9
1946	1970	25	28.6

max = 33.1 in 1917
and min = 12.3 in 1922

The period from 1896 up to and including 1970 has a mean of 23.8 x 10^9 m^3/yr and a standard deviation of 7.94 x 10^9 m^3/yr. As one might notice from the above figures, the 75-yr period, 1896-1970, can be divided into two sub-samples; the first from 1896 to 1945 and the second from 1946 up to and including 1970. The mean annual volumes of outflow are as given above and the standard deviations are 4.7 x 10^9 m^3/sec., and 12.11 x 10^9 m^3/sec. for the two sub-samples in their respective order. The first period is characterized by the small variation of the annual volumes of flow in comparison to the second one, from 1946 to 1961 with a mean of 20.1 x 10^9 m^3/yr and an excessively wet period from 1962 up to and including 1970. Applying the Fisher F-test to the variances of the flow in the periods 1896-1945 and 1946-1970 leads us to the conclusion that the two variances are non-homogeneous (significance level = 5%). The means of these two samples, however, when tested by the d-test do not appear to be significantly different one from the other (at 5% significance level).

The results in Table 8.5 show that the monthly and annual volumes of lake outflow are serially correlated. The serial correlograms of the 12 months and of the year are all quite similar. The model proposed for describing and for generating the data is the first-order auto-regressive one, as expressed by eq. 8.3. The model parameter for the yearly volumes of flow at Jinja is the first correlation coefficient, 0.8755.

8.2   THE BASIN OF THE VICTORIA NILE

8.2.1   Run-off to Lake Kyoga

The Victoria Nile Basin comprises the Upper Victoria Nile, from Lake Victoria to Lake Kyoga, the Lower Victoria Nile, from Lake Kyoga to Lake Albert, the basin of the River Kafu and the complex of Lakes Salisbury, Kyoga and Kwania. The total surface area of the Victoria Nile Basin, which is 74713 km^2, can be sub-divided into the following component areas:

Gauged areas including the catchments of Lake Salisbury
and River Kafu   46255.8 km^2

Ungauged areas   12406.6

Areas not contributing to the inflow of Lake Kyoga   11315.6

Open water lakes and swamps   4735.0

The monthly run-off from the Victoria Nile Basin to Lake Kyoga was estimated
by Hurst in Vol. V of the Nile Basin at 3% (Hurst, H.E., and Philips, P., 1938).
The report of the Hydrometeorlogical survey of the catchments of Lakes Victoria,
Kyoga and Albert gives estimates of the run-off for the years 1969 and 1970, and
for the so-called normal year (WMO, 1974). The data presented in these two
references are summarized in Table 8.6.

TABLE 8.6   Run-off from the catchment of the Victoria Nile to Lake Kyoga

Run-off to Lake Kyoga, million m^3								
Month	1932	1969	1970	Normal year	Month & year	1932	1969	1970	Normal year
Jan.	110	299	128	156	Jul.	190	161	178	193
Feb.	110	185	56	86	Aug.	220	130	356	235
Mar.	190	208	272	111	Sep.	270	175	419	249
Apr.	310	169	398	301	Oct.	270	230	528	297
May	350	368	495	284	Nov.	250	251	341	326
Jun.	260	252	284	269	Dec.	180	100	180	270
					Year	2700	2527	3638	2906

Hydrologic investigation of the basin of the River Kafu, which occupies
16700 km^2, or about 22.5% of the total catchment area of the Victoria Nile, has
shown that the run-off coefficient varies from 1.1% in the dry years to 5% in
the wet years. In these two groups of years, 1953-60 and 1961-70, the mean
annual rainfall was 1150 mm and 1240 mm, respectively. The average up to 1932
was given by Hurst at 1295 mm/yr for the whole catchment area of the Victoria
Nile.

Owing to the scarcity of the run-off data, it is not possible to give the
values of the monthly and annual run-off coefficients with a fair degree of
accuracy. The quantities listed in Table 8.6 suggest, however, a value of about
3% in a normal year increasing to 4% in a wet year for the annual coefficient.

The inflow and outflow components of the water balance of Lake Kyoga for the
years 1969, 1970 and the so-called normal year, have been included in Part 2 of
Vol. I of the report of the hydrometeorological survey. These components,
together with those obtained by us for the period 1948-1970 are as follows:

Year	P_1 mm/yr	I mm/yr	R mm/yr	O mm/yr	ΔS mm/yr	Balance	E_1, mm/yr Penman	Kohler
1969	1160	9032	558	9475	−150	1420	1867	1591
1970	1238	8732	798	9281	+110	1377	1751	1594
normal	1220	5622	639	5897	+ 67	1516	1623	−
1948-70	1220	6222	639	6331	+ 50	1700	1916	−

The above results show that the annual evaporation from the lake varies from 1700 mm, as found from the water balance for the period 1948-70, to 1916 mm as found from Penman's method. The variation of E_1 from one period to another and from one method to another is much less compared to the variation of I or O.

The values given by Hurst averaged for the period 1917-1932 were P_1 = 1292 mm/yr, I = 4351 mm/yr, R = 428 mm/yr, O = 4076 mm/yr and ΔS = 0. These values give a weighted average of 1995 mm/yr for evaporation from open water (1205 mm/yr) and potential evapotranspiration from the swamps (2228 mm/yr). Notice that the calculations made by Hurst are based on an open water surface of 1760 km^2 and on swamps of 4510 km^2 (Hurst, H.E., Black, R.P., and Simaika, Y.M., 1966). In Vol. VII of the Nile Basin, Hurst and his co-workers state that the gauging site at Masindi Port where the outflow from Kyoga used to be measured, has proved to be an unreliable site. The average discharges for the period 1940-1945 above and below Lake Kyoga were 21.0 and 19.3 milliard m^3/yr for Namasagali and Port Masindi respectively. Hurst concludes that the information available was far too little to give any certainty about the Kyoga Basine regime, except that on the whole, Lake Kyoga is probably a source of loss in the order of 1 or 2 milliards a year.

Comparison between the old measurements and/or estimates of the components of the water balance and the more recent ones shows that Lake Kyoga, instead of being a source of loss with O = 0.937 I, is a source of small gain where O = 1.045 I.

8.2.2   Water level and storage

The difference between the net inflow to the lake and the outflow from it causes the change of its surface water level and thereupon the volume of content in storage. The water level at Bugondo Pier for the periods 1928-1938 and 1942-1970 is shown in Fig. 8.7. In the same figure the annual variation of the storage depth for the same periods is shown.

Lake level observations were started in 1916 at Lale Port. In 1927 another gauge was erected at Bugondo. From then till 1938 it worked properly, and from 1939 up to and including 1941 no level measurements were taken at any of the gauging sites. From 1942 up to and including 1968 measurements of lake water level were taken again at Bugondo. Four additional gauging sites were erected in

336

the period 1965-1967. In 1968 the level gauges at Bugondo and at another site were replaced by automatic recorders.

The year 1961 witnessed the maximum gain in the monthly, as well as the annual, storage. These were 710 and 1815 mm respectively. Contrarily, the year 1918 witnessed the maximum loss in the monthly, as well as the annual, storage. These were 340 and 1505 mm respectively. The total change in storage over the period 1928-1970 (not including 1939-1941) was 805 mm, showing a rising trend of 20 mm/year.

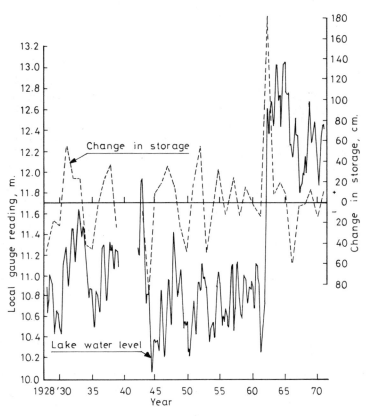

Fig. 8.7. Lake Kyoga level at Bugondo deduced from 10-day means from gauge readings 1928-39 and 1942-70. The dashed line represents the annual change in lake storage

The seasonal oscillation of the water level in Lake Kyoga is illustrated in Fig. 8.8. by two curves: one representing the level at Bugondo Pier for the years 1928-38 and 1942-70, and the second representing the level at Lale pier for the years 1917-32.

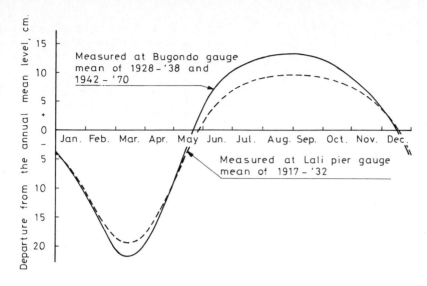

Fig. 8.8.   Seasonal oscillation of water level in Lake Kyoga

In spite of the difference in amplitude the two curves agree completely on
the times of occurrence of the maximum and minimum water levels. Similar to
Lake Victoria, the water level oscillation in Lake Kyoga can be described by the
equation

$$X_t = X_o + \sum_{j=1}^{6} (A_j \cos \frac{2 \Pi j t}{12} + B_j \sin \frac{2 \Pi j t}{12}) \qquad (8.4')$$

which is a different form of the equation 8.4 used in connection with Lake
Victoria. The harmonic coefficients A_j and B_j are 1.1613 and −16.8934, 3.7917
and 2.3205, −1.3833 and 0.2834, 0.2250 and −0.4790, 0.1221 and −0.0732 and
0.2833 and zero, respectively, for j = 1, 2, 3, 4, 5 and 6. The 12-monthly cycle
explains 91.2% of the total variation in the lake level; the 6-monthly cycle
explains 6.3% and the 4-monthly cycle, 0.6%. The remaining cycles combined, i.e.
the 3-monthly, 2.4-monthly and 2-monthly cycles, are responsible for 1.9% of the
total variation only.

The seasonal range of variation in the lake storage based on the mean monthly
depths for the period from 1942 up to and including 1970 has been analyzed statistically. The values of the basic descriptors are: mean = 54 cm/yr; standard
deviation 27.95 cm/yr; skewness = 2.209 and kurtosis = 11.154.

As in the case of Lake Victoria, the data of Lake Kyoga have shown they are
serially uncorrelated (see Fig. 8.9.). The Pearson Type III here has also proved
to be the distribution function of best fit to the observed data. The graphical

plots of the observed data and of the distribution function are shown in Fig. 8.10. The theoretical values of the range of variation of the storage depth for a number of non-exceedance probabilities are given in the same figure.

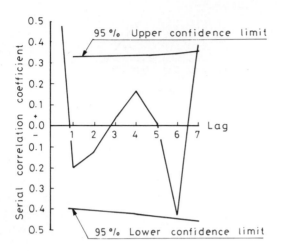

Fig. 8.9.   Serial correlogram of the seasonal variation of the storage depth in Lake Kyoga

8.2.3   Lake outflow

The flow in the lower reach of the Victoria Nile, known as Kyoga Nile, had been measured formerly at, or near, Masindi Port. The normal monthly mean discharges obtained from the measurements in 1913 and 1915-32 were reported by Hurst in Vol. V of the Nile Basin as 1.45, 1.24, 1.34, 1.32, 1.48, 1.52, 1.62, 1.64, 1.60, 1.66, 1.60 and 1.58 milliard m^3/month from January to December. The annual mean flow was thus given as 18 milliard m^3 (Hurst, H.E., and Philips, P., 1938). Later, the discharge measurements were taken at Fajao. From 1962 and onwards, the measuring site was moved to Paraa, which is situated a few kilometres downstream of Fajao. The rating curves of the Kyoga Nile at Paraa are shown in Fig. 2, Appendix E, whereas the discharges from 1948 up to and including 1970 measured at Paraa or corrected for the change of location are listed in Table 3, Appendix D.

The average hydrograph for the period 1948-70 (Fig. 8.11.) agrees well with the seasonal oscillation of the water level in Lake Kyoga, shown in Fig. 8.8., except for the steep fall of the discharge in October followed by the rise in November.

From the previous section we have already seen that the 18.0×10^9 or the 19.3×10^9 m^3 given by Hurst for the mean of the annual flow in the Kyoga Nile is quite small.

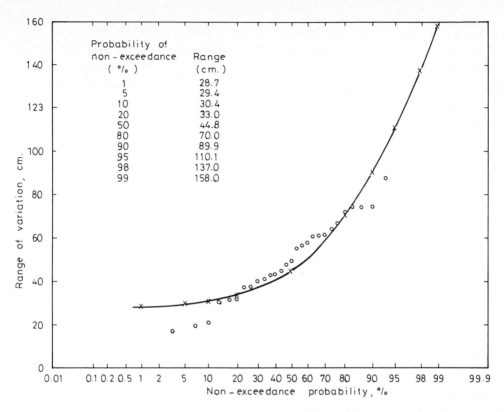

Fig. 8.10.   Fit of Pearson Type III function to the distribution of the
seasonal range of variation of the storage depth in Lake Kyoga

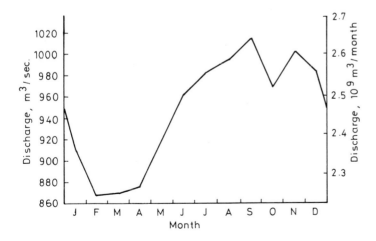

Fig. 8.11.   The average hydrograph of the Kyoga Nile at Paraa for the period
1948-1970

The statistics presented in Table 8.7 give a mean of 950 m^3/sec. or 29.9 milliard m^3/yr. This is much greater than the figures reported by Hurst. It is true that the wet spell from 1961 to 1970 had a significant effect on the general mean for the period 1948-1970, yet neither 18.0 nor 19.3 milliard m^3 can be regarded as a long-term mean. In section 8.2.2, the ratio I to O for Lake Kyoga was found as 1.045. If this ratio is multiplied by the mean flow in the Victoria Nile above Lake Kyoga which is 23.8 x 10^9 m^3/yr for the period 1900-1970, one gets an annual flow in the Kyoga Nile below the lake of 24.9 milliard m^3/yr. We shall consider this figure as the annual flow volume brought by the Kyoga Nile to Lake Albert in a normal year. The monthly and annual dis-charges of the Kyoga Nile for the period 1948-1970 appear to be strongly serially correlated. The coefficient of serial correlation drops almost linearly with the lag, from 1 to about 0.4, corresponding to lags of 0 and 5 respectively. This is the case with the twelve months and the year. We tried to fit to these data the first-order Markov model like the one we used in connection with the discharges of the Kagera and the Victoria Nile. The fitted model is that des-cribed by eq. 8.3 with parameter α = 0.9321. The residual series gave serial correlations of 0.2528, 0.1263, -0.2165, -0.2109 and 0.0534 for lags 1, 2, 3, 4 and 5, respectively. The coefficients of the original series for the five lags in their order are 0.9321, 0.8074, 0.6488, 0.5332 and 0.4230 (see Table 8.7). The serial correlations of the residuals fall well inside the 95% confidence band pointing to the independence of the residuals at this level of confidence.

8.3   THE BASIN OF LAKES GEORGE AND EDWARD AND RIVER SEMLEEKI

8.3.1   Hydrologic regime of the basin

The total surface area of this basin is 30500 km^2 of which 2500 km^2 are open water; 300 km^2 are occupied by Lake George and 2200 km^2 by Lake Edward. The catchment areas of these two lakes are 8000 km^2 and 12000 km^2, respectively. The remaining 8000 km^2 are the catchment area of the River Semliki. The two lakes are connected to each other by Kazinga Channel, which is a fairly wide and deep carrier. The situation produced by having Lake Edward joined to Lake George by this channel from one side and to Lake Albert by the Semliki River from another side qualifies the whole area to be considered as a one-hydrologic unit. The old regime reported by Hurst and his co-workers in the Nile Basin was based upon direct precipitation on the lakes of 1.365 m/yr and evaporation of 1.30 m/yr. The rainfall on the whole catchment was taken equal to 1.365 m/yr, the run-off coefficient was considered to be 12% and the change in the lakes' storage, ΔS = 0. These figures give an outflow from Lake Edward of 3.65 x 10^9 m^3/yr. In addition to this volume, the Semliki River receives a volume of 2.05 x 10^9 m^3 each year as run-off from its catchment. Our estimate for the

TABLE 8.7 The basic statistical descriptors of the outflows from Lake Kyoga at Paraa for the period 1948–1970

Month and Year	Basic statistical descriptors					Serial correlation coefficients				
	$\bar{X}$, m³/s	s, m³/s	C_v	C_s	C_k	r_1	r_2	r_3	r_4	r_5
January	912.6	438.2	0.4801	0.4585	1.7528	0.8947	0.7865	0.6852	0.5173	0.3898
February	868.7	417.1	0.4800	0.4723	1.7656	0.8849	0.7722	0.6823	0.5159	0.3919
March	870.7	444.7	0.5107	0.5545	1.9137	0.8819	0.7595	0.6068	0.4920	0.3882
April	876.9	462.1	0.5269	0.6474	2.0946	0.9183	0.7694	0.6255	0.5027	0.3923
May	921.9	488.7	0.5301	0.5927	1.8903	0.9172	0.7585	0.6161	0.5023	0.4222
June	961.9	487.3	0.5065	0.5638	1.8277	0.8987	0.7535	0.6239	0.4967	0.4303
July	981.5	481.4	0.4905	0.5197	1.7744	0.8884	0.7561	0.6308	0.5039	0.4374
August	994.7	474.9	0.4775	0.5529	1.8597	0.9013	0.7592	0.6008	0.5015	0.4263
September	1015.7	461.7	0.4546	0.5622	1.9540	0.8851	0.7683	0.5871	0.5144	0.4035
October	968.6	410.6	0.4240	0.5388	1.8225	0.8314	0.7469	0.6735	0.5455	0.4325
November	1003.6	471.0	0.4693	0.4800	1.9122	0.8930	0.8236	0.6493	0.5411	0.4046
December	984.0	475.0	0.4828	0.3411	1.6393	0.9141	0.8132	0.6735	0.5316	0.4184
Year	949.9	456.2	0.4802	0.5013	1.7251	0.9321	0.8074	0.6488	0.5332	0.4230

period 1948-1970 is based on somewhat different values for the hydrologic variables in the regime. The mean annual rainfall on the lakes was 1400 mm and the evaporation from the lake surface was 1800. These figures show a deficit of 1.0 milliard m^3 a year. The run-off coefficient is taken as 13% and this brings the annual outflow from Lake Edward to: $0.13 \times 1.4 \times 20 \times 10^9 - 1.0 \times 10^9 = 2.64 \times 10^9$ m^3/yr. The rainfall on the Semliki Catchment was about 1600 mm/yr in the period 1948-70 and the run-off about 16%. These figures produced a run-off of about 2.05×10^9, which means a total flow in the Semliki of 4.69×10^9 m^3 per year. This figure agrees quite well with the mean of the annual discharge given in Table 4, Appendix D.

8.3.2   Seasonal oscillation of lake level

The seasonal oscillation of the water level in Lake Edward is shown in Fig. 8.12. This oscillation too can be described by the same relation given as eq. 8.4'. The A and B coefficients of the harmonics 1 thru 6 are: 7.7292 and 0.8453, 10.012 and 2.3095, -0.1787 and -0.1547, 0.5000 and 0.2680, 0.4995 and zero, and -0.012 and zero respectively. As with Lakes Victoria and Kyoga, the first and second harmonics combined, i.e. the sum of the 12-monthly and 6-monthly cycles explain almost 94% of the total variation. In spite of this similarity, one finds that the 12-monthly cycle in Lake Edward explains 34% of the total variation and not 91%, as in Lakes Victoria and Kyoga, whereas the 6-monthly cycle in Lake Edward's level is responsible for 60% of the total variation and not of 6% only as is the case with Lake Victoria or Kyoga.

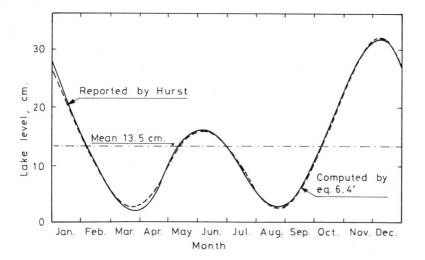

Fig. 8.12.   Seasonal oscillation of water level in Lake Edward

8.3.3   Discharges of the Semliki River

The monthly and annual discharges of the Semliki in the period 1948-1970
are given in Table 4, Appendix D.

From the scanty data that prevailed up to 1932, the average monthly discharges of the Semleeki given by Hurst were: 130, 102, 103, 106, 112, 116, 112,
116, 112, 103, 106, 121, 135 and 140 m^3/sec. for the months from January to
December, respectively.

The Semliki has been gauged regularly since 1940. The rating curves measured
at Dweramul are shown in Fig. 3, Appendix E. The average discharge hydrograph
for the period 1948-1970 is shown in Fig. 8.13., whereas Table 8.8 gives the
basic statistical descriptors of the monthly and annual discharge data for the
same period.

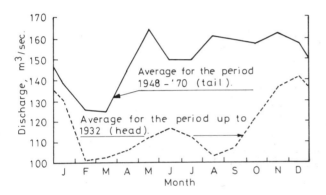

Fig. 8.13.   The average hydrograph of the River Semliki

The comparison between the annual discharge of the Semliki, Table 4, Appendix D, and the annual discharge of the Victoria Nile, Table 2, or the Kyoga
Nile, Table 3, both Appendix D, shows that the response of the Semliki Basin to
the wet spell 1961-1970 is less than that shown by either the Victoria or the
Kyoga Basins. In fact the strongly noticeable rise in the Semliki discharge
took place during the 3-year period 1962-64 only. This probably led to a rise
of not more than 10% in the general mean over the period 1948-70.

The discharge series of the Semliki River does not behave differently from
the other rivers dealt with until now, as far as their serial correlation is
concerned. Furthermore, the historical data of the Semliki can be described by
a first-order auto-regressive model like the one given by eq. 8.3, with a parameter α = 0.6206. The residual series left from fitting this model to the historical data appears to be uncorrelated with 95% confidence.

TABLE 8.8   The basic statistical descriptors of the discharge of the Semliki River in the period 1948-1970

Month and Year	Basic statistical descriptors					Serial correlation coefficient				
	$\bar{X}$, m³/s	s, m³/s	C_v	C_s	C_k	r_1	r_2	r_3	r_4	r_5
January	138.3	58.96	0.4263	1.2368	3.9216	0.5485	0.0557	−0.2181	−0.1604	0.0444
February	125.5	50.01	0.3984	1.6881	5.7768	0.5990	0.1143	−0.0874	−0.1688	−0.0589
March	124.5	50.28	0.4038	1.6870	6.7057	0.5689	0.0864	−0.0710	−0.1699	0.0885
April	144.3	58.26	0.4037	1.4727	5.4226	0.4725	0.0937	−0.0684	−0.1634	−0.0826
May	163.8	86.56	0.5286	2.9423	13.3763	0.4756	−0.0924	−0.1928	−0.1632	−0.0196
June	148.6	64.51	0.4340	2.4654	11.2517	0.4603	−0.1133	−0.2552	−0.2674	−0.0798
July	148.6	54.77	0.3686	2.3550	11.1160	0.3900	−0.0631	−0.2559	−0.2393	−0.0653
August	159.5	55.37	0.3472	1.6507	5.7890	0.5737	−0.0619	−0.4122	−0.3093	−0.1528
September	157.5	52.13	0.3309	1.4275	4.5335	0.4635	0.1424	−0.3420	−0.3380	−0.4123
October	155.9	42.77	0.2743	1.6791	6.3812	0.4481	0.0608	−0.2115	−0.4278	−0.3885
November	160.9	51.65	0.3209	1.3237	4.0309	0.5751	0.1421	−0.0751	−0.2060	−0.2021
December	155.3	56.70	0.3651	1.4672	4.3285	0.4370	−0.0268	−0.2542	−0.1980	−0.0468
Year	150.6	51.13	0.3395	2.0387	14.0376	0.6206	0.1261	−0.1865	−0.2456	−0.1021

8.4   LAKE ALBERT CATCHMENT

8.4.1   Run-off to Lake Albert

The catchment area of Lake Albert, not including the drainage basin of the Semliki, is 17000 km^2. The surface area of the lake itself is 5300 km^2. The Kyoga Nile flows into the north-eastern end of the lake and the Semleeki into its southern end. These two rivers supplied Lake Albert with 29.91 and 4.64 milliard m^3 respectively, as means for the period 1948-1970. Hurst gave the figures of 18-19.3 x 10^9 for the Kyoga and 5.7 x 10^9 m^3 a year for the Semliki. Based on a yearly rainfall of 1256 mm and a run-off coefficient of about 12%, he gave the annual run-off from the lake catchment as 2.56 x 10^9 m^3/yr. This figure, added to the 5.7 x 10^9 m^3 supplied by the Semliki, brings the annual inflow into Lake Albert to 8.26 milliard m^3/yr or 1.55 m spread over the lake surface. The ramaining figures used by Hurst to draw the regime of the Lake Albert mean for the years 1913 and 1915-1932, were 0.81 m rainfall at Butiaba and 1.2 m/yr evaporation from the lake surface. The inflow to the lake from the Lower Victoria Nile taken as 3.4 m brought the sum of the gains to the lake to 5.76 m. Since the sum of the losses was claimed to have been 5.74 m and the average rise of the lake level during that period, ΔS, was 0.06 m, the total loss was estimated at 5.80 m. The difference between the gains and losses (4 cm spread over the lake surface) was regarded as the error in the balance sheet of Lake Albert (Hurst, H.E., and Philips, P., 1938).

The hydrometeorological survey of the catchments of Lakes Victoria, Kyoga and Albert used the relation between the slope of the main drainage channel in a basin and the annual run-off coefficient, obtained from the gauged sub-basins (only 30% of the lake catchment), for estimating the total catchment run-off. Remarkable enough is that the annual run-off coefficient, estimated by this method at 12.37%, is very close to the 12% assumed by Hurst more than 40 years ago. The mean annual precipitation for the period 1948-1970 was 1285 mm. This rainfall produced an annual run-off of 17 x 10^9 x 0.1237 x 1.285 = 2.702 milliard m^3, which is equivalent to a 503 millimetre depth distributed uniformly over the surface of Lake Albert. This depth is distributed into 28, 21, 24, 36, 56, 31, 22, 26, 39, 64, 80 and 75 mm for the 12 months from January to December, respectively (WMO, 1974).

The balance sheets for the years 1969 and 1970 and for the so-called normal year, as reported by the hydrometeorological survey and for the period 1948-1970 based on our estimates are as follows:

Year	P_1, mm/yr	I, mm/yr Semliki	Kyoga	R, mm/yr	O, mm/yr	ΔS, mm/yr	E_1, mm/yr
1969	766	849	8990	540	9490	-260	1913
1970	1021	1026	8799	578	9081	120	2223
Normal	709	864	5602	503	6282	42	1546
1948-70	800	885	5642	510	6364	42	1431

8.4.2 Lake water level and storage

Records of water level at the gauge of Butiaba are available since 1912.
These levels plotted against time in years up to and including 1970 are shown in
Fig. 8.14. In this figure the annual change in the storage depth of the lake is
also shown. It is clear that the change in the level of Lake Albert and in its
storage depth is principally produced by the change in the regime of Lakes
Victoria and Kyoga and to a much less extent by the regimes of Lakes George and
Edward. The water level at the beginning of each month for the period 1912-1970
is included in the report of the hydrometeorological survey of the catchments of
Lakes Victoria, Kyoga and Albert (WMO, 1974). The difference between the maximum and the minimum level for each year, i.e. the range of oscillation has been
computed and tested statistically. This range reached a minimum of 18 cm in 1969
and a maximum of 197.5 cm in 1917 with a mean of 67.3 cm/yr, standard deviation
of 38.26 cm/yr, skewness of 1.535 and kurtosis of 5.371. The serial correlogram
of the series of range is shown in Fig. 8.15.

The empirical frequency distribution of this set of data and the fit of the
Pearson Type III function to it are shown in Fig. 8.16. The values of the range
of variation estimated from the distribution function for some of the non-
exceedance probabilities are also given in the same figure.

The mean water level in Lake Albert for the period 1912-1970 has been computed, as well as the mean level at the beginning of each month. The plot of the
monthly departure from the general mean versus time in month is represented by
curve (4) in Fig. 8.17. We have tried to fit to this curve the basic model described by eq. 8.4' and the result is shown as curve (3) in the same figure. The
A-coefficients are 16.61, 4.41, 0.34, 2.03, zero, and -0.23 and the B-coef-
ficients are -4.97, -0.42, -0.23, -1.13, 0.03 and zero for the 1, 2, 3, 4, 5 and
6 harmonics respectively. For Lake Albert, one finds that the 12-monthly cycle
is responsible for 88% of the total variation in the lake level, and the 12- and
6-monthly cycles combined explain 94% of the total variation.

The so-called actual and calculated water levels which were reported by Hurst
for the period 1913 and 1915-32 in Vol. V of the Nile Basin have been plotted in
Fig. 8.17 as curves (1) and (2) respectively. The disagreement between the two
sets of curves ((1) and (2), and (3) and (4)) is certainly too big. This situation is similar to that of the water level in Lake Victoria (see Fig. 8.3).

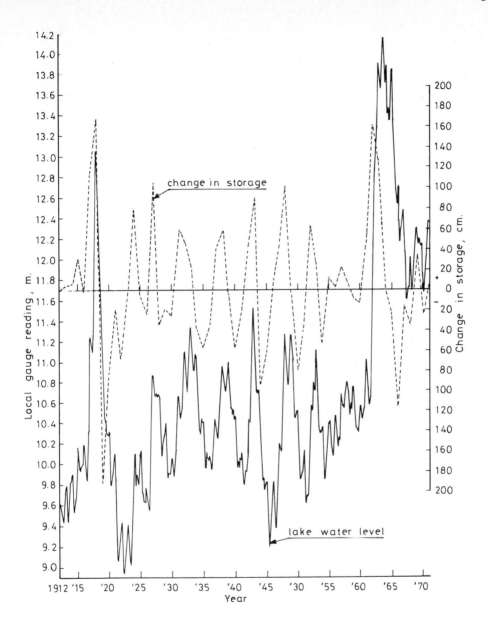

Fig. 8.14.   Lake Albert level at Butiaba deduced from 10-day means from gauge readings 1912-70. The dashed line represents the annual change in lake storage

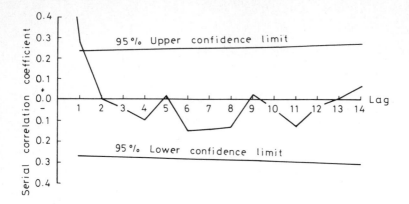

Fig. 8.15. Serial correlogram of the seasonal oscillation of the water level
in Lake Albert

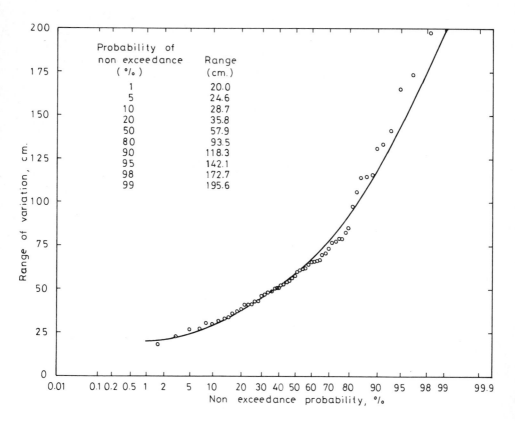

Fig. 8.16. Fit of Pearson Type III function to the distribution of the
seasonal range of variation of the water level in Lake Albert

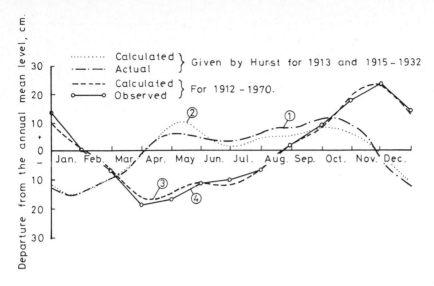

Fig. 8.17.   Seasonal oscillation of water level in Lake Albert

The annual variation in the storage depth for the period considered has been statistically tested, and the successive terms found to be serially uncorrelated at the 95% level of confidence. The series has a mean of 54.75 mm/yr and standard deviation of 652 mm/yr. The skewness of the series is quite small, 0.0715, justifying the possibility of fitting the normal function to the empirical distribution of the data. In Fig. 8.18., the data are plotted and so are the theoretical distributions, as given by the normal and the Pearson III functions.

8.4.3   Lake outflow

The discharge of the Albert Nile used to be measured for some time at Pakwach, a short distance downstream of the confluence of the Kyoga Nile with the exit flow from the lake. Unfortunately, none of the rating curves developed at Pakwach has shown stability for a sufficiently long time. Therefore, another site called Panyango, further downstream of Pakwach, has been used for gauging the lake outflow into the Albert Nile. Figure 4, Appendix E, shows the rating curve at Panyango and its relation with Pakwach. The monthly and annual discharges for the period 1948-70 are given in Table 5, Appendix D. The tabulated discharges have been analyzed statistically and their basic descriptors are given in Table 8.9. From the serial correlations given in this table, it is clear that all the monthly and the annual series undergo a strong dependence among their individuals. Furthermore, each of these series behaves more or less similarly from the others. It has accordingly been found sufficient to try to fit the first-order auto-regressive model, eq. 8.3, to the annual series only. For this discharge series the model parameter has a value of 0.9065. The

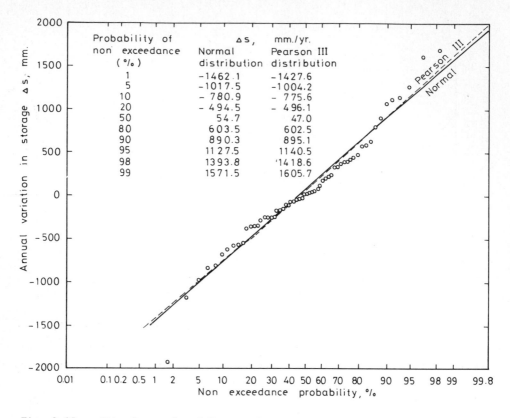

Fig. 8.18. Fit of normal and Pearson Type III functions to the distribution of the annual variation in the storage depth in Lake Albert during the period 1912-1970

residual terms in this model appear to be uncorrelated serially at the 95% level of confidence. With this simple model, one can generate any number of annual discharges as may be needed.

The mean hydrograph of the outflow of Lake Albert for the period 1948-1970 is shown in Fig. 8.19. From 1904 up to 1955 the lake outflow used to be found from a rating curve established between the discharge of the Bahr el Jebel at Mongalla and the lake level at Butiaba. The discharge at Mongalla was reduced by 5% to account for the losses between the lake exit and Mongalla (440 km). The mean of the outflow found by this method for the period 1904-1947 was about 24.6×10^9 m³/yr. The period from 1948 up to and including 1970 can be divided into two sections. In the first, from 1948 to 1955, discharges were found similar to the discharges from 1904 up to 1947. From 1956 and onward, the discharges were obtained from the Panyango-Pakwach rating relationship. The mean for the period 1948-1970 was about 33.7 milliard m³/yr, which is about 30% larger than that for the period 1904-1947. Almost the same difference exists between the

TABLE 8.9 The basic statistical descriptors of the outflows from Lake Albert at Panyango for the period 1948-1970

Month and Year	Basic statistical descriptors					Serial correlation coefficients				
	$\bar{X}$, m^3/s	s, m^3/s	C_v	C_s	C_k	r_1	r_2	r_3	r_4	r_5
January	1090.9	576.05	0.5281	0.5398	1.9144	0.9192	0.7330	0.5627	0.3973	0.2344
February	1046.9	583.10	0.5570	0.5637	1.9350	0.9055	0.7087	0.5335	0.3591	0.1929
March	1014.1	586.65	0.5785	0.6080	2.0092	0.8895	0.6835	0.5023	0.3222	0.1561
April	1004.3	589.65	0.5877	0.6109	2.0221	0.8818	0.6739	0.4862	0.3018	0.1414
May	1034.5	587.57	0.5680	0.6207	2.0545	0.8590	0.6356	0.4634	0.2951	0.1318
June	1033.8	596.19	0.5767	0.6046	2.0189	0.8562	0.6375	0.4620	0.3107	0.1759
July	1040.9	591.40	0.5682	0.6087	2.0111	0.8566	0.6409	0.4753	0.3446	0.2074
August	1062.2	586.29	0.5519	0.5969	1.9656	0.8721	0.6712	0.5006	0.3839	0.2562
September	1099.9	574.93	0.5227	0.5907	1.9785	0.8736	0.6987	0.5124	0.3949	0.2720
October	1118.6	569.40	0.5090	0.5704	1.9928	0.8931	0.7402	0.5450	0.4263	0.2746
November	1157.0	574.40	0.4965	0.4433	1.8601	0.9328	0.7945	0.5947	0.4504	0.2892
December	1163.5	589.16	0.5064	0.3701	1.7179	0.9372	0.7865	0.6009	0.4513	0.2996
Year	1071.9	577.24	0.5395	0.5902	1.9514	0.9065	0.7204	0.5384	0.3865	0.2356

corresponding monthly values of the two periods (see Fig. 8.19.). The two methods combined give an overall mean of about 27.7 million m^3/yr for the outflow of Lake Albert during the period 1904-1970. This figure is strongly supported by the estimate of the mean outflow using the long-term balance of the lake.

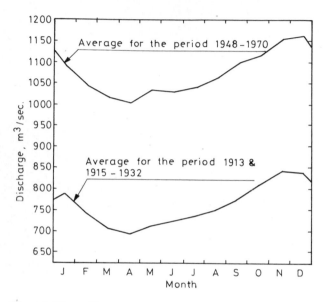

Fig. 8.19.   The average hydrographs of the Albert Nile at the exit of Lake Albert for two different periods

The inflow to the lake supplied by the Semleeki is 4.64 x 10^9 m^3/yr and by the Kyoga Nile, 24.9 x 10^9 m^3/yr (section 8.2.3). The long-term rainfall on the lake catchment is about 1250 mm/yr and the annual run-off coefficient is 12.4%. These figures produce an annual run-off to the lake equal to 2.63 x 10^9 m^3. The direct precipitation on the lake is about 800 mm/yr and the annual evaporation is in the order of 1700 mm (see Chapter 5). This means a loss of 4.77 x 10^9 m^3/yr. From section 8.4.2, we have already seen that the yearly change in the storage depth in the lake is close to 5.5 cm. This is equivalent to a loss of 0.29 milliard m^3/yr. The annual outflow then equals 4.64 + 24.90 + 2.63 - 4.77 - 0.29 = 27.11 milliard m^3/yr. The distribution of this volume between the months of the year follows that shown in Fig. 8.19.

8.5   THE BAHR EL JEBEL BASIN

8.5.1   The Bahr el Jebel from the exit of Lake Albert to Mongalla

8.5.1.1   The torrents between Lake Albert and Mongalla

The length of the Bahr el Jebel from the exit of Lake Albert to Mongalla is about 440 km and the surface area of its basin is 79000 km^2. From Lake Albert to Nimule, a distance of about 225 km, the river is a broad sluggish stream fringed with swamps and lagoons. The distance between Nimule and Rejaf is about 155 km. The river in this reach is a fast-flowing stream whose course is obstructed by some rapids. In the next stretch, which extends from Rejaf to Mongalla, a distance of about 60 km, the bed slope decreases considerably.

A number of small streams join the Bahr el Jebel between Lake Albert and Nimule. Discharge measurements at Nimule began in 1913 and the rating curve is shown in Fig 5, Appendix E. There are, however, many breaks in the available records. One may therefore take 1.5 milliard m^3/yr as a lump sum contribution of these streams to the flow in the Bahr el Jebel above Nimule.

Counterbalancing this gain there is a certain conveyance loss between Lake Albert and Nimule. This loss can be estimated at about 4% of the annual volume of the lake's outflow. A certain portion of this loss is in fact the resultant of the evaporation from the open water (260 km^2 and evaporation depth about 1800 mm/yr), the evapotranspiration from the swamps (120 km^2 and evapotranspiration rate of about 2000 mm/yr), and the annual precipitation, which is about 1300 mm/yr.

Based upon our estimate of the Lake Albert outflow, 27.1 mlrd m^3/yr, the long-term mean flow at Nimule must then be about $(27.1 \times 0.96) + 1.5 = 27.5$ mlrd m^3/yr. This amount flows a distance of 155 km before it reaches Mongalla.

The Assua River joins the Bahr el Jebel at a short distance below Nimule. It supplies the main river by about 1.5×10^9 m^3 in an average year. Fig. 6, Appendix E, shows the rating curve of the Assua River. Hurst used the rainfall and run-off data of the Assua Basin for the period 1924-1935 to develop expressions giving the monthly and the annual discharge for any given rainfall.

The annual run-off coefficient that corresponds to rainfall of about 1300 mm is close to 3%. Since the surface area of the Assua Basin is 39000 km^2, the annual run-off becomes $1.3 \times 0.03 \times 39 \times 10^9 = 1.52 \times 10^9$ m^3. Other streams join the Bahr el-Jebel in the reach between Nimule and Mongalla (see the maps, Figs. 2.4., and 2.11.). These streams are known to feed the main river by some 1.2 mlrd m^3/yr. The average hydrographs of the torrents between the exit of Lake Albert and Mongalla and between Nimule and Mongalla, River Assua being included, are shown in Fig. 8.20.

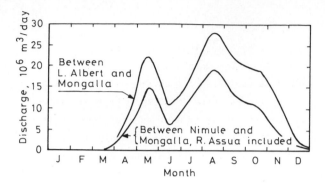

Fig. 8.20.   The average hydrographs of the torrential streams between Lake
Albert and Mongalla and between Nimule and Mongalla for the period 1923-1932

The annual rainfall on the Bahr el Jebel Basin in the reach between Nimule
and Mongalla can be taken as 1250 mm. This is the average of the rainfall at
Nimule, Stat. 142, and Kajo Kaji, Stat. 141 (see Appendix C). The evaporation
rate for the same stretch of river outside the swamps is on average 5 mm/day,
or about 1800 mm/yr. The details about this figure can be found in Chapter 5.
To account for the other transit losses such as seepage and spilling we shall
consider the flow at Mongalla as being 4% less than that at Nimule. The long-
term mean of the annual discharge at Mongalla then becomes 0.96 x 27.5 plus the
2.7 mlrd m^3 supplied by the torrents between Nimule and Mongalla. The resultant
outflow at Mongalla can thus be estimated at 29.1 mlrd m^3/yr.

8.5.1.2   Discharges at Mongalla
     The monthly and annual discharges measured at Mongalla for the period 1912-
1973 are given in Table 6, Appendix D. The gauge-discharge points used for establishing the rating curve in the period 1911-27 are shown in Fig. 7, Appendix E.
     The annual volume of flow at Mongalla in the period considered showed a
maximum of 55.8 x 10^9 m^3 in 1917 and another maximum of 60.5 x 10^9 m^3 in 1963.
The latter was, however, much broader than the former. The minimum flow was
observed in 1922 and has a value of about 15.3 x 10^9 m^3. It goes without saying
that the occurrence of the maximum and minimum flows together with all other
flows at Mongalla is primarily influenced by the hydrologic situation in the
complex of the Equatorial Lakes.
     The graphical plot of the annual discharge series of Mongalla is shown in
Fig. 8.21. The part spanning the period from, say, 1925 up to, say, 1955, fluctuates in the range between 20 and 30 mlrd m^3/yr. The ups and downs in this
part occur rather regularly - once every 5 years, on average.

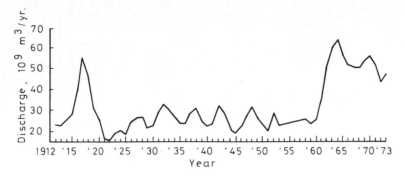

Fig. 8.21.   Graphical plot of the annual discharge series of Mongalla in the period 1912-1973

We have examined the statistical properties of the 12 months and of the annual series. The values of the mean, $\bar{X}$, standard deviation, s, the variation coefficient, C_v, the skewness, C_s and the kurtosis, have been computed from eqs. 4.1 to 4.5, respectively. The skew and the kurtosis have been corrected for the bias in their estimate. Eq. 4.6 has been used to compute the serial correlation coefficient, r_L, corresponding to lag L. Since the size of the sample N is 62 years, we have to stop with our calculation of r_L at $L = \dfrac{N}{4}$ or 15. The results obtained from these calculations are presented in Table 8.10.

The serial correlation coefficients show that all the examined series have more or less the same pattern of correlation. Each series is serially correlated up to lag L = 6 or 7. This can be seen from the example correlogram of the January discharges shown in Fig. 8.22. The autogregressive model has been fitted to the monthly series, and they all showed that the first-order model, eq. 8.3, gives the best fit.

A much better fit to the series of annual discharges was obtained from the second-order autoregressive model. This order has been described and used in connection with the analysis of the rainfall data (see Chapter 4). The model parameters α_1 and α_2 are computed from $r_1(1-r_2)/(1-r_1^2)$ and $(r_2-r_1^2)/(1-r_1^2)$, respectively. The model parameter of each of the monthly series is its first serial correlation coefficient, $\alpha = r_1$ (see Table 8.10). For the annual series, $\alpha_1 = 1.2743$ and $\alpha_2 = -0.4359$. The serial correlation coefficients of the residuals left from the fit of the first-order model to the monthly data and from the fit of the second-order model to the annual data are listed in Table 8.11.

The probability distribution functions of good fit to the monthly and annual discharge series at Mongalla are the two-parameter lognormal and the Pearson Type III functions. These two functions give almost the same variate value for a 100-year recurrence interval. Nevertheless, the two-parameter lognormal has proven to be slightly superior to the Pearson III and to provide, in general,

TABLE 8.10   The basic statistical descriptors and serial correlation coefficients of the monthly and annual discharges of the Bahr el-Jebel at Mongalla for the period 1912-1973

Item	Jan.	Feb.	Mar.	Apr.	May	June	July	Aug.	Sep.	Oct.	Nov.	Dec.	Year
Basic statistical descriptor													
$\bar{X}$, 10^6 m^3	2328	2030	2137	2187	2629	2521	2728	3178	3119	3101	2777	2584	31381
s, 10^6 m^3	1065	972	1002	978	1082	1044	1005	1143	1280	1347	1195	1162	12610
C^v	0.4574	0.4790	0.4687	0.4470	0.4116	0.4141	0.3684	0.3597	0.4105	0.4344	0.4305	0.4498	0.4018
C^s	1.2190	1.2000	1.2174	1.2620	1.2484	1.1701	1.0805	0.9634	1.2189	1.3824	1.0611	1.0529	1.1497
C^k	3.4060	3.2276	3.3746	3.6285	3.9149	3.6491	5.3332	3.2432	4.0455	4.7638	2.8798	2.7898	3.1595
Serial correlation coefficient													
r_1	0.8342	0.8439	0.8356	0.8483	0.7865	0.8052	0.7591	0.7379	0.6996	0.7165	0.8421	0.8509	0.8875
r_2	0.6569	0.6666	0.6540	0.6590	0.5681	0.5735	0.6067	0.5586	0.4953	0.5235	0.7030	0.6910	0.6951
r_3	0.4982	0.5163	0.5211	0.4984	0.4002	0.4015	0.4275	0.4428	0.2978	0.3994	0.6234	0.5992	0.5290
r_4	0.3974	0.4254	0.4253	0.3938	0.3305	0.3170	0.3876	0.3704	0.2532	0.2518	0.4674	0.4851	0.4222
r_5	0.3260	0.3644	0.3505	0.3465	0.2869	0.2971	0.3315	0.3601	0.2370	0.2293	0.4087	0.4333	0.3715
r_6	0.3077	0.3548	0.3252	0.3278	0.2619	0.2749	0.3084	0.3710	0.2992	0.3112	0.3966	0.3932	0.3441
r_7	0.2531	0.3095	0.2730	0.2865	0.2120	0.2216	0.1610	0.1812	0.2175	0.2029	0.2859	0.3037	0.2760
r_8	0.1763	0.2306	0.1851	0.2112	0.1375	0.1312	0.1812	0.1161	0.1437	0.1507	0.2119	0.2309	0.2011
r_9	0.1336	0.1720	0.1377	0.1566	0.1161	0.0917	0.0552	0.0782	0.0645	0.1702	0.2033	0.2174	0.1456
r_{10}	0.0644	0.0915	0.0562	0.0683	0.1103	$\bar{0}$.0773	$\bar{0}$.0084	0.0244	0.0399	0.0581	0.0847	0.1331	0.0716
r_{11}	0.0585	0.0445	0.0647	0.0490	0.0096	$\bar{0}$.0071	$\bar{0}$.0724	$\bar{0}$.0558	$\bar{0}$.1077	$\bar{0}$.0389	$\bar{0}$.0122	$\bar{0}$.0135	$\bar{0}$.0351
r_{12}	0.1376	0.1369	0.1450	0.1314	$\bar{0}$.1570	$\bar{0}$.1639	$\bar{0}$.1830	$\bar{0}$.1039	$\bar{0}$.1060	$\bar{0}$.0341	$\bar{0}$.0707	$\bar{0}$.0728	$\bar{0}$.1214
r_{13}	0.1176	0.1190	0.1235	0.1086	$\bar{0}$.1681	$\bar{0}$.1748	0.2157	0.1916	0.1584	0.0729	0.0799	0.1286	$\bar{0}$.1298
r_{14}	0.0616	0.0692	0.0682	0.0779	0.1173	0.1031	$\bar{0}$.1248	0.0116	0.0340	0.0469	0.0627	0.0874	$\bar{0}$.0642
r_{15}	0.0446	0.0511	0.0575	0.0577	0.0861	0.0673	0.0301	0.0740	0.1058	0.1034	0.0454	0.0645	0.0207

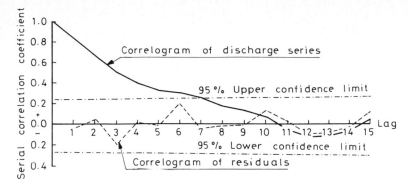

Fig. 8.22.   Serial correlogram of the January discharge series at Mongalla and of the residuals left from the fit of the first-order autoregressive model to it

a better fit to the observed data. Fig. 8.23. is an example of the fit of the lognormal function to the empirical distribution of the annual discharge series from 1912 up to and including 1973. For this series and also for the 12-month series, the 100 and 200-year discharges computed by this distribution function are:

Discharge, 10^6 m^3	Jan.	Feb.	Mar.	Apr.	May	June	July	Aug.	Sep.	Oct.	Nov.	Dec.	Year
100-yr	5693	5092	5303	5307	6059	5789	5823	6615	7164	7479	6434	6140	70724
200-yr	6290	5626	5865	5864	6687	6352	6326	7133	7881	8273	7031	6709	77408

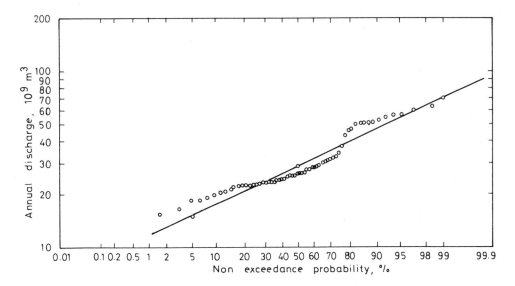

Fig. 8.23.   Fit of the lognormal function to the distribution of the annual discharges at Mongalla in the period 1912-1973

TABLE 8.11 Serial correlation coefficients of the residuals left from fitting autoregressive models to the discharge series at Mongalla

Serial correlation coefficients

Month & Year	r 1	r 2	r 3	r 4	r 5	r 6	r 7	r 8	r 9	r 10	r 11	r 12	r 13	r 14	r 15
January	-.058	.052	-.192	.005	-.014	.200	-.048	-.019	-.017	.137	.037	-.094	-.084	-.024	.128
February	.030	-.105	-.011	.119	.043	.138	.106	-.144	.114	.088	.028	-.111	-.127	.071	.104
March	.107	-.115	-.027	-.046	.002	.161	.028	-.132	.125	.069	-.069	-.135	-.080	.086	.112
April	.189	-.107	-.099	.139	.048	.162	.005	-.075	-.082	.063	.018	-.109	-.040	.034	.144
May	.156	-.106	-.126	-.018	.126	.120	.003	-.097	-.059	.159	.236	-.127	-.140	.018	.159
June	.200	-.135	-.138	.077	.150	.109	.036	-.098	-.095	.121	.227	-.183	-.167	.051	.125
July	-.048	.091	.181	.107	.032	.257	-.094	-.080	-.096	.136	.010	-.058	-.185	.002	.153
August	-.001	.053	-.035	.008	-.063	.346	.121	.021	-.007	.091	.005	.090	.301	.124	.114
September	-.003	.044	.169	.067	-.028	.225	-.036	-.041	.091	.199	.158	.101	.277	.140	.043
October	-.038	-.016	.079	-.065	-.042	.255	-.046	-.065	.138	-.033	-.122	-.070	-.173	.107	.066
November	-.009	-.070	.209	.159	.047	.225	.073	-.157	.235	.021	.057	-.127	-.006	.021	.088
December	.188	.084	.043	.063	.040	.176	.015	-.198	.177	.158	.038	.035	-.012	.082	.085
Year	.115	-.151	-.063	.048	.098	.275	-.007	-.114	.011	.244	.044	-.121	.291	.139	.226

8.5.2   The Bahr el Jebel from Mongalla to Lake No

8.5.2.1   General description of the river valley

This stretch of the Bahr el Jebel Basin is characterized by the existence of vast swamps, lagoons and side channels east and west of the main river. The length of the course of the main stream between Mongalla and Lake No is about 770 km and the area of the permanent swamps is claimed to be about 8500 km^2. This is obviously an area where severe loss of water takes place.

Some of the geographic, physiographic and hydrologic features of the area have been described in some of the volumes of the Nile Basin, especially Vol. V (Hurst, H.E., and Philips, P., 1938). The pertinent features of the river valley can, however, be summarized in the following points:

i)   The width of the swamps increases from about 3 km at Mongalla to about 7 km at Terrakeka some 30 km north of Mongalla.

ii)   The head of the Aliab River appears some 90 km north of Mongalla and joins the main river not far from Bor.

iii) At Bor, about 140 km north of Mongalla, the width of the valley becomes 9 km. North of Bor are many lagoons and open channels. On the eastern side appears the Atem-Awai system of rivers. This flows in a winding course to join and rejoin the Bahr el Jebel at and near Ghabe Shambe (see the maps Figs. 2.9., and 2.12.). There the swamps are roughly 15 km wide.

iv)   The eastern channels ultimately join to form the Upper Zeraf. Here at a latitude of $7^o30'$ the swamp occupies a width of, say, 30 km, increasing to a maximum of 35 km at a latitude of $7^o40'$.

v)   At a latitude of $7^o35'$, the lagoons begin on the west of the main river to form Peake's Channel later, which rejoins Bahr el Jebel at a short distance north of the Jebel-Zeraf cuts. These cuts are meant to maintain the flow in the Bahr el Zeraf. The southern cut is usually referred to as No. 1 and the northern as No. 2.

The upper part of the Zeraf south of the heads is usually blocked, but from the cuts northward, the Lower Zeraf has a more or less defined channel. In its lower course from about latitude $8^o30'$, the Zeraf is a stream with firm banks, rising about the normal water level and separated from the Jebel by a wide strip of dry land.

vi)   From Hillet Nuer (Adok) on the Jebel northwards, the permanent swamp is a few kilometres wide. The Jebel ends its course downstream Lake No. Eighty kilometres further downstream, the Lower Zeraf joins the White Nile through the Abu Tong cut.

Fig. 8.24 is a sort of longitudinal profile of the Bahr el Jebel from Mongalla to Lake No, on which the locations of the principal channels leaving or joining the main stream are shown. A very general picture of the discharge

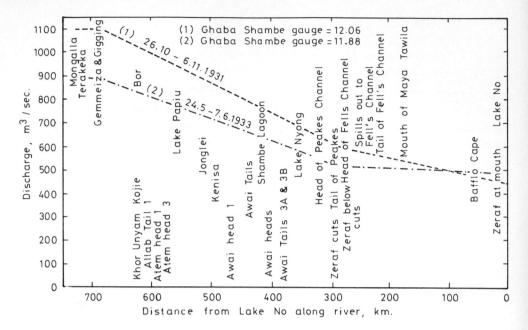

Fig. 8.24. The total discharge passing any cross-section of the valley of the Bahr el Jebel between Mongalla and Lake No (Hurst, H.E., and Philips, P., 1938)

passing any cross-section of the Bahr el Jebel valley in the reach considered can be obtained from curves (1) and (2). These lines show clearly that, in general, a heavy loss takes place from one section to another. The loss per kilometre length of the main stream is, however, heavier from Gemmeiza/Gigging to the Zeraf cuts (300 km from Lake No) than in the lowermost 300 km. In the upper 400 km reach the high water looses its discharge at the rate of about 1.225 m^3/km´ compared to about 0.925 m^3/km´ for the average water. For these two waters the loss in the last 300 km is 0.4 m^3/km and 0.1 m^3/km, respectively. The 1100 m^3/sec. discharge at Gemmeiza reaches the outlet at Lake No at 470 m^3/sec., after loosing about 57.3%, and from the initial 900 m^3 at Gemmeiza, 44.4% is lost in its way to the escape. In very general terms, a figure of 50% for the loss of water in the Bahr el Jebel Basin downstream of Mongalla is not unreasonable.

The stage-discharge measurements used formerly for preparing the rating curves of the main river and some of its principal channels are plotted graphically in a set of figures. These figures, which are included in Appendix E, are:

Fig. 8 - Khor Unyam Kojie, 2.5 km south of Bor,

Fig. 9 - River Aliab, Tail 1, 16 km north of Bor,

Fig. 10 - Bahr el Jebel at Bor,

Fig. 11 - El Jebel-Zeraf cut 1 at tail and at head,

Fig. 12 - El Jebel-Zeraf cut 2 at tail and at head,

Fig. 13 - Bahr el Jebel downstream of the Jebel-Zeraf cut 2,

Fig. 14 - Bahr el Jebel at a distance of 281 km from Lake No, downstream of the
tail of Peake's

Fig. 15 - The White Nile in the neighbourhood of Lake No,

Fig. 16 - Bahr el Zeraf, 3 km from the mouth, and

Fig. 17 - The White Nile at Abu Tong in the neighbourhood of the Zeraf mouth.

The Jonglei canal planned to convey part of the Bahr el Jebel water from near
Jonglei to the mouth of the Zeraf with the aim of saving some of the water lost
in the swamps was originally thought of by A.D. Butcher. We shall describe this
diversion canal in some detail in connection with the storage and conservation
works of the Nile Water.

8.5.2.2   The loss of water in the Bahr el-Jebel swamps

The Bahr el Jebel swamps comprise the swamps in the basins of the Bahr el·
Jebel itself and its off-shoot the Bahr el Zeraf. The area of the permanent
swamps has been frequently given as 8300 km^2. It is not possible to give any
figure for the temporary swamps, as their area varies from one season to another
and from one year to another. The essential particulars about the loss of water
in the swamps can be found in the Nile Basin Volumes V, VII and X. A few additional figures can be found in the Jonglei canal report. The available data show
that the highest correlation between the discharges above and below the swamps
can be obtained with a lag of three months. From 1912 to 1922 the below swamps
discharge was taken as the difference between the discharges of the White Nile
at Malakal and of the Sobat, both based on the discharge curves constructed from
the discharges measured during the same year. From 1923 onwards, the swamp discharge was the sum of the discharges of the Zeraf and of the White Nile at Abu
Tong, obtained by linear interpolation between measured discharges.

The relation between the quarterly mean discharges at Mongalla and below the
swamps from the data in the period 1912-1945 is shown graphically in Fig. 8.25.
From the annual flow volumes at Mongalla and the annual flow volumes below the
swamps, the curve in Fig. 8.26., has been constructed so as to read the percentage water lost given the flow at Mongalla. This curve can be described by the
equation

$$L = 2.25 \ V_M - 0.0175 \ V_M{}^2 \tag{8.5}$$

where L is the loss in percent and V_M is the annual flow volume at Mongalla, in
milliard m^3.

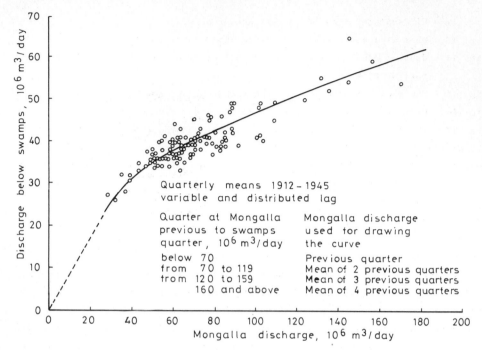

Fig. 8.25.   The relation between the discharge at Mongalla and below the swamps

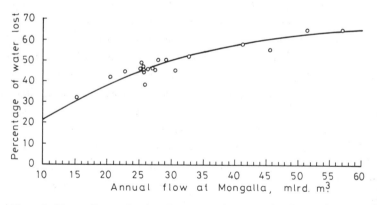

Fig. 8.26.   The relation between the annual flow volume at Mongalla and the percentage water lost in the swamps

The long-term mean of V_M has been estimated at 29.1×10^9 m³/yr (see section 8.5.1). Eq. 8.5 gives a loss of 49.1% for this value, or about 14.3×10^9 m³/yr. In other words, from an annual volume of 29.1 milliard m³ at Mongalla, only 14.8 milliard m³ reach the White Nile and the rest is lost. These discharge values are in perfect agreement with the 29 and the 14.7 m³/yr at Mongalla and Malakal respectively, for the period 1905-66, as given in the Phase I - Jonglei Project Report (Jonglei Area Executive Organ for Development Projects, 1975).

One can therefore proceed with the calculation using these figures confidently. The annual loss expressed as depth over the swamp area is 1723 mm/yr. The annual rainfall still has to be added, so as to obtain the total loss.

The thirty-year mean rainfall is 868 mm at Tonga, station 119; 1076 mm at Fangak, station 121; 795 mm at Ghabet Shambe, station 130; 903 mm at Bor, station 133 and 905 mm at Terrakekka, station 135 (see Appendix C). The mean of these five stations, which is 909 mm/yr, brings the total depth of water lost in the swamps up to 2632 mm/yr or 7.2 mm/day. The depths given by Hurst in Vol. V of the Nile Basin are 2300, 2360 and 2400 mm/yr for the periods 1927-31, 1927-37 and 1932-37 respectively (Hurst, H.E., and P., 1938). The average of these three figures is 2353 mm/yr or 6.5 mm, which is 10% less than our estimate. Both 7.2 mm/day or 6.5 mm/day as loss of water from the Bahr el-Jebel swamp are significantly larger than the loss estimated from any of the Olivier, Thornthwaite, Penman or the energy-balance methods (see Chapter 5). If we keep the 5 mm loss per day and the 8300 km^2 area of the swamp, the annual loss should therefore be (365 x 5 - 910) x 10^{-3} x 8300 x 10^6 = 7.6 x 10^9 m^3, which is 6.7 mlrd m^3 less than the 14.3 x 10^9 m^3 already given. This difference can be attributed to the inaccuracy in the discharge measurement, the escape of water into some channels without having it measured, and most important of all, the inundation of the temporary swamps.

The division of the flow between the Bahr el Jebel and the Bahr el Zeraf can be seen from the following example:
Period: from 1927 to 1931
Bahr el-Jebel upstream the cuts = 11140 x 10^6 m^3/yr
Cut 1 at head = 2490 x 10^6 m^3/yr reduced to 2370 x 10^6 m^3/yr at tail
Upper Zeraf just above tail of cut 1 = 1050 x 10^6 m^3/yr
Cut 2 at head = 805 x 10^6 m^3/yr reduced to 797 x 10^6 m^3/yr at tail
Zeraf below cut 2 = 4490 x 10^6 m^3/yr (gain = 273 x 10^6 m^3/yr)
Jebel below cut 2 = 7440 x 10^6 m^3/yr (loss = 405 x 10^6 m^3/yr)

This division of the flow can be seen from the map in Fig. 8.27. The proportion of the flow into the Bahr el Zeraf to that in the Bahr el Jebel is variable, depending mostly on the level of water in the swamp. On average, the flow is divided such that two-thirds of it finds its way into the Bahr el Jebel and the remaining one-third into the Bahr el Zeraf.

364

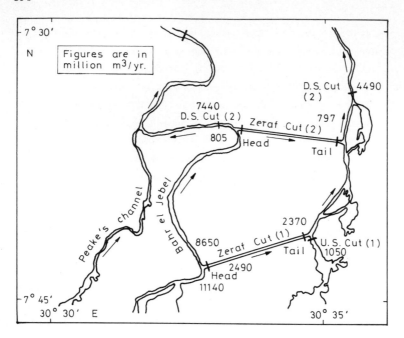

Fig. 8.27. Map showing the division of the flow between the Bahr el Jebel and the Bahr el-Zeraf

8.6 THE BAHR EL-GHAZAL BASIN

The Bahr el-Ghazal Basin is situated west of the Bahr el Jebel Basin and extends to the divide Congo-Nile. The surface area above the swamps and including the sub-basins of the western tributaries of the Bahr el Jebel below Mongalla is 528 000 km^2. The basic particulars point to an annual run-off from this area of 18.4 mlrd m^3 (1.1 m/yr rainfall and 3.2% run-off coefficient). Neglecting those tributaries which discharge their water into the Bahr el Jebel, the annual flow entering the swamps has been estimated at 16.0 mlrd m^3. Ahmed gave the following discharge values (Ahmed, A.A., 1960):

River Lol (at Nyamlel)	4000 x 10^6	m^3/yr
River Punjo (at Aluk)	500 x 10^6	m^3/yr
River Jur (at Wau)	5000 x 10^6	m^3/yr
River Tonj (at Tonj)	1500 x 10^6	m^3/yr
River Gel (at Chui Bet)	550 x 10^6	m^3/yr
River Naam (at Mvolo)	520 x 10^6	m^3/yr
River Yei (at Monderi)	2000 x 10^6	m^3/yr
	14070 x 10^6	m^3/yr

Hurst and his co-workers have shown that the average flow at Wau on the River Jur for the period 1928-1935 was 5929 x 10^6 m^3/yr. Of this amount, 1117 x 10^6 m^3 per year reached Ghabat el-Arab and finally 656 x 10^6 m^3/yr reached the mouth of the Bahr el-Ghazal at Lake No. Fig. 8.28. is a map showing the rivers and the location of the gauging points mentioned here.

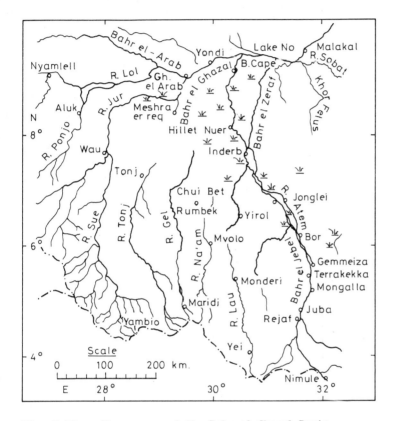

Fig. 8.28.   The swamps of the Bahr el-Ghazal Basin

The 14.07 milliard m^3/yr flow into the swamps and only 0.6 milliard m^3/yr reach the mouth at Lake No. The 80 kilometre reach from Meshra er Req to Ghabat el-Arab has a slope of 1 cm/km. This slope increases to 2.2 cm/km for the 54 km reach from Ghabat el-Arab to Yondi. But from Yondi to Lake No, a distance of 76 km, the swamp has no slope at all. This too-little slope does not produce any run-off, so water spreads on the swamps and is left to be evapotranspired. The total loss in the swamps of the Bahr el-Ghazal is (14.1 - 0.6) x 10^9 + (0.95 x 14.5) x 10^9 = 26.8 x 10^9 m^3/yr. The annual rainfall on the swamps is taken as 950 mm/yr (see Figs. 4.3.-4.5.). The total loss when divided by 14 500 km^2 gives a depth of water of about 1880 mm/yr, or 5.15 mm/day. This figure

represents the rate of evapotranspiration from a swamp, as already discussed in
Chapter 6 and in the previous section, in connection with the Bahr el Jebel
swamps.

The discharge of the Bahr el Ghazal at its mouth at Lake No has been measured
regularly since 1923. The gauge-discharge measurements of the Bahr el Ghazal
and tributaries near Lake No are shown graphically in Fig. 18, Appendix E.

The mean annual volume of flow is close to 650 million m^3. This volume is
distributed over the months of the year, as shown in Fig. 8.29. In the same
figure the water level in Lake No is shown. This water level, like the other
lakes discussed in the previous sections, undergoes a seasonal fluctuation and
can be described by eq. 8.4'. The mean level can be taken at 13.68 metres on the
gauge. The harmonic coefficients A_1 thru' 6 are 21.87, -0.25, 0.33, -0.58, 0.40
and 0.25, respectively and the coefficients B_1 thru' 6 are -16.95, -0.17, -0.42,
0.22, 0.12 and zero, respectively. The coefficients A_1 and B_1 describing the
first harmonic (12-month cycle) explain almost 99% of the total variance of the
monthly levels about the mean.

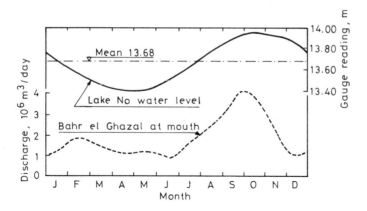

Fig. 8.29.   The discharge hydrograph of the Bahr el-Ghazal at mouth and the
water level in Lake No

8.7   THE WHITE NILE BASIN

8.7.1   The White Nile from Lake No to the mouth of the Sobat

In the reach from Lake No to the mouth of the Sobat the White Nile has a
small slope and its course is fringed with swamps. A number of tributaries pour
their water into the White Nile in the said reach. These tributaries are shown
schematically in Fig. 8.30. The gains and losses in the stretch from Lake No to
the Abu-Tong cut have been calculated for a number of years and the net loss was
found at 0.4 million m^3/day.

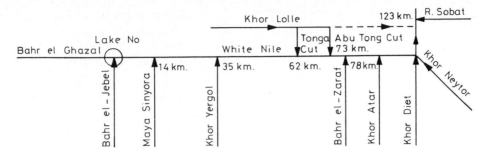

Fig. 8.30.   The tributaries of the White Nile between Lake No and the mouth of
the Sobat

The average loss in million m^3/day for the months from January to December
is: -1.2, -0.4, 0.3, 0.3, 0.7, 1.3, 2.0, 1.3, 0.5, -0.4, 0.6 and -0.1, respectively. Hurst reported that the loss in this stretch of the river should be
increased by a minimum amount of 0.5 million m^3/day to allow for the contributions of the Khors (Hurst, H.E., and Philips, P., 1938). The loss in the 50 km
stretch from Abu-Tong cut to the mouth of the Sobat was estimated at 1.5 million
m^3/day. The total net loss from the reach extending from Lake No to the mouth of
the Sobat is 2.4 x 10^6 m^3/day or 875 x 10^6 m^3/yr or about 35% more than the
inflow of the Ghazal into Lake No. The annual rainfall here can be taken as the
mean of the rainfall at Malakal, Station 118, Tonga, Station 119, and at
Meshra'er Req, Station 125. The long-term means at these stations are 819, 868
and 830 mm/yr, respectively, and the overall mean is 840 mm/yr. Since the evapotranspiration in this stretch is 1650 mm/yr or 4.5 mm/day, the net loss must be
810 mm/yr or 2.22 mm/day. From these figures the size of the wet area from which
the net loss takes place must be about 875 ÷ 0.81 or 1080 km^2. This loss brings
the 14.7 mlrd m^3 flowing in the Bahr el Jebel and el-Zeraf in a normal year to
14.45 mlrd m^3 just above the mouth of the Sobat. The report on Phase I of the
Jonglei Project gives the sum of the discharges of the Jebel and Zeraf as 14.74
mlrd m^3/yr at Malakal for the period 1905-1965. The maximum during this period
was 33.0 x 10^9 m^3/yr and it took place in 1964. The minimum was 10.3 x 10^9 m^3/yr
in 1922. The statistical testing of this set of data shows that the standard
deviation is 3.356 mlrd m^3/yr and the skewness is 3.324. Furthermore, these data
appear to be strongly correlated. The residuals of the first-order autoregressive model, AR1, do not, however, appear to be significantly dependent at the
95% level of confidence. The serial correlation coefficients of the original and
of the residual series for the first 15 lags are as follows:

368

Lag No.	Serial correlation of original series	Serial correlation of residual series	Lag. No.	Serial correlation of original series	Serial correlation of residual series
1	0.7303	0.1985	9	0.0301	0.0574
2	0.3512	0.1165	10	0.0088	0.0501
3	0.0997	-0.0902	11	-0.0365	-0.0337
4	-0.0237	-0.0808	12	-0.0463	-0.0184
5	-0.0394	-0.0261	13	-0.0335	-0.1290
6	-0.0282	-0.0442	14	0.0570	0.0525
7	-0.0034	-0.0037	15	0.1291	0.1342
8	0.0223	0.0282			

8.7.2   The White Nile At Malakal

The flow at El-Malakal on the White Nile is produced by the flows in the Bahr el Jebel, el-Zeraf, el-Ghazal and by the Sobat and its tributaries. The hydrology of all these streams, except the Sobat and its tributaries, has been presented in the previous sections. One needs, however, to investigate the hydrology of the Sobat before drawing any complete picture of the flow at Malakal.

8.7.2.1   The hydrology of the Sobat and its tributaries

This was described in Vol. VIII of the Nile Basin. In Vol. X of the Nile Basin one also finds additional information about the Machar swamps. The Sobat is formed by two streams: the Baro and the Pibor.

8.7.2.1.1   The hydrology of the Baro

The Baro springs from the Ethiopian Plateau and flows from east to west. The surface area of its basin is 41400 km^2, of which 23500 km^2 are considered mountainous and the rest is a low-elevated country. A good deal of this basin lies above 1500 m and portions are above 2000 m. In the mountainous part, the annual rainfall is at, or more than, 1500 mm, at Gambeila (between 500 m and 1000 m a.m.s.l.) the rainfall is about 1290 mm/yr and in the plain (Station 123 at Nasser, Station 127 at Akobo Post and Station 132 at Pibor Post) it drops to 880 mm/yr. The evaporation varies from slightly less than 4.0 mm/day to more than 5 mm/day, depending on the location.

In the plain the Baro is joined by a few streams coming from the Plateau. The most important of these streams are the Khor Jokau, the River Adura, the Khor Mokwai "2" and the Khor Machar. These streams are shown on the map, Fig. 2.15., and their locations measured from the Baro-Pibor junction can be seen from the schematic drawing, Fig. 8.31. This figure also shows the inflow-outflow cycle of the main river and its tributaries, as estimated for a normal year (June - November).

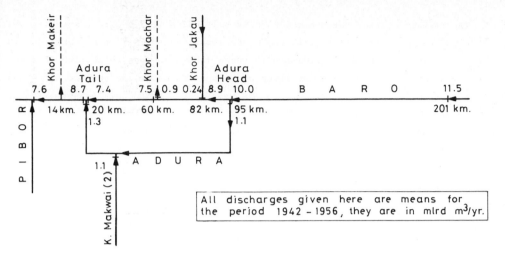

Fig. 8.31.   The inflow-outflow chart of the River Baro and its tributaries

The data published in the Nile Basin, Vol. X, give a mean flow volume of
11.6 x 10^9 m^3/yr for the period 1928-56 and of 11.5 x 10^9 m^3/yr for the period
1942-56 (Hurst, H.E., Black, R.P., and Simaika, Y.M., 1966). Either figure leads
to an annual run-off coefficient for the mountainous part of the basin of about
35% (area = 23500 km^2, annual precipitation of 1400 mm).

The surface area of the Baro Basin down Gambeila is about 18000 km^2. It
receives an annual rain depth of, say, 1.05 m, on average. Consider an annual
run-off coefficient of 14%, a a figure that was found for some parts of the
basin. The Baro should then receive an additional flow of 2.6 x 10^9 m^3/yr at its
mouth. Accordingly, one should expect an annual flow volume of about 14 mlrd
m^3/yr at the mouth of the Baro. Instead, a volume of 7.6 mlrd m^3/yr was found as
a long-term mean. The loss from Gambeila to the Baro-Pibor junction, without
taking account of all the run-off that should be discharged into the Baro, was
11.5 - 7.6 = 3.9 mlrd m^3/yr, 4.0 mlrd m^3/yr and 4.1 m^3/yr for the periods 1942-
56, 1929-62 and 1929-57, respectively. The 4 mlrd m^3/yr representing the average loss between Gambeila and the Baro-Pibor junction are distributed such that
1.5 mlrd m^3 is lost between Gambeila and the head of the Adura, 1.5 mlrd m^3
between the head and the tail of the Adura and almost 1.0 mlrd m^3 between the
tail of the Adura and the mouth of the Baro.

The drainage basin of the River Baro is characterized by the loss of not less
than 35% of its annual yield, due to spillage over the plain and also the feeding of the Khors, which flow through the Machar swamps. The area of these swamps
has been estimated at 6700 km^2 (Hurst, H.E., Black, R.P., and Simaika, Y.M.,
1966).

The average outflow through the Khor Machar, which is 0.9 mlrd m^3/year, flows through the swamps and is lost there. Other losses on the Baro are mainly due to spillage over its banks. In addition to these facts, one has to mention that the Machar swamps do not contain any lakes or lagoons to store the spilled water and thereupon to let part of it return to the main stream in the dry season, as in the case of the Jebel swamps.

This state of affairs has resulted in an almost uniform stage hydrograph which repeats itself regularly each year, whether the flood is moderate or rich. The stage hydrograph of the Baro at Gambeila for the period 1929-1933 is shown in Fig. 8.32 (Hurst, H.E., 1950). To these hydrographs one may fit the equation 8.4' with a mean level X_o = 11.034 metres on the local gauge. The harmonic coefficients A_1 thru' A_6 are: -0.5720, -0.4125, -0.1567, 0.1733, 0.0508 and 0.0217 and the coefficients B_1 thru' B_6 are: -1.946, 0.2007, 0.2500, 0.1413, -0.0710 and zero, respectively. The fitted hydrographs are also shown in Fig. 8.32.

8.7.2.1.2   The Machar Swamps

The inflow to the Machar swamps is supplied from three sources. The first source is the direct rainfall, which, on average, amounts to 0.9 m/yr. Considering the area covered by the swamps as 6700 km^2, the total volume of rain comes to 6.03 mlrd m^3/yr. The second source is the run-off from the eastern tributaries coming from the Ethiopian foothills. These tributaries are the Khor Ahmar, Tombak, Yabus, Daga and Lau and the areas of their basins are 600, 900, 4300, 2900 and 1600 km^2, respectively. Considering an annual run-off coefficient of 14% and an annual rainfall depth of 1.0 m, the annual run-off reaches 1.44 mlrd m^3/yr (Hurst, H.E., Black, R.P., and Simaika, Y.M., 1966). The third source of inflow to the swamps is the spilling of the Baro. This has already been mentioned in the previous section and found to be, on average, 4 mlrd m^3/yr. Assuming that two-thirds of this volume reaches the swamps, the total inflow amounts to (6.03 + 1.44 + 2.67) x 10^9 or 10.14 mlrd m^3/yr. This amount disappears totally by evapotranspiration in the swamps. Assuming this is all correct, the evapotranspiration from the Machar swamps should be in the order of 1515 mm/yr, or 4.15 mm/day. The same figure has been given while discussing the evaporation from the Sobat Basin (Chapter 5).

Preventing or minimizing the loss of water from the basin of the Baro was discussed in some of the Nile Basin volumes, especially Vol. X. An account of the water conservation schemes in this river basin shall be presented in the next chapter.

371

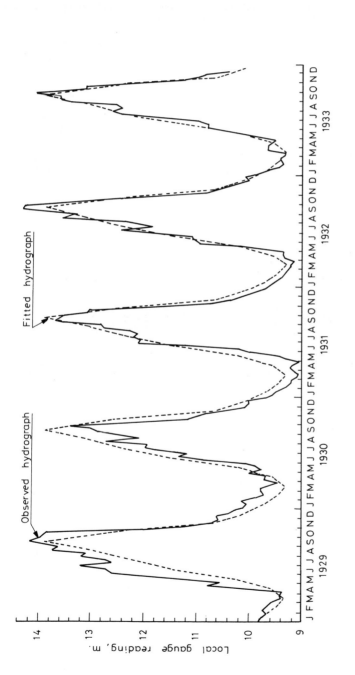

Fig. 8.32.   The observed and the fitted stage hydrographs of the River Baro at Gambeila for the period 1929-33

8.7.2.1.3   The hydrology of the Pibor

The main stream of the Pibor flows from south to north, whereas several of
its tributaries rise in the mountains of Ethiopia. The basin of the Pibor and
its tributaries has an area of 109 000 km^2, i.e., about 2½ times the area of the
Baro Basin. Nevertheless, the contribution of the Pibor to the flow in the Sobat
is much less than that of the Sobat.

Fig. 8.33. is the inflow-outflow chart of the Pibor and its tributaries. The
discharge values written on this chart are about the average for the 1929-1933
period, a rather low-flow period. According to Hurst and his co-workers, the
2.84 mlrd m^3/yr should be increased up to, say, 3.1 mlrd m^3/yr due to the uncer-
tainties in some of the measurements (Hurst, H.E., 1950). The annual rainfall at
Akobo Post, station 127, was 940 mm and at Pibor Post, station 132, 909 mm, both
means of the period 1938-1967 (see Chapter 4). If we consider 0.925 m as the
average rainfall, the annual run-off coefficient must then be in the order of 3%,
a very low figure indeed.

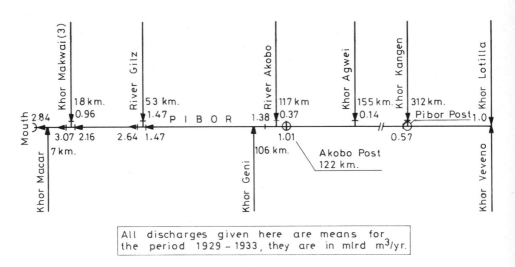

Fig. 8.33.   The inflow-outflow chart of the River Pibor and its tributaries

Hurst also gave an estimate of the gross volume of flow from Pibor Port to
the mouth on the Sobat, a distance of 312 km, at 3.6 mlrd m^3/yr and the net at
3.1, as already mentioned. This means that the loss is about 0.5 mlrd m^3/yr, or
about 14% of the gross flow.

8.7.2.1.4   The Sobat below the Pibor-Baro Junction

Below the Pibor-Baro junction, the main stream is known as the Sobat. It
flows a distance of about 350 km in a north-westerly direction before it joins

the White Nile. The Sobat below the junction has a sub-basin of about 36 800 km^2 in area, which brings the total drainage basin area to 187 200 km^2.

The Sobat has an average slope of 3 cm/km in the low-flow season and 4 cm/km in the high-flow season. The main tributaries are the Khor Fullus, Nyading, Twalor and Wakau. They join the main river at distances of 16, 239, 290 and 307 km from the mouth, respectively. In a normal year the Sobat at a head carries 12.4 mlrd m^3/yr. Of this amount, 3.1 mlrd m^3 are supplied by the Pibor and the rest by the Baro (see section 8.7.2.1.3). The average rainfall on the Sobat sub-basin can be taken as 780 mm/yr. This figure is the average of the annual rain depths at Kodok, station 117, Malakal, station 118, Abwong, station 120, and Nasser, station 123. For these stations in their order, the mean rainfall for the period 1938-1967 was 738, 819, 763 and 894 mm/yr. Assuming the annual run-off coefficient at 4%, the run-off should then be 0.04 x 0.78 x 36.8 or 1.15 mlrd m^3 per year. This figure is slightly higher than the 1.08 mlrd m^3/yr reported in Vol. VIII of the Nile Basin as a mean for the period 1934-1947 (Hurst, H.E., 1950). The sum of the flow at the Sobat head and the run-off is 13.55 mlrd m^3/yr. Reducing this amount by about 5% for the net conveyance loss, the amount that finally reaches the mouth on the White Nile is 12.9 mlrd m^3/yr.

The discharges of the Sobat at Hillet Doleib, 9 km above the mouth, have been measured since 1911. The discharge-gauge measurements for the rating curve can be seen from Fig. 19, Appendix E. The average hydrograph at Hillet Doleib corresponding to an annual volume of 12.9 mlrd m^3/yr is shown in Fig. 8.34.

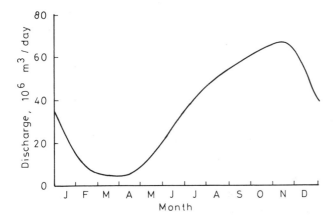

Fig. 8.34.   The discharge hydrograph of the Sobat at Hillet Doleib, near mouth

The relation between the discharges measured at the Sobat mouth, in fact at Hillet Doleib, Y, and the difference between the discharge at Malakal less the discharges of the Bahr el Ghazal, Jebel and Zeraf as at Malakal, X, has been

examined for the purpose of checking the consistency of the results. The linear
regression relation found is

$$Y = - 0.4255 + 0.9827 X \tag{8.6}$$

with a correlation coefficient of 0.9619. Since the mean flow at Malakal for the
period 1905-1966 was 28.82 mlrd m^3/yr and the sum of the Jebel and Zeraf was
14.74 mlrd m^3/yr, the mean of Y must be 1408 mlrd m^3/yr. The mean of X over the
same period, as obtained from eq. 8.6, has to be 14.76 $\pm$ 1.84 or in the range of
between 16.60 and 12.92 mlrd m^3/yr (confidence level = 95%). This result shows
that the figure 29.2 mlrd m^3/yr, though on the low side, yet needs not be
rejected.

8.7.2.2   The discharges at Malakal

The monthly and annual discharges of the White Nile at Malakal for the period
1912-73 are given in Table 7, Appendix D (Cairo University - Massachusetts
Institute of Technology, 1977). The gauge-discharge measurements used for preparing the rating curve of the White Nile at Malakal are shown in Fig. 20,
Appendix E.

The annual flow volume at Malakal in the period investigated showed two
maxima; the earlier one took place in 1918 and was sharp, and the second in 1964
and was broader. The two flow volumes in their order of occurrence were 44.35
and 48.64 mlrd m^3/yr respectively. Each of them is almost 80% of the recorded
maximum at Mongalla on the Bahr el-Jebel (see section 8.5.1.2). The lowest minimum observed at Malakal was 23.32 mlrd m^3/yr and it took place in 1940. This is
slightly less than the minima which were observed in 1913, 1922 and 1950 and
their values were 23.83, 23.59 and 23.75 mlrd m^3/yr. The average of these four
minima is slightly more than 1.5 times the minimum at Mongalla on the Bahr el
Jebel. The mean flow volume at Malakal, being 29.44 mlrd m^3/yr, is about 6% less
than the mean flow volume at Mongalla for the same period 1912-1973. These figures show the interaction between the contributions of the Bahr el-Ghazal, Jebel
and Zeraf on one hand and the contributions of the Baro, Pibor and the Sobat on
the other. Of special interest is that the mean annual flow at Malakal is
slightly less than the mean annual flow at Mongalla. This result means that the
gain from the 750 000 km^2, which comprise the sub-basins of the Ghazal, Jebel,
Zeraf, Baro, Pibor and the Sobat, all downstream of Mongalla, are a few percent
less than the loss which takes place in the swamps and the low-lying areas. An
important feature in the annual flow series at Malakal is that the ratio of the
maximum to the minimum is about 2:1 (see Fig. 8.35), whereas this ratio, except
for Mongalla, is about 4:1 (see Fig. 8.21).

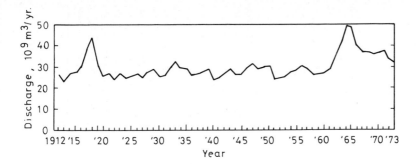

Fig. 8.35.   Graphical plot of the annual discharge series of Malakal in the
period 1912-1973

The statistical properties of the 12-monthly series and of the annual series
have been examined and the results presented in Table 8.12. These results show
that the monthly discharge series, except that of March, contain a rather strong
serial correlation between their elements; the serial correlation in the annual
series is even stronger. The correlation coefficient which is significantly
different from the zero, at a confidence level of 95%, is limited here to lag 2
or 3. It has been found, however, that one can remove a good deal of the dependence from any of the series by fitting a linear autoregressive model to it. The
residuals left after fitting the model have shown they are uncorrelated at the
same level of significance. This can be seen from Table 8.13. The order of the
model well fitting the monthly series is the first[+], whereas the second order[++]
is a better fit to the annual series. In this respect the discharge series of
Malakal resemble those of Mongalla.

The probability distribution function of good fit to the monthly and annual
series of Malakal is the Pearson Type III only. Fig. 8.36. shows the fit of this
distribution function to the annual discharge series. For this series and for
the 12-monthly ones, the 100 and the 200- year discharges computed by this function are:

Discharge, 10^6 m^3	Jan.	Feb.	Mar.	Apr.	May	June	July	Aug.	Sep.	Oct.	Nov.	Dec.	Year
100-yr	5563	4465	4914	2940	2723	2949	3514	4125	4763	5533	5574	5699	49060
200-yr	6094	5014	5652	3221	2910	3094	3684	4268	5082	2928	6009	6125	52660

+   Model parameter is the first serial correlation coefficient r_1 (see Table
    8.12)
++ Model parameters are $\alpha_1 = 1.0813$ and $\alpha_2 = -0.3499$

TABLE 8.12   The basic statistical descriptors and serial correlation coefficients of the monthly and annual discharges of the White Nile at Malakal for the period 1912-1973

Item — Basic statistical descriptor	Jan.	Feb.	Mar.	Apr.	May	June	July	Aug.	Sep.	Oct.	Nov.	Dec.	Year
$\bar{X}$, 10^6 m^3	2446	1735	1704	1484	1650	2015	2505	2871	3080	3402	3310	3158	29438
s, 10^6 m^3	939	732	786	404	350	309	312	370	470	609	628	789	5671
C^v	0.3839	0.4219	0.4614	0.2723	0.2120	0.1525	0.1246	0.1287	0.1525	0.1789	0.1898	0.2500	0.1926
C^s	1.4868	2.2417	3.0920	2.0042	1.1324	1.0071	1.3408	1.5958	1.9511	1.7958	2.0020	1.3147	1.7325
C_k	5.7771	8.6520	13.7243	7.9193	4.3800	3.7999	4.4741	5.9292	8.9434	8.3041	9.5429	6.9154	5.9970

Serial correlation coefficient

	Jan.	Feb.	Mar.	Apr.	May	June	July	Aug.	Sep.	Oct.	Nov.	Dec.	Year
r_1	0.4934	0.6058	0.2808	0.6661	0.6498	0.6193	0.7472	0.7649	0.6785	0.6987	0.6708	0.5145	0.8010
r_2	0.2832	0.3385	0.0966	0.3905	0.4005	0.3590	0.4872	0.5115	0.4623	0.5010	0.4449	0.2916	0.5162
r_3	0.2296	0.2464	0.0540	0.2622	0.2977	0.2877	0.3361	0.3029	0.2718	0.3794	0.2809	0.2321	0.3218
r_4	$\overline{0}$.0390	0.0321	0.0064	0.1548	0.1755	0.0453	0.1710	0.2172	0.1660	0.2955	0.0901	0.0450	0.1675
r_5	$\overline{0}$.0285	0.0348	$\overline{0}$.0058	0.2050	0.1894	0.0215	0.1435	0.1778	0.1009	0.1624	0.0620	0.0592	0.1035
r_6	0.1015	0.1166	0.0214	0.2133	0.2395	0.1474	0.2142	0.1595	0.0773	0.0563	0.0399	0.0986	0.1257
r_7	0.0720	0.1201	0.0459	0.2212	0.2442	0.1696	0.2016	0.1470	0.0718	0.0237	0.0271	0.0422	0.1347
r_8	0.1442	0.0989	0.0202	0.2071	0.2261	0.1309	0.1338	0.1023	0.0473	$\overline{0}$.0644	$\overline{0}$.0320	$\overline{0}$.0074	0.0997
r_9	0.1305	0.0794	0.0186	0.1226	0.1808	0.1534	0.0930	0.1014	0.0157	0.0803	0.0441	0.0378	$\overline{0}$.0268
r_{10}	$\overline{0}$.0046	$\overline{0}$.0342	$\overline{0}$.0593	0.0445	0.1317	0.1054	0.0398	0.0402	0.0440	0.0763	0.0660	$\overline{0}$.0871	$\overline{0}$.0397
r_{11}	0.1468	0.0861	$\overline{0}$.1119	$\overline{0}$.0600	0.0234	$\overline{0}$.0536	0.0782	0.0829	0.1056	0.1283	0.1279	0.1721	$\overline{0}$.0957
r_{12}	$\overline{0}$.0174	$\overline{0}$.0861	$\overline{0}$.1119	$\overline{0}$.0987	0.1392	0.1214	0.1368	0.1822	0.1655	0.1330	0.1156	$\overline{0}$.0200	0.1482
r_{13}	0.0642	$\overline{0}$.0708	$\overline{0}$.0856	0.0987	0.1347	0.1616	0.1681	0.2101	0.1689	0.1076	0.1237	0.0494	$\overline{0}$.1324
r_{14}	0.0230	0.0310	0.0500	0.0259	0.1629	0.1879	0.1571	0.1128	0.0399	0.0233	0.0009	0.0424	$\overline{0}$.0146
r_{15}	0.2570	0.0980	0.0851	0.0725	0.0257	0.0702	0.0788	0.0145	0.0490	0.0604	0.0721	0.1849	0.0826

Month of the year

TABLE 8.13   Serial correlation coefficients of the residuals left from fitting autoregressive models to the discharge series at Malakal

Month & Year	Serial correlation coefficients														
	r_1	r_2	r_3	r_4	r_5	r_6	r_7	r_8	r_9	r_{10}	r_{11}	r_{12}	r_{13}	r_{14}	r_{15}
January	-.018	-.025	.189	-.196	-.038	.128	-.076	.109	.026	-.029	-.210	.055	.076	-.071	.182
February	.020	.090	.153	-.192	-.035	.108	.018	.035	-.050	-.075	-.096	-.001	-.053	-.070	.134
March	-.011	-.006	.026	-.010	-.009	.039	.028	.010	-.018	.034	-.070	-.063	-.041	-.008	-.048
April	.053	-.106	.015	-.151	.103	.093	.022	.099	.010	.002	.090	-.011	.111	.026	.150
May	.065	.111	.062	.101	.027	.143	.053	-.073	.024	.040	.005	-.152	-.017	-.133	.124
June	.009	-.124	.238	.217	.059	.225	.041	-.037	.081	.091	-.035	-.069	-.048	-.139	.037
July	-.096	-.134	.071	-.189	.110	.192	.088	-.051	-.002	.071	.083	-.082	-.063	-.107	-.028
August	-.013	-.061	-.140	-.005	.000	-.029	.052	.106	-.116	-.025	-.066	-.042	-.225	.001	.108
September	-.008	-.039	-.053	-.024	-.033	-.015	.037	-.052	-.039	.034	-.023	-.077	-.149	.035	.103
October	-.030	-.018	.009	.109	.011	-.080	.064	-.097	-.039	.052	-.066	-.054	-.120	.096	.088
November	-.016	.004	.076	-.149	.019	.013	.051	-.054	-.021	-.065	-.067	.000	-.131	-.049	.056
December	.016	-.006	.162	-.096	.027	.079	.008	-.068	.116	-.054	.213	.112	-.102	-.010	.154
Year	.019	-.152	.152	.032	-.043	.086	.022	.079	-.060	-.046	.064	-.121	-.210	.031	.239

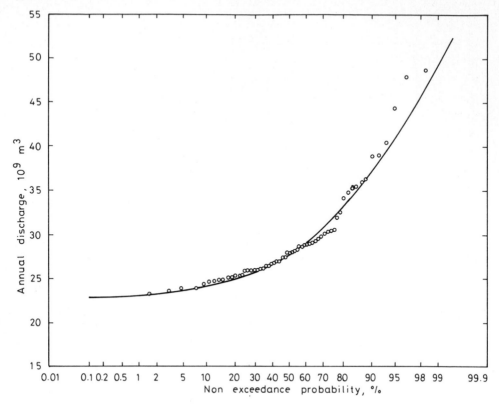

Fig. 8.36.   Fit of the Pearson Type III function to the distribution of the
annual discharges at Malakal in the period 1912-1973

8.7.3   The White Nile from Malakal to just above the junction with the Blue
         Nile

Below the confluence of the Sobat, the White Nile flows a distance of, say,
840 km, without being joined by any important tributary except by the Blue Nile
at the downstream end of this reach. The discharges have for some time been
measured at Mogren upstream of the junction of the Blue Nile and near Khartoum.

The gauge-discharge measurements for the period 1912-27 are shown in Fig. 21,
Appendix E. The gauging site at Mogren was described as being good enough at low
stage, but not good in a flood. This might explain the terrible scatter of the
points in the stage-discharge diagram.

For the first 358 km, the river has a waterway of from 300 to 500 metres in
width with numerous islands. The mean width may be taken as 425 metres in low
supply when the river is within the banks. For the next 490 km to the tail at
Khartoum, the mean width of the water surface is 850 metres in low supply. The
general depth of the water at low stage is 4 metres, and 7 metres in flood. On
either side of the waterway of the upper reach is a low ridge swamped in flood

and beyond that is a deep depression, deep in the centre and rising to the ridge on one side and to the high land and forest on the other. Each depression may be 3 km in width where it is wide and a few hundred metres where it is narrow, so that the flooded valley may have a width of 6 km in some places. The ridges are broken by openings through which the water passes in and out of the marshy depressions. These depressions are ocvered by a dense growth of needs. At Begelein the side depressions contract and the forests approach the river. Fifty kilometres further to the north, the extent of the swamps decreases and the river width varies from 700 to 900 metres. Some 30 km further to the north the sudd grasses disappear, and though there is flooding, there are no swamps. The summer channel in the 490 km upstream of Khartoum is 850 metres in width and the flood channel is 4.3 km. The summer depth of water is about 4 m. The rainfall in the 850 km reach of the White Nile dies out almost linearly with distance from Malakal to Khartoum. For the 30-year period 1938-1967, the mean rainfall over the different stations was:

Station No.	Location	Rainfall mm/yr	Station No.	Location	Rainfall mm/yr
118	Malakal	819	100	Kosti/Rebeck	403
117	Kodok	738	91	Dueim	315
116	Melut	644	85	Geteina	202
111	Renk	541	81	Jebel Awlia	199
106	Gebelein	431	79	Khartoum	156

For the same river reach, the open water evaporation has been estimated at about 1900 mm/yr at Malakal to about 2920 mm/yr at Khartoum (see Chapter 5: Evaporation). From these figures it is clear that the loss per year varies from about 1080 mm at Malakal to 2760 mm at Khartoum; if we take 1.9 m as an average loss for the whole reach and the average width of the water surface at, say, 1 km, it is therefore not difficult to realize that the White Nile from below the confluence of the Sobat to just above the White Nile junction looses, on average, 1.6 mlrd m^3/yr. This figure departs slightly from the average loss for the period 1914-1937, which was 1.9 mlrd m^3/yr with a standard deviation of 1.2 mlrd m^3/yr. Hurst reported that most of the variation in the loss was due to the fact that the loss was the difference between two much larger quantities, thereby containing the errors of both. Not much weight can be attached to the loss in any particular year, but the mean loss is probably correct at half a milliard.

The Jebel el-Aulia reservoir was put into use for the first time in 1937. The estimated loss in the post-reservoir period, 1937-1948, was 2.9 mlrd m^3/yr. The aim and function of this storage work shall be discussed, however, in Chapter 9. If we now adhere to a mean annual flow of 29.44 mlrd m^3/yr at Malakal, the natural river discharge above the Blue Nile junction can be taken as 27 to 28 mlrd m^3/yr.

8.8   THE BLUE NILE BASIN

8.8.1   Lake Tana

The Lake Tana and its catchment have been described in Chapter 2. The catchment area excluding the lake, which in itself is 3000 km^2, is 13750 km^2. The hydrologic variables are so that the average rainfall and evaporation balance each other at about 1300 mm/yr; the run-off from the catchment to the lake, assuming no change in storage, must then equal the lake outflow. Taking the annual run-off coefficient as 22%, the annual lake outflow is about 3.93 mlrd m^3/yr. This figure is nearly the same as the mean of the outflows in the period 1920-1933, which was 3.85 mlrd m^3/yr. These outflows are given in Table 8, Appendix D. For a given lake level the discharge can be read from the rating curve, Fig. 22, Appendix E. Table 8, Appendix D, has been used for preparing the average hydrograph of Lake Tana outflow, which is shown in Fig. 8.37.

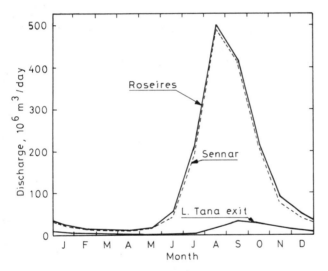

Fig. 8.37.   The average hydrograph of the Blue Nile at the exit of Lake Tana, Roseires, and Sennar

8.8.2   The Blue Nile from Lake Tana to Roseires

The Blue Nile receives tributaries a short distance away from its exit from Lake Tana. The first tributary is called Chimbil and is said to bring as much as 10 m^3/sec. in flood. Below the junction of Chimbili with the Blue Nile the tributaries increase in size and importance as the river enters the canyon in which it remains until within a few kilometres from the Sudan boundary.

The Blue Nile in its upper reach is joined by the River Bashile and River Jamma (see the map, Fig. 2.18.). The important tributaries of the lower reaches are the Didessa, Dabus and Balas. Hurst and his co-workers estimated the flow at about 2.2 x 10^6 m^3/day in low-flow season and at about 220 x 10^6 m^3/day in flood season. This means that after flowing for about 330 km, the discharge becomes about 10 times as much as its initial value at the exit of the lake. The Blue Nile discharge at Roseires, 935 km below the exit of Lake Tana, is about 7 million m^3/day, emphasizing that the contribution of the reach below Kutai (Kilo 330) is more than the contribution of the upper reach. It seems, however, that the discharge from Lake Tana up to Roseires increases with distance from the lake.

The rating curve of the Blue Nile at Roseires is shown in Fig. 23, Appendix E, and the monthly and annual discharges in the period 1912-1973 are given in Table 9, Appendix D. The statistical descriptors and the serial correlation coefficients of these data have been calculated and given in Table 8.14. From this table one can notice that the monthly and the annual discharges at Roseires are very much less dependent than the monthly and the annual discharges at either Mongalla or Malakal. It is only the months of low flow, January thru' May, and October that show significant correlation at the 95% level of confidence. The discharges of January, February and March are seen to be serially correlated up to lag 3, whereas April is serially correlated at lag 1 only and the first serial coefficient is just significantly different from zero at the 95% confidence level. The May and October series show serial correlation at lag 3 and at lag 1 or 2. The months June thru' December, with the exception of May, and the year, are not serially correlated.

One can get some rough impression about the lack of dependence in these series by comparing the graphical plot of, for example, the annual series at Roseires (Fig. 8.38.) to the annual series at Malakal or Mongalla. The dependent component in the series of January thru' April can be described, however, by a first-order autoregressive relation having the first correlation coefficient as a parameter (see Table 8.14). The residuals left after removing the dependent component appear to be uncorrelated at 95% confidence level. The serial correlation coefficients of the residuals of these series are presented in Table 8.15.

The probability distribution functions that may serve as good fit to the monthly and annual data are the Pearson Type III and the normal functions. The fit of the latter to the annual discharge data is as shown in Fig. 8.39. For all the discharge series the 100 and 200-year discharges computed from the distribution function best fitting each series are as follows:

Discharge, 10^6 m^3	Jan.	Feb.	Mar.	Apr.	May	June	July	Aug.	Sep.	Oct.	Nov.	Dec.	Year
100-yr	1642	1414	1385	1116	1851	3426	10902	22918	19941	13088	4639	2495	73091
200-yr	1752	1586	1577	1267	2096	3780	11300	23595	20619	13876	4886	2588	75270

TABLE 8.14   The basic statistical descriptors and the serial correlation coefficients of the monthly and annual discharges of the Blue Nile at Roseires for the period 1912-1973

Month of the year

Item	Jan.	Feb.	Mar.	Apr.	May	June	July	Aug.	Sep.	Oct.	Nov.	Dec.	Year
Basic statistical descriptor													
$\bar{X}$, 10^6 m³	878	543	445	369	601	1632	6548	15499	12515	6813	2672	1470	49216
s, 10^6 m³	268	235	246	199	341	533	1691	2881	2884	2386	748	398	9272
C_v	0.3050	0.4335	0.5520	0.5378	0.5679	0.3265	0.2582	0.1859	0.2305	0.3502	0.2799	0.2709	0.1884
C_s	0.7348	2.1822	2.4359	2.2961	2.1238	1.6445	0.3363	-0.2392	-0.1443	0.4263	0.4241	0.0479	-0.1747
C_k	4.0851	8.6791	9.7202	10.5525	8.3400	10.0111	3.4673	5.6194	4.1538	3.4400	3.5667	3.4878	3.9756
Serial correlation coefficient													
r_1	0.5059	0.7513	0.5126	0.2489	0.0269	0.1702	0.1173	-0.0705	0.1891	-0.0237	-0.0613	0.2171	0.1383
r_2	0.4093	0.5161	0.4183	0.0902	0.0381	-0.0468	0.0898	0.0381	0.0527	0.0690	0.0563	0.1603	0.0166
r_3	0.3560	0.3503	0.3555	0.1394	0.3773	-0.0312	-0.0059	-0.0248	0.0991	0.3098	0.0882	0.1329	0.1574
r_4	0.1784	0.1801	0.1091	0.1305	0.0754	0.1119	0.1735	0.0987	0.1384	0.1597	0.2219	0.1290	-0.1441
r_5	0.3179	0.1226	0.0646	0.0301	0.0528	-0.1254	-0.2370	0.0248	-0.0237	0.0514	0.0201	0.1724	-0.0448
r_6	0.1772	0.1404	0.0574	-0.0989	0.2903	0.2221	0.1646	0.1481	0.0259	0.2608	0.0369	0.0616	0.0963
r_7	0.2330	0.1080	0.0394	-0.1533	-0.0249	-0.0445	0.1590	0.0557	-0.0597	-0.0409	-0.0206	0.1086	-0.0237
r_8	0.2221	0.1086	-0.0463	-0.1586	-0.0338	0.0017	-0.1283	0.0057	-0.1337	-0.0815	0.0606	-0.0749	-0.0232
r_9	0.1424	0.1483	0.1040	-0.0659	0.1250	-0.1117	0.0147	-0.0846	-0.0238	0.0739	0.0126	-0.0140	-0.0778
r_{10}	0.2617	0.1309	0.0960	0.2179	0.0337	-0.0500	-0.0876	-0.1179	-0.1651	-0.0971	-0.1538	-0.1764	-0.0466
r_{11}	0.0815	0.1146	0.0879	0.2505	-0.0483	-0.1069	0.0782	0.0035	0.1826	0.1670	0.1889	0.1365	0.2977
r_{12}	0.1939	0.1227	0.2886	0.3045	0.0710	0.0515	0.1616	0.3042	0.2214	0.0225	0.1009	0.1467	0.1970
r_{13}	0.1665	0.1230	0.0445	-0.0059	0.0094	-0.1430	0.1510	-0.2180	-0.0178	0.0429	0.0307	-0.0065	-0.0573
r_{14}	-0.0153	0.0410	0.0880	0.0277	0.2652	-0.3026	0.2122	-0.0826	-0.0284	-0.0663	-0.0958	-0.1136	-0.1566
r_{15}	0.0572	-0.0219	0.1392	0.0055	0.0161	0.1991	0.1783	-0.0007	-0.1043	0.1357	0.0238	0.1581	-0.0275

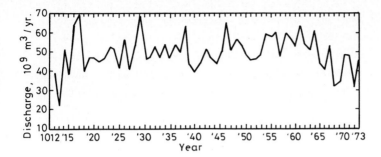

Fig. 8.38.   Graphical plot of the annual discharge series of Roseires in the period 1912-1973

TABLE 8.15   Serial correlation coefficients of the residuals left from fitting autoregressive models to some of the discharge series at Roseires

Month	Serial correlation coefficients														
	r_1	r_2	r_3	r_4	r_5	r_6	r_7	r_8	r_9	r_{10}	r_{11}	r_{12}	r_{13}	r_{14}	r_{15}
Jan.	.082	.067	.167	-.183	.231	-.074	.042	.108	-.094	.216	-.127	.097	.136	-.152	.042
Feb.	.073	-.047	.041	-.163	-.111	.077	-.031	-.039	.095	.013	-.019	.021	.121	-.024	-.008
Mar.	-.079	.076	.166	-.073	-.085	.026	.042	-.020	.054	.029	-.082	.299	-.131	.007	.192
Apr.	.016	-.005	.091	.087	-.017	.073	.081	.131	-.156	.160	.131	.231	-.081	.026	.017

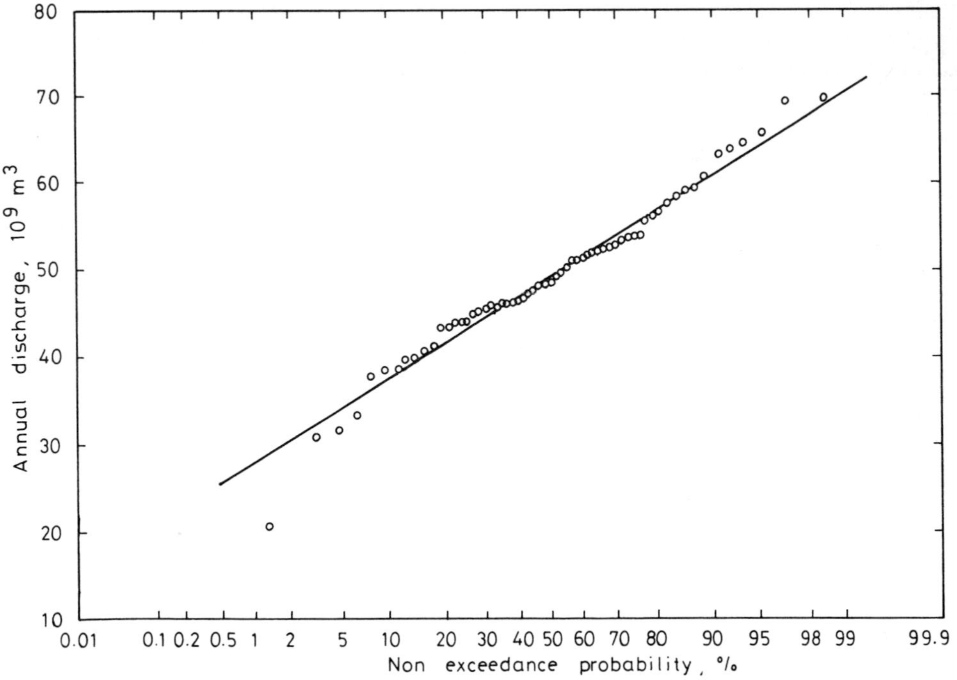

Fig. 8.39.   Fit of the normal function to the distribution of the annual discharges at Roseires in the period 1912-1973

The first phase of the construction of a concrete dam on the Blue Nile at Roseires was completed in 1966. The primary purpose of this dam is to store water and release it downstream in the shortage season to supply the Gezinah Managil Extension and the river bank pump schemes with water as may be needed. We shall describe this storage work in more detail in the next chapter.

8.8.3 The Blue Nile from Roseires to Sennar

The Blue Nile below Roseires is a mild stream with a slope of about 0.12×10^{-3} which is about one-tenth the slope of the torrential stream which prevails all the way from the exit of Lake Tana to Roseires. There is a gauging station at Wadi el-Aies, near Singa, about 180 km downstream of Roseires, and another at Makwar, near Sennar, some 270 km below Roseires. The rating curves of these two stations in the given order are shown graphically in Figs. 24 and 25, Appendix E, respectively.

The mean annual rainfall of the 30-yr period 1938-67 at Roseires, station 110, is 785 mm; at Singa, station 101, 580 mm and at Sennar, station 97, 463 mm. These data suggest the figure of 600 mm as an average depth of rain in the reach from Roseires to Sennar. For the same reach of the Blue Nile the average annual evaporation is about 2450 mm. The net loss can therefore be estimated at 1.85 m/yr. If we take 1 km as an average value for the river width, the evaporation minus precipitation loss comes to 0.5 mlrd m^3/yr. Hurst and his co-workers gave the average volume of flow at Roseires for the period 1912-1950 as 49.6×10^9 m^3/yr and that at Sennar for the same period as 48.7×10^9 m^3/yr with 0.9 mlrd m^3/yr as the total transmission losses. They considered the loss in a normal year as 2% and in a high year as 4%, both of the flow volume at the upstream station, i.e. Roseires. The results of our calculation for the mean flow at the two stations in the period from 1912 up to and including 1973 show that whereas the mean at Roseires was 49216×10^6 m^3/yr, that at Sennar was 47185×10^6 m^3/yr. These figures bring the total loss to 2 mlrd m^3/yr, which is somewhat bigger than the previous results, pointing to the possibility of increasing withdrawal of water from the river between Roseires and Sennar.

The statistical descriptors and the serial correlations of the monthly and annual data at Sennar are given in Table 8.16. Although the mean monthly flows at Roseires and Sennar are very nearly equal (see Fig. 8.37.), the pair of series for each month at these two stations do not always behave similarly. Of the Sennar series, which show dependence at lag 1 with 95% confidence, are those of February, September and December; at lag 2, November; lag 3, October and lag 4, March. For all these series, except that of March, one can describe the dependence in the series by the first-order linear Markov model. The coefficients of serial correlation of the residual series have been computed and

TABLE 8.16 The basic statistical descriptors and the serial correlation coefficients of the monthly and annual discharges of the Blue Nile at Sennar for the period 1912-1973

Item	Jan.	Feb.	Mar.	Apr.	May	June	July	Aug.	Sep.	Oct.	Nov.	Dec.	Year
Basic statistical descriptor													
$\bar{X}$, 10^6 m^3	682	449	404	373	573	1402	5892	15222	12383	6437	2202	1186	47185
s, 10^6 m^3	260	160	170	153	304	573	1852	2771	3159	2632	1003	476	10012
C^v	0.3817	0.3565	0.4204	0.4094	0.5313	0.4083	0.3143	0.1820	0.2551	0.4089	0.4554	0.4016	0.2122
C^v	1.1508	1.0651	1.1641	0.6165	1.9219	1.4886	0.1255	0.0882	0.5352	0.6046	1.2769	0.9890	0.5123
C^s_k	5.4273	4.4911	5.6132	2.9148	8.2095	8.6921	3.9041	5.1156	6.4598	3.8337	5.8684	5.7073	5.0149
Serial correlation coefficient													
r_1	0.1811	0.3267	0.1574	0.1699	0.0846	0.2333	-0.0274	-0.0845	0.2805	0.1353	0.1973	0.3017	0.1872
r_2	0.0996	0.0450	0.0108	0.0327	0.0326	0.0599	-0.1135	-0.0529	0.0655	0.0640	0.2722	0.2635	0.0301
r_3	0.1784	0.0561	-0.0538	0.0012	0.1991	0.0771	-0.1331	-0.0951	0.0979	0.2551	0.2628	0.1619	0.1176
r_4	-0.0421	-0.2558	-0.3774	-0.2155	-0.0713	0.0902	-0.0552	-0.1391	0.1118	-0.0728	0.0144	-0.0717	-0.0110
r_5	0.2182	0.1209	0.2031	-0.0765	0.0424	0.0880	-0.0815	-0.0497	0.0480	-0.0250	0.0915	0.0793	-0.0110
r_6	0.0753	0.0251	0.1041	-0.0142	0.1976	0.1817	0.1002	0.1225	0.0425	0.1917	0.1574	0.0648	0.1281
r_7	0.1118	-0.0638	-0.1558	-0.0669	-0.0517	0.1005	0.1756	-0.0580	0.0513	-0.0240	0.0790	0.0253	0.0630
r_8	-0.1109	-0.0303	-0.0492	-0.1085	-0.0553	-0.0205	0.1384	0.0652	-0.1631	-0.1433	0.1010	0.0916	-0.0150
r_9	0.0212	0.0773	-0.1523	-0.1633	-0.0474	0.1754	-0.0120	0.0440	-0.0179	-0.0189	0.0049	0.0128	-0.0374
r_{10}	0.1525	0.0718	0.0612	0.0398	0.0160	0.0678	0.0276	0.0269	-0.0783	0.0879	0.0910	0.0924	-0.0059
r_{11}	-0.1070	0.0516	0.0366	0.1117	0.0547	0.1207	-0.1403	0.0153	-0.1638	0.2224	-0.1724	0.1445	-0.1759
r_{12}	0.1571	0.2162	0.3133	0.0715	0.0565	0.1078	-0.1298	0.2387	0.1748	-0.0716	-0.0475	0.0425	0.1837
r_{13}	0.0595	0.0719	-0.0118	-0.0559	-0.0155	-0.1611	-0.0776	-0.2285	0.0585	0.0186	0.0237	0.0558	-0.0054
r_{14}	-0.1780	-0.0523	-0.0806	-0.2350	0.2107	-0.1245	-0.1453	-0.1220	-0.0298	-0.1617	-0.1117	-0.1096	-0.1207
r_{15}	-0.0560	-0.0523	-0.0463	-0.3877	-0.0137	0.2046	-0.0846	0.0294	0.0819	-0.0904	0.0101	-0.0244	-0.0368

listed in Table 8.17. The tabulated values do not justify rejecting the null hypothesis that the serial correlation coefficients are not significantly different from zero at the 95% level of confidence.

TABLE 8.17   Serial correlation coefficients of the residuals left from fitting autoregressive models to some of the discharge series at Sennar

Month	Serial correlation coefficient														
	r_1	r_2	r_3	r_4	r_5	r_6	r_7	r_8	r_9	r_{10}	r_{11}	r_{12}	r_{13}	r_{14}	r_{15}
Feb.	.075	-.009	.036	-.275	-.168	-.054	-.031	-.007	.076	.057	.001	.179	.040	-.064	-.034
Sep.	-.021	-.014	.122	-.131	.120	.008	.089	-.178	.054	-.033	-.179	.223	.111	-.022	.041
Oct.	-.047	.096	.078	-.084	.033	.057	-.053	.063	-.110	.106	-.146	.014	-.029	-.132	-.039
Nov.	-.032	.203	.210	-.048	.076	.140	.026	.098	-.020	.099	-.147	-.008	.031	-.088	.026
Dec.	-.024	.179	.131	-.120	.100	.075	-.023	.074	-.053	.121	-.173	.096	.019	-.086	.008

The Sennar Dam (Makwar) was built across the Blue Nile in 1925 a few kilometres above Sennar. This dam was built exclusively in the interest of the Sudan. It serves two purposes: it raises the level sufficiently high for the water to flow into the main Gezirah canal, and it stores water from the flood to be used during the period January to April when there is no surplus water in the Blue Nile over Egypt's requirements. More detailed description of this storage will appear in Chapter 9.

8.8.4   The Blue Nile below Sennar to Khartoum

Below Sennar, the Blue Nile flows north-west for a distance of 350 km before it joins the White Nile at Khartoum. Between Sennar and Wad Medani the Blue Nile receives the Rahad. This tributary rises on the Ethiopian Plateau a few kilometres west of Lake Tana under the name of the Sidd. In its course of 750 km the river flows in the Ethiopian Plains as the Aima and changes its name in the Sudan to the Dinder. The drainage basin of the Dinder has an area of 160 000 km^2. The average areal rainfall is about 0.80 to 0.85 m/yr and the annual run-off coefficient reaches 22%. These figures suggest a total run-off of about 3 mlrd m^3 in a normal year. Vol. IX of the Nile Basin gives 2.97 mlrd m^3 as an average volume of flow per year in the period 1912-1950 and 3.83 mlrd m^3 as the flow volume in a high year (1946). The gauge-discharge measurements for establishing the rating curve of the Dinder are taken at Hillet Idris. The data for the period 1924-1927 are shown in Fig. 26, Appendix E.

The Blue Nile below Wad Medani receives the Rahad, which springs nearly from the same place as the Dinder. This tributary has a length of 800 km and a drainage basin of about 8000 km^2. For the same rainfall as on the catchment of the

Dinder, i.e. 0.80-0.85 m/yr, and an annual run-off coefficient of, say, 16%, the total flow reaching the mouth of the Rahad comes to about 1.1 mlrd m^3/yr. Vol. IX of the Nile Basin gives 1.08 mlrd m^3 as the yearly average for the period 1912-1950. The rating curve of the Rahad for the period 1922-1927 at Abu-Haraz, near the mouth, is shown in Fig. 27, Appendix E.

The Dinder and the Rahad hardly carry any water in the period from January to May. The hydrograph of each of these two tributaries has a more or less triangular shape with a base width of about 200 days. The peak discharges have been found to be about 480 and 160 m^3/sec. for the Dinder and the Rahad, respectively.

The mean annual rainfall varies from 463 mm at Sennar, station 97, to 340 mm at Managil, station 89 to 385 at Wad Medani, station 88, to 312 at Ruffa, station 86 to 254 mm at Kamlin, station 83, to 160 mm at Khartoum stations 78/79. In this reach of 350 km, average depths of 320 mm and 2740 mm/yr can be used for rainfall and evaporation, respectively. Assuming an average width of the river of 800 m, the loss becomes (2.74 - 0.32) x 350 x 10^6 x 0.8 = 0.68 mlrd m^3/yr to be rounded to, say, 0.85 mlrd m^3/yr to account for some seepage loss. The balance at the mouth of the Blue Nile near Khartoum is 47.185 + 2.970 + 1.080 - 0.850 = 50.385 mlrd m^3/yr. This figure is about 2% less than that given by Hurst in Vol. IX of the Nile Basin (Hurst, H.E., Black, R.P., and Simaika, Y.M., 1959). The monthly and annual discharge series of the Blue Nile at Khartoum are presented in Table 11, Appendix D. The data used for preparing the rating curve are shown in Fig. 28, Appendix E. The change of the gauge site alternatively between Buri, Soba and Khartoum could be one of the reasons responsible for the heavy scatter of the plotted points.

The already derived annual flow volume at Khartoum is nearly in perfect agreement with the mean annual flow for the period 1912-1973, which was 50.369 mlrd m^3/yr (see Table 8.18). Either figure, derived or computed, shows that the flow at Khartoum is about 6.8% larger than that at Sennar. One should not forget, however, that this is an average percentage for 62 years, around which individual years fluctuate positively and negatively. The graphical plot of the annual series at Sennar and Khartoum, Fig. 8.40. shows that the flow volume in some years at Sennar was equal to, or even larger than, that at Khartoum.

Some of the monthly series and the annual discharge series at the mouth of the Blue Nile appear to be slightly significantly different from zero at the 95% level of confidence. The remaining series, which are the May, July, August, September and October series, are not significantly different from the zero at the same level of confidence. The series showing significant dependence at lag 1 are those of February, April, June, December and the year. Those at lags 3, 4, 5 and 6 are the series of November, March, January and June, respectively. The dependence in all these series can be adequately described by a first-order

TABLE 8.18 The basic statistical descriptors and the serial correlation coefficients of the monthly and annual discharges of the Blue Nile at Khartoum for the period 1912-1973

Item	Jan.	Feb.	Mar.	Apr.	May	June	July	Aug.	Sep.	Oct.	Nov.	Dec.	Year
Basic statistical descriptor													
$\bar{X}$, 10^6 m^3	740	455	408	386	480	1137	5371	15853	13945	7657	2487	1264	50369
s, 10^6 m^3	263	159	157	180	225	500	1730	2835	3243	2932	1061	486	10258
C_v	0.3551	0.3491	0.3847	0.4657	0.4685	0.4394	0.3221	0.1788	0.2326	0.3829	0.4267	0.3843	0.2037
C_s	1.0112	1.0912	1.3087	1.7888	1.7085	2.2234	0.3456	0.5503	0.1190	0.5897	0.1503	0.8370	0.1919
C_k	5.5867	4.8671	6.6164	7.8484	7.3397	11.8552	3.1507	4.2886	4.5216	4.0881	5.2131	5.0576	3.9845
Serial correlation coefficient													
r_1	0.1420	0.2836	0.2268	0.2848	0.1060	0.2403	0.1562	-0.1054	0.1593	0.1458	0.1345	0.3182	0.2521
r_2	0.0902	0.0295	-0.0426	0.1544	0.0865	0.0230	-0.0039	0.0913	0.1411	0.0216	0.1660	0.2251	0.1146
r_3	0.1664	0.0427	-0.1317	0.0540	0.1790	0.1538	-0.0680	-0.1318	0.1638	-0.1908	0.2458	-0.1552	-0.0675
r_4	0.0247	-0.2222	-0.3582	-0.2050	-0.0574	-0.0580	-0.1442	-0.0479	0.1046	-0.1095	-0.0630	-0.0411	-0.0950
r_5	0.2745	0.0090	-0.1651	-0.1528	-0.2076	-0.0206	-0.1104	-0.1228	0.1783	-0.0287	0.0724	0.0570	-0.0089
r_6	0.0672	0.1211	-0.1057	-0.1588	0.0940	0.3831	0.1682	0.0627	-0.0342	0.2068	0.0721	0.0687	0.1187
r_7	0.0450	-0.0100	-0.1141	-0.1233	-0.0238	0.0188	0.1670	0.0317	0.1139	-0.0271	0.0171	0.0226	-0.0746
r_8	-0.0796	-0.0100	-0.0155	-0.0043	-0.2028	-0.0390	-0.1587	-0.1298	0.1811	0.1240	0.0970	0.0850	-0.0439
r_9	-0.0269	-0.0567	0.1768	0.0458	-0.0848	0.1915	-0.1043	0.0258	-0.1033	-0.0015	-0.0010	0.0384	-0.0760
r_{10}	-0.1097	-0.0066	0.0200	0.2177	0.1029	0.1051	0.0908	0.0096	0.0663	-0.1174	0.1105	0.0725	-0.0671
r_{11}	0.1097	0.0058	0.1111	0.0896	0.0512	0.1125	0.0584	0.0254	0.1801	-0.1772	0.1609	0.1999	0.1999
r_{12}	0.1336	0.1659	0.2714	0.2490	0.0508	0.2121	0.1279	0.1158	0.1679	0.0182	0.0818	0.0412	0.0913
r_{13}	0.0280	0.1313	0.0500	0.1506	0.0325	0.1338	0.0797	0.2944	0.0902	0.0033	0.0037	0.0238	0.0162
r_{14}	-0.2083	-0.0363	-0.0710	-0.0117	-0.2270	-0.2470	-0.1616	-0.1556	-0.0071	-0.1239	-0.1255	-0.1113	-0.1343
r_{15}	0.0379	0.0754	0.0331	0.0258	0.0420	0.1809	0.1759	0.0787	0.0152	0.1162	0.0154	0.0367	-0.1344

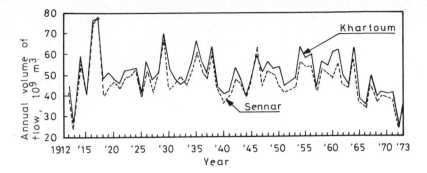

Fig. 8.40.   The graphical plot of the annual series at Sennar and Khartoum on the Blue Nile in the period 1912-1973

autoregressive model for which the first serial correlation coefficient is a parameter. The residuals left from fitting this model to each of the dependent series when tested appear not to be serially correlated at the chosen level of confidence. The only exception can be found at lag 6 for the month of June, which is significantly different from zero. The serial correlation coefficients of the residual series are given in Table 8.19.

TABLE 8.19   Serial correlation coefficients of the residuals left from fitting autoregressive models to some of the discharge series at Khartoum

Month & Year	Serial correlation coefficient														
	r_1	r_2	r_3	r_4	r_5	r_6	r_7	r_8	r_9	r_{10}	r_{11}	r_{12}	r_{13}	r_{14}	r_{15}
Jan.	.057	.095	.071	.081	.162	.058	.040	.116	.091	.120	.091	.064	.061	.194	.033
Feb.	.107	.030	.037	.238	.158	.031	.009	.051	.058	.003	.042	.116	.077	.063	.003
Mar.	.053	.031	.066	.278	.202	.089	.046	.004	.191	.046	.123	.251	.012	.080	.022
Apr.	.014	.045	.077	.183	.087	.108	.090	.029	.073	.209	.054	.183	.053	.082	.088
June	.020	.089	.171	.113	.084	.396	.067	.063	.199	.137	.087	.228	.110	.133	.062
Nov.	.016	.155	.229	.080	.095	.067	.005	.106	.018	.099	.123	.035	.016	.091	.016
Dec.	.042	.154	.135	.091	.080	.066	.033	.062	.003	.115	.222	.117	.002	.080	.015
Year	.033	.087	.062	.079	.013	.110	.058	.031	.031	.017	.183	.133	.003	.098	.049

The monthly series of the low-flow season are distributed like a lognormal, whereas the monthly series of the high-flow season are distributed more or less like a Pearson Type III distribution.

The series of annual discharges is nearly normally distributed (see Fig. 8.41.).

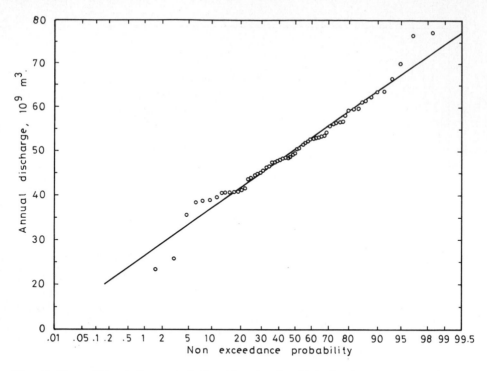

Fig. 8.41. Fit of the normal function to the distribution of the annual discharges at Khartoum on the Blue Nile in the period 1912-1973

For all discharge series at Khartoum, the 100 and 200-year discharges computed from the distribution function best fitting each series are as follows:

Discharge 10^6 m^3	Jan.	Feb.	Mar.	Apr.	May	June	July	Aug.	Sep.	Oct.	Nov.	Dec.	Year
100-yr	1537	946	912	1016	1255	2995	9835	21268	21482	15720	5797	2686	74424
200-yr	1670	1041	998	1133	1396	3367	10388	21693	22658	16834	6328	2895	77194

8.9   THE MAIN NILE BELOW KHARTOUM TO JUST ABOVE THE JUNCTION OF THE ATBARA

8.9.1   Tamaniat discharges

Regular discharge measurement of the Main Nile began at Tamaniat in 1912.
This station is situated 41 km below the confluence of the Blue and White Niles
at Khartoum. Since 1934 the flood measurements have been taken at Shambat, which
is 6 km further below. The gauge-discharge measurements for the period 1912-1973
are plotted in Fig. 29, Appendix E. The monthly and annual discharges for the
same period are given in Table 12, Appendix D. These discharge data have been
statistically analyzed and the results presented in Table 8.20.

TABLE 8.20   The basic statistical descriptors and the serial correlation coefficients of the monthly and annual discharges of the Main Nile at Tamaniat for the period 1912-1973

Item	Jan.	Feb.	Mar.	Apr.	May	June	July	Aug.	Sep.	Oct.	Nov.	Dec.	Year
Basic statistical descriptor													
$\bar{X}$, 10^6 m³	3219	2387	2397	2352	2328	2889	6645	16585	16339	10906	5406	4031	75575
s, 10^6 m³	832	639	650	903	893	724	1586	3103	3379	3136	1366	822	11341
C^v	0.2585	0.2678	0.2710	0.3840	0.3838	0.2506	0.2388	0.1871	0.2068	0.2075	0.2527	0.2040	0.1506
C^s	0.6503	1.3988	0.3981	0.6738	0.9868	1.2860	0.5751	0.8162	0.0114	0.8752	1.2665	0.5521	0.4132
C_k	3.8837	6.1783	3.2296	2.7410	3.2408	5.1900	3.8961	4.9800	3.6400	4.8392	5.7600	4.1110	4.3247
Serial correlation coefficient													
r_1	0.2789	0.3731	0.6303	0.7757	0.6064	0.3581	0.2470	-0.0886	0.2839	0.1636	0.1431	0.3896	0.2998
r_2	0.1997	0.1855	0.4002	0.6804	0.4715	0.1138	0.0055	0.1776	0.1672	0.0327	0.1943	0.2659	0.0869
r_3	0.1381	0.2437	0.4779	0.6463	0.4051	-0.0609	0.1191	0.0279	0.1653	0.1862	0.1992	0.1655	0.0043
r_4	-0.0930	-0.0707	0.3513	0.5397	0.2572	0.2451	-0.1707	0.0265	0.1067	0.1114	0.0858	0.0791	-0.2272
r_5	0.0624	-0.0560	0.2665	0.5138	0.3301	-0.1449	-0.1613	-0.0229	0.0629	0.0526	0.0175	-0.0690	-0.1608
r_6	0.0287	0.1120	0.3933	0.5044	0.2965	-0.0493	-0.0942	0.0898	-0.0372	0.1676	0.0775	0.0987	-0.0685
r_7	-0.0683	0.0351	0.4059	0.5175	0.3474	0.0265	0.0651	0.0081	0.0642	0.0410	0.0417	0.1471	-0.1356
r_8	0.0225	0.0119	0.3795	0.4840	0.3770	0.0586	0.1061	0.0645	0.0469	-0.1441	0.0627	-0.0874	-0.1054
r_9	0.0129	0.0617	0.4047	0.4621	0.2576	0.1402	-0.0450	0.0182	0.0724	-0.0074	-0.0030	-0.0576	-0.0829
r_{10}	-0.0401	-0.0181	0.3090	0.3999	0.2468	0.0570	0.0629	-0.0757	0.0561	0.0900	0.1366	0.0174	-0.0312
r_{11}	0.1861	-0.0932	0.2545	0.3479	0.1141	-0.0340	-0.0068	-0.0801	0.1577	0.1566	-0.0918	0.1027	0.1376
r_{12}	0.0024	-0.0178	0.2720	0.3387	0.0379	-0.0437	0.0688	0.0810	0.1733	0.0419	0.0260	0.0981	0.0995
r_{13}	-0.0498	0.0820	0.1691	0.2713	0.0158	0.2559	0.1081	0.2072	0.0084	0.0822	0.0523	0.0638	0.0982
r_{14}	-0.0432	0.0626	0.1381	0.2433	0.0130	0.2786	0.1969	0.1799	0.0222	0.0755	0.0822	0.0462	0.1396
r_{15}	0.1759	0.1269	0.2128	0.2760·	0.0820	0.1287	0.2367	0.1238	0.0803	0.0353	0.1304	0.0631	0.0412

Month of the year

The discharge of the Main Nile at Tamaniat is equal to the sum of the discharges of the Blue Nile and the White Nile, both at Khartoum. Since the latter has been estimated from the Malakal discharges, one can say that the discharge at Tamaniat is equal to the discharge of the Blue Nile at Khartoum plus that of the White Nile at Malakal minus the losses from Malakal to Khartoum. Because of the small distance between Khartoum and Tamaniat, 40 km, the conveyance loss between these two stations can be ignored without any marked effect on the final result. Since the outflow of the Blue Nile at mouth is nearly twice as much as the flow of the White Nile at Khartoum, it is then not strange that the discharge series at Tamaniat, especially the annual series, are more affected by the Blue Nile discharges than by the White Nile ones. This can be made evident by comparing the plot in Fig. 8.42. with those in Figs. 8.40. and 8.35.

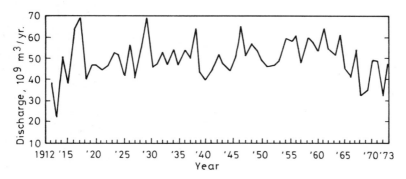

Fig. 8.42. Graphical plot of the annual discharge series at Tamaniat on the Main Nile in the period 1912-1973

In the above discussion, no account has been taken for the effect of the time lag between one station and the other, the effect of the backwater produced by the junction of the two Niles and of the storage in reservoirs, the effect of reservoir regulation and the storage losses and inaccuracy in measuring the water stage and discharge.

The overall effect of these items can be seen from the results of the regression and correlation analysis of the difference in the flow between Tamaniat on the Main Nile and Khartoum on the Blue Nile, Y, and the flow at Malakal on the White Nile, X.

The analysis was performed on the monthly and the annual series. The X's are readily available in Table 7, Appendix D, whereas the Y's are the differences between Tables 12 and 11, Appendix D.

Let

$$Y = a + bX \tag{8.7}$$

where a and b are the regression constant and coefficient, respectively. The correlation coefficient r_{XY} can be computed from the formula

$$r_{XY} = \frac{n \ \Sigma XY - \Sigma X \Sigma Y}{\{ \ n \ \Sigma X^2 - (\Sigma X)^2 \quad n \ \Sigma Y^2 - (\Sigma Y)^2 \ \}^{\frac{1}{2}}} \tag{8.8}$$

The values of a, b, and r_{XY} for the 13 series have been calculated and put in Table 8.21. The coefficient of linear correlation is at a fairly high value in January and February then fluctuates between moderate to fairly stong till July. The high flood discharges of the Blue Nile are probably the reason behind the almost zero to poor correlation during August, and September and October, respectively. From November onwards till the end of the year, the correlation between X and Y improves considerably.

TABLE 8.21   Regression of the flow difference between Tamaniat and Khartoum on the flow at Malakal

Month & Year	Regression Equation	a, 10^6 m^3	b	r_{XY}
January	Y = 969.375 + 0.618756 X	969.375	0.618756	0.84493
February	Y = 756.845 + 0.675871 X	756.845	0.675871	0.84838
March	Y = 1214.280 + 0.454712 X	1214.280	0.454712	0.57058
April	Y = - 275.977 + 1.512053 X	275.977	1.512053	0.73911
May	Y = -1551.564 + 2.061950 X	1551.564	2.061950	0.84765
June	Y = -1190.822 + 1.459891 X	1190.822	1.459891	0.79768
July	Y = -1419.763 + 1.075955 X	1419.763	1.75955	0.60137
August	Y = 1449.095 - 0.201233 X	1449.095	0.201233	0.06804
September	Y = - 257.030 + 0.855047 X	257.030	0.855047	0.33322
October	Y = 1658.424 + 0.467589 X	1658.424	0.467589	0.38881
November	Y = 903.017 + 0.608749 X	903.017	0.608749	0.71579
December	Y = 1113.904 + 0.524161 X	1113.904	0.524161	0.82904
Year	Y = 792.214 + 0.834794 X	792.214	0.834794	0.91140

X and Y are expressed in million m^3

The strongest correlation between X and Y can be found, as can be expected, in the annual series. The regression line for these data is shown in Fig. 8.43. The estimate of the mean difference of flow between Tamaniat and Khartoum is 25367 $\pm$ 4310 or from 21057 to 29677 million m^3/yr (95% confidence level). As our estimate of the annual flow at Khartoum on the White Nile just above the confluence of the Blue Nile is at 27 to 28 mlrd m^3 (see section 8.7.3), one is not

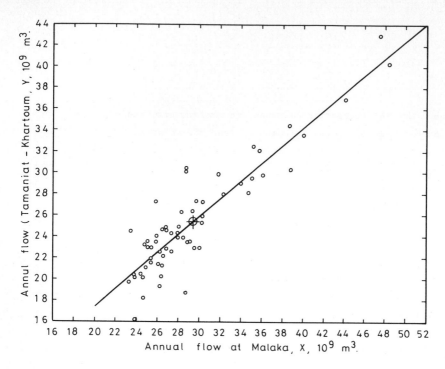

Fig. 8.43. Simple linear regression of the difference in annual flow between Tamaniat and Khartoum, Y, and the annual flow at Malakal, X, for the period 1912-1973

justified in rejecting the hypothesis that the two estimates are not signifi-cantly different one from the other at the given level of confidence.

The results presented in Table 8.20 indicate that all the discharge series at Tamaniat, except those of August, October and November, are serially correlated. The dependent component in the series with correlation can be described by an autoregressive model of the first or second order. The series of the residuals have been tested and found not serially correlated. This can be seen from the results included in Table 8.22.

The probability distribution functions that have been found as good fit to the discharges at Tamaniat are the Pearson Type III and the 2-parameter lognormal. Fig. 8.44. shows the fit of the latter to the annual volumes of flow during 1912-1973.

The 100- and 200-yr discharges obtained from the theoretical functions for all the 12 months and the year are as follows:

Discharge 10^6 m^3	Jan.	Feb.	Mar.	Apr.	May	June	July	Aug.	Sep.	Oct.	Nov.	Dec.	Year
100-yr	5540	4477	4097	4902	5027	5213	10983	21953	24212	20110	9757	6267	105402
200-yr	5863	4833	4314	5262	5443	5600	11577	22263	25043	21506	10474	6572	109144

TABLE 8.22   Serial correlation coefficients of the residuals left from fitting autoregressive models to some of the discharge series at Tamaniat

Month & Year	r_1	r_2	r_3	r_4	r_5	r_6	r_7	r_8	r_9	r_{10}	r_{11}	r_{12}	r_{13}	r_{14}	r_{15}
Jan.	.011	.084	.120	$\overline{.218}$	.103	.016	$\overline{.133}$	.044	$\overline{.076}$	$\overline{.036}$	.163	.026	$\overline{.012}$	$\overline{.059}$	.137
Feb.	.013	$\overline{.028}$	.213	$\overline{.153}$	$\overline{.060}$	.099	$\overline{.030}$	$\overline{.003}$	.028	$\overline{.067}$	.129	.056	$\overline{.062}$	$\overline{.062}$	.141
Mar.	.014	$\overline{.254}$	.294	.025	$\overline{.111}$	.163	.098	.064	.093	$\overline{.038}$	.015	.117	$\overline{.052}$	.078	.143
Apr.	$\overline{.168}$	$\overline{.055}$	$\overline{.204}$	$\overline{.098}$	.055	.121	.036	.014	.112	$\overline{.011}$	$\overline{.031}$	.126	$\overline{.057}$	.061	.138
May	$\overline{.089}$	.027	.185	$\overline{.160}$	.201	.087	$\overline{.052}$	.173	.000	.156	.010	.008	$\overline{.039}$	$\overline{.028}$	.062
June	.003	$\overline{.058}$	.121	$\overline{.287}$	$\overline{.096}$	.181	$\overline{.085}$	.041	.128	.015	.069	.044	$\overline{.171}$	$\overline{.151}$	$\overline{.056}$
July	.034	$\overline{.004}$	$\overline{.094}$	$\overline{.070}$	$\overline{.128}$	.109	.030	$\overline{.103}$	$\overline{.058}$	.036	$\overline{.057}$	.108	$\overline{.094}$	.098	$\overline{.083}$
Sep.	$\overline{.060}$	.098	.150	$\overline{.121}$	.163	$\overline{.073}$	.113	$\overline{.065}$	$\overline{.029}$	.006	$\overline{.176}$	.219	.098	.021	.025
Dec.	$\overline{.055}$	.137	.129	$\overline{.123}$	.004	$\overline{.026}$	$\overline{.114}$	$\overline{.037}$	.038	.047	$\overline{.167}$	.129	.027	$\overline{.074}$	.049
Year	$\overline{.037}$	.061	.032	$\overline{.151}$	$\overline{.047}$	.039	$\overline{.106}$	$\overline{.040}$	.050	.006	$\overline{.156}$	.110	.068	$\overline{.132}$	$\overline{.005}$

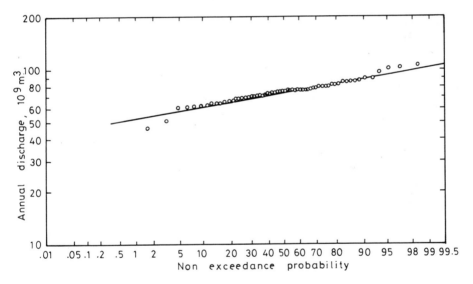

Fig. 8.44.   Fit of the 2-parameter lognormal function to the distribution of the annual discharges of the Main Nile at Tamaniat in the period 1912-1973

8.9.2   Hassanab discharges

The Main Nile flows a distance of 277 km below Tamaniat before it reaches Hassanab station, which is located about 5 km above the junction of the Atbara with the Nile. In this reach of the Main Nile there is no gain at all. The annual rainfall decreases from about 160 mm/yr at Khartoum to about 65 mm/yr at Atbara and Zeidab. The potential evaporation in all this reach has an average of 8 mm/day or 2920 mm/yr. The net loss can be taken as 2800 mm/yr. Assuming the average width of the area from which the losses take place as 1.5 km, the annual loss should then be in the order of 277 x 1.5 x 2.8 x 10^6 or about 1.16 mlrd m^3 a year. This is about the same as 1.2 mlrd m^3/yr given for the period 1912-52 in Vol. IX of the Nile Basin (Hurst, H.E., Black, R.P., and Simaika, Y.M., 1959).

The gauge-discharge measurements for the period 1924-1927 at Hassanab are shown graphically in Fig. 30, Appendix E. The monthly and annual discharges in the period 1912-1973 are given in Table 13, Appendix D, and the results of their statistical analysis are in Table 8.23.

From this table and from Table 8.20, one can compute the monthly and yearly volumes representing the change in the contents of the river trough plus the losses or gains averaged over the same period, i.e. 1912-1973. The results of computation are given below together with the figures averaged for the period 1912-1952, for comparison.

Change of contents plus gains or losses, 10^6 m^3, averages for

Period	Jan.	Feb.	Mar.	Apr.	May	June	July	Aug.	Sep.	Oct.	Nov.	Dec.	Year
1912-52	-50*	0	40	70	150	280	650	600	100	-400	-260	-30	1200
1912-73	-51	-23	32	69	125	295	698	603	93	-331	-192	-58	1299

These results show clearly that the sum of gains is smaller than the sum of losses. The net yearly loss is about 1.2 to 1.3 mlrd m^3/yr. The gain between Tamaniat and Hassanab takes place in the period from October to February, whereas the loss takes place in the remaining months.

The change in the flow from Tamaniat to Hassanab caused by the gains, losses and change in river trough contents has affected the structure of some of the discharge series. Compared to Tamaniat, the July, September and the yearly series of Hassanab are serially uncorrelated. The low-flow series, i.e. January thru' June and December remain, as do those of Tamaniat, serially correlated. Like the series of the upstream stations, the dependence in the correlated discharge series of Hassanab can be described by an autoregressive model. The residual series have been tested and found not to be correlated (confidence level = 95%). Table 8.24 gives the values of the serial coefficients of the residual series. The majority of the discharge series at Hassanab can be well fitted by the Pearson Type III function and the rest by the normal and lognormal functions. The 100-yr and the 200-yr discharges obtained from the theoretical distributions of all the series are as follows:

Discharge 10^6 m^3	Jan.	Feb.	Mar.	Apr.	May	June	July	Aug.	Sep.	Oct.	Nov.	Dec.	Year
100-yr	5320	4386	4024	4511	4587	4351	10132	20522	22890	18523	9400	5913	98210
200-yr	5581	4700	4236	4799	4916	4606	10692	20803	23498	19417	9947	6121	100931

* all the minus signs here mean gain

TABLE 8.23 The basic statistical descriptors and the serial correlation coefficients of the monthly and annual discharges of the Main Nile at Hassanab for the period 1912-1973

Item	Jan.	Feb.	Mar.	Apr.	May	June	July	Aug.	Sep.	Oct.	Nov.	Dec.	Year
Basic statistical descriptor						Month of the year							
$\bar{X}$, 10^6 m^3	3270	2410	2365	2283	2203	2594	5947	15982	16432	11237	5598	4089	74276
s, 10^6 m^3	784	638	632	847	840	621	1533	2602	3039	2864	1320	757	9931
C_v	0.2399	0.2648	0.2674	0.3709	0.3814	0.2393	0.2578	0.1628	0.1849	0.2548	0.2358	0.1851	0.1337
C_s	0.3969	1.1142	0.4132	0.4204	0.7249	0.7058	0.5622	0.7796	0.2655	0.2945	0.7769	0.1384	0.1379
C_k	3.5739	5.8476	4.0487	2.4931	2.6125	3.1923	4.1176	4.0537	3.7169	3.5484	4.4776	3.6457	4.1719
Serial correlation coefficient													
r_1	0.2873	0.3053	0.5758	0.7643	0.6911	0.3636	0.1626	$\overline{0}.0772$	0.1237	0.1608	0.1173	0.2747	0.2194
r_2	0.0539	0.1360	0.3374	0.6791	0.5679	0.3156	$\overline{0}.0495$	0.0206	0.1007	0.0179	0.0632	0.1578	$\overline{0}.0144$
r_3	$\overline{0}.1465$	0.2461	0.4184	0.6500	0.4898	0.0960	$\overline{0}.1053$	$\overline{0}.0752$	0.1229	0.2022	0.1232	0.0630	$\overline{0}.0220$
r_4	$\overline{0}.1117$	$\overline{0}.0916$	0.3218	0.5617	0.3491	0.2080	$\overline{0}.1198$	$\overline{0}.0356$	$\overline{0}.1156$	$\overline{0}.1083$	$\overline{0}.0942$	$\overline{0}.0960$	$\overline{0}.2246$
r_5	$\overline{0}.0354$	0.1068	0.3637	0.6114	0.4237	0.0035	0.1018	0.0385	0.0694	0.0384	0.1448	0.0378	0.1460
r_6	$\overline{0}.0152$	0.2651	0.4761	0.6035	0.4329	0.0428	0.1686	0.1348	0.0984	0.1996	0.0012	0.1935	0.0098
r_7	$\overline{0}.0901$	0.0670	0.4200	0.5762	0.4501	0.2141	0.1506	0.0465	0.0332	0.0365	0.0316	$\overline{0}.1105$	0.1094
r_8	$\overline{0}.0367$	$\overline{0}.2068$	0.4044	0.5329	0.4135	0.2329	0.1119	$\overline{0}.0553$	$\overline{0}.1550$	$\overline{0}.1493$	$\overline{0}.1315$	$\overline{0}.0034$	$\overline{0}.1353$
r_9	0.0860	0.2060	0.3867	0.4821	0.3026	0.2518	$\overline{0}.1292$	$\overline{0}.0467$	$\overline{0}.1062$	$\overline{0}.0253$	$\overline{0}.0085$	0.0088	$\overline{0}.1150$
r_{10}	$\overline{0}.0500$	$\overline{0}.0066$	0.2597	0.4157	0.2592	0.0612	0.0737	$\overline{0}.0145$	$\overline{0}.0396$	$\overline{0}.0730$	$\overline{0}.1245$	0.0280	$\overline{0}.0711$
r_{11}	$\overline{0}.1967$	$\overline{0}.0689$	0.2527	0.3772	0.1445	$\overline{0}.0823$	0.0151	$\overline{0}.0609$	0.1459	0.2051	0.0811	$\overline{0}.1206$	$\overline{0}.0961$
r_{12}	0.0888	0.0308	0.2850	0.3661	0.0960	0.0887	0.1835	0.1280	0.2487	0.0987	0.0329	0.0955	0.1633
r_{13}	$\overline{0}.0625$	$\overline{0}.0897$	0.1510	0.3019	0.0921	$\overline{0}.1614$	$\overline{0}.0638$	$\overline{0}.0750$	0.0161	0.1107	0.1170	0.0670	0.0008
r_{14}	$\overline{0}.0731$	$\overline{0}.0823$	0.1189	0.2869	0.1035	$\overline{0}.1369$	$\overline{0}.1422$	0.0216	0.1102	$\overline{0}.0288$	0.0214	$\overline{0}.0483$	0.0228
r_{15}	0.0971	0.1244	0.1701	0.3064	0.1404	0.1285	$\overline{0}.2713$	0.0099	0.1608	0.0324	0.1623	0.0029	0.0048

TABLE 8.24   Serial correlation coefficients of the residuals left from fitting
autoregressive models to some of the discharge series at Hassanab

Month	Serial correlation coefficient														
	r_1	r_2	r_3	r_4	r_5	r_6	r_7	r_8	r_9	r_{10}	r_{11}	r_{12}	r_{13}	r_{14}	r_{15}
Jan.	.052	-.035	.115	-.151	-.018	-.039	-.109	-.019	.006	-.068	-.199	.103	-.034	-.059	.076
Feb.	.020	-.060	.212	-.161	-.025	.128	.044	.015	.002	-.075	-.116	.080	.045	-.079	.140
Mar.	-.012	-.171	.240	-.032	.033	.237	.071	.086	.183	-.020	.029	.278	.035	-.029	.112
Apr.	-.171	.028	.169	-.165	.158	.130	.052	.057	.060	-.011	.001	.102	-.047	-.020	.127
May	-.125	.022	.165	-.215	.152	.096	.105	.138	-.033	.107	-.042	-.026	-.010	-.004	.048
June	.088	.191	.073	<u>-.292</u>	.071	-.020	.131	.095	.154	-.018	-.070	-.035	-.087	.107	.128
Dec.	.037	.092	.052	-.159	.035	-.175	-.073	-.007	-.024	.041	-.178	.102	.026	-.053	.027

8.10   THE ATBARA RIVER

The total surface area of the drainage basin of the Atbara is about 100 000 km^2, of which about 68000 km^2 comprise the basin of the Setit, which is the major tributary of the Atbara. The rest of the area belongs to the lower Atbara below the junction of the Setit. It is the rainfall on the catchment area of the Setit that is responsible for the major part of the flow in Atbara. The annual mean rainfall on the catchments of the Setit and Lower Atbara can be taken as 800 mm/yr and 300 mm/yr respectively. The annual run-off coefficients for these two catchments in their order can be taken as 0.20 and 0.10 respectively. These figures produce an annual run-off of 68 x 10^9 x 0.8 x 0.2 + 32 x 10^9 x 0.3 x 0.1 = 11.84 x 10^9 m^3/yr.

The gauge-discharge measurements of the Atbara at mouth are shown graphically in Fig. 31, Appendix E. The monthly and annual discharges in the period 1912-73 are given in Table 14, Appendix D, and the results of the statistical analysis of these discharge data are included in Table 8.25. The mean flow at Atbara, near the mouth of the river, for the period considered was 11.88 mlrd m^3/yr. This volume of flow distributes itself over the months, in a normal year, as shown in Fig. 8.45. The first five months of the year are practically dry. The effective base width of the hydrograph is from June up to and including December. The flood season covers August and September and the remaining months represent the low-flow season. The Atbara, in this respect, resembles the Blue Nile; both are torrential streams. This feature leads us to review the results in Table 8.25 carefully. The period from January up to and including May is practically dry. There are a few years in which the river can bring some little water in this period. This does not, however, improve the long-term mean sensibly.

Instead it produces a considerable scatter leading to a high coefficient of variation and rather meaningless coefficients of skew and kurtosis.

TABLE 8.25 The basic statistical descriptors and the serial correlation coefficients of the monthly and annual discharges of the River Atbara at Atbara, near mouth, for the period 1912-1973

Item	Jan.	Feb.	Mar.	Apr.	May	June	July	Aug.	Sep.	Oct.	Nov.	Dec.	Year
Basic statistical descriptor													
$\bar{X}$, 10^6 m^3	20.3	7.2	1.1	3.6	7.7	73.2	1616	5582	3496	812	176	57.6	11885
s, 10^6 m^3	31.1	19.7	4.1	18.8	31.8	101	812	1842	1421	415	112	47.2	3913
C^v	1.5319	2.7352	3.6348	5.2828	4.1288	1.3748	0.5026	0.3299	0.4065	0.5115	0.6332	0.8186	0.3292
C^s	2.2993	5.0538	5.7714	5.7813	4.6593	1.5918	1.9888	1.4805	0.4947	0.8970	1.1367	0.6843	1.0933
C_k	10.1766	32.8253	40.6174	36.8470	24.6862	4.9233	9.7805	7.1945	3.6017	4.6077	5.7258	2.9403	5.6846
Serial correlation coefficient													
r_1	0.3363	0.1957	0.0886	0.4284	0.3328	-0.2172	-0.2936	0.2240	0.2212	0.4174	0.2469	0.4918	0.0342
r_2	0.3139	0.1248	0.0784	0.0388	0.0122	0.1892	0.0429	0.0149	0.0159	0.2319	0.3424	0.3979	0.0400
r_3	0.3972	0.1756	0.1637	0.0401	0.0358	0.1145	0.0944	0.1476	0.0318	0.1443	0.2688	0.3274	-0.0062
r_4	0.1523	0.0069	-0.0072	0.0414	0.0186	0.1876	0.0504	-0.0483	-0.0637	0.0701	0.1054	0.2067	-0.0427
r_5	0.3116	0.1404	0.1384	0.0428	0.0701	0.0185	0.0793	-0.0479	0.2231	0.1295	0.2966	0.2861	0.0956
r_6	0.1995	0.0640	0.0435	0.0442	0.0700	0.1052	0.0079	0.1144	0.0124	0.1486	0.0774	0.1414	0.0772
r_7	0.1824	0.0740	0.1557	0.0311	0.0477	0.1247	0.0311	0.1125	0.0920	0.0979	0.0165	0.1690	0.0590
r_8	0.3030	0.3010	0.0752	0.0084	0.0304	0.0462	0.2671	0.2373	0.0740	0.0196	0.1580	0.1940	0.1909
r_9	-0.0171	0.0327	0.0816	0.1007	0.0125	0.0659	0.2832	0.1087	0.1521	0.0256	0.1589	0.0322	0.1857
r_{10}	0.0152	0.0074	0.0621	0.0593	0.0130	0.1261	0.1466	0.0488	0.1084	0.1576	0.0237	0.1379	0.0571
r_{11}	0.1243	0.0759	0.0905	0.0111	0.0338	0.0823	0.0181	0.1270	0.1111	0.1224	0.2330	0.2048	0.1697
r_{12}	0.0051	0.0728	0.0892	0.0120	0.0134	0.2258	0.1616	0.0459	0.0197	0.1538	0.2523	0.1300	0.0695
r_{13}	0.0919	0.0434	0.0812	0.0130	0.0376	0.0834	0.0231	0.0521	0.0700	0.1043	0.1108	0.0515	0.0464
r_{14}	0.0372	0.0470	0.0916	0.0140	0.0396	0.0059	0.0985	0.1179	0.1699	0.3209	0.2930	0.2120	0.1643
r_{15}	0.0204	0.0523	0.0843	0.0151	0.0417	0.1900	0.0406	0.0508	0.1220	0.3084	0.1448	0.1217	0.0702

Month of the year

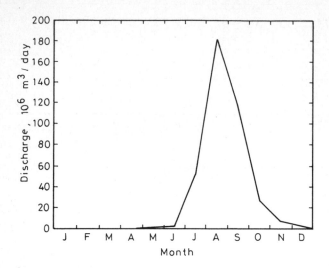

Fig. 8.45.   The average hydrograph of the Atbara for the period 1912-1973 at Atbara, near mouth

We shall, therefore, consider the monthly series from June to December only and the annual series. Of these series those belonging to June, October, November and December are the ones whose individuals appear to be serially correlated. First-order autoregressive models have been fitted to the historic data of these series and the residuals examined statistically. The serial correlations of the residual series are given in Table 8.26. They show that they are not significantly different from zero with 95% confidence, except for November at lags 2 and 4, and for December at lag 8.

TABLE 8.26   Serial correlation coefficients of the residuals left from fitting autoregressive models to some of the discharge series at Atbara

Month	Serial correlation coefficient														
	r_1	r_2	r_3	r_4	r_5	r_6	r_7	r_8	r_9	r_{10}	r_{11}	r_{12}	r_{13}	r_{14}	r_{15}
July	.136	.058	.023	.217	.121	‾.031	‾.097	.165	‾.023	.162	‾.201	.049	‾.170	‾.016	‾.087
Oct.	‾.188	‾.003	‾.097	‾.019	.097	.154	‾.076	.010	‾.132	.078	.058	.069	‾.021	‾.134	‾.228
Nov.	.067	.267	.189	.000	.274	.021	‾.031	.204	‾.158	.060	‾.150	.154	.005	.193	.039
Dec.	.121	.166	.232	‾.153	.250	.010	.042	.283	‾.115	.029	‾.049	‾.161	.091	‾.124	‾.063

The probability distribution functions that serve as good fit to the discharge data of the Atbara are the 2-parameter lognormal and the Pearson III functions. The fit of the former to the annual flow volumes in the period 1912-1973 is shown in Fig. 8.46.

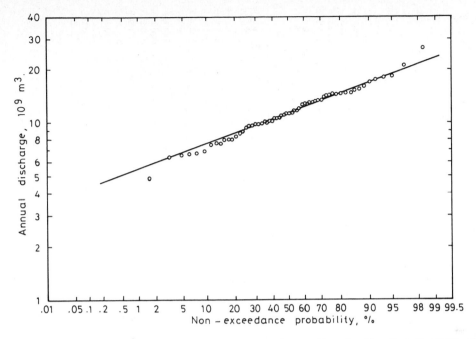

Fig. 8.46.   Fit of the 2-parameter lognormal function to the distribution of the annual discharges of the Atbara near mouth for the period 1912-1973

The computed 100-yr and 200-yr discharges for the months from June up to and including December and for the year are as follows:

Discharge 10^6 m^3	June	July	Aug.	Sep.	Oct.	Nov.	Dec.	Year
100-yr	415	4543	11697	7311	2039	524	191	23961
200-yr	476	5106	12747	7817	2223	571	210	25874

8.11   THE MAIN NILE BELOW THE MOUTH OF THE ATBARA

8.11.1   The discharge of the Main Nile at Dongola

The monthly and annual discharges of the Main Nile at Dongola for the period 1912-1973 is given in Table 15, Appendix D. The Main Nile flows a distance of about 760 km below the confluence of the Atbara before it reaches Dongola. A picture of the average width of this reach of the river during the high-flow season can be seen from Fig. 2.23. The overall average width between high and low flow seasons can be taken as 400 m. This river reach runs in a real arid zone with about 8 mm/day free water evaporation. This figure leads us to the conclusion that the annual flow reaching Dongola is about 0.88 mlrd m^3/yr less than that flowing just below the Atbara junction. From Table 8.27, which contains the values of the basic statistical descriptors and the serial correlation coefficients of the monthly and annual discharges at Mongalla, one can find the

TABLE 8.27   The basic statistical descriptors and the serial correlation coefficients of the monthly and annual discharges of the Main Nile at Dongola, for the period 1912-1973

Item	Jan.	Feb.	Mar.	Apr.	May	June	July	Aug.	Sep.	Oct.	Nov.	Dec.	Year
Basic statistical descriptor						Month of the year							
$\bar{X}$, 10^6 m³	3539	2501	2281	2207	2108	2153	5188	19071	21321	13904	6857	4473	85570
s, 10^6 m³	714	692	636	789	859	765	1476	3948	4230	3558	1858	890	12874
C^v	0.2018	0.2767	0.2789	0.3575	0.4076	0.3551	0.2845	0.2070	0.1984	0.2559	0.2710	0.1989	0.1505
C^s	0.5865	1.2652	1.5769	0.6144	0.8619	1.0336	0.5971	0.3176	0.1462	0.2891	0.6995	0.4226	0.2381
C^k	4.4003	5.6025	7.8229	3.1226	2.8286	3.8752	3.7711	3.3608	4.1119	3.7548	3.6472	3.9191	4.0081
Serial correlation coefficient													
r_1	0.2859	0.3166	0.4137	0.7311	0.6982	0.4567	0.2706	-0.1964	0.1525	0.3266	0.1025	0.2228	0.1863
r_2	0.1550	0.0823	0.0887	0.5691	0.5610	0.3018	0.0740	0.0293	0.0213	0.1538	0.1743	0.2480	0.0310
r_3	0.0820	0.1364	0.1344	0.5426	0.5113	0.2569	-0.1822	0.0984	0.0938	0.2262	0.2654	0.1867	0.0157
r_4	-0.2000	-0.1932	-0.0402	0.4567	0.4107	-0.0792	-0.1028	-0.1098	-0.0439	-0.0278	-0.0854	-0.0402	-0.1500
r_5	-0.0521	0.1275	0.0652	0.4300	0.4320	0.0762	0.0879	-0.1103	0.1033	0.0404	0.1326	0.0486	-0.1593
r_6	0.0512	0.0608	0.0605	0.4524	0.4090	0.1541	0.0865	0.1118	0.0170	0.2545	0.0764	0.0483	0.0292
r_7	-0.1095	0.0058	0.1217	0.4959	0.4568	0.1265	0.0348	0.0428	0.0520	0.0602	0.0019	0.0882	0.0967
r_8	-0.0484	0.0163	0.0648	0.4898	0.4570	0.2379	0.1066	0.1401	0.0252	0.0579	0.0647	0.0268	-0.0326
r_9	0.0263	0.1233	0.1487	0.4598	0.3731	0.1856	-0.0723	0.0666	0.0860	0.0381	0.0361	0.0209	0.0472
r_{10}	-0.0797	-0.0505	0.0024	0.3763	0.3052	0.1553	0.1142	0.0413	0.0491	0.0883	0.1278	0.1365	0.0405
r_{11}	-0.1556	-0.2203	-0.0057	0.3224	0.2000	0.0038	-0.0309	0.1870	0.1612	0.1322	0.1280	0.0648	0.1369
r_{12}	0.0100	0.0152	0.1082	0.3314	0.1783	0.0457	0.0782	0.1182	0.1293	0.0709	0.0057	0.1189	0.1317
r_{13}	0.1222	0.1160	-0.0107	0.2601	0.1602	0.1912	0.1888	0.1641	0.1276	0.0336	0.0885	0.1409	0.0163
r_{14}	-0.0126	0.1156	-0.0321	0.2275	0.1354	-0.2375	-0.2386	0.1468	0.1399	-0.1038	0.1160	-0.1046	0.2023
r_{15}	0.1391	0.1847	0.1197	0.2530	0.1889	0.0054	-0.3107	-0.0630	0.0365	-0.0221	0.1454	0.1293	-0.0598

annual flow volume averaged over the period 1912-1973 at 85.57 mlrd m^3. This
figure is 0.58 mlrd m^3/yr less than the sum of the mean flows at Hassanab and
Atbara for the same period. Fig. 8.47. shows a graphical plot of the annual flow
volumes at the mouth of the Atbara, at Hassanab and at Dongola. Because of the
largeness of the flow at Hassanab in proportion to the flow at the Atbara mouth,
almost 6 to 1, it is quite understandable that the plot of the flow series at
Dongola is very much parallel to that of Hassanab and not of the Atbara.

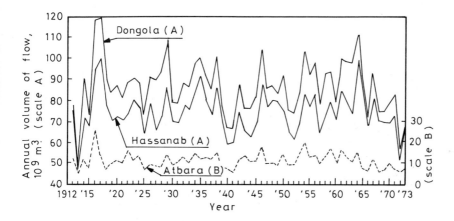

Fig. 8.47.   Graphical plot of the annual flow volume series at Atbara, Hassanab
and Dongola in the period 1912-1973

The scatter diagram of the sum of the flows at Hassanab and at the mouth of
the Atbara, Y, against the flow at Dongola, X, can be seen from Fig. 8.48. The
regression of Y on X can be represented by the equation

$$Y = 3.4125 + 0.96703 X \qquad\qquad (8.9)$$

in which Y and X are given in mlrd m^3/yr. The correlation coefficient, r_{XY},
being equal to 0.96632, is certainly strong.

From the values of the serial correlation coefficients listed in Table 8.27,
one can observe that all the series, except those of the high flood discharges,
August and September, are significantly correlated, at least with lag 1. The
first correlation coefficient shows, in general, a considerable rise with
decreasing mean discharge. The deterministic component in the serially correlated discharge series can be fitted each by a first-order autoregressive scheme
with the first serial coefficient as a parameter. The stochastic or the residual
component, which was left from the fit, was tested and the null hypothesis that

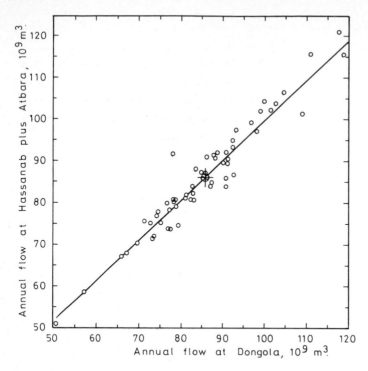

Fig. 8.48. Simple linear regression between the sum of the flows at Hassanab and Atbara mouth, Y, and the flow at Dongola, X, for the period 1912-1973

that its serial correlation coefficients are not significantly different from zero (95% confidence level) could not be rejected. The only exception was the residual series of June where the coefficient at lags 3 and 4 differs slightly from zero at 95% level of confidence. The serial correlations of the residual series are listed in Table 8.28.

The probability distribution functions which are good fit to the monthly and annual discharge series are the Pearson III, the 2-parameter lognormal and the normal functions. The fit of the Pearson Type III function to the distribution of the annual flow volumes at Dongola is shown graphically in Fig. 8.49. The theoretically computed 100-yr and 200-yr discharges are as follows:

Discharge 10^6 m^3	Jan.	Feb.	Mar.	Apr.	May	June	July	Aug.	Sep.	Oct.	Nov.	Dec.	Year
100-yr	5517	4708	4685	4389	4622	4486	9524	27322	30690	22923	12103	6814	117755
200-yr	5788	5079	4806	4688	5005	4846	9811	28053	31600	24027	12845	7114	121617

TABLE 8.28   Serial correlation coefficients of the residuals left from fitting autoregressive models to some of the discharge series of the Nile at Dongola

Month	Serial correlation coefficient														
	r_1	r_2	r_3	r_4	r_5	r_6	r_7	r_8	r_9	r_{10}	r_{11}	r_{12}	r_{13}	r_{14}	r_{15}
Jan.	.050	.083	.086	.208	.037	.038	.137	.012	.062	.092	.120	.045	.066	.003	.079
Feb.	.067	.069	.135	.220	.070	.061	.053	.006	.001	.062	.208	.046	.042	.112	.161
Mar.	.067	.161	.154	.081	.065	.050	.087	.016	.047	.096	.056	.104	.045	.076	.114
Apr.	.052	.155	.148	.042	.008	.118	.039	.076	.092	.015	.050	.118	.027	.039	.094
May	.119	.036	.154	.109	.155	.081	.051	.154	.038	.077	.048	.027	.012	.050	.077
June	.044	.019	.250	.326	.082	.173	.060	.162	.061	.111	.005	.041	.127	.174	.093
July	.024	.040	.222	.003	.087	.033	.026	.128	.153	.101	.095	.138	.166	.098	.147
Oct.	.037	.033	.232	.077	.026	.239	.009	.063	.105	.068	.107	.098	.022	.096	.006
Nov.	.004	.169	.243	.080	.159	.076	.011	.082	.036	.104	.108	.021	.063	.094	.096
Dec.	.018	.212	.154	.079	.070	.020	.090	.026	.053	.113	.122	.115	.094	.111	.135

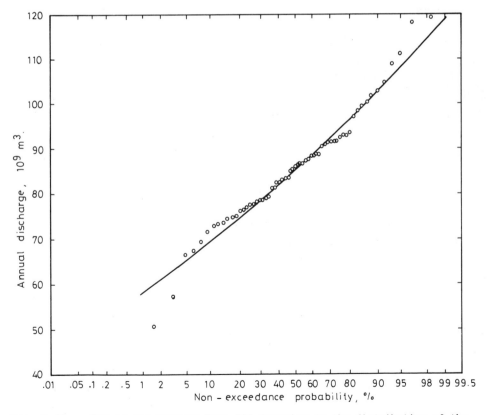

Fig. 8.49.   Fit of the Pearson Type III function to the distribution of the annual discharges in the Main Nile at Dongola for the period 1912-1973

8.11.2   The discharge of the Main Nile at Aswan

Downstream of Dongola the Main Nile flows to the north then bends to the east and to the west and once more to the east then north-east to Wadi Halfa. This reach is about 450 km in length. Discharge measurement at Wadi Halfa began in 1911, and with a gap during the First World War, continued until 1931. The gauge-discharge measurements, 1911-1927, which were used for constructing the rating curve for this station are shown in Fig. 32, Appendix E. The measuring site at Wadi Halfa was inundated by the backwater caused by the second heightening of the Aswan Dam about 1934. The site was moved to Kajinarti, which is about 50 km south of Halfa.

The loss in the reach below the junction of the Atbara with the Main Nile and Wadi Halfa is about 1210 km x 0.4 km x 2.7 m, or about 1.3 mlrd m^3/yr. Vol. IX of the Nile Basin gives 86.1 mlrd m^3/yr as the sum of the flows at Hassanab and the Atbara mouth. The flow in the Main Nile at Kajinarity or Halfa was 85.3 mlrd m^3/yr. These two figures are the averages for the period 1912-1952. This means that the loss averaged over the same period is 0.8 mlrd m^3/yr. This is relatively much less than the 1.3 mlrd m^3/yr already given by us. One should not forget, however, that the absolute figure is in itself quite small, and having it estimated as the difference between two much larger quantities strongly influences its accuracy.

The Main Nile flows about 345 km in a real arid zone, with rainfall of less than 10 mm/yr, before reaching Aswan. The rating curve points measured in the period 1918-1927 at Khannaq station are shown in Fig. 33, Appendix E. The monthly and annual discharges are listed in Table 16, Appendix D. These data differ from the corresponding ones at Dongola by the conveyance loss between the two stations and the storage losses at Aswan. The total loss in a year averaged over the period 1912-1973 was 3.368 mlrd m^3, or about 3.94% of the annual flow at Dongola. The close resemblance between the discharges at Aswan and Dongola can be seen, for example, from a comparison between Fig. 8.50. and Fig. 8.47., both showing the plot of the annual flow volumes.

The monthly and annual discharge series at Aswan, like all main stations on the Nile, have been analyzed statistically and the results obtained are presented in Table 8.29. Examination of these results shows that all discharge series at Aswan, monthly and yearly, undergo some significant dependence in their structure. The dependent component in each series can be described by a first-order autoregressive equation in which the first serial correlation coefficient is a parameter. The residuals left after removing the dependent component from each discharge series have been examined and the hypothesis that the serial correlation between the items of the residual series is not significant (at 95% confidence level) could not be rejected for most of the discharge series. It is

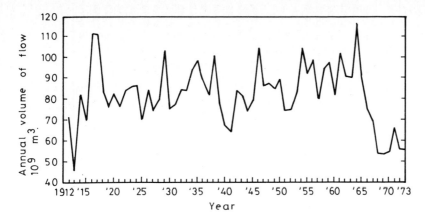

Fig. 8.50.   Graphical plot of the annual flow volume series at Aswan on the
Main Nile in the period 1912-1973

possible that the dependence in some of the remaining series needs to be described by higher-order autoregressive equations so as to render the residual
components to be uncorrelated. The serial correlation coefficients of the residual series can be found in Table 8.30.

The average hydrograph at Aswan for the period 1912-1973 is shown in Fig. 8.51.
This hydrograph does not represent, in the first place, the natural condition.
The low-flow supply of the river in the period from November up to and including
June is somewhat modified by the existing storage works on the Nile. The storage
reservoir at Aswan alone has resulted in the increase in the average volume of
flow during the mentioned season from 26.12 mlrd m^3 at Dongola to 27.69 mlrd m^3
at Aswan, i.e. a net gain of 1.57 mlrd m^3. A detailed discussion of the storage
on the Nile shall be presented, hwoever, in the next chapter.

The probability functions that serve as good fit to the monthly discharge
data are the Pearson Type III and the 2-parameter lognormal. The normal function
is a very good fit to the distribution of the annual discharge series. This can
be seen from Fig. 8.52. The monthly and annual flows with 100 and 200 years
return period have been computed from these distribution functions and are as
given below:

Discharge 10^6 m^3	Jan.	Feb.	Mar.	Apr.	May	June	July	Aug.	Sep.	Oct.	Nov.	Dec.	Year
100-yr	5546	4898	5162	4659	6337	8018	8355	26416	28792	21265	13145	7624	117618
200-yr	5809	5208	5587	5025	7031	9056	8816	27114	29144	21696	13640	8001	121416

The above analysis of the discharges at Aswan is based on the assumption that
they are homogeneous. The validity of this assumption seems questionable.

TABLE 8.29 The basic statistical descriptors and the serial correlation coefficients of the monthly and annual discharges of the Nile at Aswan, for the period 1912-1973

Item		Jan.	Feb.	Mar.	Apr.	May	June	July	Aug.	Sep.	Oct.	Nov.	Dec.	Year
Basic statistical descriptor														
$\bar{X}$, 10^6 m³		3578	2704	2480	2153	2214	2399	4670	16623	19466	13742	7376	4788	82202
s, 10^6 m³		740	777	868	861	1261	1610	1420	5168	6388	4311	2241	1020	15224
C^v		0.2069	0.2874	0.3501	0.3998	0.5695	0.6713	0.3040	0.3109	0.3282	0.3137	0.3038	0.2131	0.1852
C^s		0.4676	0.6999	1.0892	0.8272	1.3802	0.7765	0.3694	0.5681	−0.1680	0.7856	0.3376	0.6138	−0.1610
C^s_k		4.3987	3.8095	3.7018	2.9905	4.0417	5.1028	3.0563	2.9880	4.1945	4.0836	3.2447	4.1724	3.1683
Serial correlation coefficient														
r_1		0.2623	0.4792	0.6342	0.7779	0.8389	0.7964	0.4750	0.4028	0.6773	0.5825	0.2422	0.3128	0.4838
r_2		−0.0012	0.3085	0.5246	0.6737	0.7537	0.7179	0.2813	0.4415	0.5779	0.4198	0.2059	0.2150	0.2979
r_3		−0.0265	0.3268	0.4820	0.6242	0.6772	0.6410	0.0503	0.3200	0.4575	0.3146	0.2193	0.1412	0.1679
r_4		−0.2332	0.0584	0.3157	0.5071	0.5665	0.4976	0.0802	0.2886	0.3050	0.1217	−0.0769	−0.1717	−0.0033
r_5		−0.1661	0.1387	0.3027	0.4893	0.5169	0.4611	0.0975	0.1232	0.2561	0.1336	0.0817	0.0609	0.0239
r_6		−0.1288	0.1851	0.3187	0.4737	0.4455	0.3325	0.2759	0.1201	0.0435	0.0754	−0.0023	−0.1121	−0.0362
r_7		−0.1844	0.1207	0.2627	0.4597	0.4047	0.2246	0.1627	−0.0411	−0.0300	−0.0169	−0.1147	−0.1706	−0.1360
r_8		−0.0753	0.1917	0.2659	0.4613	0.3662	0.2296	0.1331	−0.0605	0.1222	−0.0695	0.0579	−0.0423	−0.0973
r_9		0.1196	0.1302	0.1591	0.3885	0.2990	0.1062	0.0604	0.1825	0.1917	0.0969	0.0967	0.0354	0.1543
r_{10}		−0.0425	0.0568	0.0497	0.3437	0.2463	0.0775	0.0417	0.1010	0.1563	0.1163	0.0451	0.0575	−0.0974
r_{11}		0.2115	0.2149	0.0226	0.3166	0.2038	−0.0079	−0.0129	0.1659	0.1708	0.1198	0.1707	0.1302	0.1478
r_{12}		0.0185	−0.0536	0.0232	0.2938	0.1759	−0.0198	0.0877	0.0629	0.1021	0.0512	0.1166	0.0295	0.0282
r_{13}		−0.0043	−0.0352	−0.0005	0.2389	0.1637	−0.0268	0.1041	0.2363	0.1945	0.0742	0.0208	0.0927	0.1212
r_{14}		0.0589	−0.0422	0.0104	0.2256	0.1439	−0.0524	0.2234	0.1882	0.1713	0.1647	0.1612	0.0592	0.2148
r_{15}		0.1374	0.1113	0.0873	0.2340	0.1496	−0.0045	0.2515	0.1817	0.0873	0.1277	0.0213	0.1157	0.1268

TABLE 8.30   Serial correlation coefficients of the residuals left from fitting first-order autoregressive models to the discharge series of the Main Nile at Aswan

Month & Year	Serial correlation coefficient														
	r_1	r_2	r_3	r_4	r_5	r_6	r_7	r_8	r_9	r_{10}	r_{11}	r_{12}	r_{13}	r_{14}	r_{15}
Jan.	.480	448	.458	.313	.389	.362	.287	.382	.240	.247	.252	.221	.189	.178	.189
Feb.	‾.126	‾.117	.076	‾.176	.004	.009	‾.070	‾.040	.106	‾.057	.203	.071	‾.007	‾.001	.080
Mar.	‾.221	‾.114	.349	‾.268	.063	.135	.088	.097	.158	‾.057	.208	.104	‾.027	.080	.147
Apr.	‾.255	‾.040	.239	‾.150	.042	.178	.006	.002	.151	.069	.030	.061	.010	.129	.136
May	.143	.046	.296	‾.067	.195	.138	.066	.130	.018	.061	.008	.050	.004	‾.056	.107
June	.229	.083	.278	‾.209	.339	.087	‾.003	.165	‾.160	.141	.045	.086	.010	.138	.155
July	‾.049	.123	‾.118	.080	‾.024	.254	.021	.094	‾.033	.043	‾.059	.173	‾.054	.112	‾.134
Aug.	‾.074	.283	.141	.203	.029	.175	‾.079	.188	‾.141	.023	.151	.113	‾.194	‾.044	‾.053
Sep.	.110	.197	.173	‾.007	.289	.086	.143	‾.034	‾.058	‾.032	‾.115	.182	‾.227	.015	‾.041
Oct.	.043	.134	.202	‾.060	.161	.141	‾.038	‾.006	.029	‾.095	.041	.054	‾.031	.103	.069
Nov.	‾.026	.165	.231	‾.083	.151	.031	.090	.037	.102	.044	.129	‾.061	.043	.124	.026
Dec.	.051	.154	.163	‾.170	.059	.036	‾.137	.032	‾.049	.070	.159	.055	.068	‾.086	.092
Year	.041	.114	.096	‾.015	.034	.082	‾.102	.059	‾.093	‾.040	‾.134	.106	‾.096	.117	‾.024

The underlined values are those serial coefficients which are significantly different from zero at 95% confidence level

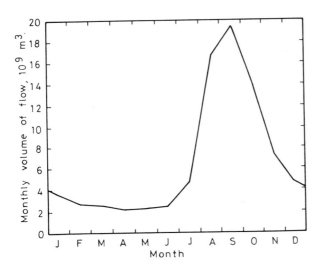

Fig. 8.51.   The average hydrograph of the Main Nile at Aswan for the period 1912-1973

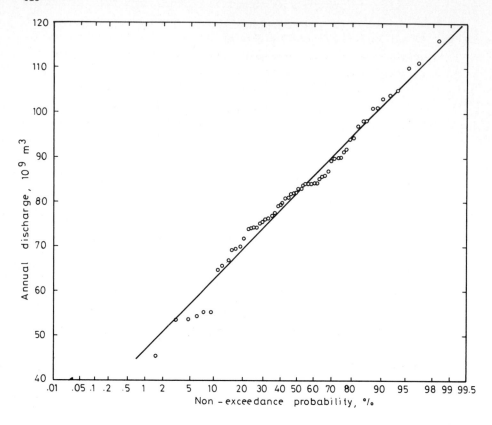

Fig. 8.52.   Fit of the normal function to the distribution of the annual discharges in the Main Nile at Aswan for the period 1912-1973

A mathematical model of the River Nile from its exit at Lake Albert to its
entrance into Lake Nasser upstream Aswan has very recently been designed and
calibrated (Fahmy, A., Panattoni, L., and Todini, E., 1982).

The different components of the Nile system were all modelled using the so-
called Constrained Linear Systems model (CLS). This type was chosen as a consequence of an analysis of the purposes of the model and the amount and nature
of the available data. The first component covers the reach from Pakwach (exit
of the Nile at Lake Albert) to Mongalla; the second covers the reach from
Mongalla to Malakal and Hillet Doleib on the White Nile and the third, which is
divided into two parts, covers the River Sobat. The fourth component or submodel
covers the reach from Malakal (downstream the mouth of the Sobat) to Mogren; the
fifth deals with the Blue Nile, and the sixth covers the reach from Mogren to
Wadi Halfa.

After the individual submodels were identified, they were assembled to give a
model of the entire system from Lake Albert to Lake Nasser. A fortran programme
was used to simulate the behaviour of all the system and the different reaches.

The model has been used by its designers for computing the discharges at Wadi Halfa after feeding it with the discharges at Pakwach in the period from 1953 to 1972. The comparison between the observed and computed hydrographs at Wadi Halfa show that the residuals have a mean and a standard error of 2.55 and 28.6 million m^3/day, respectively. The latter figure corresponds to 12.2% of the daily discharge at Wadi Halfa.

8.11.3   From Aswan to the Mediterranean Sea

The water that had been leaving the old Aswan Dam every year used to flow in the Nile and its branches on its way to the Mediterranean Sea. A certain part of this flow had been used for land irrigation and for domestic purposes and the rest had been discharged into the sea. The annual quantity passing downstream of Aswan Dam in such a normal year as 1947 was 84.3 mlrd m^3. Of this amount, 6.6 mlrd m^3 were used for basin irrigation, 38.1 mlrd m^3 for perennial irrigation and less than 1 mlrd m^3 for domestic and industrial purposes. The rest, almost 39 mlrd m^3 was thrown into the sea. The consumption of water in Egypt that year, which reached 45.7 mlrd m^3, was very near to the full share of Egypt in the Nile water at that time. This was limited to 48 mlrd m^3/yr. Irrigation was, and still is, accomplished via an intricate canal system. The general layout of the main canals is shown in Fig. 8.53.

From Aswan to Cairo there is hardly any rainfall worth mentioning and the discharge downstream of Aswan is practically the only source of water. In this reach, gauge discharge measurements are taken more or less regularly at a number of places. Examples of these are given for Hawatka station near Assiut in 1926-1927, Fig. 34, Appendix E, and for Beleida station near Koraimat in 1920-1922, Fig. 35, Appendix E.

Before the construction and operation of the storage works on the Nile, agriculture in Egypt depended almost entirely on the natural supply of the river. The annual inundation of the Nile Valley in the late summer generally supplied enough moisture to the soil to ensure fair crops in the fall and winter. No wonder, then, that the most important annual event in Egypt was the Nile flood and, therefore, records were engraved on the cliff walls in various places, notably at Semna, a section of the second cataract (see map, Fig. 2.25.). Several sections of nilometers, at various points along the stream channel, have been discovered, and their inscriptions have been deciphered and correlated.

C.S. Jarvis combined the discoveries and memoires of Sir H. Lyons, M. LeGrain, Prince Omar Toussoun, Aboul Mehasin and many others who kept themselves busy with the Nile water levels. This led Jarvis to publish his marvellous paper on the flood-stage records of the River Nile in Egypt (Jarvis, C.S., 1935). From

412

this paper we have copied the maximum and minimum annual levels of the river at the Roda nilometer, Cairo, in the period from 622 up to 1933 and presented it here as Fig. 8.54.

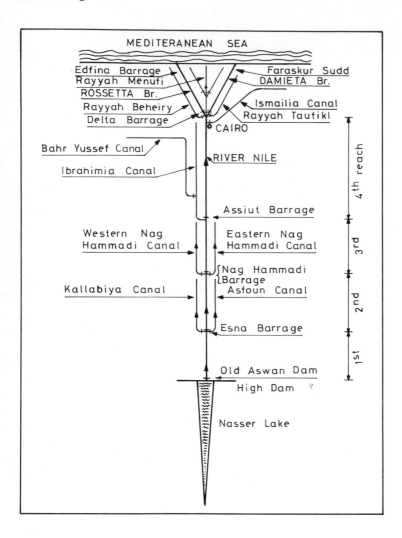

Fig. 8.53.

The discussions on this paper by Hurst and K.O. Ghaleb, both well equipped with outstanding information gained from actual experience with the river, have pointed to a number of sources of error in these data. With full recognition of these errors and other possible discrepancies in the data, one cannot deny the value of the length of this series and its fair completeness, especially from about 622 to about 1450 A.D.

From these data Jarvis found the rise in the Nile flood level ascribable to sedimentation to range from 0.10 to 0.15 m per century. J.C. Stevens repeated this calculation using the simple 10-yr averages and the progressive 50-yr averages (Jarvis, C.S., 1935). He concluded that the average rise was practically 4 inches (0.10 m) per century for both the maximum and minimum stages. He further concluded that no periodic cycle was in evidence, but high cycles alternated with low cycles of irregular duration. From that nearly continuous record and from the sporadic records of isolated periods running back over 5000 years, there appears to have been little or no climatic changes that can be detected. The conclusions drawn by T.H. Means from his discussion on the same paper were almost identical to those given by Stevens which have already been presented (Jarvis, C.S., 1935).

We shall present the change in the river bed as a result of the construction of the High Aswan Dam in Chapter 9.

The evaporation loss between Aswan and Cairo in the pre-High Dam period could be figured out approximately as the length of the reach, 900 km, times the average weighted width, 500 m, times the annual evaporation depth, 2 m. This gives 0.9 mlrd m^3/yr to be rounded off to 1.5 mlrd m^3/yr to account for the losses from the network of irrigation canals. Whether the resultant of the seepage from the river and the return flow to it could be considered as a net gain or net loss is not precisely known.

The relationship between both the loss or gain and the factors affecting it in the post-High Dam period, was studied by Saleeb, S.I. (1977). To implement this study, the river reach from downstream of Aswan to Cairo was divided into four reaches (see map, Fig. 8.53.). The outflow from an upstream reach is considered as inflow to the next downstream reach, and so on. Taking the town of Assiut as the centre of gravity of the reach from Aswan to Cairo, the loss or gain was found to be affected by the discharge just downstream of Aswan, the mean air temperature and humidity at Assiut 5 days lagging behind Aswan and the groundwater level at Assiut. The regression models for the daily and monthly loss or gain have been developed and the optimum monthly discharges found. These discharges are included in Table 8.31. The monthly means of the 6-yr period 1968-73 and of the 56-yr period 1912-67 are also included in this table for the purpose of comparison. The optimum discharges have been found on the grounds that they correspond to total loss or gain equal to zero.

A short distance below Cairo, the river bifurcates into its two branches: Damietta and Rosetta. These branches are the main source of water feeding the irrigation canals in Lower Egypt. They were also used in the pre-High Dam period to convey the excess flood water to the Mediterranean Sea. This is no longer the case after exercising full control on the Nile water by means of the High Aswan Dam.

414

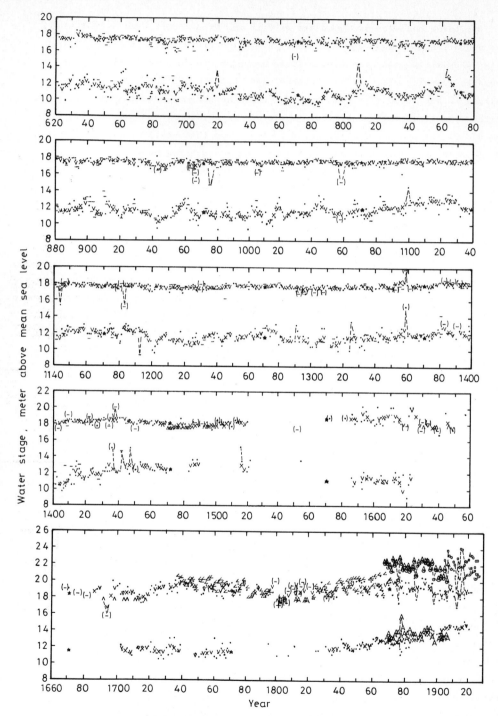

Fig. 8.54.   Maximum and minimum stages of the Nile River at Roda, Egypt

In Fig. 8.54., the various plotted points may be identified as follows:

Applying to all the records:

v = five-year average for the data shown by dots, sup-
    plemented by - or (-) when the location of the dot
    is not given;

x = ten-year average for data shown by dots, supple-
    mented by - or (-) when the location of the dot is
    not given; and,

★ = one hundred-year average

Applying to the Roda Gauge at Cairo:

. = records compiled by Omar Toussoun[+] covering the
    1300-yr period from 622 to 1921 A.D.;

(.) = confirmation from textual notes for the records
    compiled by Omar Toussoun: when - is lacking, an
    agreement is indicated between all three sources;

- = records compiled by Aboul Mehasin[++], covering the
    period, 20 to 855 of the Hegira, or 641 to 1451
    A.D., a total of 811 years. A small + indicates
    extra data, representing 25 surplus years of
    Mohammedan reckoning;

(-) = records from notes compiled by Ibn Iyas and
    others[+++], for the period 769 to 1878 A.D.

∠ = records by Lyons[*]; and

∧ = records indicating "wafa", or the stage that
    assures plenty, at which the canals were opened
    to supply the basins; the maximum flood stage,
    ordinarily, was somewhat higher.

Applying to the gauge downstream from Aswan Dam:

A = maximum annual river stages at the Aswan gauge
    above the assumed datum, 71.0 m, or 232.9 ft,
    above mean Mediterranean Sea level;

ʊ = maximum ten-day average gauge heights, and, there-
    fore, somewhat below the actual maximum.

Applying to the El-Leisi gauge, 37 miles upstream from Cairo:

o = maximum ten-day average gauge heights and, therefore,
    somewhat below the actual minimum.

The gauge heights plotted as ordinates in Fig. 8.54. are
readings from the Roda gauge on the Nile River at Cairo.
The exceptions, marked A or J, refer to readings of the
Aswan gauge and those marked O refer to readings of the
El-Leisi gauge, 37 miles above Cairo. In all cases the
readings are the mean Mediterranean Sea level elevations
at Alexandria.

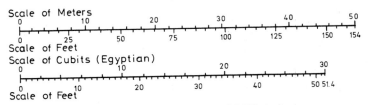

Scale of Meters

Scale of Feet

Scale of Cubits (Egyptian)

Scale of Feet

1 Cubit = 23.9417 Digits, 10 Digits = 8.5875 Inches

[+]Memoirs, Inst. of Egypt, 1925 Vol. 9.; [++]Loc. Cit., 1923, Vol. 4.; [+++]Loc. cit.,
1923 and 1925, Vol. 4 and 9.; [*]"The Nile Flood in 1905", by Capt. H.G. Lyons

TABLE 8.31   Comparison between the actual and the proposed optimum discharges
downstream of Aswan Dam

Month & Year	Average for the periods, mlrd m^3		Optimum discharge from multiple regression model mlrd m^3
	1912-67 (56 yrs)	1968-73 (6 yrs)	
January	3625	3142	2713
February	2579	3870	3646
March	2335	3836	3900
April	1968	3877	3588
May	1894	5200	5180
June	1959	6507	6552
July	4448	6742	6699
August	17743	6167	5943
September	21095	4265	3891
October	14808	3788	3419
November	7777	3634	3069
December	4948	3296	2880
Year	85149	54324	51480

REFERENCES

Ahmed, A.A., 1960. Recent developments in Nile control. Proceedings Institution of Civil Engineers, London, Vol. 17, Paper No. 6102: 137-180.

Berg, C.L., 1953. Detailed analysis of a discharge measurement on the Victoria Nile. Proceedings Institution of Civil Engineers, London, Part III, Vol. 2, Paper No. 5935: 609-613.

Cairo University/MIT, 1977. The River Nile project; stochastic modelling of the Nile inflows to Lake Nasser. Cairo University/MIT technological planning programme, Cairo, Egypt.

Fahmy, A., Panattoni, L., and Todini, E., 1982. A mathematical model of the River Nile. Engineering applications of computational hydraulics (Abbott, M.B., and Cunge, A.J.: editors), Pitman, London: 111-130.

Hurst, H.E., and Philips, P., 1932. The Nile Basin, Vol. II, measured discharges of the Nile and its tributaries, Physical Department Paper No. 28, Government Press, Cairo, Egypt.

Hurst, H.E., and Philips, P., 1933. The Nile Basin, Vol. IV, ten-day mean and monthly mean discharges of the Nile and its tributaries. Physical Department Paper No. 30, Government Press, Cairo, Egypt.

Hurst, H.E., and Philips, P., 1938. The Nile Basin, Vol. V, the hydrology of the Lake Plateau and the Bahr el-Jebel. Physical Department Paper No. 35, Schnidler's Press, Cairo, 235 pp.

Hurst, H.E., Black, R.P., and Simaika, Y.M., 1951. The Nile Basin, Vol. VII, the future conservation of the Nile. Physical Department Paper No. 51, Eastern Press (reprinted), Cairo, 157 pp (with appendices).

Hurst, H.E., 1950. The Nile Basin, Vol. VIII, the hydrology of the Sobat and White Nile and the topography of the Blue Nile and Atbara, Physical Department, Paper No. 55, Government Press, Cairo, 125 pp.

Hurst, H.E., Black, R.P., and Simaika, Y.M., 1959. The Nile Basin, Vol. IX, the hydrology of the Blue Nile and Atbara and of the Main Nile to Aswan with some reference to projects. Nile Control Department, Paper No. 12, General Organization for Government Printing Offices, Cairo, 206 pp.

Hurst, H.E., Black, R.P., and Simaika, Y.M., 1966. The Nile Basin, Vol. X, the major Nile projects. Nile Control Department, Paper No. 23, General Organization for Government Printing Offices, Cairo, 217 pp.

Jarvis, C.S., 1935. Flood-stage records of the River Nile. Transactions ASCE, Paper No. 1944 (with discussions by H.P. Gillete, R.W. Davenport, H.E. Hurst, T.H. Means, J.W. Breadsley, J.C. Stevens, J.W. Shuman, K.O. Ghaleb and C.S. Jarvis): 1012-1071.

Jonglei Area Executive Organ for Development Projects, 1975. Jonglei Project, report on phase I, Tamaddun Press, Khartoum, 100 pp.

King, J.W., 1975. Solar phenomena, weather and climate. European Space Agency, ESA Bulletin No. 3, Neuilly-sûr-Seine, France: 24, 49-51.

Kite, G.W., 1981. Recent changes in level of Lake Victoria, Bulletin of hydrological sciences, Vol. 26, No. 3: 233-243.

Rodriguez, I. and Yevjevich, V., 1967. Sunspots and hydrologic time series. Proceedings of the International Hydrology Symposium, Ft. Collins, Colorado, Vol. I: 397-405.

Saleeb, I.S., 1977. River Nile water systems analysis using digital computers. Proceedings of the International Conference on Computer Applications in Developing Countries. AIT, Bangkok, Vol. II: 777-791.

Shahin, M., 1983. Effect of storage works in the Nile River system on the homogeneity in the annual flow series. Proceedings of the Symposium on Water Resources, Varna, Bulgaria: 11-23.

WMO, 1974. Hydrometeorological survey of the catchments of Lakes Victoria, Kyoga and Albert, Vol. I, Parts I & II: meteorology and hydrology of the basin, WMO, Geneva.

WMO, 1974. Hydrometeorological survey of the catchments of Lakes Victoria, Kyoga and Albert, Vol. III: preliminary reports on the index catchments, WMO, Geneva.

WMO, 1974. Hydrometeorological survey of the catchments of Lakes Victoria, Kyoga and Albert, Vol. IV: hydrological studies of selected river basins, WMO, Geneva.

Chapter 9

WATER STORAGE AND CONSERVATION

INTRODUCTION

While analyzing and discussing the stream flow in the River Nile system, Chapter 8, a brief mention was made to water storage and conservation works. This chapter will deal with a somewhat detailed description of such works; those completed and functioning and those which are underway.

The first storage work built, at least in contemporary history, is the old dam at Aswan, Egypt (1898-1902). The operation of this work was based on storing a small volume of flood water and releasing it in the next low-flow season to the downstream reach of the river to improve the natural discharge. Both storage and release were accomplished in the same water year. One may thus describe the reservoirs formed by the old Aswan Dam and its heightenings as annual storage works.

Subsequent storage works have been designed and operated on the same basis. The only exception is the High Dam at Aswan, Egypt (1956-1964) which was designed according to the theorem of long-term or century storage. The principal lines of the annual and long-term storage theorems are briefly reviewed and discussed in the next sections.

9.1   WATER STORAGE IN THE NILE BASIN

9.1.1   Annual storage

The method used to determine the live-storage capacity of a reservoir was developed in 1882 by Rippl. Since this method can, preferably, be worked out graphically, it is commonly referred to as the Rippl-diagram or mass curve method.

9.1.1.1   Simple case of seasonal storage where demand and supply are equal and
           there are no storage losses

Suppose that the data in columns 2 and 4 of Table 9.1 represent the monthly natural flow reaching a reservoir and the demand downstream of it, respectively. The graphic plot of these data versus the time in months shows clearly that there are months of excess (natural supply > demand) and months of deficit (natural supply < demand). Fig. 9.1a. shows that the season from August to January is a period of excess whereas the rest of the year, except February, is one of deficit. February is the only period without excess or deficit. This case implies that the reservoir must be filled in the period of excess, neither filled nor emptied in February, and emptied from March to July to compensate for

420

the deficit of the natural supply in this period.

The Rippl-diagram method offers a simple tool for determining the required reservoir capacity. The mass curves of the supply and demand can be obtained by plotting the figures in columns 3 and 5, Table 9.1, versus time. The difference between the ordinates of the two mass curves at any instant equals the volume of reservoir contents at that particular instant. In the present example the maximum storage is attained at the end of January and remains constant till the end of February, after which the reservoir is emptied till the end of the year. The same result can be seen in the last column in Table 9.1.

TABLE 9.1   Monthly and cumulative supply and demand (Shahin, M., 1971)

Month	Monthly supply, 10^6 m^3	Cumulative supply, 10^6 m^3	Monthly demand, 10^6 m^3	Cumulative demand, 10^6 m^3	Cumulative supply-demand, 10^6 m^3
August	80	80	30	30	50
September	70	150	15	45	105
October	35	185	15	60	125
November	25	210	10	70	140
December	20	230	05	75	155
January	15	245	05	80	165
February	10	255	10	90	165
March	10	265	20	110	155
April	10	275	30	140	135
May	10	285	40	180	105
June	05	290	50	230	60
July	10	300	70	300	0

Sometimes it could be more convenient to draw the so-called "differential mass curve", which is simply a graphical plot of the ordinate difference between the two mass curves versus time, using a horizontal datum. This curve facilitates the reading of the reservoir contents at any time during the year (see Fig. 9.1b.). In this example, the maximum capacity of the reservoir as read from the last column of Table 9.1 is 165 x 10^6 m^3. The same figure can be obtained either from the maximum ordinate difference between the two curves or simply as the maximum ordinate of the differential mass curve.

9.1.1.2   Case of supply greater than demand

If, in this case, we use the same mode of computation as in section 9.1.1.1, a certain amount of water will remain unutilized up to the end of the year (see Fig. 9.2a.). This amount is simply the excess of the yearly supply over the yearly demand. Were it assumed that the reservoir must be empty by the end of the year, the last ordinates of the supply and demand mass curves must then be equal. The method of computation can be modified as follows: with reference to

Fig. 9.2a., instead of starting the computation of the mass demand from the beginning of the year at O, the order of the procedure is reversed, and the computation starts from point B corresponding to the end of the year. The mass demand is now drawn from right to left until it intersects the mass supply at C. From C down to O the mass demand should coincide with the mass supply. In other words, the reservoir has to be kept empty in the period OC and filling starts only on the date corresponding to C. This method results in a maximum reservoir capacity R_2 smaller than R_1 (the difference being equal to that between the mass supply and the mass demand). As the discharge of a river can hardly be forecast accurately, one may, instead of waiting till the time represented by point C, begin the filling at O and continue till the required amount is stored. After that, the reservoir has to be maintained full, until demand exceeds the natural supply.

Another way to accomplish the same purpose is shown in Fig. 9.2b. The diagram shown can be constructed by first plotting the mass supply curve. The ordinates of the mass demand are computed and the reservoir size, R_2, found. The mass demand is then drawn starting from the origin. On this diagram the ordinate DE, equal to R_2, can be found. From E a curve parallel to the mass supply curve is drawn, as is a curve parallel to the mass demand curve starting from point B. The two mass curves intersect at point G and the line OEGB should represent the mass curve of the outflow. This curve can be divided into three distinct parts: O-E, in which the outflow equals the demand and the filling of the reservoir takes place; E-G, in which the outflow exceeds the demand the the reservoir contents are kept constant, and G-B, in which the outflow equals the demand the and the reservoir emptying takes place.

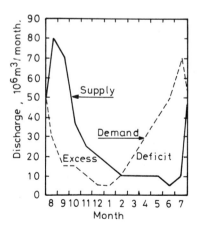

Fig. 9.1a.   Hydrographs of natural
supply and demand

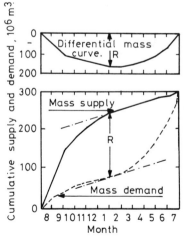

Fig. 9.1b.   Mass curves of supply
and demand. Differential mass curve
and reservoir capacity, all for equal
cumulative supply and demand

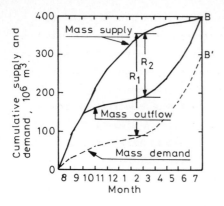

Fig. 9.2a. Mass curves of supply, demand and outflow and minimum reservoir size for the case of cumulative supply greater than cumulative demand

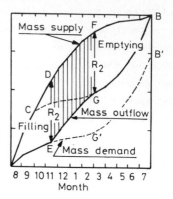

Fig. 9.2b. A possible solution for the minimum reservoir capacity in the case of cumulative supply greater than cumulative demand (notice the change in mass outflow than in 9.2a.)

9.1.1.3   Case of supply equal to, or greater than, demand, initial storage required

In the above cases supply and demand have been arranged so that the ordinate of the mass supply at any time is equal to, or greater than, the ordinate of the mass demand. This is, however, not always the case and the two mass curves often intersect at least once. The storage calculation here can better be explained using the data in Table 9.2 (Shahin, M., 1971).

TABLE 9.2   Monthly and cumulative supply and demand and difference between mass supply and demand with and without initial storage

Month	Monthly supply, 10^6 m^3	Cumulative supply, 10^6 m^3	Monthly demand, 10^6 m^3	Cumulative demand, 10^6 m^3	Cumulative supply minus cumulative demand, 10^6 m^3	
					without initial storage	with initial storage
August	10	10	45	45	−35	20
September	15	25	35	80	−55	0
October	30	55	30	110	−55	0
November	60	115	15	125	−10	45
December	80	195	10	135	60	115
January	50	245	05	140	105	160
February	20	265	10	150	115	170
March	10	275	15	165	110	165
April	10	285	25	190	95	150
May	05	290	30	220	70	125
June	05	295	35	255	40	95
July	05	300	45	300	0	55

We start by plotting the supply and demand hydrographs as shown in Fig. 9.3a. This plot shows one period of excess, whereas the deficit is split into two shorter periods; one at the beginning of the year and the other from March till the end of the year. Since the yearly demand and supply are equal, the volumes of excess and deficit mus be equal too.

As before, we start by drawing the mass supply and demand curves. From Fig. 9.3b. and Table 9.2 one can see that the required demand cannot be fulfilled, unless a certain initial storage is provided. Furthermore, the maximum negative cumulative difference which Table 9.2 gives as 55×10^6 m^3 should be considered as the minimum volume needed for the initial storage. To adjust the mass supply curve to this situation one needs to add an amount of 55×10^6 m^3 to its ordinates. Fig. 9.3b., as well as Table 9.2, show that the maximum difference between the adjusted mass supply and the adjusted mass demand is 170×10^6 m^3. This represents the full capacity required for the reservoir. The same volume can be found from the differential mass curve and from the last column in Table 9.2. The mass diagram can be split into four parts. The partially full reservoir (55×10^6 m^3) in August is depleted gradually and becomes empty at the end of September. The demand is equal to the natural supply in October and so the reservoir remains empty. In November the supply exceeds the demand and the storage continues till the reservoir becomes completely full at the end of February (volume of contents = 170 million m^3). From the beginning of March till the end of the year the reservoir is partly emptied till the volume of contents reaches the initial storage (55×10^6 m^3) by the end of July.

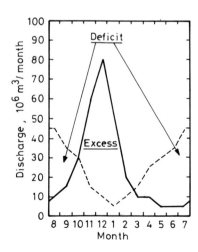

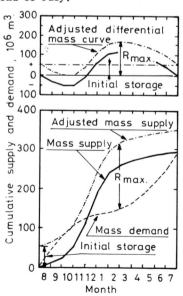

Fig. 9.3a.   Hydrographs of natural supply and demand

Fig. 9.3b.   Mass curves and differential mass curves. Initial storage is needed

9.1.1.4   Case of a sequence of years, the cumulative supply in each year being
           equal to, or greater than, the cumulative demand

It is common in reservoir operation to consider a sequence of years rather
than a single year. The demand is fixed according to the purpose for which water
is used: irrigation, hydro-electric power development, flood control, etc., and
for at least some time, it remains the same, or nearly the same, each year. The
quota for Egypt in the pre-High Aswan Dam condition was limited to 48 mlrd m^3/yr
(1929-1964). This quota has been increased to 55.5 mlrd m^3/yr in the post-dam
condition which began in 1965. From 1869-1870 up to 1979-1980 the annual supply
at Aswan always exceeded either figure, except in the water year 1913-1914 when
the supply fell to about 42 mlrd m^3.

A number of methods for determining the "safe yield" given a reservoir was
reviewed by Bernier J. (1966). He defined the safe yield as that yield corresponding to a certain probability of failure in filling the reservoir. The method he
developed for computing the said probability is based on the theory of the
Markov processes which allow taking into account the dependence between the inputs to the reservoir. For a more extensive presentation of the available techniques and methods related to reservoir capacity and yield, the reader is referred to the work of McMahon, T., and Mein, R. (1978). Fig. 9.4. shows a short
sequence composed of three years. In the first year supply equals demand and the
reservoir capacity needed to guarantee the fixed demand is 165×10^6 m^3. This
figure can be obtained as described in section 9.1.1.1. In the second year the
same reservoir capacity, i.e. 165×10^6 m^3 is necessary to satisfy demand. Since
the net supply in this year (380×10^6 m^3) is, however, in excess of the demand
(300×10^6 m^3), the reservoir at the end of the water year is not totally empty.
Instead, there remains a volume of contents of 80×10^6 m^3. If the reservoir
should be empty by the end of the year, the procedure described in section
9.1.1.2 (see Fig. 9.2b.) has to be used. As a result of the excess of the net
supply over the demand, the outflow curve does not coincide in its full length
with that of the demand. The outflow curve here consists of two external arms
each similar to the corresponding parts of the demand curve, and of a central
part which is parallel to the supply curve and in which the volume of the reservoir contents is maintained constant. The third year has a cumulative supply of
420×10^6 m^3 at its end which is greater than the fixed demand (300×10^6 m^3).
Moreover, the monthly values of the supply and the demand are so that a new
reservoir capacity R_2 smaller than R_1 would be adequate to guarantee the
required demand. If the supply could be precisely forecasted one needs to fill
the reservoir partly up to R_2 (115×10^6 m^3). The reservoir contents should then
be kept constant at this volume till the natural supply begins to fail to
satisfy the demand. From this moment onwards the reservoir contents are depleted

gradually till the end of the year.

The above-described case is not really too different from that of the annual storage on the Main Nile or its tributaries. The annual supply is, in general, much greater than the annual demand. Nevertheless the daily supply is, in some months, less than the daily demand and in other months, more. The storage capacities are fairly small, so most of the flood water is released downstream without being used, and a small amount only is stored to help improve the natural supply during the low-flow season to meet with the demand.

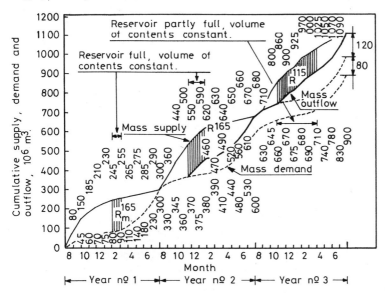

Fig. 9.4.   Reservoir capacity and operation for a sequence of years in which the yearly net supply equals or exceeds the yearly demand

9.1.2   Annual storage works on the Nile and its tributaries

9.1.2.1   The old Aswan Dam

Up to the beginning of the beginning of the twentieth century the amount of water that could be used for irrigating the summer crops in Egypt was almost limited to the natural supply of the river. Such an amount in a low-flow year was hardly sufficient for irrigating 1.5 million acres.

Under the thrust of increasing population it was decided to extend perennial irrigation to a vast area. To fulfill the necessary irrigation requirement the Aswan Dam was first built in 1898-1902 to store just 1 mlrd m^3 of the flood water and to use it together with the natural supply of the river in the following low-flow season.   The dam was first heightened in 1912 and thereupon the storage capacity of the reservoir increased to about 2.3 mlrd m^3. The second heightening of the dam took place in 1934, which brought the capacity to a

figure between 5.0 and 5.1 mlrd m^3 (notice that in the period 1902-1964 the mean
annual supply was about 84 mlrd m^3 and the annual demand 48 mlrd m^3).

In 1944 the idea of a third heightening of the dam by about 11 metres, which
would have increased the reservoir capacity up to 10 mlrd m^3, was proposed. The
idea was given up owing to the danger of silt deposition in the reservoir when
used for flood protection and to the difficulty of filling the enlarged reservoir with non-silty water, and also to reluctance to complicate the structure
any further (Hurst, H.E., Black, R.P., and Simaika, Y.M., 1959).

The cross-section of the dam and its two heightenings is shown in Fig. 1,
Appendix F. Some of the technical data relevant to the dam and the storage
reservoir are included in the same Appendix. The rule curve illustrating the
usual operation of the reservoir is shown in Fig. 9.5. for 1953. Towards the end
of July the reservoir was rather empty and its level had fallen to about 98
where it remained for a short time until the flood wave raised the level to
something between 101 and 103 (notice that the water level in the Nile downstream the dam had increased by about 6 m). The reservoir level fell to about 99
at the end of September and by the tenth of October the normal filling usually
began and was complete at level 121 sometime in December. It remained approximately at that level until the supply reaching the reservoir failed to cope with
the requirements for irrigating the summer crops. This usually happened at the
beginning of February. Water from the reservoir was then used to supplement the
supply and the level fell until the reservoir was almost empty towards the end of
July. It was customary to begin the filling phase when the river stage downstream
Aswan was 90.5 m above sea level or when the falling limb of the supply hydrograph reached 530 million m^3/day. This always took place between the tenth and
the 20th of October every year. These figures were the outcome of experiments
which showed that at level 90.50 the suspended matter in the Nile water had
practically no influence on the storage capacity of the reservoir. The result of
the experiments for the period 1914-1927 is shown in Fig. 9.6.

The final programme of the filling phase had to be planned as soon as the
reading of the local gauge at Atbara reached 14.00, which meant that the level
chosen at Aswan would be reached there about 8 days later (distance = 1555 km
and velocity of propagation of flood = 2.25 m/sec.).

The reservoir was used a number of times as an emergency flood escape to
reduce the danger of breaching the river banks in Middle and Lower Egypt. The
volume of silty water impounded in the reservoir during the disastrous flood of
1954 was about 3 mlrd m^3 with a silt content of about 9 million m^3.

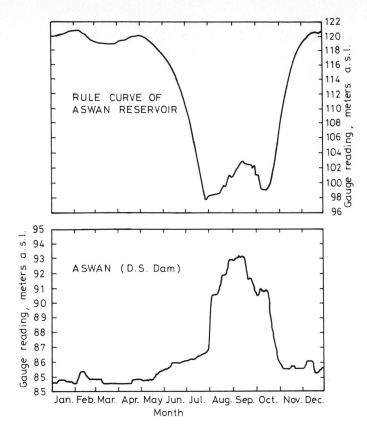

Fig. 9.5.   The stage hydrograph of the Nile downstream of Aswan and the rule curve of the reservoir, both for 1953 (Hurst, H.E., Black, R.P., and Simaika, Y.M., 1959)

Some of the results obtained from investigating the silt regime in the Nile, especially at Kajnarti, Wadi Halfa and Gaafra (closest to Aswan), have been published, among others, by Hurst and co-workers in Vol. IX of the Nile Basin (Hurst, H.E., Black, R.P., and Simaika, Y.M., 1959), by Simaika and El-Sherbini (1957) and by Simaika alone (1961). The main result obtained from that last investigation is that the silt concentration, expressed in parts per million per weight, was approximately 400, 3100, 2500, 1000 and 300 for July, August, September, October and November, all measured at Wadi Halfa. The measurements at Gaafra seemed to be influenced by the operation of the Aswan reservoir. Long-term averages at Gaafra were, however, 300, 3000, 2000, 700 and 170 parts per million by weight for the months July up to and including November successively. It was further reported that the amount of the suspended sediment in the rising stage of the flood was much higher than in the falling stage. With the advancement of the flood wave in time from the end of July towards the end of October a significant increase in the percentage of the coarse fraction (sand) and a

corresponding decrease in the fine fraction (silt and clay) had been noticed.
The detailed results obtained from investigating the silt in the Nile have been
employed in the estimation of the capacity to be allotted to dead storage (storage of sediments) in the reservoir created by the Aswan High Dam.

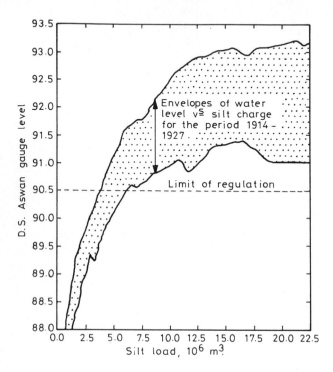

Fig. 9.6.   Water level at Aswan and the silt load in the river water

9.1.2.2   The Sennar (Makwar) Dam

The dam construction was completed in 1925. It is built on the Blue Nile some
350 kilometres south-east of Khartoum for the benefit of the Sudan. Its purpose
is to store part of the Blue Nile water for irrigating the cotton raised in the
Gezirah area and to raise the water intake upstream of the Gezirah canal up to
the required level. The reservoir capacity created by the Sennar Dam is about
0.8 mlrd m^3. The cross-section of the dam and some of the related technical data
are presented in Fig. 2, Appendix F.

The Sennar reservoir has been designed to operate in such a manner that during the low stages of the river, i.e. from January to July, the discharge downstream of the dam remains the same as it would have been had there been no dam
at all. This means that the discharge passing through the sluices in this period
is about equal to that which enters the reservoir from the upstream; but as the

Gezirah canal still continues to draw its fully supply, the amount of water
leaving the reservoir during its emptying period must be greater than the amount
that enters it. Therefore, the reservoir level during the said period drops
gradually with time (see Fig. 9.7., El-Zein Sagheyroon, S.S., 1965).

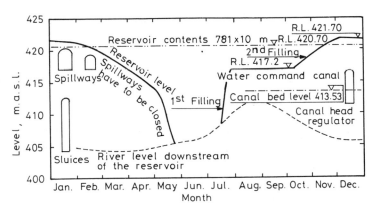

Fig. 9.7.   Operation of the Sennar reservoir and the downstream water level

Bearing in mind that irrigation of the Gezirah land must begin at the end of
July, the first filling is accomplished in the second half of July so that on the
the first of August the level upstream of the dam is raised to 417.20 to enable
the canal to draw its full share from the river. The criterion to start filling
the reservoir is that the daily natural flow of the Blue Nile equals or exceeds
160 million m^3/day. On the first of August the river flow becomes big enough to
permit further raising of water to levels higher than 417.20. However, this cannot be done until the Nile water is sufficiently clear of silt, to ensure that
solids are not deposited in the reservoir. Sediment deposition below the level
of 417.20 is not, however, too harmful as the corresponding reservoir contents
merely act as a water cushion and have nothing to do with live storage. To avoid
accumulation of silt deposits at higher levels the reservoir level should remain
at 417.20 till the Blue Nile water is sufficiently free of silt, usually around
mid-October. At this time the second filling of the Sennar reservoir starts and
is considered complete when the level is 421.70, at the end of November.

The reservoir level is kept at, or higher than, 417.20 as long as the Gezirah
land has to be irrigated and the Gezirah canal has to draw its full share. This
continues till the first of April, after which the level is allowed to fall
below 417.20. The canal then serves in supplying water for domestic purposes.
The fall goes on until the reservoir attains its minimum level between the end
of May and mid-July (see Fig. 9.7.), and this date can therefore be taken as the

end of the emptying phase of the operation cycle.

9.1.2.3   The Jebel el-Aulia Dam

This dam was constructed in 1937 on the White Nile some 44 kilometres south
of Khartoum to store water for the benefit of Egypt. The storage capacity of the
reservoir at the time the dam was completed was 3.5 mlrd m^3. Due to the continuous silting up of the reservoir basin the live capacity has shrunk gradually to
2.2 mlrd m^3 at Aswan by 1960.

The cross-section of the Jebel el-Aulia Dam and some technically-related data
are given in Fig. 3, Appendix F.

The operation of the Jebel el-Aulia reservoir depended to a large extent on
conditions at Aswan. At the beginning of February both reservoirs used to be
full. Since demand for irrigation in Egypt at that time exceeded natural supply,
the emptying of the Aswan reservoir used to begin in February; when it reached
a certain level, the emptying of the Jebel el-Aulia reservoir started. By the
time the released water arrived at Aswan, the volume of the reservoir contents
had further decreased, so that enough room was available for receiving the
released water as well as the natural river supply.

The volume of reservoir contents is that volume of water overlying the surface level of the natural river. For a certain level in a reservoir, the contents drop as the level of the natural river rises, and vice-versa. The estimate
of the reservoir contents therefore depended on the upstream gauge level (Wadi
Halfa in the case of the Aswan Dam). Fig. 9.8. shows the reservoir content
curves in terms of the reservoir level and the gauge reading at Halfa.

The emptying phase of operation of the Jebel el-Aulia reservoir started between the first of February and the first of March according to the natural income of the river. This phase used to last about two-and-a-half months, while
that of the Aswan reservoir continued until the end of July, when the natural
income began to surpass the quantity of water required. Similar to the Sennar
reservoir, the Jebel el-Aulia was filled in two stages; the first from the end
of July till the twentieth of August, when the reservoir level reached 376.50 m
above mean sea level and the second stage from the first of September and continued till the level reached 377.20. The difference between the two levels,
i.e. 376.50 and 377.20, left during the break between the two filling stages
corresponds to 1 mlrd m^3. It was meant to act as a safety valve when the Blue
Nile showed an extremely high flood. If this occurred, all the sluices of the
dam were fully opened and part of the flood water flowed backwards into the
White Nile.

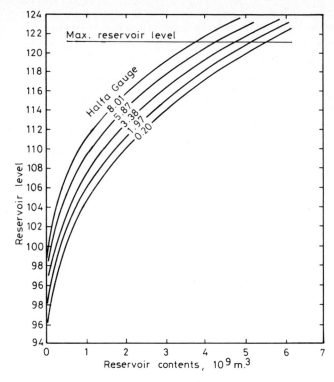

Fig. 9.8.   Aswan reservoir contents and their relation to Halfa gauge reading and the reservoir level (Hurst, H.E., Black, R.P., and Simaika, Y.M., 1959)

In the pre-High Aswan Dam period, the old Aswan and the Jebel el-Aulia reservoirs used to contribute about 7.8 mlrd m^3/yr at Aswan to the supply in the low-flow season, February to July.

9.1.2.4   Khashm el-Girba Dam

The main objective of the reservoir created by this dam is to regulate some of the Atbara water in order to supply the irrigation canals of the Atbara scheme with the necessary flow. In the technical data in Fig. 4, Appendix F, it is mentioned that the initial storage capacity was 1.3 mlrd reduced to about 0.95 mlrd m^3 by 1971, with the possibility of a further reduction by 40 million m^3 each year.

The flood of the Atbara begins in the latter half of June, depending on the date of fall of the torrential rains. The first filling of the reservoir up to level 462.00 m above mean sea level begins in the period from the first of July to the tenth of July every year in order to operate the turbine pumps to lift water to the main canal. This canal branches at kilo 26 off to three branch canals and at kilo 14 there is a pumping station for direct irrigation of some land. It may be of interest to mention that all the land irrigated by these canals is destined for the people of Wadi Halfa and district who were displaced by the High Aswan Dam project in Egypt.

During the first part of the filling phase the main sluices of the dam are left open to let the silt-laden water flow to downstream. This goes on till the end of August or when the river discharge near the dam site falls to 110×10^6 m^3/day. The second part of the filling phase then begins and continues till the beginning of October when the reservoir level reaches 473.20. During and after this period all the excess water is allowed to flow downstream until the flood ceases. In the second part of the filling phase the main sluices of the dam are shut off.

The stored water serves to supply the main canal from the beginning of November till the end of May.

9.1.2.5   The Roseires Dam

The storage reservoir formed by this dam was designed to retain water up to level 480 m above mean sea level in its first phase and up to 490 m above mean sea level in its second phase. These two levels correspond to volumes of 3 mlrd m^3 and 6.8 mlrd m^3, respectively.

The primary purpose of the Roseires Dam is to store water and to pass it downstream when required by the Gezirah, Managil extension and the river bank pump schemes (see Fig. 2.22.), all for the benefit of the Sudan.

The Roseires reservoir is operated in conjunction with Sennar with the purpose of satisfying the irrigation requirements upstream and downstream of the dam, and generating the maximum possible power. The filling of the reservoir during the rising flood including the peak, when the silt content is at its maximum, is avoided and filling is delayed to the latest possible time during the falling flood. Therefore the filling date is enforced by either the first of September, if the flow has never risen above 325 million m^3/day, or by the day later than the first of September when the discharge has fallen to 325 million m^3/day.

Cross-sections and some of the technical details of the dam are given in Fig. 5, Appendix F.

9.1.3   Over-Annual storage

Consider the succession of mass inflow curves shown in Fig. 9.9. Assume that line AB has a slope Q that should not be exceeded to avoid the flooding of the area downstream of the storage reservoir. For this purpose the maximum size R_2 needs to be empty before the arrival of the flood in the second year to store all inflow discharge in excess of Q. From the same figure it is clear that in

year number 3 one needs an empty space $R_3 < R_2$ for regulating the flow from the
reservoir to the downstream. The assumed Q, however, implies that the reservoir
cannot be emptied completely before the flood in the fourth year comes. A reservoir as such is said to have an over-year or over-annual storage.

This type of storage work does not exist in the Nile Basin. All the storage
works there belong to either the annual storage, or the long-term storage types.

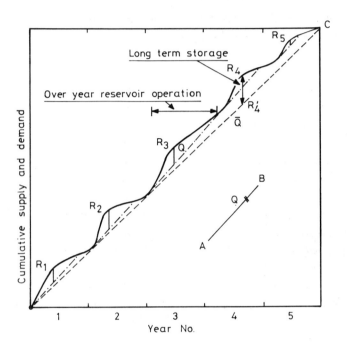

Fig. 9.9.   Over-annual and long-term reservoir operation

9.1.4   Long-term storage

Assume that the inflow and demand mass curves in Fig. 9.9., instead of being
for five years only, represent a much longer sequence. Let the line connecting
OC have a slope equal to the mean net inflow of the sequence, $\bar{Q}$. The storage
required to provide the downstream with demand $\bar{Q}$ is R'_4. This is known as long-
term storage and sometimes as century storage.

9.1.4.1   Design of reservoir capacity

The Rippl-diagram method can also be used for determining the capacity of a
long-term storage reservoir. Let the annual volumes of inflow to such a reser-
voir be X_1, X_2, ... X_n and the annual demand be $\alpha \bar{X}$, where $0 < 1$ (α = draft
rate) and $\bar{X}$ is the long-term mean inflow.

According to Kottegoda, N. (1980), one needs to determine the earliest year, j, which satisfies the condition

$$X_j \geq \alpha \bar{X} > X_{j+1} \quad , \quad j = 1, 2, \ldots, n - 1 \tag{9.1}$$

Suppose this corresponds to year $j = K_1$, it is suggested that two computations have to be made:

i) The length l of the period in which the reservoir level lies below the level at time K_1. From all such depletion periods select the maximum l_1 for which the following constraint is satisfied

$$\alpha\bar{X} \cdot 1 - \sum_{i=1}^{1} X_{K_1+i} \geq 0 \tag{9.2}$$

ii) The deficit d given by the equation

$$d = \max_{1 \leq m \leq l_1} (\alpha\bar{X}m - \sum_{i=1}^{m} X_{K_1+i}) \tag{9.3}$$

Let the maximum value of d be called d_1 and the value of m which maximizes d to d_1 be m_1.

In order to find the next value $j = K_2$, after time $l_1 + K_1$, equation 9.1 has to be applied for the years $j = K_1 + l_1 + 1$, $K_1 + l_1 + 2$, ..., $n - 1$, until the constraint it represents is satisfied. Equations 9.2 and 9.3 are used to determine l_2 and d_2 and m_2. This procedure is repeated for the whole sequence and the following statistics evaluated:

the maximum deficit $\quad : d_{max}^{(\alpha)} = d_j = \max (l_i)$, $i = 1, 2, 3, \ldots.,$

the maximum duration $\quad : m_{max} = m_j,$ and

the longest depletion period: $l_{max}^{(\alpha)} = l_j = \max (l_i)$, $i = 1, 2, 3, \ldots,$

all for a withdrawal rate $\alpha\bar{X}$.

The maximum deficit for any given α, $d_{max}^{(\alpha)}$, is the minimum reservoir capacity required to supply $\alpha\bar{X}$ every year. This capacity depends on the historical sequence of length n, which does not recur over a future design period of length n.

To avoid this shortcoming in the Rippl-diagram method, Hurst, H.E. (1951), approached the problem using the adjusted range R_n of a sequence of X_i values (i = 1, 2, ..., n) obtained from the cumulative departure from the mean $\bar{X}$. This can be expressed as

$$R_n = \max_{1 \le i \le n} \left\{ \sum_{i=1}^{i} X_i - i\bar{X} \right\} - \min_{1 \le j \le n} \left\{ \sum_{j=1}^{j} X_j - j\bar{X} \right\} \tag{9.4}$$

To proceed with the question of long-term storage, Hurst developed the variable known as the adjusted rescaled range R_n/s, where s is the estimate of the standard deviation from the sample data. This variable was computed for 720 phenomena and the values obtained led to the conclusion that R_n/s varies with the size of the data sample n according to the relation

$$\frac{R_n}{s} = \left(\frac{n}{2}\right)^K \tag{9.5}$$

where K is a variable index, often referred to as Hurst coefficient. The overall mean and standard deviation of K are 0.72 and 0.092, respectively. Moreover, the frequency distribution of the K values is claimed to be the normal one, as shown in Fig. 9.10. (Hurst, H.E., 1951).

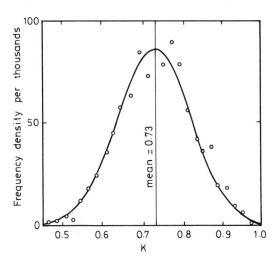

Fig. 9.10.   Frequency density graph of index

Equation 9.5 can be rewritten as

$$\frac{R_n}{s} = 0.61 \, (n)^{0.72} \tag{9.6}$$

It should be noted, however, that for a purely random time series process K tends towards 0.50 (Feller, W. 1951). The fact that natural time series yield an average K greater than 0.50 is known as the Hurst phenomenon.

For a detailed review, discussion and interpretation of the work of Hurst and his co-workers, the reader may be referred to Hurst (1951 and 1956), Fathy and Shukry (1956), Klemes (1974), Boes and Salas (1978), and Hipel and McLoed (1978).

Practical considerations may sometimes make it impossible to provide a storage capacity as large as R_n. Consequently, a different problem arises when it is required to guarantee a draft less than the mean. Hurst and his co-workers investigated this problem and developed a number of formulas to be used for computing the new capacity S given a new draft B, which is less than the long-term mean inflow in the reservoir. These formulas are

$$\log_{10} (\frac{S}{R_n}) = - 0.08 - 1.00 (\frac{\bar{X}-B}{s}) \qquad (9.7)$$

$$\log_{10} (\frac{S}{R_n}) = - 0.11 - 0.88 (\frac{\bar{X}-B}{s}) \qquad (9.8)$$

and

$$\frac{S}{R_n} = 0.91 - 0.89 (\frac{\bar{X}-B}{s})^{\frac{1}{2}} \qquad (9.9)$$

For the 51-year sequence (1906-1956) of the annual flow volumes of the River Niger at Koulikoro, Mali, if the mean $\bar{X}$ is taken at unity, $s = 0.2398$, $B = \alpha\bar{X}$, $\alpha = 0.85$. R_n will be approximately 2.8 (this corresponds to K = 0.75). Equations 9.7, 9.8 and 9.9 yield values of S of about 0.55, 0.61 and 0.58, respectively. Though these values do not compare exactly to $d_{max}^{(\alpha)} = 0.67$ obtained from the mass curve analysis, yet they are of about the same order of magnitude (Kottegoda, N.T., 1980). The important conclusion is that by taking $\alpha = 85\%$, i.e. assuming the new draft to be 15% less than the original mean inflow and subsequently the reduction in draft in proportion of the standard deviation to be at 62.5%, the new storage capacity is only 20 to 24% of the original capacity R_n.

The reduction in the storage capacity obtained from the different formulas as a result of reducing the draft below the long-term mean is represented by the curve in Fig. 9.11.

9.1.4.2   Regulation of long-term storage reservoirs

Hurst investigated a number of regulation schemes for long-term storage reservoirs (1956). In the first regulation, 32 phenomena were considered, the starting content of reservoir was $\frac{1}{2} R_{100}$ (where $R_{100} = 16.7s$) and the draft was taken equal to the mean of the historical data $\bar{X}$. The reservoir showed to be filled in 66% of the cases and emptied in 59% of them. In the second regulation the starting content was R_{100} and the reservoir emptied in 41% of the cases. The next seven regulations were divided by Hurst into 4 main types: (a) has a

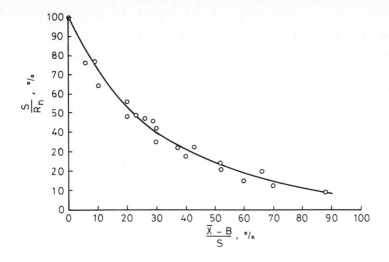

Fig. 9.11.   Relationship between reduced draft B and maximum deficit S

constant draft, (b) has a draft that varies with inflow to reservoir, (c) has a draft that varies with volume of reservoir contents and (d) has a draft that varies both with inflow and volume of reservoir contents.

The results of regulations based on initial data of 30 years are given in Table 9.3. The subscript denotes the length of period, $s' = 1.1\ s_{30}$ and the capacity $R'_{100} = 16.7\ s'$. From these results it is clear that regulation 7 is superior to 4 as the starting content in the two regulations is equal, $\frac{1}{2}\ R'_{100}$, and a reduction of draft of $0.1\ s'$ would have prevented the reservoir emptying in 94% of cases (14 out of 15). If it were permissible to start with the reservoir full, as in regulation 6, a draft of $\bar{X}_{10}$ changing every year would have met all cases except 1, i.e. all but 2%.

TABLE 9.3   Results of reservoir regulation types (a) and (b)

Regulation Type	Number Number	Number of phenomena	Starting content	Draft	% of cases reservoir fills	empties
(a)	3	51	$\frac{1}{2}\ R'_{100}$	$\bar{X}_{30}$	44	38
(b)	4	51	$\frac{1}{2}\ R'_{100}$	$\bar{X}_{10}$ changing every 5 years	23	19
	5	51	$\frac{3}{4}\ R'_{100}$	$\bar{X}_{10}$ changing every 5 years	56	5
	6	51	R'_{100}	$\bar{X}_{10}$ changing every 5 years	–	1
	7	51	$\frac{1}{2}\ R'_{100}$	$\bar{X}_{10}$ changing every year	12	15

Regulation 8, type (c) was tried in one case only, i.e. the Nile discharge at Aswan, and a drastic scale of reduction of draft with decreasing content was needed to prevent emptying. Furthermore, it was reported that the scheme was not considered good enough, compared with others, to warrant examination (Hurst, H.E., 1956).

The last regulation, number 9, belongs to type (d). In this regulation the reservoir started half-full and the draft began with $\bar{X}_{10}$ then changed every year as in regulation 7. The draft was further reduced or increased by a sliding scale in which the reservoir content was divided into nine parts. When the content was in the middle ninth, the draft was maintained at $\bar{X}_{10}$; when in the fourth one-ninth (from the bottom) it was reduced to $(\bar{X} - g)$, and so on down to $(\bar{X} - 4g)$, where g is the step of the scale. As a measure of safety against floods the scale could be applied to the reservoir contents in the upper four-ninths of the reservoir to increase the draft.

It was concluded that regulation 9 looked as if it were the type which would be most generally useful.

The rest of this section shall be devoted to a numerical example in which regulation number 7 is applied to the annual flow volume of one of the Central European rivers included in Table 9.4a.

TABLE 9.4a   Historical flow data used in connection with regulation 7

Year	Serial No.	Flow, 10^9 m^3/yr	Year	Serial No.	Flow, 10^9 m^3/yr	Year	Serial No.	Flow, 10^9 m^3/yr
1851	-29	10.88	1861	-19	7.88	1871	-9	10.28
1852	-28	9.27	1862	-18	7.00	1872	-8	5.85
1953	-27	10.38	1863	-17	5.39	1873	-7	5.17
1854	-26	12.24	1864	-16	5.82	1874	-6	4.92
1855	-25	13.18	1865	-15	7.73	1875	-5	7.16
1856	-24	8.76	1866	-14	5.46	1876	-4	11.95
1857	-23	6.94	1867	-13	14.19	1877	-3	8.77
1858	-22	6.09	1868	-12	10.50	1878	-2	8.45
1859	-21	8.20	1869	-11	7.41	1879	-1	9.05
1860	-20	10.91	1870	-10	8.86	1880	0	12.30

Consider the 30-year sequence from 1851 up to and including 1880. The year 1851 shall be referred to as year -29, year 1852 as year -28, ..., 1880 as year 0, 1881 as year 1, 1882 as year 2, ... etc. This set of data has a mean $\bar{X}_{30}$ = 8.70 x 10^9 m^3/yr, standard deviation S_{30} = 2.55 x 10^9 m^3/yr, skewness = 0.37, kurtosis = 2.55, K = 0.758 and the serial coefficients from lag 1 to lag 7 as 0.3322, -0.0380, -0.1134, -0.0734, -0.1202, -0.1936 and -0.1632, respectively. The reservoir content at the beginning of regulation 7 is $\frac{1}{2}$ R'_{100} = $\frac{1}{2}$ x 16.7 x 1.1 x s_{30} = 23.67 x 10^9 m^3. The annual flow volumes in the period from 1881 up

to and including 1964 are included in Table 9.4b. Using these data in regulation number 7 the following results can be obtained: reservoir contents at the
end of year $1 = \frac{1}{2} R'_{100} + I_1 - O_1 = \frac{1}{2} R'_{100} + I_1 - \bar{X}_{(-9\to0)}$ = (23.67 + 10.09 -
8.39) 10^9 = 25.37 x 10^9 m^3, reservoir contents at the end of year 2 = 25.37 x
$10^9 + I_2 - O_2$ = 25.37 x $10^9 + I_2 - \bar{X}_{(-8\to1)}$ = (25.37 + 10.56 - 8.37) x 10^9 =
27.56 x 10^9 m^3. Notice that I_1, I_2, ... are the inputs to reservoir in years
1, 2, ..., $\bar{X}_{(-9\to0)}$ is the mean inflow for the ten-year period 1871-1880,
$\bar{X}_{(-8\to1)}$ is the mean for the ten-year period 1872-1881, ... etc. The procedure
has been repeated to determine the yearly draft and the volume of reservoir contents at the end of each period (see Table 9.4b). Unfortunately, we have to stop
the regulation after year number 61 because the reservoir is filled. To resume
the regulation the draft in that year should have been increased by at least
1.787 mlrd m^3, or about 0.64s.

9.1.5   Long-term storage works in the Nile Basin

9.1.5.1   Owen-Falls Dam

This dam may be regarded as the first long-term storage work constructed on
the Nile. The storage reservoir was designed with the idea of raising the level
of Lake Victoria to 1134.75 m above sea level, i.e. raising it by about 3.0 m,
thereby increasing the lake storage by 200 mlrd m^3.

Because of the largeness of the surface area of the reservoir (about 67 x 10^9
m^2) it needs too long a time before it attains the design level. From Chapter 8
we have already learnt that the losses in the Sudd region and in the Bahr el
Jebel Basin increase considerably with the flow leaving Lake Victoria. Moreover,
such an outflow has to travel such a long distance before reaching Malakal or
Aswan that the usefulness of any improvement in the lake outflow to the Sudan or
Egypt becomes questionable unless the conveyance conditions of the river channel
are equally improved.

A cross-section of this dam and some of its leading technical particulares
are given in Fig. 6, Appendix F.

Since its construction, the dam has been used for hydro-electric power development for the benefit of Uganda only.

9.1.5.2   The High Dam at Aswan

The population of Egypt was estimated by the French mission of Napoleon
Bonaparte at less than 2.5 million. In 1821 Mohammed Ali found a somewhat higher
figure: 2.536.000. In 1846 the population had reached 4.5 million; in 1882
nearly 7 million; by 1927 this had doubled to 14.2 million. Since then the population has almost trippled. By 1970 it was 33 million and in 1980 it was estimated at 40 million. These figures demonstrate the remarkable increase in the

TABLE 9.4b   Volume of water in storage and annual draft

Year	Serial No.	Inflow, 10^9 m^3/yr	Draft 10^9 m^3/yr	Initial Storage, 10^9 m^3	Year	Serial No.	Inflow, 10^9 m^3/yr	Draft 10^9 m^3/yr	Initial Storage, 10^9 m^3
1881	1	10.09	8.39	23.67	1923	43	10.88	9.95	30.47
1882	2	10.56	8.37	25.37	1924	44	10.78	10.20	31.40
1883	3	9.71	8.84	27.56	1925	45	9.05	10.35	31.98
1884	4	8.79	9.30	28.43	1926	46	16.97	9.75	30.68
1885	5	6.47	8.79	27.92	1927	47	11.23	10.28	37.89
1886	6	9.15	9.62	25.60	1928	48	8.51	10.37	38.84
1887	7	6.37	9.33	25.13	1929	49	6.62	10.65	36.98
1888	8	12.40	9.09	22.16	1930	50	8.42	10.29	32.94
1889	9	9.52	9.49	25.47	1931	51	10.53	9.91	31.07
1890	10	13.69	8.96	25.50	1932	52	8.89	10.37	31.69
1891	11	9.40	9.10	30.22	1933	53	5.39	10.19	30.21
1892	12	9.30	9.03	30.52	1934	54	4.79	9.64	25.41
1893	13	7.41	8.91	30.79	1935	55	7.85	9.04	20.57
1894	14	8.92	9.25	29.29	1936	56	8.70	8.92	19.38
1895	15	11.29	9.26	28.97	1937	57	9.87	8.09	19.15
1896	16	11.89	9.74	31.00	1938	58	11.04	7.96	20.93
1897	17	13.62	10.02	33.14	1939	59	14.98	8.21	24.01
1898	18	8.64	10.77	36.75	1940	60	15.12	9.05	30.78
1899	19	9.02	10.37	34.61	1941	61	21.98	9.72	36.85
1900	20	14.29	10.32	33.26	1942	62	9.97	filled	49.12
1901	21	9.65	10.38	37.24	1943	63	4.89		
1902	22	8.10	10.40	36.51	1944	64	12.21		
1903	23	8.14	10.29	34.21	1945	65	9.84		
1904	24	7.05	10.36	32.06	1946	66	9.84		
1905	25	9.24	10.17	28.76	1947	67	8.07	Regulation stopped	Reservoir filled
1906	26	9.27	9.97	27.83	1948	68	10.66		
1907	27	10.31	9.70	27.13	1949	69	7.13		
1908	28	7.08	9.37	27.74	1950	70	5.87		
1909	29	9.84	9.22	25.45	1951	71	6.62		
1910	30	14.29	9.30	26.08	1952	72	7.62		
1911	31	8.14	9.30	31.07	1953	73	6.78		
1912	32	10.81	9.15	29.90	1954	74	7.03		
1913	33	8.42	9.42	31.57	1955	75	11.48		
1914	34	9.24	9.45	30.58	1956	76	10.85		
1915	35	15.04	9.66	30.37	1957	77	10.41		
1916	36	11.70	10.24	35.75	1958	78	12.84		
1917	37	10.38	10.49	37.20	1959	79	6.94		
1918	38	5.58	10.49	37.09	1960	80	8.85		
1919	39	10.28	10.34	32.18	1961	81	8.96		
1920	40	12.24	10.39	32.11	1962	82	8.45		
1921	41	5.93	10.18	33.96	1963	83	5.68		
1922	42	10.72	9.96	29.71	1964	84	6.29		

population of Egypt in the last century-and-a-half.

Such a state of affairs had to be met with, at least in part, by increasing
the share of Egypt in the Nile's water and, more important, reducing the immense
variability of the flow at Aswan between the flood and the low-flow season.

The future conservation of the Nile as portrayed by Hurst, Black and Simaika
in Vol. VII of the Nile Basin, consisted of long-term reservoirs in Lakes
Victoria and Albert, a regulator below Kyoga to avoid the delay in the passage
of water below the Upper Victoria Nile, the Jonglei diversion canal to reduce
the losses in the Bahr el Jebel Basin, an over-year storage reservoir in Lake
Tana, and another reservoir at the 4th Cataract on the Main Nile near Merowe for
annual storage and regulation of the discharge coming from the Upper Nile reservoirs (Hurst, H.E., Black, R.P., and Simaika, Y.M., 1946).

By 1950, of all the above-mentioned works, only the Owen Falls Dam had been
built. From the previous section we have seen that the benefit of this storage
work to either Egypt or the Sudan is limited. In 1952 Hurst and his co-workers
suggested a combination of the Upper Nile projects and a high dam to be built at
Aswan for the full utilization of the Nile water.

The hydrologic design of the live storage in the High Aswan reservoir was
based on equations 9.6 and 9.7 with $\bar{X}$ = 92 mlrd m^3/yr, B = 84 mlrd m^3/yr and
s = 18 mlrd m^3/yr. These values give a maximum deficit S = 85.9 mlrd m^3, rounded
off to 90 mlrd m^3. In other words a reservoir with a live storage volume of 90
mlrd m^3 can guarantee a gross draft of 84 mlrd m^3/yr at Aswan. The maximum
reservoir losses were estimated at 10 mlrd m^3/yr, thus bringing the annual net
draft to 74 mlrd m^3.

The annual rate of sedimentation at Aswan has been estimated at about 60 to
62 million m^3/yr and the life age of the reservoir at 500 years. The volume
allocated to dead storage is thus 31 mlrd m^3.

An additional room of 41 mlrd m^3 has been left as a safety measure for protection against high floods when the reservoir is full. The total size of the
reservoir formed by the High Dam is thus 162 x 10^9 m^3.

The net draft, which is 74 mlrd m^3/yr, is greater than the sum of the shares
of Egypt, 48 mlrd, and the Sudan, 4.5 mlrd, i.e. 52.5 mlrd m^3/yr, in the pre-
dam condition. The gain of 21.5 mlrd m^3/yr is divided in such a way that Egypt
gets 7.5 mlrd, bringing its new share to 55.5 mlrd m^3/yr and the Sudan gets 14.0
mlrd, bringing its new share to 18.5 mlrd m^3/yr. The ratio between 55.5 and 18.5
is equal to the ratio between the Egyptian and the Sudanese populations at the
time of amending the Nile Water Treaty in 1959. The discharges passing downstream of the dam were 53.12, 54.85, 55.36, 55.96, 55.29, 56.29, 56.25, 54.40,
54.71, 57.66, 62.18, 58.03 and 56.7, 56.72 mlrd m^3 for the years 1968-69, 1969-
70, 1970-71, ..., 1979-80 and 1980-81, respectively, with an average of

56.2 mlrd m^3 (Abul-Atta, A., 1975).

The surface area of the reservoir and the volume of its contents corresponding to the different water levels are given in Table 9.5. When the storage in
reservoir is at full level, i.e. 175 m above mean sea level the surface area
reaches 5168 km^2 and the corresponding evaporation loss amounts to about 14 mlrd
m^3 in a normal year. This figure is nearly 40% greater than that assumed by the
Ministry of Irrigation in Egypt.

TABLE 9.5   Area of water surface and volume of reservoir contents (Abul-Atta,
             A., 1978)

Water level, m.a.s.l.	Surface area, km^2	Volume of content 10^9 m^3	Water level, m.a.s.l.	Surface area, km^2	Volume of content 10^9 m^3	Water level, m.a.s.l.	Surface area, km^2	Volume of content 10^9 m^3
120	450	5.2	142	1380	23.8	164	3454	74.3
121	480	5.7	143	1449	25.2	165	3581	77.9
122	510	6.2	144	1511	26.7	166	3726	81.5
123	540	6.8	145	1589	28.3	167	3871	85.3
124	570	7.3	146	1663	29.9	168	4016	89.2
125	600	7.8	147	1737	31.6	169	4162	93.3
126	634	8.5	148	1812	33.4	170	4308	97.6
127	668	9.2	149	1887	35.3	171	4480	101.9
128	702	9.9	150	1962	37.2	172	4652	106.4
129	736	10.6	151	2052	39.2	173	4824	111.1
130	749	11.3	152	2142	41.3	174	4996	116.1
131	796	12.1	153	2232	43.5	175	5168	121.3
132	844	12.9	154	2323	45.7	176	5358	126.5
133	892	13.7	155	2414	48.1	177	5548	131.9
134	940	14.6	156	2521	50.5	178	5738	137.5
135	988	15.6	157	2628	53.1	179	5928	143.4
136	1038	16.6	158	2735	55.7	180	6118	149.5
137	1089	17.6	159	2842	58.5	181	6329	155.8
138	1140	18.7	160	2950	61.5	182	6540	162.3
139	1191	19.9	161	3076	64.5	183	6751	168.9
140	1242	21.2	162	3202	67.6	184	6962	175.7
141	1311	22.5	163	3328	70.9	185	7174	182.7

The reservoir has been in operation since mid 1964. Since then, up to and
including 1976, the change in the volume of its contents is shown in Fig. 9.12.
The operation rule adopted by the Ministry of Irrigation, Egypt, is to release
the quota assigned to Egypt each average year, and to release more water to the
downstream in years with high floods. At any rate the reservoir level on the
first of August each year should not exceed 175 m above mean sea level to be
able to accommodate the coming flood.

The water balance of the reservoir is drawn each year, and the actual losses
and the natural river supply are computed. Table 9.6 gives the results obtained
from such a balance for the period 1964-76. The periodic figures of the mean

supply and storage losses can be used in the future to amend the Nile Water
Treaty between Egypt and the Sudan whenever this proves necessary.

The natural supply at Aswan in 1973 was quite small and the reservoir had to
be emptied by 30.36×10^9 m^3 in the period from first of January to eighth of
July to cope with the demand. This happened in a single year, i.e. 1973. However,
there is always the chance of having a succession of low-flow years, which will
eventually lead to the failure of supplying the full demands of Egypt and the
Sudan. The application of a reduced draft is then necessary to cover part of the
water requirements of each country. This has to be strictly applied with their
consent to avoid complete depletion of the live storage in the reservoir. These
operation rules appear in detail in the items included in the Nile Water Treaty
of 1959.

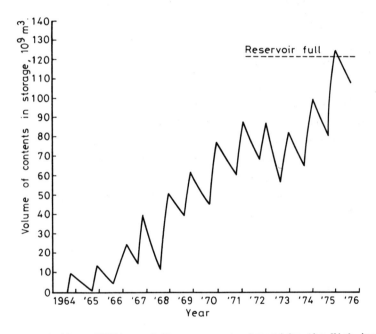

Fig. 9.12.   Filling of the reservoir formed by the High Aswan Dam

In spite of the fact that only a very small portion of the room allocated for
the dead storage has been occupied by the sediments deposited in the last 15
years, the total volume designed for both dead and live storages has been nearly
full with water only since 1975-76. The reservoir surface and volume of contents
are, therefore, at their maxima (see Fig. 9.12. and Table 9.6)

TABLE 9.6 Water balance of the reservoir formed by the High Dam at Aswan, Lake
Nasser (Abul-Atta, A., 1978)

Year	Maximum storage level, m.a.s.l.	Estimated losses, mlrd m^3			Water balance method, mlrd m^3			
		Seepage and absorption	Evaporation	Total losses	Inflow to lake	Outflow + change in storage	Total losses	Loss by seepage + absorption
1964	126.00							
1965	133.61	0.279	1.872	2.151	88.411	87.611	0.800	-
1966	140.74	1.022	2.308	3.330	71.422	69.662	1.760	-
1967	142.40	0.448	4.003	4.451	90.185	86.535	3.650	-
1969	156.50	6.836	5.466	12.302	73.768	66.598	7.170	1.704
1969	161.23	4.363	6.782	11.145	74.047	65.977	8.070	1.288
1970	164.87	4.251	7.823	12.074	77.258	68.324	8.934	1.111
1971	167.62	3.994	9.158	13.152	77.152	66.517	10.635	1.477
1972	165.26	-	9.587	9.587	58.050	45.145	12.905	3.318
1973	166.24	-	8.763	8.763	79.527	60.502	9.025	0.262
1974	170.61	4.878	9.694	14.572	84.934	70.465	14.469	4.775
1975	175.70	10.468	11.167	21.635	97.988	81.629	16.359	5.192
1976	176.51	1.929	12.443	14.372	68.964	54.820	14.144	1.701

For a better insight into the input to the reservoir, with the aim of establishing a better operating policy and helping to evaluate future projects, syn-
thetic stream flow data have been generated by means of stochastic models. These
are the multivariate autoregressive Markovian model, the disaggregation model
and the Broken line model. The three types of models have been applied to four
stations: Malakal, Khartoum, Atbara and Aswan and the results tested. It has
been concluded that if the Broken line method is first invoked for generating
annual stream flow, and the obtained data are disaggregated to monthly levels
using the disaggregation model, the combination will represent a more highly
powerful tool than the implementation of the Markovian model by itself (Cairo
University - Massachussets Institute of Technology Technological Planning Program, 1977).

It is not known yet how the High Dam Authority or the Ministry of Irrigation,
Egypt, will be able to incorporate the generated stream flows in the operation
of the reservoir (Lake Nasser).

In addition to the six tunnels with their twelve openings which feed twelve
turbines of the power station (see Fig. 7, Appendix F), there is a number of
spillways. The main spillway has twelve sluices; the auxiliary spillway situated
underneath the main one has twelve sluices also, and the emergency spillway has
thirty sluices. The latter is designed to operate when the reservoir level
reaches 178 m above mean sea level. All these sluices combined can discharge
11000 m^3/sec., or about 950 million m^3/day, to the downstream. Needless to say,

the High Aswan Dam Authority has laid down the rules to operate the different
spillways and developed the rating relations (stage-discharge) of their sluices
(Abul-Atta, A., 1978).

In order to limit the release to the downstream of Lake Nasser in the high-
flow years to the fixed demands of Egypt only, the Ministry of Irrigation there
has thought of releasing the excess water to Toshka depression. This depression
is located south-west of the High Dam and its level varies from 121 to 180 m
above mean sea level. At the level of 180 m the depression has a surface area of
6000 km^2 and a storage volume of 120 mlrd m^3. The reservoir is connected to the
depression by an old channel (Khor) 72 km in length (see Fig. 9.13.). With some
improvement the channel can serve as a diversion canal from Lake Nasser to the
escape depression.

It has been estimated that if there is an abnormally high year, similar to
1878-1879, with an annual volume of flow in the order of 150 mlrd m^3, the
release to the depression may reach 54 mlrd m^3. A sequence of high-flow years
similar to that in the period from 1870 to 1902 will need about 25 years to fill
the depression after satisfying the loss by evaporation and all other losses. On
the other hand, in a sequence of normal years, as that which happened from 1940
to 1970, there is no need to release water to the escape depression for periods
as long as 20 years or more.

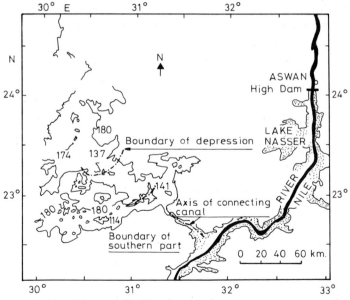

Fig. 9.13   Plan of Toshka depression

9.2   CONTROL WORKS ON THE MAIN NILE BETWEEN ASWAN AND THE MEDITERRANEAN SEA

9.2.1   Introduction

The idea of having a control on the Nile water levels and the amounts diverted to the canals taking from the river in the reach between Aswan and the Mediterranean dates back to the time Mohammed Ali was ruling Egypt. The difficulties encountered with the construction of the first barrages at the apex of the Delta caused a delay in the completion of that work until 1861. The original design was to raise the water level by 3.5 m. When in 1863 the head was raised from 1 to 1.5 m, cracks appeared in the barrage and further deterioration took place later. Repairs were undertaken and only upon their completion in 1890 was the barrage able to hold its designed head for the first time.

The barrages are, in fact, open type weirs or dams where the water flows through the vents and does not spill over a crest. The principal reason behind the choice of such a type of construction was to reduce the chance of sedimentation upstream of the barrage and the scour downstream of it, common features associated with the construction of weirs on alluvial streams. The flow through the vents as well as the water levels upstream and downstream of the structure are regulated by means of vertical steel gates resting either directly on the concrete floor or on a weir crest built monolithically with the barrage floor. The latter design helps in reducing the height of the gates, their own weight and the capacity of the lifting organs. Moreover, it was found more economical to use two overlapping short principal gates instead of a tall principal one. Emergency gates are placed upstream and downstream of the principal one. The former are usually made of timber. The upstream emergency gates act as a first line of defence to the barrage in case of complete closure for any sort of repair. The downstream emergency gates when set in position isolate the barrage from the downstream water and it thus becomes possible to evacuate the barrage from the volume of water impounded in it. This condition has always been taken into consideration while designing the floor as it is then subjected to the maximum difference in head. To reduce the floor thickness without endangering it, repair work requiring complete closure of the barrage should be carried out during the lowest possible river levels.

The protection of the downstream reach against scour is usually accomplished by extending the floor horizontally and by placing vertical cut-off walls and in some instances by end sills. In addition to all these measures, further downstream protection of the river bed and sides against scour is achieved by stone pitching for a distance between 10 and 20 m.

9.2.2   The barrages on the Nile from Aswan to the Mediterranean Sea

9.2.2.1   Esna barrage

The construction of the original barrage at Esna was completed in 1908. It helped in raising the upstream water level by 2.5 m to guarantee the requirements for basin irrigation in the Province of Qena (see Fig. 9.14.). The barrage consists of 120 vents each 5 m wide. Every two adjacent vents are separated by a 2 m thick pier, and every ten vents form a unit separated from the adjacent units by a 4 m thick pier. The barrage is combined with a lock 80 m long and 16 m wide. The Esna barrage was remodelled in the period between 1945 and 1947 to withstand a head difference of 4.5 m.

9.2.2.2   Nag-Hammadi barrage

This barrage was constructed in the period 1927-1930 to help irrigate the basins situated between Nag-Hammadi and Dayrut (about 210 000 hectares). The designed maximum difference in head is 4.5 m. The barrage consists of 100 vents each 6 m wide separated by piers each 2 m thick. Every ten vents form a unit separated from the adjacent units by a 4 m thick pier. Similar to the Esna barrage, this one is also combined with a lock, the basin of which is 80 x 16 m.

9.2.2.3   Assiut barrage

This barrage was constructed in 1902 to help in the summer irrigation of an area of about 400 000 hectares situated in Middle Egypt and El Fayum Province (see Fig. 9.14.) through the Ibrahimia canal. In 1938 the barrage was remodelled to withstand a difference of head of 4.3 m. This barrage consists of 110 vents each 5 m wide. The pier is, as usual, 2.0 m thick, but here every nine vents form a unit separated from the adjacent units by a 4 m thick pier.

9.2.2.4   The Delta barrages

In 1935 it became obvious that the old Mohammed Ali barrages had reached a poor state and could not stand any further remodelling. The new Delta barrages were thus constructed in 1937 and completed in 1939 to withstand a maximum head difference of 3.8 m during the summer season. The new barrages are situated a short distance downstream of the old ones and upstream of the submerged sills.

The barrage on the Damietta branch consists of 34 vents whereas that on the Rosetta branch consists of 46 vents. The width of each vent is 8 m and the pier separating one vent from the other is 2.5 m thick. Each barrage is combined with a lock of 80 x 12 m.

448

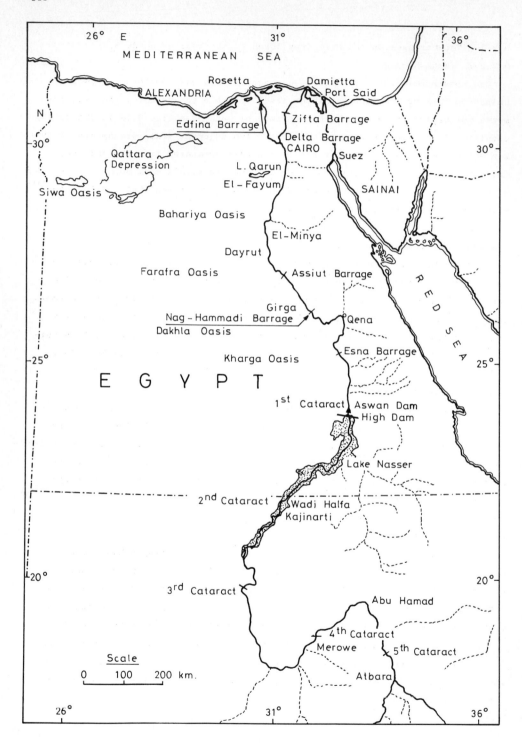

Fig. 9.14. Map showing the location of the Nile control works in Egypt

9.2.2.5 Zifta barrage

This barrage is situated on the Damietta branch some 87 km downstream of its mouth. The barrage was first constructed in 1903 then remodelled between 1949 and 1953. It consists of 50 vents each 5.0 m wide. The pier thickness is 2.0 m and every 10 vents form a separate unit having a pier 3 m thick. The design difference of head is 4.0 m and the summer upstream water level is enough for the canals feeding the eastern part of the Nile Delta. Combined with the barrage is a lock whose chamber is 55 x 12 m.

9.2.2.6 Edfina barrage

This barrage was constructed on the Rosetta branch in the period 1948-1951. The barrage was designed with the aim of avoiding the discharge of excessive amounts of fresh-water into the sea during the summer season. This was done to prevent the intrusion of the salt water from the sea into the river branch. This control work has no doubt helped to improve irrigation conditions in a vast area in the north-western part of the Nile Delta. The summer design head on the Edfina barrage was 2.8 m. The number of vents is 46, each 8 m wide; every two vents are separated by a 2.5 m thick pier. The lock combined with the barrage has a chamber of 80 x 12 m.

Fig. 8, Appendix F, shows some cross-sections of the above-described barrages.

9.3 WATER CONSERVATION SCHEMES

9.3.1 Conservation schemes of Bahr el Jebel and Bahr el Zeraf water

Next to storage and control of Nile water, conservation schemes aiming at saving the tremendous amounts which are lost in certain sub-basins of the river have been planned and the execution of one of them is already under way.

In many parts of the book, especially in Chapter 8, we have emphasized that the loss in the Bahr el Jebel and Bahr el Zeraf sub-basins amounts to not less than 14 mlrd m^3/yr. In order to save all, or some, of this water, two basic schemes have been thought of: the diversion canal scheme or the Jonglei canal and the embankment scheme. The layouts of the different schemes have been combined in one map (Fig. 9.15.) and analyzed by Salih, A.M.A. (1981).

The diversion canal has always been considered superior to the embankment scheme and its excavation and the construction of the related regulation works are under way. The first phase, which is about to be completed, comprises the excavation of one canal along the direct line (Fig. 9.15.), with a cross-section that allows the flow of 20 million m^3/day, and the construction of three regulators: one at the outlet and one on the River Atem at Jonglei latitude. In this phase, out of the average discharge of 75 million m^3/day at Mongalla, 66 million m^3/day reach Jonglei; 20 flow through the canal and the rest through Bahr el

Jebel and Bahr el Zeraf. The Jonglei canal will deliver 19 million every day at the outlet, and the natural rivers (Jebel and Zeraf), 32. The total volume of flow delivered in an average day at the outlet is 51 million m^3. The amount of water reaching Malakal without having a diversion canal is about 39 million m^3/ day (see Fig. 8.25.). So the first phase of the Jonglei canal saves about 4.4 mlrd m^3/yr at Malakal or 3.6 mlrd m^3/yr estimated at Aswan, Egypt. This gain shall be divided equally between Egypt and the Sudan. The detailed calculation of the gain to be expected in such a low-flow year as 1912 and in another year such as 1960 are included in Table 9.7 (Executive Organ for Development Projects in Jonglei Area, 1975).

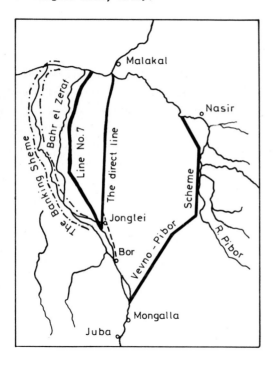

Fig. 9.15.   Plan of the proposed schemes

The second phase is planned on the basis of having a long-term storage in the Equatorial Lakes to equalize their natural outflows and improve the carrying capacities of the channels of Bahr el Jebel north of Mongalla as well as of Bahr el Zeraf to enable them to transmit the normal flow of 75 million m^3/day. If this improvement can be realized, then 71 million will reach Jonglei, of which 51 will be carried by the Jebel and the Zeraf and 20 by the canal. These will deliver 45 and 19 million m^3/day at the outlet, respectively. The daily flow reaching Malakal on an average day after implementing phase two is expected to

TABLE 9.7   Expected saving of Nile water produced by phase 1 of Jonglei scheme

Discharge in million m³/day for the months of the year 1912

Description of item	Jan.	Feb.	Mar.	Apr.	May	June	July	Aug.	Sep.	Oct.	Nov.	Dec.
Monthly mean discharge at Mongalla	52.6	51.3	54.3	61.1	67.3	61.9	71.2	80.7	88.1	89.6	78.3	65.9
Monthly mean discharge at Jonglei	53.4	47.8	46.8	49.4	55.5	60.7	56.2	64.0	71.0	76.2	77.2	69.2
Canal discharge reaching Malakal	19.0	19.0	19.0	19.0	19.0	19.0	19.0	19.0	19.0	19.0	19.0	19.0
Discharges of Jebel and Zeraf reaching Malakal	32.0	27.6	23.0	22.0	24.4	28.9	31.8	29.2	33.1	35.2	36.4	36.6
Total discharge at Malakal												
– after having the canal	51.0	46.6	42.0	41.0	43.3	47.9	50.8	48.2	52.1	54.2	55.4	55.6
– before having the canal	37.5	35.8	34.4	34.1	34.9	36.3	37.5	36.4	38.1	39.6	40.6	40.7
Net gain at Malakal	13.5	10.8	7.6	6.9	8.4	11.6	13.3	11.8	14.0	14.6	14.8	14.9
Net gain, 10⁶ m³/month	418.0	304.0	236.0	207.0	260.0	348.0	412.0	366.0	420.0	453.0	444.0	462.0

Net gain, 10⁹ m³/yr at Malakal   4.33
Net gain, 10⁹ m³/yr at Aswan   3.50

Discharge in million m³/day for the months of the year 1960

Description of item	Jan.	Feb.	Mar.	Apr.	May	June	July	Aug.	Sep.	Oct.	Nov.	Dec.
Monthly mean discharge at Mongalla	53.9	47.7	44.7	47.7	51.8	45.5	74.1	94.1	100.0	70.9	68.1	63.5
Monthly mean discharge at Jonglei	63.8	49.0	43.4	41.6	43.3	47.1	47.0	66.0	79.9	83.5	69.8	61.4
Canal discharge reaching Malakal	19.0	19.0	19.0	19.0	19.0	19.0	19.0	19.0	19.0	19.0	19.0	19.0
Discharges of Jebel and Zeraf reaching Malakal	32.1	29.8	24.1	19.5	18.0	19.4	22.4	22.3	33.8	37.3	38.0	35.0
Total discharge at Malakal												
– after having the canal	51.1	48.8	43.1	38.5	37.0	38.4	41.4	41.3	52.8	56.3	57.0	54.0
– before having the canal	41.1	31.7	25.4	34.5	33.5	32.7	34.4	36.8	38.6	37.0	34.6	44.0
Net gain at Malakal	10.0	17.1	17.7	4.0	3.5	5.7	7.0	4.5	14.2	19.3	22.4	10.0
Net gain, 10⁶ m³/month	310.0	493.0	548.0	120.0	108.0	170.0	217.0	140.0	436.0	598.0	672.0	310.0

Net gain, 10⁹ m³/yr at Malakal   4.32
Net gain, 10⁹ m³/yr at Aswan   3.50

be 64 million m^3, whereas without phases one and two it is limited to 40 million
m^3. The saving at Malakal will then be 8.8 mlrd m^3/yr estimated at 7.8 mlrd m^3/
yr at Aswan. The alternative solution is to double the carrying capacity of the
diversion canal either by widening phase-one canal or excavating another line.
Of the 40 million m^3/day carried by these canals, 38 million are expected to
reach Malakal and from the 31 million m^3 carried by el Jebel and el Zeraf, 25
million m^3 are expected to reach Malakal.

The saving of 8.8 mlrd m^3/yr at Malakal or of 7.8 mlrd m^3/yr at Aswan will be
equally divided between the Sudan and Egypt. Accordingly, the shares of the two
countries in the Nile water will increase to about 23 and 59 mlrd m^3/yr, respectively.

In his review of the Jonglei development project, Ibrahim, A.M., mentioned
that the total loss in the Bahr el Jebel and Bahr el Zeraf Basins amounts in a
normal year to more than 20 mlrd m^3. Of this amount, around 7 mlrd m^3 are lost
by direct evaporation from the rainfall (1977). He added that if this amount is
properly used in irrigated agriculture the annual revenue will be not less than
200 million pounds (based on 1977 prices).

9.3.2   Conservation scheme of Bahr el Ghazal water

The only conservation schemes laid down for the reclamation of the Bahr el
Ghazal swamps were those planned by Ahmed, A. el Aziz (1960). According to him
the total known loss from the complex of the sub-basins of the Bahr el Jebel, el
Zeraf and el Ghazal is in the order of 28 mlrd m^3/yr. From Chapter 8, the discharges of the streams in the Bahr el Ghazal Basin at the sites indicated on
Fig. 8.28. are about 14.07 mlrd m^3 in a normal year. Of this amount only 0.6
mlrd m^3 reaches the White Nile at Lake No and the rest, i.e. 13.47 mlrd are lost
in the swamps. Obviously the difference between the last figure and the 28 mlrd
m^3 is lost in the swamps of the Bahr el Jebel and Bahr el Zeraf Basins.

The idea of excavating a main collector running around the marshy land leading directly to the White Nile (see Fig. 9.16) was considered. This proposal,
however, was not appreciated. It was thought impracticable because of the
extreme flatness of the country; even if it were practicable, the difficulties
inherent in the maintenance of the channel and banks of the collector drain in
such a territory would be bound to entail enormous running expenses.

Ahmed, A. el Aziz (1960) proposed a hydro-electric scheme which he claimed
could deal with all the losses incurred in the swamps of both the Bahr el Jebel
and the Bahr el Ghazal. The scheme involved the generation of power from a
series of waterfalls in the Bahr el Jebel between Nimule and Rejaf, of which
Fola rapids (a drop of about 16 m) are the best known. The power would be transmitted several hundred kilometres to two points on the White Nile, where

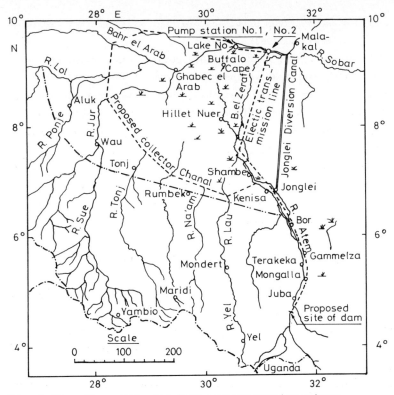

Fig. 9.16.   Upper Nile and Sudd regions pumping scheme

electrically driven pumping stations would be installed at the outlets of the
Bahr el Jebel and Bahr el Zeraf respectively for lifting water from the swamps
into the White Nile. The object of the pumping scheme is to create an artificial
slope in a really flat land, whose flatness is the reason for the formation of
the swamps. The drop created by the pumps, however small, would greatly enhance
the flow into the depression thus formed.

In his discussion of the paper of Ahmed, A., Hurst criticized the conservation scheme on the grounds that of the complicated and costly scheme, only the
embanking of the lower parts of the Bahr el Jebel and Bahr el Ghazal made a
small reduction of the losses in the Sudd region. The dam and power station at
Bedden (see Fig. 9.16.), the transmission line and the two pumping stations con-
tributed nothing (Ahmed A. el Aziz et al, 1961). In his discussion of the same
paper, Snelson, K.E. proposed to build reservoirs in the rivers above the swamps,
and to lead to the White Nile the share of their water which used to flow northward by a gravity channel with an acceptable slope. The proposed alignment of
this collector drain is shown by the dash-dot line in Fig. 9.16. As the hydrology of the Bahr el Ghazal swamp was not accurately known, the thought that the
conservation scheme proposed by Ahmed would lead to a saving of 12 mlrd m^3/yr

on the basis of 1 mlrd m^3/month, from being lost in that swamp was not widely
accepted (Ahmed, A. el Aziz et al, 1961).

It is remarkable that since that time, 1960-61, nothing has been planned or
undertaken to conserve the water lost in the Bahr el Ghazal swamps. It seems
that Snelson, K. was right when he said that this would be the last and almost
certainly the most expensive stage of Nile development and would present an
engineering challenge of magnitude to the engineers of the future who would
undertake it (Ahmed, A. el Aziz et al, 1961).

Any of these schemes, however, is expected to yield a saving of 12 mlrd m^3/yr
at Malakal or roughly 10 mlrd m^3/yr at Aswan. If this net gain is divided
equally between Egypt and the Sudan their shares in the Nile water at Aswan will
rise to about 64 and 28 mlrd m^3/yr respectively.

9.3.3 Conservation schemes in the basin of the Sobat and tributaries
9.3.3.1 Conservation of the Baro

Until now there are no definite plans laid down for conserving the water lost
in the basin of the Sobat and tributaries, similar to those of the Bahr el Jebel,
el Zeraf and el Ghazal. The reader will only find a number of guidelines discussed here and there, especially in Volumes VII thru' X of the Nile Basin (1946,
1950, 1959, 1966).

The Baro discharge undergoes losses from June to November each year. In this
period, notwithstanding the inflow to the secondary streams, the river stretch
from Gambeila down to the Pibor-Baro junction looses about 4 mlrd m^3 in a normal
year. If the inflow to these streams is considered, the loss simply reaches
5 mlrd m^3 each year. The details of these figures have already been presented in
Fig. 8.31, Chapter 8.

The conservation scheme consists of two components; a storage reservoir and
river embankment wherever necessary. The estimated reservoir capacity should be
between 5 and 7 mlrd m^3. Since the precipitation up the valley of the Baro
exceeds evaporation, any site suitable from constructional and geotechnical conditions would be suitable as far as hydrology is concerned. Two major difficulties arise in relation to the storage scheme. Firstly, the possibility of heavy
siltation in the reservoir and secondly, the dam site, if one exists, is bound
to be deep in Ethiopia and thereupon its construction, maintenance and operation
shall need all the support and co-operation from the Ethiopian authorities
(Hurst, H.E., Black, R.P., and Simaika, Y.M., 1966). The embankment of some
stretches of the Baro and its tributaries like Khor Mokwai and Khor Adura have
to be planned and designed in conjunction with the storage reservoir.

9.3.3.2   Conservation of water lost in the Machar swamps

The water balance of the Machar swamps has already been discussed in section
8.7.2.1.2, Chapter 8. The inflow consists of about 6.03 mlrd m^3/yr rainfall,
1.44 mlrd m^3 run-off from the eastern tributaries; Khor Lau, Daga, Yabus, Tombak
and El-Ahmar, and 2.67 mlrd m^3 from the Baro spill. The total sum of 10.14 mlrd
m^3/yr is lost altogether by evapotranspiration from the swamp.

The conservation scheme so far considered is that based on the diversion of
the Baro. This requires the construction of a barrage at the head of Khor Machar,
enlarging the cross-section of Khor Machar, excavating a canal through the
marshes to connect Khor Machar to Khor Adar and improving the carrying capacity
of the Adar up to its mouth on the Sobat (see Fig. 2.15.). This conservation
scheme is expected to yield a net gain of 3 mlrd m^3/yr.

Another line of thought is to combine the diversion of the Baro with the
reclamation of the Machar swamps. This includes the canalization of the eastern
tributaries which feed the marshes and the Khor Machar and Khor Wol. The
remaining part of the scheme is as already mentioned above, i.e. enlarging the
Khor Machar and excavating the canal connecting the same Khor to Khor Adar. This
scheme, known as the combined scheme, is expected to yield an average gain of
4.4 mlrd m^3/yr at the White Nile, or about 4 mlrd m^3/yr at Aswan. An equal division of this gain will bring the share of Egypt and the Sudan in the Nile water
up to, say, 66 and 30 mlrd m^3/yr respectively.

The sum of these two figures, 96 mlrd m^3/yr, coincides with the grand total
of 96.57 mlrd m^3/yr obtained by Dekker, G. (1972) from the data published by
Hurst for the period 1912-1947. The details of the latter are quite different
from ours. Dekker assigns 62 and 23.8 mlrd m^3/yr for the requirements in Egypt
and the Sudan respectively. Of the rest, which is about 10.8 mlrd m^3/yr, 7.31
mlrd m^3 are for Ethiopia; 2.46 mlrd m^3 for Kenya, Uganda and Tanzania; 1.0 mlrd
m^3 for Rwanda and Burundi. Obviously, the largest difference between the two
estimates lies in the 7.3 mlrd m^3 assigned by Dekker to Ethiopia and not to
Egypt and the Sudan as it should be.

9.4   SOME ENVIRONMENTAL IMPACTS OF THE NILE STORAGE, CONSERVATION AND CONTROL
      WORKS

9.4.1   Introduction

The construction of water storage, conservation and control works for stream
flow and water level regulation is mostly unavoidable to satisfy the interests
of land irrigation and drainage, hydro-power development, inland navigation and
other purposes. Such works, which are often referred to as man-made works, eventually lead to environmental changes. These changes need to be carefully investigated and certain measures implemented before any detrimental effect arises.

In this section we shall mention only some of the environmental changes obviously caused by man-made works, mostly in Egypt. Of these works the High-Aswan Dam is one that played, and probably still plays, a major role.

9.4.2   Environmental changes
9.4.2.1   Water quality

The Main Nile in Egypt receives its water, as we already know, from two principal sources: the Equatorial Lakes Plateau and the Abbysinian Plateau. The contributions of these two sources are about two-sevenths and five-sevenths, respectively.

The water leaving Lake Victoria has as much low concentration of dissolved compounds as the water of the lake. The alkalinity is a sodium balanced alkalinity, with rather low Ca^{++} concentrations. The contribution from Lake Albert has a pronounced effect. The high concentration of total salts, alkalinity, pH, phosphate, sulphate and chloride in the lake lead to strong increases in these quantities downstream in the Albert Nile. The influence of papyrus swamp in the Sudd region is remarkable, for here sulphate decreases and iron increases sharply due to the low oxygen concentrations. The accumulation of dissolved salts in the Lake No swamp is remarkably high, but the water which flows from it is apparently insufficient to alter the conductivity of water in the main stream. The River Sobat has a diluting effect. It resembles the Blue Nile in having a greater calcium than sodium content, high nitrate and low chloride contents during flood, for they both spring from one plateau. In the Blue Nile alkalinity is mainly balanced by calcium, while in the White Nile Na^+ predominates over Ca^{++} because the White Nile derives its water from the Equatorial Lakes Plateau, where this is a common phenomenon. The algal growth, which in African waters is more limited by the availability of nitrogen than by phosphate, dictates the levels of PO_4-P and NO_3-N (Golterman, H., 1975).

The salt concentration in the Nile water, in parts per million, as obtained from old analyses of samples collected from different locations, is as follows:

Year	Lake Victoria	Lake Albert	Lake George	Lake Edward	Bahr el Jebel	Bahr el Zeraf	River Sobat
1884	120	540	270	360	–	–	–
1902-06	134	–	–	–	164	220	70

Year	White Nile	Lake Tana	Blue Nile	River Atbara	Main Nile (Cairo)
1884	170	170	120	170	–
1902-06	140-200	–	103-106	–	124-260

The lowest and highest concentrations of the Main Nile at Cairo in the period 1919-1927 were 128 ppm (average for September: flood season) and 232 (average for April: near the end of the low-flow season). The corresponding figures for 1963 were 167 ppm for October and 200 ppm for March.

The Ministry of Irrigation, Egypt, has reported that the concentrations of the Nile water near Cairo in 1972 and 1975, i.e. 8 and 11 years after the construction of the High Dam, were 198 and 170 ppm, respectively. The main results obtained from the 1972 analysis are as given in Table 9.8.

TABLE 9.8   Results of analysis of Nile water near Cairo in 1972 (Abul-Atta, A., 1978)

Month	Total dissolved solids, ppm	Ca O	Mg O	SO_4^{--}	Cl^-
January	211	12	58.0	18.0	26.0
February	198	16	53.0	22.5	21.5
March	190	18	55.0	22.0	20.0
April	207	10	53.0	26.0	18.0
May	207	10	63.0	15.0	20.0
June	181	18	52.5	11.2	18.0
July	168	12	46.0	14.5	16.0
August	177	12	42.0	18.0	15.5
September	181	14	47.0	17.5	19.0
October	237	16	61.0	17.0	15.0
November	207	16	62.0	17.5	27.5
December	221	16	61.0	20.0	28.0

In addition to the quality of the Nile water near Cairo, the dissolved solids in the water of Lake Nasser was measured in May (before the arrival of the flood) and found to be 175, 151 and 146 ppm for the surface layer in the years 1975, 1976 and 1977, respectively. Moreover, the concentration did not show any considerable variation with depth up to 60 m below the surface.

Between Aswan and Cairo the concentration of the salts in water can be seen from the longitudinal section, Fig. 9.17.

The Nile water characterization prior to and after the construction of the High Aswan Dam, especially at the Cairo area, has been under observation by the Water Pollution Department of the Egyptian National Researches Centre since 1963.   Studies have indicated that after complete impoundment, 1970-1973, a drop in bacterial counts occurred below the dam, although a rise in coliform densities were recorded in the impounded water. Moreover, the faecal streptococcus group in numbers generally exceeding 10^2 100 ml^{-1} were observed in the samples collected in the Cairo district. The usual faecal streptococci counts fall between 10^1 100 ml^{-1} to 10^2 100 ml^{-1} for water samples collected both upstream and downstream of Cairo. The rise in the Cairo area could possibly be

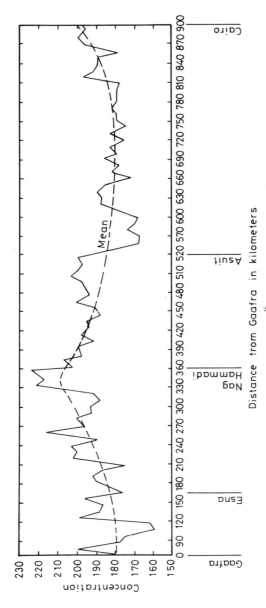

Fig. 9.17. Salinity concentration, ppm, of the Nile water between Aswan and Cairo

due to human interference on a large scale and to a possible rise in the pollu-
tant load reaching this part of the river (Saleh, F., 1980).

One may fairly conclude that some biological, as well as chemical, changes in
the quality of the Nile water have been, and shall be, caused by the long
impoundment of water in the reservoir. One of the noticeable effects of these
changes is the thermal stratification of water in Lake Nasser, especially during
the summer season. The stagnant water layer at and near the lake bottom has led
to the situation where biological decomposition of organic matter completely
removes dissolved oxygen, which cannot be replenished. The absence of oxygen
might well threaten the lake as a future potential fishery.

9.4.2.2   Sedimentation

The sediments transported by the Nile are made up of bed and suspended loads.
The investigations carried out in the pre-High Dam period have shown that the
bed-load transport was only 1 to 2% of the total transport of the river. It
accordingly became customary to consider the total sediment load equal to the
suspended load. The latter was regularly measured (1928, 29, 30, 31, 38, ..., 55,
..., 63, ...) especially during the flood (August-November) at Wadi Halfa,
Gaafra (35 km downstream of Aswan), El-Hanady, El-Samata, Sallam and El-Ekhsas.
The main conclusions drawn from those investigations were as follows:

i)   98% of the annual sediments are brought by the Nile during the flood season
     (August-November). The observed sediment varies from 100 to 5800 ppm cor-
     responding to discharges of 200 to 1000 million m^3/day (see Fig. 9.18.).
     The annual volume of sediments reaching Aswan amounts to about 60 million
     m^3, weighing about 125 million metric tons.

ii) The percentages of clay, silt and sand fractions in the suspended load
     change with time during the flood season. The average percentages are as
     follows:

Month	Clay < .002 mm	Silt .002 - .02 mm	Sand .02 - .2 mm
August	35	45	20
September	30	45	25
October	30	45	25
November	35	35	30

iii) The graphic plot of the discharge Q reaching Aswan against the concentra-
     tion C looks like a hysterisis loop, indicating that the sediment concen-
     tration for a given discharge is bigger during the rising flood than during
     the falling flood.

     The relationship between the sediment concentration in ppm and the dis-
     charge in million m^3/day was found to be of the type (Quélennec, R., and
     Kruk, C.B., 1974):

460

$$C = a \cdot Q^{(b-1)} \qquad\qquad\qquad (9.10)$$

where a = 1.525 x 10^{-1} and b = 2.59 for the rising flood

and   a = 1.823 x 10^{-2} and b = 2.79 for the falling flood

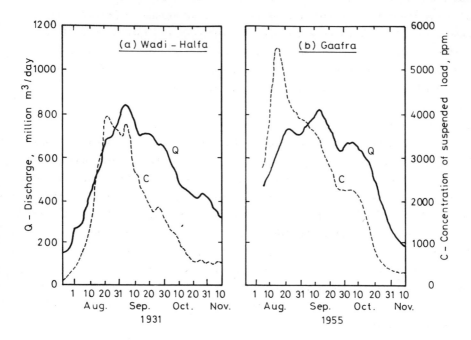

Fig. 9.18.   Discharge and concentration of suspended matter in the Nile water during the flood seasons of 1931 and 1955

The High Aswan Dam has been in operation since 1964. When the flood reaches Lake Nasser the velocity of water drops from more than 1.0 m/sec. to about 0.02 m/sec. This considerable fall in velocity leads to the deposition of about 98% of the suspended matter in the lake formed by the High Dam. The sediment load which passed through the dam to the downstream was estimated at 26.3, 5.7, 3.8, 3.2, 2.3, 1.9, 2.8, 2.4, 27. and 2.7 million metric tons for the years 1964, 65, 66, 67, 68, 69, 1970, 71, 72 and 73, respectively. In other words, the average weight of sediments passing downstream of Aswan in the post-dam condition is about 2.5 million metric tons compared with 125 million metric tons in the pre-dam condition (Shalash, S., 1974).

Generally, the tendency is greater for deposition of silt in the upper reaches of the reservoir rather than in the vicinity of the dam. Observation on concentration of solid materials on sections of the reservoir showed that it decreases from year to year at the sections near the dam. During the flood of

1971 the concentration of silt became negligible at the section situated 250 km upstream of the dam. In the first years of the initial filling of the dam (1965-1967) silt was deposited just upstream of it (Abu Zeid, M., 1979).

The Nile is a silt-bearing river and the delta and valley in Egypt are made up of its sediments (Chapter 7). The rate of sedimentation in the pre-storage period was estimated at 6 to 15 cm/century, with an average of 0.8 mm/yr, used to cover the surface of agricultural soil in Egypt. Since 1964 this amount ceased to reach the soil, with the following consequences:

i)   loss of a unique source of natural nourishment to the soil, and

ii) loss of supply of fresh soil, thereby eliminating any chance of natural improvement of depth to water table.

The losses described by i) and ii) may be overcome by artificial fertilizers and soil amendments. Both seem to be energy-consuming and expensive.

9.4.2.3   Degradation of river channel

The degradation of the river channel from Aswan to the Mediterranean coast was observed long before the construction of the High Aswan Dam.

The depth of the scour hole downstream of the Nag-Hammadi barrage reached almost 70 cm shortly after it was put into operation. The channel scour is regarded as a phenomenon associated with such construction of storage and control works as the old Aswan Dam, the Esna and the Nag-Hammadi barrages, etc., since they are chiefly responsible for disturbing the flow and silt regime in the river. Further degradation of the Nile as a result of the High Dam was anticipated prior to its construction. Many scientific and technical authorities have been, and probably still are, involved in estimating the rate and ultimate depth of degradation at different points on the Nile between Aswan and the sea.

The bed degradation from 1964 up to 1971 reached 0.22, 0.41, 0.24 and 0.27 m downstream Aswan Dam, Esna barrage, Nag-Hammadi barrage and Assiut barrage, respectively (Shalash, S., 1974). For the same sites the maximum drop in the water level of the Nile was 0.40, 0.80, 0.58 and 0.50, respectively (Kenawy, I., et al, 1973). A more complete presentation of the drop in water level in the period from 1963 up to and including 1977 at the said four stations plus the station of El-Ekhsas is shown in Fig. 9.19 (Hartung, F., 1978). This figure shows clearly that degradation downstream of the dam and farther to the downstream of Esna barrage has been recovering since 1973 at a rate of 3 to 4 cm/yr. Contrary to this observation, the channel below the Nag-Hammadi barrage has been eroding at about the same rate, and a short distance below the Assiut barrage at about 1.5 cm/yr.

The graph of El-Ekhsas shows a rise in the water level of 43 cm in 14 years or 3 cm/yr pointing to bed agradation in this reach of the river. This means

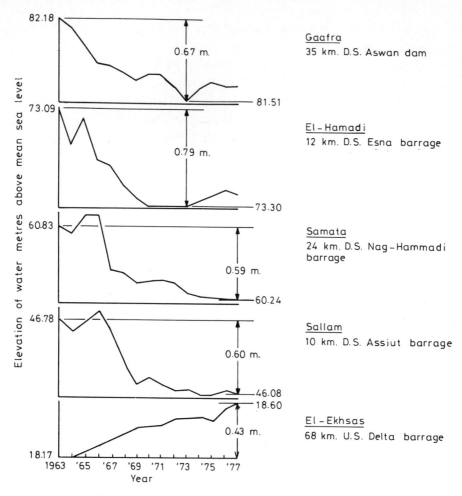

Fig. 9.19.   Change of Nile water level since 1963 at five different sites corresponding to discharge of 100 million m³/day)

that bed degradation has not taken place for some distance upstream of the Delta barrages and for the whole reach from these barrages down to the sea.

A number of theorems and approaches have been developed and worked out for predicting the ultimate bed degradation and the time the river needs before its equilibrium is reached. The wide discrepancy between the results so far obtained has been argued on the following grounds:

i)   the assumptions used in developing the theorems describe neither the actual characteristics of the river channel nor the flow conditions adequately;

ii)   the measures adopted by the Ministry of Irrigation, Egypt, seem to be effective, at least locally, in deccelerating the degradation, and

iii) the number of years of observation is so far too short to be able to draw any definite conclusion about the extent of agreement between theory and observation.

It does not seem possible, however, to predict future degradation with a high degree of accuracy. Nevertheless, the data collected point clearly to the need for strengthening or reconstructing the barrages on the Nile from Aswan to the sea and for providing a better protection to the bed and slopes of the river in this reach.

9.4.2.4   Beach erosion

Since ancient times, the sea coast defended itself against wave attack, storms, and sea and coastal currents. A quasidynamic equilibrium was achieved and the balance between the sea forces, the sediments carried away from the shores and the sediments supplied by the Nile seemed to swing in favour of accretion on the coast. All hydraulic structures on the Nile between Aswan and the sea and which led to instability in the amount, rate and temporal distribution of the silt transport to the sea, resulted in disturbance to the said equilibrium. Beach erosion near the river mouths at Damietta and Rosetta was observed as early as 1902. Since then, the shore line has been frequently mapped both before and after the construction of the High Aswan Dam. According to Wassing, F. (1964), the rate of retreat of the promontories at Damietta and Rosetta had an average of 29 m/yr in the period from 1898 to 1960. The gradual development of the contour line of the shore at the mouths of the river branches is as shown in Fig. 9.20.

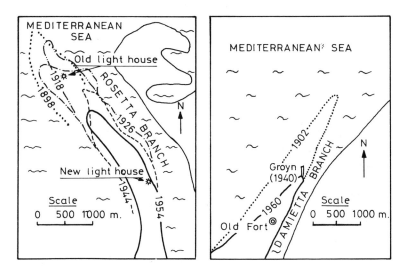

Fig. 9.20.   Retreat of the shore line near the mouth of each of the Damietta and Rosetta branches from 1898 to 1960

Before the construction of the High Dam, an annual mean of about 35 mlrd m^3 of Nile water was discharged to the Mediterranean throughout the flood period from August to November. This volume of water used to carry not less than 90 million metric tons of sediments along to the sea. In the pre-dam condition, the surface current velocity at the mouths of the branches was 4 knots and less than 0.5 knot at the bottom, both during the flood season. After the construction of the dam these velocities dropped considerably. The fact that the Nile has no longer discharged its sediments to the sea since 1964 is producing an imbalance in the near-coast sediment budget, thereby making the coast vulnerable to considerable erosion.

The Egyptian authorities have, for quite some time, been investigating the beach erosion phenomenon, both theoretically and in the field, with the help of some of the international organizations. Some measures aiming at both erosion prevention and accretion on the coast are under way.

The High Dam is not only responsible for the accelerated instability of the shore line, but also for the change in the estuariane circulation pattern from two-layered to one-layered flow. Such a change in the circulation pattern has led to increased surface salinity in the post-dam period. This can easily be seen from Figs. 9.21a. and 9.21b. The salinity of the deep water observed during September and October in the post-dam years is similar to that during the flood season in the pre-dam years (Sharaf el-Din, S., 1976).

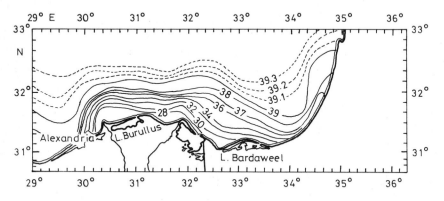

Fig. 9.21a.   Map showing the surface salinity distribution during the flood season (August-October) before the construction of the High Dam

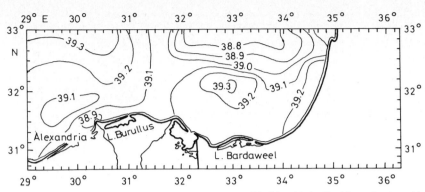

Fig. 9.21b.   Map showing the surface salinity distribution during the period
(August-September) after the construction of the High Dam

9.4.2.5   Groundwater level

As a rule, when land is put under irrigation, the water table rises, and if
no proper drainage is applied the rise continues till the soil mass becomes
water-logged and practically produces no yield. The situation in Egypt is no
different than elsewhere. Balls, W.L. (1953), listed the minimum levels of the
water table in the neighbourhood of Cairo in the period from 1894-1951. The
levels in wells and/or pits situated at 1 km, 2 km and between 1 and 2 km from
the Nile on the Giza farm are given in Table 9.9. Balls commented on these
measurements saying, "... it is evident that the minimum level of the water
table is conditioned by seepage water from canals, and by drainage water from
surface irrigation, superposed on a foundation of mass inflitration from the
barrage (Delta) pond".

TABLE 9.9   Minimum levels of the water table in the neighbourhood of Cairo
             (Balls, L., 1953)

Year	Water level		Year	Water level		Year	Water level
	1 km	2 km		1 km	2 km		
1894	14.29	14.67	1907	15.39	15.48	1928	16.40
1895	14.84	14.95	1908	15.43	15.47	1929	16.45
1896	14.73	14.81	1909	–	–	1930	16.30
1897	14.75	14.81	1910	–	–	1931	16.30
1898	14.50	14.55	1911	–	–	1932	16.30
1899	14.85	15.03	1912	16.00		1933	16.80
1900	14.24	14.40	1913	16.44		1934	16.35
1901	14.70	14.74	1914–22	No data		1935	16.35
1902	14.74	14.89	1923	<16.70		1936	16.65
1903	14.90	15.03	1924	16.30		1937	16.80
1904	15.41	15.55	1925	16.25		1938	16.50
1905	15.39	15.48	1926	16.20		1939	16.90
1906	15.43	15.47	1927	16.20		1950	16.80
						1951	16.75

In the pre-High Dam period the hydrograph of the Nile at Aswan consisted of a long low-flow season (December-July) followed by a flood season (August-November). As a result of the hydraulic connection between the Nile and the groundwater, the river flood was transmitted to the groundwater reservoir in the form of a reduced wave. The reverse happened during the low-flow season, i.e. the wave was transmitted from the groundwater reservoir to the river and the base flow used to sustain the low-flow in the river. In the locations where the semi-confining layer has a limited resistance to vertical flow, it was possible to observe the effects of canal rotations and irrigation applications on the groundwater level (Shaḥin, M., 1955).

Since the construction of the High Dam, the discharges passing downstream of Aswan are regulated in such a way as to satisfy the demand. As a result, the discharge in what used to be a low-flow season has been improved and the flood season does not exist any longer. These conditions have caused the water in the semi-confining layer to reach a level almost higher than that of the piezometric head of the groundwater. The latter, in turn, is higher than the corresponding water level in the Nile. This can be attributed to the application of irrigation water in amounts far in excess of the actual need of the soil and the seepage from the canals running at high levels for a long time.

The hydrographs of the groundwater levels at some locations in Upper and Middle Egypt before and after the construction of the High Dam are shown in Fig. 9.22. (Attia, F., et al, 1983). From this figure it is clear that all the hydrographs show much more uniformity with time after the construction of the dam than before it. Secondly, most of the hydrographs show a rise in the water level in what used to be the low-flow season during the pre-dam condition. This is mostly coupled with a fall in the level in what used to be the flood season before the dam construction. Thirdly, the piezometric heads at many places seem to have approached a stable level during the last few years.

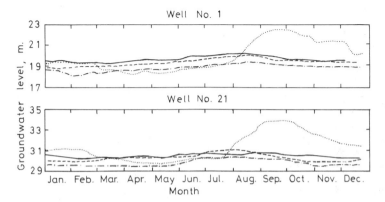

Fig. 9.22.   Groundwater heads before and after the High Aswan Dam

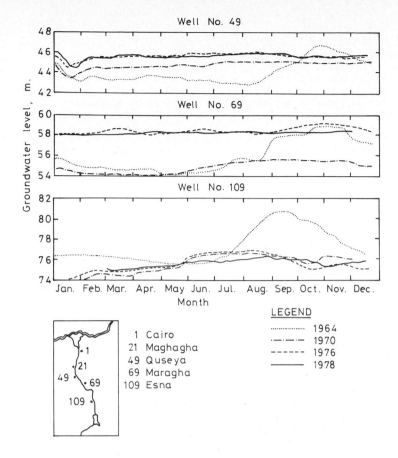

Fig. 9.22.   (continued)

9.5   REFERENCES

Abul-Atta, A., 1975. The water policy of Egypt. Ministry of Irrigation, Cairo,
    Egypt, 15 pp (with drawings).
Abul-Atta, A., 1978. Egypt and the Nile in the post-dam epoc. Ministry of
    Irrigation and Land Reclamation, Cairo, Egypt, 145 pp (in Arabic, with draw-
    ings).
Abu Zeid, M., 1979. Short and long-term impacts of the River Nile projects.
    Water Supply and Management, Vol. 3: 275-283, UK.
Ahmed, A. el-Aziz, 1960. Recent developments in the Nile control. Proceedings
    of the Institute of Civil Engineers, Vol. 17, Paper no. 6102: 137-180,
    London, UK.
Ahmed, A. el-Aziz, 1961. Discussion of 'Recent developments in the Nile con-
    trol', Paper no. 6102, and of 'An analytical study of the storage losses in
    the Nile Basin with special reference to Aswan Dam Reservoir and the High
    Dam Reservoir' Paper no. 6370. Proceedings of the Institute of Civil Engi-
    neers. Vol. 19 (special excerpt): 337-415, London, UK.

Attia, F., Amer, A., and Hefny, K., 1983. Effect of High Aswan Dam on groundwater conditions in Upper Egypt. Proceedings of the International Conference on Water Resources Development in Egypt, Cairo, Egypt: 99-119.

Balls, L.W., 1953. The yields of a crop, E. & F.N. Spon Limited. London, UK, 144 pp.

Bernier, J., 1966. Sur la garantie assurée par un réservoir de régulation saisonnière. Int. Assoc. Sci. Hydro., Symposium of Garda. Pub. no. 71, Vol. 2: 575-589.

Bertlin, D.P., and Olivier, H., 1954. Owen Falls, constructional problems. Proceedings of the Institute of Civil Engineers, 3, Part I, Paper no. 6008: 670-699, London, UK.

Boes, D., and Salas, J., 1978. Nonstationarity of the mean and the Hurst phonomenon. Water Resources Research, Vol. 14: 135-143.

Cairo University/Massachussets Institute of Technology, 1977. Stochastic modelling of Nile inflows to Lake Nasser. Published by Cairo University, Giza, Egypt.

Corney, J.V., 1966. The construction of Roseires Dam. The Sudan Engineering Society Journal, no. 11: 7-13, Khartoum, the Sudan.

Dekker, G., 1972. A note on the Nile. Water Resources Research, Vol. 8 no. 4: 818-828.

Executive Organ for Development Projects in Jonglei Area, 1975. Jonglei project, phase one. Tamaddon Press, Khartoum, the Sudan.

Fathy, A., and Shukry, A.S., 1956. The problem of reservoir capacity for longterm storage. Proceedings of the American Society of Civil Engineers, Paper no. 82, Hy 5: 1-27.

Feller, W., 1951. The asymptotic distribution of the range of sums of independent random variables. Annales of Mathematical Statistics, 22: 427-432.

Fitt, R., Marwick, R., and Whitaker, F., 1967. The Roseires Dam, Sudan: planning and design. Proceedings of the Institute for Civil Engineers, Vol. 38, Paper no. 7047: 21-51, London, UK.

Golterman, H.L., 1975. River ecology (edited by B.A. Witton). Chapter 2: Chemistry, 39-80. The University of California Press, 725 pp.

Hartung, F., 1978. 75 Jahre Nilstau bei Assuan, Entwicklung und Fehlentwicklung. Bericht no. 40, Versuchanstalt für Wasserbau der Techinschen Universität München-Oskar v. Miller Institut, Munich, W. Germany.

Hipel, K., and McLoed, A.I., 1978. Preservation of the re-scaled adjusted range, Part one - a reassessment of the Hurst phenomenon, Water Resources Research, Vol. 14: 491-508.

Hurst, H.E., Black, R.P., and Simaika, Y.M., 1946. The Nile Basin, Vol. VII, the future conservation of the Nile. Physical Department Paper no. 51, Eastern Press, Cairo, Egypt, 159 pp (with drawings).

Hurst, H.E., 1950. The Nile Basin, Vol. VIII, the hydrology of the Sobat and the White Nile and the topography of the Blue Nile and Atbara. Physical Department Paper no. 55, Government Press, Cairo, Egypt, 125 pp.

Hurst, H.E., 1951. Long-term storage capacity of reservoirs (with discussion). Transactions of the American Society of Civil Engineers, Vol. 116, Paper no. 2447: 770-808.

Hurst, H.E., 1956. Methods of using long-term storage in reservoirs (with discussion). Proceedings of the Institution of Civil Engineers, Part I, Paper no. 6059: 519-590. London, UK.

Hurst, H.E., Black, R.P., and Simaika, Y.M., 1959. The Nile Basin, Vol. IX, the hydrology of the Blue Nile and Atbara and the Main Nile to Aswan, with some reference to projects, Nile Control Department Paper no. 12, Government Printing Offices, Cairo, Egypt, 206 pp.

Hurst, H.E., Black, R.P., and Simaika, Y.M., 1966. The Nile Basin, Vol. X, the major Nile projects, Nile Control Department Paper no. 23, Government Printing Offices, Cairo, Egypt, 247 pp.

Ibrahim, A.M., 1977. Jonglei development project. ICID Bulletin, Vol. 26, no. 2: 70-72 and 81, New Delhi, India.

Kinawi, I.Z., and El-Ghamry, O., 1971. Some effects of the High Dam on the
    Environment. 11th Congress on Dams and Grand Barrages, Madrid, Spain, Q40,
    R59: 959-973.
Klemeš, V., 1974. The Hurst phenomenon - a puzzle? Water Resources Research,
    Vol. 10: 675-688.
Kottegoda, N.T., 1980. Stochastic water resources technology. The MacMillan
    Press Limited, London, UK, 384 pp.
Kruk, C.B., and Quélennec, R., 1975. Suspended sediment transport of the Nile.
    Paper submitted to a seminar on Nile sedimentology, Alexandria, Egypt, 7 pp
    (with drawings).
McMahon, T.A., and Mein, R.G., 1978. Reservoir capacity and yield. Elsevier
    Scientific Publishing Company, Amsterdam, 213 pp.
Saleh, F.A., 1980. Isolation and enumeration of Faecal Streptococci from Nile
    water (1975-1976). Water Research, Vol. 14: 1669-1678.
Salih, A.M., 1981. Reclamation of water from el-Jebel Swamps. Water Interna-
    tional, Vol. 6 no. 2: 71-74.
Shahin, M., 1955. Effect of Nile and canal water levels on subsoil water move-
    ment in irrigated farms. Thesis submitted to the Faculty of Engineering,
    Cairo University for M.Sc degree, Egypt.
Shahin, M., 1971. Hydrology of the Nile Basin, Lecture notes, 2nd edition,
    International Courses in Hydraulic and Sanitary Engineering, Delft, The
    Netherlands, 140 pp.
Shalash, S., 1974. Facts about degradation. Progress report of the Hydraulic
    Research Station of the Ministry of Irrigation, Egypt, 32 pp (with drawings).
Sharaf el-Din, S., 1976. Effect of the Nile flood on the estuarine and coastal
    circulation pattern along the Mediterranean Egyptian coast. Hydrological
    Sciences Bulletin XXI, 3: 451-461.
Simaika, Y.M., and El-Sherbini, H., 1957. Some aspects of erosion in Egypt.
    Proceedings of the International Association of Scientific Hydrology,
    Toronto, Canada, Vol. I: 381-386.
Simaika, Y.M., 1961. Degradation of the Nile due to interception of the silt in
    the High Aswan Reservoir. Proceedings of the 7th Congress on Dams and Grand
    Barrages, Rome, Italy, C-7: 639-656.
Wassing, F., 1964. Coastal engineering problems in the Delta region of UAR
    (Egypt). Memoranda W1-W6, Reports of UN expert to the Department of Ports
    and Lighthouses, Alexandria, Egypt.
Westlake, C., Mountain, R., and Paton, T., 1954. Owen Falls, Uganda, hydro-
    electric development. Proceedings of the Institution of Civil Engineers 3,
    Part I, Paper no. 6007 (with discussion): 630-669, London, UK.
El-Zein Sagheyroon, S., 1965. Proposed projects for the utilization of the
    Sudan share of Nile water. The Sudan Engineering Society Journal, no. 10:
    5-12, Khartoum, the Sudan.

APPENDIX A$_1$ - Annual depth of rainfall, in mm

No.	Station Name	Latitude, N	Longitude, E	Altitude, M	1938	1939	1940	1941	1942	1943	1944	1945	1946	1947
	EGYPT													
1	Sidi Barrani	31° 38′	25° 58′	27	178	127	–	–	–	–	–	–	–	–
2	Borollos (L.H.)	31 36	31 05	10	200	84	170	143	(49)	260	263	187	256	113
3	Sallum	31 33	25 11	07	85	84	–	–	–	–	–	–	50	307
4	Damietta (L.H.)	31 31	31 51	02	180	91	120	114	97	170	184	133	71	48
5	Damietta	31 25	31 49	03	93	60	38	79	74	158	108	79	49	43
6	Rosetta (L.H.)	31 24	30 25	02	132	–	–	–	–	–	–	–	–	116
7	Rosetta	31 24	30 25	02	–	–	–	–	–	–	–	–	248	188
8	Mersa Matruh	31 22	27 14	07	191	126	–	–	–	–	89	121	66	149
9	Tolombat El-Boseili	31 20	30 24	02	172	(56)	143	116	59	153	194	151	171	110
10	Tolombat El-Tolombat	31 18	30 04	01	215	57	155	88	58	167	266	320	114	100
11	Edfina	31 18	30 31	03	252	93	132	153	96	277	216	223	170	113
12	Port Said (A.P.)	31 17	32 15	01	64	57	25	68	67	80	122	88	45	–
13	Sirw	31 14	31 39	02	54	42	26	63	60	95	123	61	46	(30)
14	Kom el Tarfaia	31 14	30 09	02	204	97	144	101	46	185	130	117	95	54
15	Alexandria	31 12	29 53	32	228	75	199	155	93	194	218	198	144	113
16	Mex	31 09	29 50	05	160	77	184	124	78	171	270	225	92	71
17	Kafr El-Dawar	31 08	30 08	03	167	61	142	137	53	261	242	200	69	87
18	Kafr El-Sheikh	31 07	30 57	07	77	46	62	58	67	116	90	130	56	41
19	El-Arish	31 07	33 46	10	(126)	104	94	108	98	121	210	–	–	37
20	Sakha	31 07	30 57	06	88	35	66	74	75	136	89	83	62	29
21	Ras El Dabaa	31 06	28 28	15	161	101	72	–	–	–	–	–	–	–
22	Mansura	31 03	31 23	07	(47)	34	30	56	73	81	43	81	61	27
23	Fuka	31 02	27 56	26	153	–	–	–	–	–	–	–	–	–
24	Damanhur	31 02	30 28	06	64	22	47	99	58	151	76	160	74	48
25	Amria	31 01	29 48	11	110	26	92	–	–	–	–	–	–	–
26	Borg El-Arab	30 55	29 32	10	180	20	108	–	–	–	–	–	–	–
27	Hammam	30 50	29 25	15	80	46	73	–	–	–	–	–	–	–
28	Kafr El-Zayyat	30 49	30 49	10	17	37	14	60	70	83	45	92	24	17
29	Tanta	30 47	31 00	14	(51)	(27)	51	48	(49)	98	36	95	86	(08)
30	Faqus	30 44	31 48	10	41	29	50	69	65	50	54	56	25	36
31	El-Quseima	30 40	34 22	330	–	07	54	47	62	38	75	112	28	102
32	Zagazig	30 35	31 30	11	29	16	38	36	08	37	37	38	21	33
33	Shebin El-Kom	30 33	31 00	11	–	–	–	–	–	–	–	38	32	11
34	El-Hassana	30 28	33 48	250	–	00	45	29	43	29	43	64	06	16
35	Benha	30 28	31 11	14	06	03	10	18	25	43	39	13	13	05
36	Wadi El-Natrun	30 23	30 21	01	–	–	–	–	–	–	–	57	28	06
37	Fayed	30 20	32 17	13	–	20	–	–	15	(02)	51	(52)	16	12
38	Delta Barrage	30 11	31 08	20	11	18	24	17	19	29	33	(17)	12	01
39	Giza	30 02	31 13	21	09	22	15	16	17	38	36	30	30	37
40	Kuntella	30 00	34 41	540	57	–	–	–	–	–	–	76	07	08
41	Suez	29 56	32 33	08	18	58	27	16	30	26	38	–	14	(07)
42	El-Nekhel	29 55	33 45	406	–	–	–	16	37	08	13	68	10	06
43	Attaqa	29 54	32 28	03	–	–	–	–	–	(10)	18	25	42	12
44	Helwan	29 52	31 20	112	21	25	21	35	31	44	64	81	17	13
45	El-Themed	29 40	34 22	616	56	06	06	07	47	18	13	72	06	07

Annual depth of rainfall, in mm

1948	1949	1950	1951	1952	1953	1954	1955	1956	1957	1958	1959	1960	1961	1962	1963	1964	1965	1966	1967
(352)	241	166	98	106	133	144	98	189	174	128	132	143	229	110	92	341	182	143	171
268	221	170	166	137	240	224	116	157	216	140	163	149	245	76	146	313	230	203	138
102	123	(144)	(44)	34	324	(54)	33	173	112	41	84	95	160	53	39	139	98	128	105
189	191	159	118	77	182	169	89	133	121	39	82	75	180	45	90	161	149	127	133
124	137	126	(109)	94	228	168	116	161	287	45	106	59	171	40	108	115	169	129	124
363	311	82	119	131	162	204	130	104	342	-	-	-	-	-	343	169	239	104	
379	307	146	95	-	99	-	-	-	-	-	126	154	310	155	-	373	212	233	166
274	163	153	82	83	85	92	129	186	273	138	77	102	138	71	78	215	215	145	101
272	287	125	132	88	114	260	120	150	220	93	106	120	226	165	103	321	173	182	132
379	329	126	170	121	255	274	298	213	281	144	214	111	214	130	98	318	103	266	57
291	274	178	179	123	175	-	-	-	-	-	-	-	-	-	-	-	-	-	-
66	101	75	61	81	96	75	53	(33)	100	13	46	39	167	26	60	88	173	81	66
76	120	95	79	72	146	70	48	66	87	-	47	55	(67)	19	73	133	99	73	74
298	283	92	128	79	128	142	75	115	267	115	243	93	277	112	61	254	159	-	-
284	294	150	130	139	164	241	185	250	307	116	203	178	127	169	124	290	165	269	175
299	278	107	114	108	92	246	179	220	301	103	275	144	344	225	151	373	189	237	164
306	244	86	100	70	99	216	261	175	189	86	166	97	167	122	79	341	176	149	114
163	122	50	84	49	64	61	43	105	78	46	54	59	141	31	36	142	122	107	102
87	80	127	45	64	126	113	130	(64)	-	-	108	46	150	55	155	128	193	63	-
122	111	53	64	42	48	65	36	57	50	39	59	33	112	32	30	107	108	29	64
153	132	117	119	197	120	132	197	168	244	(36)	-	(80)	155	73	87	242	169	133	90
64	98	30	38	66	62	59	52	66	95	44	36	45	87	23	31	95	66	36	86
-	90	86	88	130	81	83	123	221	244	59	67	75	184	50	37	158	-	-	-
113	127	53	92	42	67	60	67	104	142	63	98	116	128	55	74	183	138	93	65
-	235	148	146	84	104	-	-	-	(144)	-	(324)	-	-	-	-	-	-	-	-
321	225	133	136	118	104	207	179	(163)	295	95	188	183	256	87	100	230	100	-	-
-	139	81	113	99	86	142	183	284	203	25	113	83	124	40	70	183	118	170	53
17	27	26	28	09	30	29	30	49	50	23	24	16	58	17	12	85	66	13	13
32	51	44	47	39	39	68	47	48	76	05	45	47	80	23	34	65	81	40	63
32	73	47	62	25	42	47	74	34	40	05	17	26	31	11	38	23	35	17	26
-	-	-	-	-	31	41	18	-	-	59	50	(51)	-	50	100	-	-	-	-
(26)	34	35	27	26	36	21	40	44	59	11	31	31	184	06	13	33	55	30	42
35	31	29	51	38	42	28	29	54	68	07	34	26	-	19	27	43	55	43	58
-	-	-	-	-	11	26	03	-	-	(33)	-	-	-	-	-	-	31	33	-
16	43	17	31	12	29	13	11	18	52	07	52	12	-	-	13	31	44	30	28
-	15	11	48	20	281	27	101	-	(25)	-	37	31	50	00	20	17	21	39	18
17	-	-	-	-	29	30	24	36	-	00	12	17	17	01	19	27	65	13	21
17	29	(22)	61	29	16	24	17	35	43	00	07	13	-	08	-	07	20	19	-
25	24	12	55	54	09	30	16	28	34	03	24	11	34	08	07	10	19	19	22
-	-	-	-	-	(14)	(31)	13	-	-	12	17	(02)	-	00	10	18	61	16	-
19	03	06	33	57	46	41	14	55	43	03	19	14	22	03	12	11	73	08	-
-	-	-	-	-	11	39	02	-	-	02	12	01	06	00	24	33	37	36	-
04	07	08	(06)	16	16	14	18	03	11	07	04	12	-	00	27	16	55	08	13
14	76	15	62	11	07	23	18	18	33	01	25	09	38	08	13	16	15	17	14
-	-	-	-	-	-	07	(14)	-	-	03	09	14	-	08	18	02	27	37	-

No.	Station Name	Lati- tude, N	Longi- tude, E	Alti- tude, M	1938	1939	1940	1941	1942	1943	1944	1945	1946	1947
	EGYPT													
46	Shakshuk	29° 28′	30° 42′	43	02	20	02	17	04	05	27	07	16	11
47	Fayoum	29 18	30 51	28	09	05	27	24	20	09	(64)	45	06	11
48	Ras El Negb	29 36	34 52	760	-	-	-	-	26	09	28	35	15	08
49	Siwa	29 12	25 19	17	11	07	00	00	21	08	04	41	04	02
50	Beni Suef	29 04	31 06	28	-	-	-	-	-	-	-	-	-	-
51	Bahariya	28 20	28 54	128	06	01	14	00	00	13	07	02	00	00
52	Tor	28 14	33 37	03	30	15	10	43	11	04	(04)	-	00	02
53	Minya	28 05	30 44	39	19	02	00	00	02	00	00	01	00	00
54	Hurghada	27 17	33 46	01	(01)	43	00	09	00	01	11	06	00	-
55	Assiut	27 11	31 06	71	15	03	03	03	06	06	13	15	00	00
56	Farafra	27 03	27 58	90	-	-	-	-	-	-	-	-	-	-
57	Qena	26 10	32 43	73	03	01	13	00	03	01	18	10	13	00
58	Qusseir	26 08	34 18	06	00	02	03	01	00	00	20	(11)	00	00
59	Nag Hammadi	26 03	32 15	69	-	-	-	-	-	00	18	02	01	00
60	Luxor	25 40	32 42	95	00	04	04	02	00	01	00	00	00	00
61	Dakhla	25 29	29 00	110	00	00	00	00	11	-	-	-	-	-
62	Kharga	25 26	30 34	70	01	00	00	00	00	08	01	08	05	00
63	Deadalus Isl.	24 55	35 52	03	02	00	03	04	06	09	13	08	06	43
64	Kom Ombo	24 29	32 56	100	-	-	-	-	-	-	-	-	-	-
65	Aswan	24 02	32 53	108	01	00	05	03	00	01	05	08	00	00
	SUDAN													
66	Wadi Halfa	21 55	31 20	125	-	-	-	01	01	00	09	09	00	00
67	Port Sudan	19 37	37 13	06	138	63	136	72	57	74	104	41	-	-
68	Abu-Hamed	19 32	33 20	312	00	57	08	00	04	00	14	04	04	01
69	Gebeit	18 57	36 51	797	68	233	82	93	213	36	201	153	122	93
70	Sinkat	18 50	36 50	859	78	169	118	(10)	167	25	277	116	133	20
71	Kareima (Merowe)	18 33	31 51	253	22	(20)	(04)	12	60	11	15	32	56	07
72	Tokar	18 25	37 45	18	83	(73)	232	166	-	-	171	35	100	50
73	Tahamiyam	18 20	36 34	647	45	168	62	00	41	41	85	69	31	00
74	Talguharai	18 17	35 55	539	107	319	98	07	52	42	74	72	37	77
75	Atbara	17 42	33 58	348	31	42	74	15	133	27	45	61	75	55
76	Zediab	17 23	33 55	365	26	00	00	00	00	-	64	72	82	26
77	Abu Deleig	15 55	33 49	400	240	180	149	143	318	211	225	196	196	210
78	Khartoum G.C.	15 37	32 33	385	382	119	149	165	213	175	118	106	222	69
79	Khartoum (A.P.)	15 36	32 33	380	364	96	73	91	275	204	152	203	248	67
80	Kassala	15 28	36 24	501	310	358	-	(340)	452	215	212	350	386	334
81	Jebel Aulia	15 14	32 30	380	336	119	76	159	241	142	151	244	154	185
82	Wadi Turabi	15 08	33 08	393	357	185	218	197	300	70	234	316	431	204
83	Kamlin	15 04	33 11	387	484	285	253	278	384	259	328	427	328	281
84	Khashm El Girba	14 59	35 55	468	-	-	-	-	-	-	-	-	-	-
85	Geteina	14 52	32 22	379	294	289	183	199	205	113	202	192	144	169
86	Rufaa	14 45	33 22	403	342	(106)	326	265	366	201	295	257	363	238
87	Wadi-Shair	14 42	33 17	400	309	248	323	232	354	336	310	246	334	221
88	Wad Medani	14 24	33 30	407	355	548	300	79	-	-	477	251	426	251
89	Managil	14 15	33 00	390	335	260	379	378	252	254	336	182	311	236
90	Gedaref	14 02	35 24	610	394	704	563	611	427	570	495	554	861	497

Annual depth of rainfall, in mm

1948	1949	1950	1951	1952	1953	1954	1955	1956	1957	1958	1959	1960	1961	1962	1963	1964	1965	1966	1967
08	17	01	12	13	10	25	06	10	07	03	–	–	09	02	06	05	01	09	06
08	28	06	26	18	08	17	03	12	12	00	03	00	11	08	01	00	06	05	01
–	–	–	–	–	37	54	16	–	–	01	73	10	–	09	14	06	52	73	–
23	06	02	07	22	03	03	08	29	17	05	00	10	08	01	24	00	02	01	10
–	–	–	–	–	03	27	01	–	10	01	08	02	17	00	04	07	07	04	07
18	00	00	00	01	06	05	00	02	06	00	00	00	02	01	06	00	02	09	00
00	00	04	01	–	(06)	29	34	–	00	04	08	00	15	00	21	05	10	06	–
14	02	01	03	06	01	10	02	(02)	18	01	02	02	02	00	02	00	00	01	00
–	–	–	–	–	–	–	–	–	–	01	00	02	02	00	00	00	02	02	04
00	00	00	00	00	00	04	00	03	11	00	00	00	00	00	00	00	00	02	02
–	–	–	–	–	03	10	00	00	00	00	00	00	02	00	00	02	00	03	00
00	57	00	00	02	12	–	00	00	01	00	00	00	00	00	01	00	00	00	00
00	06	00	00	00	00	40	00	01	00	00	04	01	00	00	01	00	00	02	12
00	03	00	00	00	–	–	–	–	–	00	00	00	00	00	00	00	00	00	00
–	–	–	–	–	–	–	–	–	–	00	00	00	00	00	00	01	04	02	00
00	00	00	00	00	00	00	00	01	00	00	00	00	00	00	00	00	00	00	00
00	00	00	00	00	00	06	01	03	05	00	00	00	00	00	02	02	00	01	00
09	10	06	04	00	00	49	13	01	02	00	08	00	00	01	00	00	00	25	37
–	–	–	–	–	–	00	00	00	–	00	00	04	–	00	00	03	01	00	00
00	02	04	00	00	03	01	–	00	01	00	00	00	00	01	00	00	00	01	00
00	02	30	00	01	00	00	00	00	00	00	00	20	02	07	00	–	–	–	–
76	179	213	209	67	28	15	46	46	146	85	68	133	117	197	165	47	144	77	30
00	23	139	02	15	48	04	00	34	01	18	19	03	09	20	00	27	16	00	08
42	77	231	65	189	106	103	85	168	53	89	327	66	171	113	77	129	69	107	142
90	114	216	80	358	213	80	43	150	51	146	405	44	82	138	58	195	54	88	146
03	86	137	01	119	106	07	02	51	43	38	58	00	48	22	00	47	29	02	133
28	42	83	145	26	17	01	51	85	166	112	45	11	250	57	200	75	117	69	102
20	39	282	–	–	–	–	–	79	66	110	129	31	169	87	88	102	101	104	35
26	99	157	65	131	100	–	47	155	25	45	100	15	75	75	15	180	30	–	07
11	39	166	68	115	107	58	27	104	41	149	85	8	134	53	16	51	87	16	56
09	17	136	15	77	123	–	–	72	47	–	135	19	176	112	21	–	87	04	67
36	102	261	194	236	(389)	(340)	240	316	334	348	470	138	293	202	108	274	179	75	140
92	45	196	126	132	–	–	165	249	190	146	246	72	278	209	133	277	138	171	331
76	47	178	117	117	224	176	179	239	178	149	229	70	242	209	143	294	135	155	316
206	388	442	337	285	401	359	265	429	317	451	336	293	267	286	207	465	234	131	265
126	206	292	165	190	292	232	149	281	209	235	–	–	253	252	50	151	100	106	139
167	210	392	184	270	496	214	317	283	213	169	375	71	276	273	250	294	239	112	168
200	151	321	185	177	188	260	183	335	235	195	357	126	327	230	184	341	263	99	223
(262)	360	585	245	297	495	599	296	369	343	397	250	300	360	–	222	232	241	231	340
196	164	419	173	204	193	185	119	154	148	189	248	228	227	385	108	108	298	157	207
265	236	342	228	269	212	265	194	459	249	317	283	220	348	425	245	543	416	272	359
398	174	369	204	223	474	407	243	500	343	–	323	156	309	336	241	334	338	236	229
445	247	430	237	382	276	463	318	442	327	439	556	270	367	435	319	421	312	269	504
384	108	225	247	165	455	295	197	477	161	270	291	207	306	585	352	378	373	310	246
518	615	713	469	575	571	557	588	698	474	590	589	439	599	574	743	555	600	616	661

No.	Station Name	Lati-tude, N	Longi-tude, E	Alti-tude, M	1938	1939	1940	1941	1942	1943	1944	1945	1946	1947
	SUDAN													
91	Dueim	13° 59′	33° 20′	379	480	446	439	138	326	403	319	230	318	159
92	Haj Abdalla	13 58	33 35	415	422	479	415	360	511	466	473	438	565	340
93	Wad Haddad	13 49	33 33	417	487	442	461	432	343	451	346	427	652	299
94	Bara	13 42	30 22	490	363	422	304	267	372	251	334	329	270	207
95	El-Fasher (A.P.)	13 38	25 20	790	310	409	201	232	167	273	252	393	-	-
96	Mafaza	13 37	34 32	435	-	-	-	-	-	-	-	-	-	-
97	Sennar (D.S.)	13 33	34 37	419	403	504	514	496	580	516	393	471	399	449
98	Geneina	13 29	22 27	779	520	504	375	419	477	496	504	490	814	677
99	El-Obeid (A.P.)	13 10	30 14	570	388	419	312	274	351	451	313	599	716	394
100	Kosti	13 10	32 40	378	388	383	334	484	374	359	353	384	416	309
101	Singa	13 09	33 57	433	626	704	531	599	771	625	686	658	549	561
102	Tendelti	13 02	31 55	413	430	611	483	371	358	375	339	443	505	289
103	Um Rawaba	12 53	31 13	446	436	373	358	348	439	416	351	419	983	309
104	Al Rahad	12 43	30 39	498	514	610	389	473	301	364	352	383	539	414
105	El-Nahud	12 42	28 26	540	537	490	482	251	352	353	412	506	514	414
106	Jebelein	12 37	32 50	386	515	542	497	419	379	369	465	428	318	311
107	Nyala	12 04	24 53	634	606	498	295	244	504	478	477	497	611	-
108	Dilling	12 02	29 38	699	925	899	600	808	649	488	691	664	867	701
109	Rashad	11 52	31 03	885	728	844	613	750	656	745	798	704	978	774
110	El-Roseires	11 51	34 23	467	1034	824	723	697	744	666	706	657	728	899
111	Renk	11 45	32 47	382	439	466	392	552	451	413	508	904	869	848
112	Abri	11 38	30 55	746	793	705	838	673	642	730	499	633	833	664
113	Kadugli	11 00	29 43	513	1074	960	758	855	678	735	712	669	805	797
114	Talodi	10 37	30 21	503	856	825	664	916	627	718	940	589	832	1059
115	Kurmuk	10 33	34 17	702	1022	1039	-	-	-	-	906	904	1175	950
116	Melut	10 27	32 12	383	681	710	387	542	648	580	684	791	792	747
117	Kodok	09 53	32 07	384	708	595	473	746	793	886	841	756	664	811
118	Malakal	09 32	31 39	389	684	900	725	825	710	568	733	900	944	775
119	Tonga	09 28	31 03	390	701	591	683	823	831	864	833	1079	1167	856
120	Abwong	09 07	32 12	389	751	472	695	836	695	795	835	923	702	826
121	Fangak	09 04	30 53	388	1163	1033	1033	1305	1245	1162	1138	1178	1461	1047
122	Aweil	08 46	27 24	400	1164	1053	798	1035	592	808	974	945	1265	836
123	Nasir	08 37	33 04	397	970	870	532	753	881	690	711	788	932	740
124	Raga	08 28	25 41	460	1398	1709	1261	1005	1332	1305	1104	1010	1343	1065
125	Meshra El Rek	08 25	29 16	427	871	942	710	652	663	973	1124	921	859	756
126	Gambeila	08 15	34 35	450	-	1162	-	943	1174	1057	1292	1414	1726	1592
127	Akobo Post	07 48	33 03	403	1010	974	419	923	1080	814	724	1221	1432	1217
128	Wau	07 42	28 01	433	1318	1156	1081	1165	944	892	1389	957	1189	1352
129	Tonj	07 16	28 45	430	-	-	-	-	-	-	1139	1024	953	1187
130	Ghaba Shambe	07 07	30 46	405	642	774	764	816	754	582	859	847	764	847
131	Rumbek	06 48	29 42	420	1070	1001	631	929	1009	966	1173	1103	979	1292
132	Pibor Post	06 48	33 08	410	890	692	655	699	987	705	986	516	682	456
133	Bor	06 12	31 33	422	932	822	755	872	823	670	1074	.804	1068	797
134	Amadi	05 31	30 20	500	1351	1607	1195	1289	1150	797	1359	1140	1260	1140
135	Terrakekka	05 27	31 45	437	891	815	740	911	1117	875	1256	886	922	990

Annual depth of rainfall, in mm

1948	1949	1950	1951	1952	1953	1954	1955	1956	1957	1958	1959	1960	1961	1962	1963	1964	1965	1966	1967
299	289	270	325	271	407	284	287	312	196	331	366	154	349	323	262	261	456	185	217
432	472	335	390	416	249	442	340	656	340	413	575	414	377	399	450	481	443	411	560
359	428	-	330	421	430	276	467	442	400	353	423	351	356	442	366	509	433	141	-
333	212	239	207	375	207	310	168	308	339	241	353	255	353	412	169	304	193	192	235
237	172	586	228	294	417	718	245	311	232	228	326	285	250	324	279	297	332	247	234
-	630	495	629	736	345	870	760	728	453	406	465	528	612	509	710	814	450	561	535
533	412	548	441	396	609	286	349	576	572	685	548	397	385	419	521	580	567	302	600
476	387	567	483	623	514	789	716	651	516	495	441	554	731	586	715	737	435	502	396
521	256	324	513	433	353	487	402	665	307	396	375	317	466	512	315	544	359	216	267
454	256	431	347	306	533	557	390	521	464	464	548	308	547	395	344	371	368	319	409
605	336	768	524	610	545	629	666	705	491	610	490	498	532	534	526	635	603	568	566
410	206	217	367	276	343	431	478	456	388	477	373	210	374	361	209	374	317	246	367
388	205	342	572	502	381	396	369	546	344	388	494	322	383	524	496	389	441	291	312
627	351	489	454	433	509	389	391	485	491	470	513	342	423	660	909	388	408	389	349
306	361	462	440	403	659	346	334	486	487	431	404	486	307	442	299	500	483	312	279
516	285	414	315	442	522	(406)	435	556	294	478	378	332	524	469	548	347	353	391	472
296	320	543	456	371	520	524	570	567	485	486	469	429	458	448	621	620	335	471	421
634	542	573	573	647	634	666	478	752	751	800	799	396	509	502	697	587	562	303	526
782	735	812	696	664	(972)	825	781	739	855	703	697	595	884	644	898	796	687	940	1029
589	673	778	749	765	828	924	691	665	610	525	588	728	812	803	829	751	575	741	658
765	731	456	476	592	702	683	526	647	439	534	617	358	590	438	752	418	419	540	494
583	618	616	618	579	751	773	800	719	765	573	554	606	812	633	-	-	-	-	-
543	738	1040	826	690	1019	1046	554	724	733	650	642	728	625	835	788	1046	868	760	741
723	800	1075	784	702	1054	821	772	869	853	519	836	827	1042	837	860	999	788	775	673
805	-	913	853	945	1083	1205	1066	1196	809	1042	872	943	965	1135	1073	1090	627	1296	1077
605	568	403	534	734	664	1290	428	676	(388)	652	601	660	734	575	596	631	548	769	646
683	642	683	813	554	804	776	602	801	811	726	721	626	726	903	882	554	-	-	1120
1032	1176	774	792	495	948	887	808	676	736	805	725	874	821	705	778	771	519	875	873
1031	919	859	871	698	963	1333	1222	1052	930	1097	920	861	1008	686	756	776	-	1032	-
856	879	917	874	725	966	749	1068	849	-	1031	602	622	767	-	430	203	-	-	-
806	689	1267	578	792	742	981	798	918	848	808	811	812	749	435	621	744	356	576	535
703	976	1195	660	990	1179	1071	944	970	947	895	558	948	671	755	544	757	-	-	-
727	619	1167	756	681	987	728	949	789	919	654	875	698	847	-	806	877	420	842	873
1035	1125	1144	1354	1132	1218	1116	1308	1150	1381	1280	1257	973	1331	1200	1441	1042	1396	-	-
919	799	1135	634	716	-	856	808	-	-	-	957	1420	1031	542	720	-	-	-	-
1270	1495	1261	1447	1170	1240	1756	1430	1427	916	1214	1374	1179	1784	1460	1581	1790	1092	1494	1339
1163	1238	1111	989	899	997	945	870	1096	944	919	918	1037	1162	1113	786	975	665	957	957
1068	1159	1487	985	1094	1181	1117	1328	1031	1210	1366	1211	1197	1100	1387	1275	1365	1004	1429	1033
822	754	1456	1040	878	829	1491	1067	937	973	1173	895	1458	1147	1413	1101	-	921	-	-
1033	960	947	638	554	684	533	655	746	595	947	661	371	990	708	463	485	-	-	-
987	833	1156	885	1197	845	1164	1070	1172	530	526	638	591	989	1004	957	877	-	-	-
1163	687	869	606	1083	981	1064	941	1197	1174	799	855	836	(1395)	1445	1091	996	800	882	1174
712	835	848	830	961	751	787	777	1006	962	787	825	632	1101	965	1142	1310	1085	953	881
1098	1112	1277	964	1085	(850)	859	1005	1256	927	1356	1317	1084	1205	1016	1278	1356	-	-	-
918	904	840	670	901	704	826	-	-	-	1312	806	1152	1259	1592	1196	-	-	856	846

APPENDIX A$_1$ (continued)

No.	Station Name	Lati-tude, N/S	Longi-tude, E	Alti-tude, M	1938	1939	1940	1941	1942	1943	1944	1945	1946	1947
	SUDAN													
136	Maridi	04° 55′	29° 28′	748	1153	1363	1249	1394	1110	1076	1056	1338	1265	1316
137	Juba	04 51	31 37	462	1366	1078	832	975	742	693	1194	1317	1200	1044
138	Yambio	04 34	28 34	724	1650	1359	1438	1642	1328	1351	1424	1723	1381	1321
139	Loka	04 16	31 01	965	999	1062	1408	1368	1264	1089	1273	1311	1382	1355
140	Yei	04 05	30 40	830	1703	1424	1245	1613	1276	1143	1476	1207	1488	1291
141	Kajokaji	03 53	31 40	1030	1645	1250	1167	1253	1476	1107	1198	1369	1657	1244
142	Nimule	03 36	32 03	620	1320	1180	1174	1376	1270	931	1099	1337	1333	1235
	UGANDA													
143	Kitgum	03 17	32 53	937	1188	1094	1212	1248	1138	994	1457	1164	1651	1306
144	Arua	03 01	30 55	1280	1474	1259	1327	1866	1509	1241	1376	1176	1432	–
145	Gulu	02 45	32 20	1106	1121	1200	1657	1768	1463	1002	1483	1522	1457	1809
146	Moroto	02 33	34 36	1524	–	697	1024	967	1111	600	542	855	1270	1080
147	Ngetta farm	02 17	32 56	1095	1397	1035	1452	1449	1331	1240	1360	1408	1543	1793
148	Aduku dispensary	01 59	32 43	1036	1271	1029	1323	1392	1118	1020	1190	1525	1578	1389
149	Katakawi dispensary	01 54	33 59	1158	1360	971	1227	1427	1055	1066	1173	1093	1077	1096
150	Butiaba	01 50	31 20	621	666	649	863	883	804	492	587	774	768	770
151	Soroti	01 43	33 37	1127	1224	920	1096	1868	1642	983	1170	1125	1390	1411
152	Masindi	01 41	31 43	1146	1045	1114	1396	1429	1252	1098	1331	1371	1192	1190
153	Ongino	01 34	34 01	1219	–	1088	1260	1430	1305	(488)	1146	1047	1194	1261
154	Serere Agr. Sta.	01 31	33 27	1139	1205	938	1457	1496	1458	997	1481	1449	1485	1156
155	Kyere	01 29	33 37	1067	1088	1209	1394	1431	1298	1041	1359	1324	1554	1511
156	Bulindi	01 28	31 28	1036	1066	1349	1136	1520	1354	1177	1548	1206	1514	1382
157	Ngora C.M.S.	01 27	33 46	1128	1389	1277	1247	1539	1396	1128	1505	1351	1421	1592
158	Nakasangola	01 19	32 28	1274	710	1110	980	1161	948	707	1295	1126	1099	1168
159	Bukedia	01 19	34 03	1113	1144	865	1457	1283	1492	1018	1196	1228	1350	1361
160	Kachumbala R.S.	01 15	34 06	1146	1079	1152	1325	1242	1037	981	1004	1299	–	–
161	Kibale V.T.C.	01 12N	33 47	1097	1323	1088	1584	1612	1432	938	1558	1322	1316	1272
162	Bugaya	01 06	33 15	1067	1145	1263	1499	1679	1217	956	1440	1180	1276	1433
163	Mbale	01 06	34 11	1220	1006	1340	1133	1029	1400	972	963	1309	1101	1481
164	Namasagali	01 01	32 57	1036	1231	1064	1261	1315	1106	916	1066	1194	1161	1392
165	Vukula	00 57	33 36	1097	1126	1045	1555	1403	1382	1040	1336	1162	1375	1355
166	Kiboga	00 55	81 46	1219	1249	1105	1096	–	1359	1274	1296	870	970	1045
167	Bulopa	00 51	33 15	1097	940	935	1494	1474	1216	1220	1752	1438	1264	1509
168	Ntenjeru	00 44	32 53	1158	1009	984	1378	1361	1342	1247	1443	1156	–	1382
169	Bukalasa Agr. St.	00 43	32 31	1128	1119	1283	1385	1326	1336	834	1244	1206	1136	1329
170	Kahangi Est.	00 42	30 28	1524	–	1355	1323	1491	1166	1164	1310	–	1327	1392
171	Tororo	00 42	34 10	1226	1137	1079	1421	1527	1470	1214	1331	1228	1653	1468
172	Fort Portal	00 40	30 17	1539	1164	1393	1429	1624	1384	1360	1456	1162	1426	1642
173	Kalagala Agr. St.	00 37	32 37	1097	1294	968	1572	1273	1335	997	1348	853	1181	1468
174	Iganga	00 36	33 28	1161	1056	1076	1345	1501	1431	900	1321	1096	1329	1376
175	Mubende	00 35	31 22	1553	1051	1163	1272	1178	974	981	–	1271	1248	1283
176	Nawanza	00 33	33 30	1189	1371	1316	1406	1532	1305	946	1377	1023	1284	1459
177	Dabani	00 28	34 05	1219	1398	1370	2001	1427	1712	1123	1324	1315	1407	1868
178	Nagoje Estate	00 27	32 53	1152	1140	1191	1310	1312	1120	893	–	1215	992	1363
179	Masafu Dispensary	00 24	34 02	1219	1344	1339	1297	1190	1417	1021	1380	1519	1627	1331
180	Lugala Estate	00 24	33 02	1280	1183	1155	1606	1364	1424	1115	1629	1390	1191	1378

Annual depth of rainfall, in mm

1948	1949	1950	1951	1952	1953	1954	1955	1956	1957	1958	1959	1960	1961	1962	1963	1964	1965	1966	1967
1550	1680	1401	841	1309	1368	–	–	1340	1767	1504	–	1440	1385	1665	1396	1444	1724	1435	1694
968	768	980	784	946	695	1054	1041	1019	1164	933	923	968	931	1300	1006	961	820	983	1294
1499	1485	1455	1351	1657	1168	1505	1322	1457	1399	1460	1508	1713	1830	1403	1646	1554	1152	1502	1502
1240	1015	1161	1293	1326	1294	1243	1409	1383	1181	1347	1398	(1249)	–	1473	1456	–	–	–	–
1503	1133	1178	1500	1225	1250	1415	(1294)	1507	1267	1138	1463	1291	1733	1566	1552	1276	–	–	1111
1183	1203	1065	1269	1577	658	1487	2252	(1511)	1201	1151	1344	1392	1606	1174	1555	1669	–	–	–
1342	1290	1005	1397	1573	1216	1238	–	–	–	1685	1035	–	–	(1019)	1333	1033	–	–	–
1067	1508	914	1262	1007	953	–	–	1021	1400	1034	–	1150	1421	1408	1352	1358	1293	1237	1252
1513	1513	1334	1491	1367	1393	1632	1879	1756	1594	1417	–	1481	–	1316	–	1453	–	–	–
1949	1726	1379	1519	1544	1377	1518	1629	1563	1505	1650	1368	1772	1894	1897	1874	1622	1484	1389	1452
902	903	685	1023	610	733	1259	870	917	730	975	820	816	1408	716	1067	923	574	974	–
1560	1654	1256	1533	1326	1229	1300	1583	1427	1217	1484	1509	1520	1698	1438	1930	1528	1374	1447	1537
1326	1154	1318	1289	981	1116	1030	1247	1438	1167	1086	–		1561	1437	1306	1395	1197	1304	1709
1337	979	1072	1464	1063	806	1058	1157	1268	936	1006	–	1274	1531	989	1251	1086	878	1127	1023
636	584	493	685	400	730	610	746	608	630	592	714	636	939	904	1076	–	570	695	549
1347	1010	1011	1570	1266	1261	1222	1312	1440	873	–	–	1405	1643	1498	1363	1044	1145	1336	1684
1118	1030	1336	1459	–	1263	1426	1141	1080	–	1153	1262	–	1629	1634	1488	1445	1316	1235	1185
1136	1195	1200	1449	1133	1106	1123	1474	1663	1556	968	–	–	1803	1550	1083	1330	1284	–	1169
1480	1235	1147	1507	1437	1165	1223	1306	1422	1306	1273	1387	1501	1822	1509	1807	1474	1225	1671	1334
1253	969	1135	1588	1035	1288	1237	1281	1450	1425	999	–	–	1554	1074	1288	1386	1114	1407	1258
1249	1265	1048	1331	1067	1137	1069	1342	1234	1072	1243	1317	–	1774	1495	1589	1578	1163	1297	1054
1343	1141	1222	1910	1411	1109	1178	1268	1154	1104	1455	–	–	1605	–	1359	–	1296	1431	1509
–	903	998	1008	1013	949	856	1159	1057	913	1013	–	–	1495	1056	1333	1041	967	–	1057
1081	865	1175	1388	1126	1073	1051	1239	1023	1097	1023	–	–	1616	1483	1561	1336	1115	1282	1485
1064	799	997	1507	1349	1070	881	1071	–	1073	–	–	–	1778	1469	–	1245	–	–	–
1437	1211	1362	1720	1278	1082	1306	1163	1247	1245	1310	–	–	2073	1427	1960	1347	1250	1596	1516
1243	1150	1096	2012	995	1219	1158	1196	1125	1208	1103	1487	1352	1746	1382	1474	1346	1180	1360	1532
1022	890	1365	1775	1207	923	1144	946	887	921	976	1146	1042	1687	1149	1632	1377	951	1040	1256
1299	958	1194	1828	1048	1102	1353	1235	1362	1212	1258	1257	1423	1806	–	–	1111	1081	–	–
1384	1367	1305	1932	1470	1133	1282	1321	1115	1041	1113	1131	1476	1876	1432	2075	1503	1351	1249	1569
1104	803	1367	2198	930	1115	1238	845	1289	1193	979	1315	984	–	1074	1327	1313	1033	1256	1387
1380	1341	1206	1504	1048	1303	1334	1556	1292	1211	1187	1399	1095	1691	1093	1692	1328	1544	1684	1457
1230	926	986	1361	1043	1080	1100	966	1117	1279	1111	1151	1095	1782	1072	1627	1447	1279	1417	1071
1121	995	1250	1630	1051	1180	1276	1289	1106	1456	1246	–	1134	2125	1780	1255	1293	1383	1332	1548
1284	1344	1353	2025	1321	1236	1396	1409	1233	1110	1350	1563	1395	1692	1370	1654	1465	1044	1582	1410
1247	1115	1330	1637	1539	1054	1283	1385	1419	1426	1267	1345	1637	1981	1753	1795	1635	1304	1528	1580
1604	1564	1385	1896	1306	1664	2116	1452	1225	1458	1328	1564	1398	1957	1620	1805	1553	1400	1795	1408
1010	945	1305	1662	988	1080	1096	1037	980	1284	1193	–	–	–	1402	1495	1188	1101	1071	1403
1264	–	1395	1495	922	1260	–	–	1110	1079	1072	1194	848	2004	1335	1661	–	–	–	–
1233	1336	–	1581	–	1220	–			1427	1135	1133	969	1647	1228	1312	1182	1185	1117	1229
1226	1199	1056	1693	1017	951	1019	1430	1061	1292	1175	1435	1248	2518	1329	1612	1398	1281	1327	–
1411	1329	1621	2152	1304	1320	1583	1625	1430	1615	1345	1569	1644	2109	1608	2005	1410	1569	1945	1985
1338	1026	1383	1946	1159	1191	930	1072	1180	–	1168	1384	1478	–	1622	2406	1897	1689	–	–
–	–	1493	2444	1264	1342	–	1368	1399	1542	1171	–	–	1941	1563	1923	–	–	–	–
–	1061	1104	1585	1087	1159	1201	1608	–	–	–	–	1427	–	1307	2169	1689	1337	1250	1352

No.	Station Name	Lati-tude, N/S	Longi-tude, E	Alti-tude, M	1938	1939	1940	1941	1942	1943	1944	1945	1946	1947
	UGANDA													
181	Monika Estate	00° 23	32° 55′	1250	1571	1525	1646	1420	1366	1103	1513	1288	1387	1542
182	Namanve	00 21	32 41	1135	1170	1138	1393	1445	1293	1082	1433	1346	1301	1127
183	Mukono Agr. St.	00 21	32 45	1184	1274	1302	1628	1509	1287	1066	1626	1370	1401	1317
184	Bwavu Estate	00 21	33 01	1280	-	-	-	-	-	1099	1452	1544	1158	1415
185	Nasmbya	00 18	32 35	1134	-	1006	1000	1311	1285	884	1170	942	1092	970
186	Kabasanda	00 17	32 13	1158	882	1028	1373	1175	1287	1025	1516	1041	1569	1947
187	Budo King's College	00 16	32 29	1311	974	1299	1703	1306	1209	-	1325	1204	1123	1143
188	Ngogwe C.N.	00 14	32 59	1158	1110	1413	1954	1586	1743	1249	1610	1279	1420	1509
189	Buyuma Island	00 11	33 18	1158	-	1484	-	1463	1836	1260	1758	1077	1574	1710
190	Kisubi St. M.C.	00 07	32 32	1173	944	1026	1498	1416	1955	1439	-	1614	1150	1561
191	Entebbe Obs.	00 04	32 29	1182	1079	999	1600	1698	1828	1521	1837	1446	1534	1722
192	Nkozi	00 01S	32 01	1189	881	705	1136	1003	1116	817	1503	982	1002	1209
193	Kalangu	00 10	31 45	1219	785	654	985	908	1235	763	1338	1185	1044	1280
194	Katigondo	00 13	31 44	1311	738	846	1134	1111	1317	851	1522	1104	1073	1473
195	Lyantonde	00 20	31 09	1219	745	806	915	1495	875	689	1236	1287	1000	918
196	Masaka	00 20	31 44	1313	939	790	1069	993	1200	1028	1535	1196	946	1310
197	Kiwala Est.	00 20	31 48	1298	-	792	1265	942	1238	927	1523	1087	1098	1720
198	Kalangala Zaza	00 20	32 19	1158	(2652)	1515	2355	2550	3355	1729	-	-	1371	2720
199	Buwunga	00 23	31 47	1250	1049	954	1273	986	1271	870	1403	1108	995	1733
200	Kyanamukaka	00 30	31 41	1219	1231	954	1206	1032	1181	1005	1335	1186	996	1576
201	Bushenyi	00 33S	30 12	1631	786	1377	1440	1578	1430	1109	1104	1103	1157	1185
202	Mbarara	00 37	30 39	1443	689	946	852	1128	855	708	823	938	1053	731
203	Bikira	00 37	31 34	1219	1063	1006	1050	1090	1165	849	1111	931	975	1321
204	Lawasamaire	00 50	30 08	1646	928	1061	1069	1450	1321	1029	901	1040	709	977
205	Katara Songo	00 55	31 38	1189	1206	994	1307	727	1304	983	1306	1064	1178	1524
206	Kabale	01 15	29 59	1871	1166	979	966	1486	1217	777	947	1059	951	916
	KENYA													
207	Kapenguria	01 14N	35 09	2134	914	630	1229	1327	1221	1077	1029	1161	1465	1414
208	Endabess Mt. Elgon	01 10	34 30	2225	1018	923	1215	1341	1322	1186	1088	1178	1288	1607
209	Kitale Agr. Dep.	01 01	35 10	1890	1165	756	963	1271	957	1009	993	1166	1344	1253
210	Turbo	00 38	35 03	1809	1225	1012	1096	1558	1085	1273	1240	1205	1399	1500
211	Tambach	00 36	35 32	1829	1066	679	1434	1542	1129	983	1257	984	1203	1522
212	Mayanga Stat.	00 33	34 27	1248	1128	1210	1299	1308	1321	1199	1315	1275	1441	-
213	Bungoma V.S.	00 31	34 30	1372	1391	1274	1782	2035	1393	1163	1645	1385	1445	1893
214	Mumias	00 19	34 30	1340	1883	1511	2069	1938	-	1294	1925	1333	1973	1781
215	Tororo	00 18	34 09	1219	1372	1401	1444	1938	1380	1176	1420	1100	1486	1577
216	Kakamega	00 14	34 51	1676	1843	1565	2246	2049	1808	-	1801	1795	1962	2398
217	Kapsabet	00 12	35 07	1998	1395	1131	1744	1802	1498	1196	1383	1443	1607	1757
218	Rangala	00 10	34 21	1384	1375	1223	-	1385	1563	1199	1559	1390	1744	1632
219	Maseno V.S.	00 00	34 37	1463	1602	1260	1658	1740	1572	1227	1420	1368	1859	1899
220	Equator	00 01S	35 24	2012	1461	1098	1855	1816	1502	1308	1505	1673	1757	1975
221	Miwani Stat.	00 05	34 59	1219	1411	1163	1364	1437	1127	1001	1146	1046	1132	1234
222	Kisumu P.C.	00 06	34 45	1146	1155	878	1310	1335	1018	935	1229	862	1421	1100
223	Chemelil Stat.	00 06	35 07	1230	1038	-	1430	1472	1303	1081	1658	-	-	2821
224	Muhoroni Stat.	00 09	35 12	1300	1341	1474	1418	1646	1604	1103	1494	1116	1336	1981
225	Fort Ternan	00 10	35 23	1768	1357	877	1393	1478	1270	935	1259	1133	1242	1548

Annual depth of rainfall, in mm

1948	1949	1950	1951	1952	1953	1954	1955	1956	1957	1958	1959	1960	1961	1962	1963	1964	1965	1966	1967
1177	1231	1429	1797	1180	1418	1352	1802	1493	1310	1347	-	1516	2142	1692	1972	1751	1480	1464	1628
1217	1008	1307	2010	1291	1226	944	1489	1236	1086	1119	1132	1589	2226	1558	1466	1603	1188	1531	1413
1219	1031	1439	1852	1331	1298	1160	1534	1397	2059	1829	1980	2260	2097	1658	1535	1458	922	1085	1066
-	1189	1012	1981	1176	1242	1087	1348	1237	1287	1147	1458	1395	1847	-	1972	1616	1239	1371	1343
1030	743	1036	1556	1040	943	588	953	-	-	-	1078	-	1765	1192	-	-	1150	-	-
1402	1108	1189	1940	1121	1243	1021	1301	1329	1575	1333	1422	1557	1620	1320	1628	1283	1011	1199	1261
1148	1048	1444	1656	1050	1091	1146	1452	1134	1500	1055	1238	1266	1543	1332	1381	1579	-	1099	1338
1305	-	-	2074	1500	1609	1288	1670	1471	1584	1514	1628	1727	2276	1767	-	1781	1163	-	1179
1356	1099	1680	2219	1512	1536	1253	1328	1844	1990	1222	1244	1326	1967	1665	2002	1886	1461	1957	1783
1357	957	1438	2087	1374	1330	969	-	-	-	-	-	-	1289	-	1525	1577	-	-	-
1411	1128	1783	-	1567	1374	1046	1519	1754	1705	1722	1444	-	1834	1641	1944	1938	1458	1368	1569
821	767	1136	1387	797	709	689	656	840	925	1122	965	1063	1371	1166	1490	919	1049	1161	1183
969	907	-	-	-	692	1105	944	-	-	-	1056	1114	1566	-	1154	1020	919	-	-
975	904	1087	1259	834	962	949	1329	938	1153	1104	1056	1151	1199	980	1471	1082	915	-	759
742	715	794	1532	323	885	-	-	707	-	-	763	-	774	599	909	811	600	632	660
851	818	1061	1252	826	845	1029	890	811	1281	954	929	1248	1213	1341	1633	1125	853	1280	1046
1006	996	1132	1537	978	947	871	1126	1217	1783	1825	-	-	1296	1099	1242	1066	857	1226	1111
1689	1390	1890	3509	2040	1526	-	2330	2095	2144	2192	2057	1998	-	1986	2497	2070	1638	2198	-
821	934	1111	1727	865	894	927	1057	1042	1066	1101	956	1519	1489	-	-	-	-	1203	1291
1065	878	1070	1421	923	1087	983	975	-	978	979	1182	1395	1295	1063	1595	1146	975	1774	1139
1446	1253	1217	1854	1015	1281	1127	1637	1318	1556	1036	1149	1159	1365	1606	1678	1137	-	-	934
910	858	742	1520	697	969	936	1101	849	1141	897	798	846	1151	1040	1228	946	781	780	743
995	824	1118	1378	969	1078	964	943	1013	1252	993	-	-	1280	992	1599	1015	806	1070	934
1037	910	1041	1401	897	1199	1171	1106	1173	-	924	887	-	1204	1165	1315	1100	871	924	860
1057	872	1127	1531	1041	1253	1175	1150	1056	1048	1251	1272	1341	-	1197	1315	-	1037	-	935
704	914	801	1156	794	1034	1082	922	1076	1095	1233	872	800	1000	1016	1281	975	897	1032	1148
1500	1030	864	1396	1096	1143	1183	-	-	-	1127	1151	1316	2026	1195	1747	1514	949	1359	2069
1354	940	1072	1532	1096	959	1129	1310	1345	1473	1485	1264	1013	1563	1081	1384	1378	1003	1299	1705
1136	1254	1070	1462	1019	887	1210	877	1076	1043	1461	862	1050	1712	1375	1444	1213	1019	-	1340
1092	1231	1596	2056	1544	1167	1598	1851	1161	1191	1341	871	906	1126	1010	1014	1163	-	-	-
941	865	903	1496	1079	-	1469	1474	1179	1231	1457	1034	1070	1800	1135	1634	1035	808	1373	-
1384	943	-	1821	1403	788	1560	832	881	1002	1078	946	1743	3201	1504	1597	-	1977	1062	-
1326	1412	1364	2073	1917	1350	1540	1594	1310	1475	1447	1184	1650	2118	1752	1867	1537	1476	1503	1525
1511	1546	1621	2203	1989	1284	1585	1244	1219	1559	1531	1759	1887	2278	2164	2245	2059	1642	1504	1887
1194	1222	1229	1834	1251	1247	1210	-	1354	-	-	1303	1297	1821	1352	1602	1391	1319	1421	1353
1813	1612	1829	2111	1954	-	1551	2027	1831	1715	1881	1743	2440	2385	2333	2389	2308	1705	2007	2601
1579	1250	1289	1840	1818	1258	1445	1621	1529	1301	-	1277	1727	2241	1911	-	1340	1153	1605	2154
1111	1409	-	1544	1602	1261	1620	1445	1602	1942	1479	1821	1552	2103	1922	1911	1736	1529	-	-
1310	1170	1239	2037	1463	1514	1445	1483	1503	1613	1411	1510	1405	2128	1754	2267	1882	1584	1714	1671
1413	1376	1249	1831	1474	1025	1551	1454	1374	1394	1650	1560	1501	2220	1795	1963	1475	935	1571	1814
889	1000	891	1422	1201	947	1086	1309	1218	1172	1182	862	880	805	-	-	-	-	975	1291
871	952	764	1562	1126	974	979	-	1006	944	916	709	963	1498	1517	1273	1172	874	1156	1012
993	1307	1216	1689	1238	-	1158	1178	-	-	-	1180	1405	1754	1619	1476	1412	1886	-	-
1147	1101	1306	1941	1272	955	1241	1643	1482	1128	1029	1315	1457	1953	1924	1421	1467	999	1472	851
963	974	1084	1405	1044	978	1160	1281	1113	1393	1329	1098	1384	1560	1496	1499	1492	-	-	-

No.	Station	Lati- tude, N/S		Longi- tude, E		Alti- tude, M	1938	1939	1940	1941	1942	1943	1944	1945	1946	1947
	KENYA															
226	Londiani	00°	10′	35°	35′	2316	1217	878	1347	1486	1158	963	966	1238	1197	1495
227	Lumbwa Stat.	00	12	35	28	1931	1056	774	1100	1177	977	838	1050	1227	1148	1444
228	Molo Station	00	15	35	44	2458	997	679	1295	1646	1604	1196	1164	888	1228	-
229	Kericho	00	23	35	17	1981	1856	1475	1843	2230	1814	1387	1608	1714	1671	1242
230	Sotik, Monieri	00	40	35	05	1813	1209	750	1373	1495	1313	1142	1340	1362	1254	1621
231	Kissi	00	41	34	47	1768	1501	1344	1567	2062	1710	1202	1687	1473	1471	1913
	TANZANIA															
232	Bukoba	01	20	31	49	1144	2192	1991	2237	2046	2051	1640	2346	1998	1813	2676
233	Tarime	01	22	34	23	1524	1209	1332	1405	1420	1294	1193	1334	1138	1263	1354
234	Musoma	01	30	33	48	1147	624	637	774	713	852	710	852	710	998	759
235	Kagondo	01	33	31	42	1296	1867	1435	1876	1415	1677	1424	1634	1960	1315	1900
236	Kwalinda	01	34	31	43	1295	1624	1388	1716	1354	1684	1433	1478	1639	1269	2062
237	Rubya Mission	01	43	31	37	1433	1651	986	1579	1153	1070	1095	1578	1429	936	1554
238	Igabiro Estate	01	48	31	33	1524	1307	836	1261	1062	815	968	1252	1055	830	1428
239	Ikizu Mission	01	56	34	03	1524	-	788	-	1153	942	865	927	796	989	-
240	Kome Mission	02	21	32	29	1134	921	1189	1476	1541	774	776	985	1176	1087	1348
241	Ngara	02	28S	30	38	1798	805	1032	949	1322	936	888	1014	846	1080	1154
242	Mwanza	02	31	32	54	1131	855	735	985	1482	1080	737	1235	1006	932	1197
243	Biharamulo	02	38	31	19	1478	970	1054	1113	1020	913	714	948	818	972	850
244	Ukiriguru	02	42	33	01	1199	704	536	977	1025	775	438	714	843	888	1013
245	Sumuvwe	02	46	33	13	1219	305	647	1037	1149	935	598	812	892	735	978
246	Geita Gold mine	02	52	32	10	1292	837	907	972	1016	785	769	1041	1071	940	1108
247	Ngudu	02	57	33	21	1219	853	730	963	1073	761	629	846	792	692	933
248	Kijma Mission	03	04	33	07	1143	443	686	725	886	-	584	-	729	639	818
249	Shanwa D.O.	03	10	33	46	1341	526	719	669	1072	860	692	703	620	524	793
	ETHIOPIA															
250	Addis Ababa	09	02	38	44	2440	1011	1104	938	1107	1154	1054	1083	1006	1137	1261

APPENDIX B

Mean annual precipitation, in millimeters, after supplementing the missing data

Station number

Year	97	98	99	100	101	102	103	104	105	106	107	108	109	110	111	112	113	114	115
1	403	520	388	388	626	430	436	514	537	515	606	925	728	1034	439	793	1074	856	1022
2	504	504	419	383	704	611	373	610	490	541	498	899	844	824	466	705	960	825	1039
3	514	375	312	334	531	483	358	389	482	479	295	600	613	723	392	838	758	664	508
4	496	419	274	484	599	371	348	473	251	419	244	808	750	697	552	673	855	916	610
5	580	477	351	374	771	358	439	301	352	379	504	649	656	744	451	642	678	627	627
6	516	496	451	359	625	375	416	364	353	369	478	488	745	666	413	730	735	718	591
7	393	504	313	353	686	339	351	352	412	465	477	691	789	706	508	499	712	940	906
8	471	490	499	384	658	443	419	383	506	428	497	664	706	657	904	633	669	589	904
9	399	814	716	416	549	505	988	539	514	318	611	867	978	728	869	833	805	832	1175
10	449	677	394	309	561	289	309	414	414	311	732	701	774	899	848	664	797	1059	950
11	533	476	521	454	605	410	388	627	306	516	296	634	782	589	765	583	543	723	805
12	412	387	256	256	336	206	205	351	361	285	320	542	735	673	731	618	738	800	637
13	548	567	324	431	768	217	342	489	462	414	543	573	812	778	456	616	1040	1075	913
14	441	483	513	347	524	367	454	454	440	315	456	573	696	749	476	618	826	784	853
15	396	623	433	306	610	276	572	433	403	442	371	647	664	765	592	579	690	702	945
16	609	514	353	533	545	343	381	509	659	522	520	634	972	828	702	751	1019	1054	1083
17	286	789	487	557	629	431	396	389	346	406	524	666	825	924	683	773	1046	821	1205
18	349	716	402	390	666	478	369	391	834	435	570	478	781	691	526	800	554	772	1066
19	576	651	665	521	705	456	546	485	486	556	567	752	739	665	647	719	724	869	1196
20	572	516	307	404	491	388	344	491	487	294	485	751	855	610	439	765	733	853	809
21	685	495	396	464	610	477	388	470	431	478	486	800	703	525	534	573	650	519	1042
22	548	441	375	548	490	373	494	513	404	378	469	799	697	588	617	554	642	836	872
23	397	554	317	308	498	210	322	342	486	332	429	396	595	728	358	606	728	827	943
24	385	731	446	547	532	374	383	423	307	524	458	509	884	812	590	812	615	1042	965
25	419	586	518	395	534	361	524	660	442	489	448	502	644	803	438	633	835	837	1135
26	521	715	315	344	526	209	496	909	299	548	621	697	898	829	752	823	788	860	1073
27	580	737	544	371	635	374	389	388	500	347	620	587	796	751	418	717	1046	999	1090
28	567	435	359	368	603	317	441	408	483	353	335	562	687	575	419	621	868	788	627
29	302	502	216	319	568	246	291	389	312	391	471	303	940	741	532	771	760	775	1296
30	600	396	267	409	566	367	312	349	279	473	421	526	1029	658	494	824	741	673	1077

APPENDIX B (continued)

Mean annual precipitation, in millimeters, after supplementing the missing data

Station number

Year	116	117	118	119	120	121	122	123	124	125	126	127	128	129	130	131	132	133	134
1	681	708	684	701	751	1163	1164	970	1398	871	964	1010	1318	1091	642	1070	890	932	1351
2	710	595	900	591	472	1033	1053	870	1709	942	1162	974	1156	1066	774	1001	692	822	1607
3	387	473	725	683	695	1033	798	532	1261	710	521	419	1081	810	764	631	655	755	1195
4	542	746	825	823	836	1305	1035	753	1005	652	943	923	1165	967	816	929	699	872	1289
5	648	793	710	831	695	1245	592	881	1332	663	1174	1080	944	934	754	1009	987	823	1150
6	580	886	568	864	795	1162	808	690	1305	973	1057	814	892	975	582	966	705	670	797
7	684	841	733	833	835	1138	974	711	1104	1124	1292	724	1389	1139	859	1173	986	1074	1359
8	791	756	900	1079	923	1178	945	788	1010	921	1414	1221	957	1024	847	1103	516	804	1140
9	792	664	944	1167	702	1481	1265	932	1343	859	1726	1432	1189	953	964	979	682	1068	1260
10	747	811	775	856	826	1047	836	740	1065	756	1592	1217	1352	1187	847	1292	456	797	1140
11	605	683	1032	1031	856	806	703	727	1035	919	1270	1163	1068	822	1033	987	1163	712	1098
12	568	642	1176	919	879	689	976	619	1125	799	1495	1238	1159	754	960	833	687	835	1112
13	403	683	774	859	917	1287	1195	1187	1144	1135	1241	1111	1487	1456	947	1156	869	848	1277
14	534	813	792	871	874	578	660	756	1354	634	1447	989	985	1040	638	885	606	830	964
15	734	554	495	698	725	792	990	681	1132	716	1170	899	1094	878	554	1197	1083	961	1085
16	664	804	948	963	966	742	1179	987	1218	980	1240	997	1181	829	684	845	981	751	850
17	1290	776	887	1333	749	981	1071	728	1116	856	1756	945	1117	1491	533	1164	1046	787	859
18	428	602	808	1222	1068	798	944	949	1308	808	1430	870	1328	1067	655	1070	941	777	1005
19	676	801	676	1052	849	918	970	789	1150	1027	1427	1096	1031	937	746	1172	1197	1006	1256
20	388	811	736	930	846	848	947	919	1381	926	916	944	1210	973	595	530	1174	962	927
21	652	728	805	1097	1031	808	895	654	1280	1026	1214	919	1366	1173	947	526	799	787	1356
22	601	721	725	920	602	811	558	875	1257	957	1374	918	1211	895	661	638	855	825	1317
23	660	626	874	861	622	812	948	698	973	1420	1179	1037	1197	1458	371	591	836	632	1084
24	734	726	821	1008	767	749	671	847	1331	1031	1784	1162	1100	1147	990	989	1395	1101	1205
25	575	903	705	686	699	435	755	865	1200	542	1460	1113	1387	1413	708	1004	1445	965	1016
26	594	882	778	756	430	621	544	806	1441	720	1581	786	1275	1101	463	957	1081	1142	1278
27	631	554	771	776	203	744	757	877	1042	1157	1790	975	1365	1110	485	877	996	1310	1356
28	548	541	519	491	474	356	1128	420	1398	979	1032	665	1004	921	1096	971	800	1085	1502
29	769	801	875	1032	808	576	1450	842	1418	1439	1494	957	1429	1453	955	1297	882	953	1306
30	646	1120	873	732	806	535	1048	873	1027	1040	1339	957	1033	1118	890	1031	1174	881	1463

APPENDIX B (continued)

Mean annual precipitation, in millimeters, after supplementing the missing data

Station number

Year	135	136	137	138	139	140	141	142	143	144	145	146	147	148	149	150	151	152	153
1	891	1153	1366	1650	999	1703	1645	1320	1188	1474	1121	1165	1397	1271	1360	666	1224	1045	1194
2	815	1363	1078	1359	1062	1424	1250	1180	1094	1259	1200	697	1035	1029	971	649	920	1114	1088
3	740	1249	832	1438	1408	1243	1167	1174	1218	1327	1657	1024	1452	1323	1227	863	1096	1396	1260
4	911	1394	975	1642	1368	1613	1253	1376	1248	1866	1768	967	1449	1392	1427	883	1868	1429	1430
5	1117	1110	742	1328	1264	1276	1476	1270	1138	1509	1463	1111	1331	1118	1055	804	1642	1252	1305
6	875	1078	693	1351	1089	1143	1107	931	994	1241	1002	600	1240	1020	1066	492	983	1098	488
7	1256	1056	1194	1424	1273	1476	1198	1099	1457	1376	1483	542	1360	1190	1173	587	1170	1331	1146
8	886	1338	1317	1723	1311	1207	1369	1337	1164	1176	1522	855	1408	1525	1093	774	1125	1371	1047
9	922	1265	1200	1381	1382	1488	1657	1333	1651	1432	1457	1270	1543	1578	1077	768	1390	1192	1194
10	990	1316	1044	1321	1355	1291	1244	1235	1306	1160	1809	1080	1793	1389	1096	770	1411	1190	1261
11	918	1550	968	1499	1240	1503	1183	1342	1067	1513	1949	902	1560	1326	1337	636	1347	1118	1136
12	904	1680	768	1485	1015	1133	1203	1290	1509	1513	1726	903	1654	1154	979	584	1010	1030	1195
13	840	1401	980	1455	1161	1178	1065	1005	914	1334	1279	685	1256	1318	1072	493	1011	1336	1200
14	670	841	784	1351	1293	1500	1269	1397	1262	1491	1519	1023	1533	1289	1464	685	1570	1459	1449
15	901	1309	946	1657	1326	1225	1577	1573	1007	1367	1544	610	1326	981	1063	1287	1266	1047	1133
16	704	1368	695	1168	1294	1250	668	1216	953	1393	1377	733	1229	1116	806	730	1261	1263	1106
17	826	1073	1054	1505	1243	1415	1487	1238	1458	1632	1518	1259	1300	1030	1058	610	1222	1426	1123
18	1056	1118	1041	1322	1409	1294	2252	2009	1576	1879	1629	870	1583	1247	1157	746	1312	1141	1474
19	1048	1340	1019	1457	1383	1507	1511	1522	1021	1756	1563	917	1427	1439	1268	608	1440	1080	1663
20	1110	1767	1164	1399	1181	1267	1201	1255	1400	1594	1505	730	1217	1167	936	630	873	747	1556
21	1312	1504	933	1460	1347	1138	1151	1685	1034	1417	1650	975	1484	1086	1006	592	1196	1153	968
22	806	1272	923	1508	1398	1463	1344	1035	1442	955	1368	820	1509	1393	1220	714	1267	1262	1175
23	1152	1440	968	1713	1249	1291	1392	1386	1150	1481	1772	816	1520	1288	1274	636	1405	1225	1227
24	1259	1385	931	1830	1642	1733	1606	1648	1421	1057	1894	1408	1698	1561	1531	939	1643	1629	1808
25	1592	1665	1300	1403	1473	1566	1174	1019	1408	1316	1897	716	1438	1437	989	904	1498	1634	1550
26	1196	1396	1006	1646	1456	1552	1555	1333	1352	1005	1874	1067	1930	1306	1251	1076	1363	1488	1083
27	1080	1444	961	1554	1243	1276	1669	1033	1358	1453	1622	923	1528	1395	1086	1717	1044	1445	1330
28	899	1724	820	1152	798	819	781	824	1293	571	1484	574	1374	1197	878	570	1145	1316	1284
29	856	1435	983	1502	957	977	935	986	1237	669	1389	974	1447	1304	1127	695	1336	1235	1246
30	846	1894	1294	1502	1089	1111	1191	1222	1257	544	1452	1084	1537	1709	1023	549	1684	1185	1169

APPENDIX B (continued)

Mean annual precipitation, in millimeters, after supplementing the missing data

Year	Station number																		
	154	155	156	157	158	159	160	161	162	163	164	165	166	167	168	169	170	171	172
1	1205	1088	1006	1389	710	1144	1079	1323	1145	1006	1231	1126	1249	940	1009	1119	1129	1137	1164
2	938	1209	1349	1277	1110	865	1152	1088	1263	1340	1064	1045	1105	935	984	1283	1355	1079	1393
3	1457	1394	1136	1247	980	1457	1325	1584	1499	1133	1261	1555	1096	1494	1378	1385	1323	1421	1429
4	1496	1431	1520	1539	1161	1283	1242	1612	1679	1029	1315	1403	1390	1474	1361	1326	1491	1527	1624
5	1458	1298	1354	1396	948	1492	1037	1432	1217	1400	1166	1382	1359	1216	1342	1336	1166	1470	1384
6	997	1041	1177	1128	707	1018	981	938	956	972	916	1040	1274	1220	1247	834	1164	1214	1360
7	481	1359	1548	1505	1295	1196	1004	1558	1440	963	1066	1336	1296	1752	1443	1244	1310	1331	1456
8	1449	1324	1206	1351	1126	1228	1299	1322	1180	1309	1194	1162	870	1438	1156	1206	1138	1228	1162
9	1485	1554	1514	1421	1099	1340	1262	1316	1276	1101	1161	1375	970	1264	1195	1136	1327	1653	1426
10	1156	1511	1382	1592	1168	1361	1384	1272	1433	1481	1392	1355	1045	1509	1382	1329	1392	1468	1642
11	1480	1253	1249	1343	997	1113	1064	1437	1243	1022	1299	1384	1104	1380	1230	1121	1284	1247	1604
12	1235	969	1265	1141	903	865	799	1211	1150	890	958	1367	803	1341	926	995	1344	1114	1564
13	1147	1135	1048	1222	998	1175	997	1362	1096	1365	1194	1305	1367	1206	986	1250	1353	1330	1385
14	1507	1588	1331	1910	1008	1388	1507	1720	2012	1775	1828	1932	2198	1504	1361	1630	2025	1637	1896
15	1437	1035	1067	1411	1013	1126	1349	1278	995	1207	1048	1470	930	1048	1043	1051	1321	1539	1306
16	1165	1288	1137	1109	949	1073	1070	1082	1219	923	1102	1133	1115	1303	1080	1180	1236	1054	1664
17	1223	1237	1069	1178	856	1051	881	1306	1148	1144	1353	1282	1238	1334	1100	1276	1396	1283	2116
18	1306	1281	1342	1268	1159	1239	1071	1163	1196	946	1235	1321	845	1556	966	1289	1409	1385	1452
19	1422	1450	1234	1154	1057	1023	1008	1247	1125	887	1362	1115	1289	1292	1117	1106	1233	1419	1225
20	1306	1425	1072	1104	913	1097	1073	1245	1208	921	1212	1041	1193	1211	1279	1456	1110	1426	1458
21	1273	999	1243	1455	1013	1023	1026	1310	1103	976	1258	1113	979	1187	1111	1246	1350	1267	1328
22	1387	1245	1317	1231	740	1208	1195	1217	1487	1146	1257	1131	1215	1399	1151	1315	1563	1345	1564
23	1500	1517	1538	1443	932	1249	1212	1379	1352	1042	1423	1476	984	1095	1095	1134	1395	1637	1398
24	1822	1554	1774	1605	1495	1616	1778	2073	1746	1687	1806	1876	1684	1691	1782	2125	1692	1981	1957
25	1509	1074	1495	1251	1056	1483	1469	1427	1382	1149	1231	1432	1074	1093	1072	1780	1370	1753	1620
26	1807	1288	1589	1359	1333	1561	1587	1960	1474	1632	1642	2075	1327	1692	1627	1255	1654	1795	1805
27	1474	1386	1578	1345	1041	1336	1245	1347	1346	1377	1111	1503	1313	1328	1447	1293	1465	1635	1553
28	1225	1114	1163	1296	967	1115	1091	1250	1180	951	1081	1351	1033	1544	1279	1383	1044	1304	1400
29	1671	1407	1297	1431	1234	1282	1253	1596	1380	1040	1477	1249	1256	1684	1417	1331	1582	1528	1795
30	1334	1258	1054	1509	1054	1485	1445	1516	1532	1256	1359	1569	1387	1457	1071	1548	1410	1580	1408

APPENDIX B (continued)

Mean annual precipitation, in millimeters, after supplementing the missing data

Station number

Year	173	174	175	176	177	178	179	180	181	182	183	184	185	186	187	188	189	190	191
1	1294	1056	1051	1371	1398	1140	1344	1183	1571	1170	1274	1216	1047	882	974	1110	1236	944	1079
2	968	1076	1163	1316	1370	1191	1339	1155	1525	1138	1302	1234	1006	1028	1299	1413	1484	1026	999
3	1572	1345	1272	1406	2001	1310	2397	1606	1646	1393	1628	1629	1000	1373	1703	1954	1719	1498	1600
4	1273	1501	1178	1532	1426	1312	1190	1384	1420	1445	1509	1383	1311	1175	1306	1586	1463	1416	1698
5	1335	1431	974	1305	1712	1120	1417	1424	1366	1293	1287	1433	1285	1287	1209	1743	1836	1955	1828
6	995	900	981	946	1123	893	1021	1115	1103	1082	1066	1099	884	1025	1014	1249	1260	1439	1521
7	1348	1321	1223	1377	1314	1282	1380	1629	1513	1433	1626	1452	1170	1516	1325	1610	1758	1661	1837
8	853	1096	1271	1023	1315	1215	1519	1390	1288	1346	1370	1544	942	1041	1204	1279	1077	1614	1446
9	1181	1329	1248	1284	1407	992	1627	1191	1387	1301	1401	1158	1092	1569	1123	1420	1574	1150	1534
10	1468	1376	1283	1459	1868	1363	1331	1378	1542	1127	1317	1415	970	1947	1143	1509	1710	1561	1722
11	1010	1264	1233	1226	1411	1338	1389	1181	1177	1217	1219	1181	1030	1402	1148	1305	1356	1357	1411
12	945	1231	1336	1199	1329	1026	1292	1061	1281	1008	1031	1189	743	1108	1048	1185	1099	957	1128
13	1305	1395	1146	1056	1621	1383	1493	1104	1429	1307	1439	1012	1036	1189	1444	1191	1680	1438	1783
14	1662	1495	1581	1693	2152	1946	2444	1585	1797	2010	1852	1981	1556	1940	1656	2074	2219	2087	1978
15	988	922	922	1017	1304	1159	1264	887	1180	1291	1331	1176	1040	1121	1050	1500	1512	1374	1567
16	1080	1260	983	951	1320	1191	1342	1159	1418	1226	1298	1242	943	1243	1091	1609	1536	1330	1374
17	1096	1057	1220	1019	1583	930	1530	1201	1352	944	1160	1087	588	1021	1146	1288	1253	969	1046
18	1037	1453	851	1430	1625	1672	1368	1608	1802	1489	1534	1348	953	1301	1452	1670	1328	1432	1519
19	980	1110	1196	1061	1430	1180	1399	1281	1493	1236	1397	1237	1169	1329	1134	1471	1844	1571	1754
20	1284	1079	1427	1292	1615	1794	1542	1306	1310	1086	2059	1287	1493	1575	1500	1584	1990	1655	1705
21	1193	1072	1135	1175	1345	1168	1171	1195	1347	1119	1829	1147	1102	1333	1055	1514	1222	1524	1722
22	1340	1194	1133	1435	1569	1384	1540	1461	1470	1132	1980	1458	1078	1422	1238	1628	1244	1365	1444
23	1335	848	969	1248	1644	1478	1637	1427	1516	1589	2260	1395	1359	1557	1266	1727	1326	1291	1285
24	2087	2004	1647	2518	2109	1777	1941	1909	2142	2226	2097	1847	1765	1620	1543	2276	1967	1289	1834
25	1402	1335	1228	1328	1608	1622	1563	1307	1692	1558	1658	1420	1192	1320	1332	1767	1665	1536	1641
26	1495	1661	1312	1612	2005	2406	1923	2169	1972	1466	1535	1972	1444	1628	1381	2046	2002	1523	1944
27	1188	1424	1182	1398	1410	1897	1438	1689	1751	1603	1458	1616	1607	1283	1579	1781	1886	1577	1938
28	1101	1314	1185	1281	1569	1689	1533	1337	1480	1188	922	1239	1150	1011	1137	1163	1461	1373	1458
29	1071	1356	1117	1327	1945	1882	1884	1250	1464	1531	1085	1371	1187	1199	1099	1394	1951	1307	1368
30	1403	1487	1229	1457	1985	1365	1927	1352	1628	1413	1066	1343	1355	1261	1338	1179	1783	1512	1569

APPENDIX B (continued)

Mean annual precipitation, in millimeters, after supplementing the missing data

Year	_____ Station number _____														
	192	193	194	195	196	197	198	199	200	201	202	203	204	205	206
1	881	785	738	745	939	963		1049	1231	786	689	1063	928	1206	1166
2	705	654	846	806	790	762		954	954	1377	946	1006	1061	994	979
3	1136	985	1134	915	1069	1265		1273	1206	1440	852	1050	1069	1307	966
4	1003	908	1111	1495	993	942		986	1032	1578	1128	1090	1450	727	1486
5	1116	1235	1317	875	1200	1238		1271	1181	1430	855	1165	1321	1304	1217
6	817	763	851	689	1028	927		870	1005	1109	708	849	1029	983	777
7	1503	1338	1522	1236	1535	1523		1403	1335	1104	823	1111	901	1306	947
8	982	1185	1104	1287	1196	1087		1108	1186	1103	938	931	1040	1064	1059
9	1002	1044	1073	1000	946	1098		995	996	1157	1053	975	709	1178	951
10	1209	1280	1473	918	1310	1720		1733	1576	1185	731	1321	977	1524	916
11	821	969	975	742	851	1006		821	1065	1446	910	995	1037	1057	704
12	767	907	904	715	818	996		934	878	1253	858	824	910	872	914
13	1136	1130	1087	794	1061	1132		1111	1070	1217	742	1118	1041	1127	801
14	1387	1349	1259	1532	1252	1537		1727	1421	1854	1520	1378	1401	1531	1156
15	797	877	834	823	826	978		865	923	1015	697	869	897	1041	794
16	709	692	962	885	845	947		894	1087	1281	969	1078	1199	1253	1034
17	689	1105	949	999	1029	871		927	983	1127	936	964	1171	1175	1082
18	656	944	1329	999	890	1126		1057	975	1637	1101	943	1106	1150	922
19	840	989	938	707	811	1217		1042	1032	1318	849	1013	1173	1056	1076
20	925	1250	1153	1194	1281	1783		1066	978	1556	1141	1252	1362	1048	1095
21	1122	1187	1104	959	954	1825		1101	979	1036	897	924	924	1251	1233
22	965	1056	1056	763	929	950		956	1182	1149	798	887	887	1272	872
23	1063	1114	1151	1259	1248	1363		1519	1395	1159	846	1388	910	1341	800
24	1371	1566	1199	774	1213	1296		1489	1295	1365	1151	1204	1204	1302	1000
25	1166	1055	980	599	1341	1099		1196	1063	1606	1040	1165	1165	1197	1016
26	1490	1154	1471	909	1633	1242		1435	1595	1678	1228	1315	1315	1315	1281
27	919	1020	1082	811	1125	1066		1118	1146	1137	946	1100	1100	1146	975
28	1049	919	915	600	853	857		888	975	845	781	871	871	1037	897
29	1161	1320	1259	632	1280	1226		1203	1774	883	780	924	924	1257	1032
30	1183	863	759	660	1046	1111		1291	1139	934	743	860	860	935	1148

APPENDIX C - Monthly rainfall, in mm and in percentage of the annual rainfall

Station		Month of the year												Year
No.	Name	Jan.	Feb.	Mar.	Apr.	May	June	July	Aug.	Sep.	Oct.	Nov.	Dec.	
1	Sidi Barrani	35.0	26.0	14.5	03.0	02.5	0	0	0	0	14.5	24.0	37.5	157
		22.3	16.6	09.2	01.9	01.6	0	0	0	0	09.2	15.3	23.9	
2	Borollos Light House	47.0	35.0	16.0	05.0	02.0	0	0	0	0	07.0	33.0	44.0	189
		24.8	18.5	08.5	02.6	01.1	0	0	0	0	03.7	17.5	23.3	
3	Sallum	22.0	17.0	10.0	02.0	03.0	0	0	01.0	01.0	09.0	25.0	21.0	111
		20.0	15.3	09.0	01.8	02.7	0	0	00.9	00.9	08.0	22.5	18.9	
4	Damietta Light House	30.0	22.0	14.0	03.0	02.0	0	0	0	0	07.0	18.0	28.0	124
		24.3	17.7	11.3	02.4	01.6	0	0	0	0	05.6	14.5	22.6	
5	Damietta	27.0	20.0	10.0	03.0	02.0	0	0	0	00.5	06.0	15.5	21.0	105
		25.8	19.0	09.5	02.8	01.9	0	0	0	00.5	05.7	14.8	20.0	
6	Rosetta Light House	46.0	31.5	12.5	03.0	00.8	0	0	0	00.3	07.5	23.0	49.0	174
		26.4	18.2	07.2	01.7	00.5	0	0	0	00.2	04.3	13.2	28.3	
7	Rosetta	61.0	35.0	12.0	06.5	04.0	0	0	0	00.5	10.0	24.0	49.0	202
		30.2	17.3	05.9	03.2	02.0	0	0	0	00.2	05.0	11.9	24.3	
8	Mersa Matruh	38.0	22.0	12.5	03.0	02.0	00.5	0	0	01.0	14.0	26.0	33.0	152
		25.0	14.5	08.2	02.0	01.3	00.3	0	0	00.7	09.2	17.1	21.7	
9	Tolombat el Boseili	42.0	24.5	06.5	02.5	01.5	0	0	0	0	06.5	23.0	41.5	148
		28.4	16.6	04.4	01.7	01.0	0	0	0	0	04.4	15.5	28.0	
10	Tolombat el Tolombat	49.5	34.5	08.0	01.0	0	0	0	0	0	07.0	23.5	54.5	179
		27.6	19.3	04.5	00.6	0	0	0	0	0	03.9	13.1	30.4	
11	Edfina	35.5	36.0	12.0	03.5	03.0	0	0	0	0	09.0	19.0	37.0	155
		22.9	23.2	07.7	02.3	01.9	0	0	0	0	05.8	12.3	23.9	
12	Port Said	15.0	13.5	08.5	03.5	02.0	00.5	0	0	0	06.0	10.5	17.5	77
		19.6	17.5	11.0	04.5	02.6	00.6	0	0	0	07.8	13.6	22.8	
13	Sirw	16.5	11.5	06.0	03.0	03.0	0	0	00.5	0	03.0	07.5	15.0	66
		25.0	17.5	09.1	04.5	04.5	0	0	00.8	0	04.5	11.4	22.7	

APPENDIX C (continued)

No.	Name	Jan.	Feb.	Mar.	Apr.	May	June	July	Aug.	Sep.	Oct.	Nov.	Dec.	Year
14	Kom el-Tarfaia	47.5	33.5	08.0	02.0	00.5	0	0	0	0	04.5	18.0	46.0	160
		29.7	20.9	05.0	01.3	00.3	0	0	0	0	02.8	11.3	28.7	
15	Alexandria (Kom el Nadura)	49.5	25.0	11.0	03.0	02.0	0	0	0	01.0	06.5	32.5	56.5	187
		26.4	13.4	05.9	01.6	01.1	0	0	0	00.5	03.5	17.4	30.2	
16	Mex	44.0	26.0	08.5	03.0	01.0	0	0	0	0	06.5	29.0	52.0	170
		25.9	15.3	05.0	01.8	00.6	0	0	0	0	03.8	17.0	30.6	
17	Kafr el Dawar	42.5	29.0	10.0	03.5	01.0	0	0	0	0	06.5	21.0	44.5	158
		26.9	18.4	06.3	02.2	00.6	0	0	0	0	04.1	13.3	28.2	
18	Kafr el Sheikh	17.0	15.0	08.2	02.0	02.0	0	0	01.0	01.5	04.5	08.5	15.5	75
		22.7	20.0	10.7	02.7	02.7	0	0	01.3	02.0	06.0	11.3	20.6	
19	El Arish	17.5	18.0	13.0	06.0	02.5	0	0	0	0	04.5	14.5	19.0	95
		18.5	18.9	13.7	06.3	02.6	0	0	0	0	04.7	15.3	20.0	
20	Sakha	19.5	16.5	09.0	03.5	02.5	0	0	01.0	01.0	05.0	09.0	14.0	81
		24.1	20.4	11.1	04.3	03.1	0	0	01.2	01.2	06.2	11.1	17.3	
21	Ras el Dabaa	30.5	19.5	12.5	02.5	01.5	0	0	0	00.5	05.5	24.0	36.5	133
		22.9	14.7	09.4	01.9	01.1	0	0	0	00.4	04.1	18.0	27.5	
22	Mansura	12.0	10.0	07.0	03.5	03.5	00.5	0	0	0	04.0	06.5	08.0	55
		21.8	18.2	12.7	06.4	06.4	00.9	0	0	0	07.3	11.8	14.5	
23	Fuka	19.0	16.0	09.5	02.0	03.0	0	0	0	00.5	09.0	17.0	22.5	93.5
		19.3	16.2	09.7	02.0	03.0	0	0	0	00.5	09.2	17.3	22.8	
24	Damanhur	23.0	19.5	10.0	03.5	02.5	0	0	0	0	04.0	10.0	21.5	94
		24.5	20.7	10.6	03.7	02.7	0	0	0	0	04.3	10.6	22.9	
25	Amria	40.5	29.5	07.0	02.0	01.0	0	0	0	0	02.5	18.0	42.5	143
		28.4	20.6	04.9	01.4	00.7	0	0	0	0	01.7	12.6	29.7	
26	Borg el-Arab	40.5	23.0	06.0	02.0	01.0	0	0	0	0	09.0	31.5	38.0	151
		26.8	15.2	04.0	01.3	00.7	0	0	0	0	06.0	20.8	25.2	

Station / Month of the year

APPENDIX C (continued)

No.	Name	\multicolumn{12}{c}{Month of the year}	Year
		Jan.	Feb.	Mar.	Apr.	May	June	July	Aug.	Sep.	Oct.	Nov.	Dec.	
27	Hammam	28.0 / 25.4	15.5 / 14.2	05.5 / 05.0	02.0 / 01.8	0 / 0	0 / 0	0 / 0	0 / 0	02.0 / 01.8	07.0 / 06.4	22.0 / 20.0	28.0 / 25.4	110
28	Kafr el Zayyat	12.5 / 21.2	11.0 / 18.6	07.0 / 11.9	02.0 / 03.4	01.5 / 02.5	0 / 0	0 / 0	0 / 0	0 / 0	03.5 / 05.9	09.5 / 16.1	12.0 / 20.4	59
29	Tanta	10.0 / 20.8	09.0 / 18.8	04.0 / 08.3	03.5 / 07.3	04.0 / 08.3	0 / 0	0 / 0	0 / 0	0 / 0	04.5 / 09.4	05.0 / 10.4	08.0 / 16.7	48
30	Faqus	11.5 / 21.7	07.0 / 13.2	06.5 / 12.3	02.5 / 04.7	02.0 / 03.8	00.5 / 01.0	0 / 0	0 / 0	0 / 0	06.0 / 11.3	08.5 / 16.0	08.5 / 16.0	53
31	El Qusseima	08.5 / 18.0	05.0 / 10.6	11.0 / 23.4	04.5 / 09.6	02.5 / 05.3	0 / 0	0 / 0	0 / 0	0 / 0	02.0 / 04.3	06.0 / 12.8	07.5 / 16.0	47
32	Zagazig	05.5 / 17.8	06.5 / 21.0	04.0 / 12.9	02.0 / 06.4	02.0 / 06.4	0 / 0	0 / 0	0 / 0	0 / 0	02.0 / 06.4	04.0 / 12.9	05.0 / 16.2	31
33	Shebin el Kom	07.0 / 20.6	07.5 / 22.1	02.5 / 07.4	02.0 / 05.8	04.0 / 11.8	0 / 0	0 / 0	0 / 0	0 / 0	02.0 / 05.8	02.5 / 07.4	06.5 / 19.1	34
34	El-Hassana	04.0 / 14.7	03.0 / 11.1	05.5 / 20.4	02.0 / 07.4	01.5 / 05.6	0 / 0	0 / 0	0 / 0	0 / 0	01.5 / 05.6	06.5 / 24.1	03.0 / 11.1	27
35	Benha	04.0 / 18.2	03.0 / 13.6	03.0 / 13.6	01.0 / 04.6	01.0 / 04.6	0 / 0	0 / 0	0 / 0	0 / 0	01.5 / 06.8	03.5 / 15.9	05.0 / 22.7	22
36	Wadi el Natrun	04.5 / 11.8	04.5 / 11.8	02.5 / 06.6	01.0 / 02.6	05.0 / 13.3	0 / 0	0 / 0	0 / 0	00.5 / 01.3	00.5 / 01.3	10.0 / 26.3	09.5 / 25.0	38
37	Fayed	02.0 / 09.1	02.0 / 09.1	04.5 / 20.4	01.0 / 04.5	02.5 / 11.4	0 / 0	0 / 0	0 / 0	0 / 0	00.5 / 02.3	02.0 / 09.1	07.5 / 34.1	22
38	Delta Barrage	03.0 / 16.7	04.0 / 22.2	03.0 / 16.7	00.5 / 02.8	01.0 / 05.5	0 / 0	0 / 0	0 / 0	0 / 0	02.0 / 11.1	00.5 / 02.8	04.0 / 22.2	18
39	Giza	04.5 / 17.6	04.0 / 15.7	03.5 / 13.7	02.0 / 07.8	01.5 / 05.9	0 / 0	0 / 0	0 / 0	0 / 0	02.5 / 09.8	02.5 / 09.8	05.0 / 19.7	25.5

APPENDIX C (continued)

Station		Month of the year												Year
No.	Name	Jan.	Feb.	Mar.	Apr.	May	June	July	Aug.	Sep.	Oct.	Nov.	Dec.	
40	Kuntella	06.0 27.3	02.5 11.4	02.0 09.1	01.5 06.7	02.0 09.1	0 0	0 0	0 0	0 0	02.0 09.1	02.0 09.1	04.0 18.2	22
41	Suez	02.0 09.5	03.0 14.3	03.0 14.3	01.0 04.8	01.0 04.8	0 0	0 0	0 0	0 0	02.5 11.9	04.5 21.4	04.0 19.0	21
42	El-Nekhl	05.0 22.7	05.0 22.7	04.0 18.3	02.0 09.1	01.5 06.8	0 0	0 0	0 0	0 0	00.5 02.3	01.0 04.5	03.0 13.6	22
43	Attaqua	01.0 06.1	02.0 12.1	01.5 09.1	00.5 03.0	05.5 33.3	0 0	0 0	0 0	0 0	01.5 09.1	01.5 09.1	03.0 18.2	16.5
44	Helwan	05.5 19.3	04.0 14.0	04.0 14.0	02.5 08.8	02.5 08.8	0 0	0 0	0 0	0 0	01.0 03.5	03.5 12.3	05.5 19.3	28.5
45	El-Themed	03.0 09.1	06.0 18.2	04.5 13.6	01.5 04.5	02.0 06.1	0 0	0 0	0 0	0 0	03.0 09.1	10.0 30.3	03.0 09.1	33
46	Shakshuk	01.5 15.0	01.5 15.0	01.0 10.0	01.0 10.0	01.0 10.0	0 0	0 0	0 0	0 0	00.5 05.0	01.0 10.0	02.5 25.0	10
47	Fayoum	01.0 07.1	02.0 14.4	02.0 14.4	01.0 07.1	01.0 07.1	0 0	0 0	0 0	0 0	01.0 07.1	01.0 07.1	05.0 35.7	14
48	Ras el Nagb	04.0 08.7	04.0 08.7	04.5 19.6	02.0 08.7	02.0 08.7	0 0	0 0	0 0	0 0	01.0 04.3	02.0 08.7	03.5 15.2	23
49	Siwa	01.0 10.0	01.5 15.0	01.0 10.0	01.0 10.0	01.5 15.0	0 0	0 0	0 0	0 0	0 0	01.0 10.0	03.0 30.0	10
50	Beni-Suef	01.0 12.5	02.0 25.0	01.0 12.5	00.5 06.3	00.5 06.3	0 0	0 0	0 0	0 0	0 0	0 0	03.0 37.4	08
51	Baharia	0 0	01.7 33.3	0 0	00.5 10.0	0 0	0 0	0 0	0 0	0 0	0 0	01.3 26.7	01.5 30.0	05
52	Tor	01.5 12.5	02.0 16.7	02.0 16.7	0 0	0 0	0 0	0 0	0 0	0 0	01.0 08.3	02.0 16.7	03.5 29.1	12

492

APPENDIX C (continued)

Station

No.	Name	Month of the year												Year
		Jan.	Feb.	Mar.	Apr.	May	June	July	Aug.	Sep.	Oct.	Nov.	Dec.	
53	Minya	01.0 33.3	01.0 33.3	0 0	00.2 06.6	00.2 06.6	0 0	0 0	0 0	0 0	00.1 03.6	0 0	00.5 16.6	03
54	Hurghada	0 0	0 0	0 0	0 0	0 0	0 0	0 0	0 0	0 0	0 0	02.0 80.0	00.5 20.0	02.5
55	Assiut	00.5 14.2	01.0 28.6	0 0	0 0	0	0 0	0 0	0 0	0 0	0 0	01.0 28.6	01.0 28.6	03.5
56	Farafra	0 0	0 0	0 0	0 0	0 0	0 0	0 0	0 0	0 0	01.5 100.0	0 0	0 0	01.5
57	Qena	0 0	01.0 20.0	0 0	0 0	0 0	0 0	0 0	0 0	0 0	01.0 20.0	02.0 40.0	01.0 20.0	05
58	Qusseir	0 0	0 0	0 0	0 0	0 0	0 0	0 0	0 0	0 0	01.0 25.0	02.0 50.0	02.0 25.0	04
59	Nag Hammadi	0 0	0 0	0 0	0 0	0 0	0 0	0 0	0 0	0 0	0 0	0 0	02.0 100.0	02
60	Luxor	0 0	0 0	0 0	0 0	0 0	0 0	0 0	0 0	0 0	00.5 100.0	0 0	0 0	00.5
61	Dakhla	0 0	00.5 100.0	0 0	0 0	0 0	0 0	0 0	0 0	0 0	0 0	0 0	0 0	00.5
62	Kharga	0	0	0 0	0 0	0	0 0	0 0	0	0	0	0	0	0
63	Deadalus Island	01.0 10.5	0 0	0 0	0 0	00.5 05.3	0 0	0 0	0 0	0 0	00.5 05.3	05.0 52.6	02.5 26.3	09.5
64	Kom Ombo	0 0	0 0	0 0	0 0	0 0	0 0	0 0	0	0 0	01.0 100.0	0 0	0 0	01
65	Aswan	0 0	0 0	0 0	0 0	01.0 100.0	0 0	0 0	0 0	0 0	0 0	0 0	0 0	01

APPENDIX C (continued)

No.	Name	Station		Month of the year												Year
		Jan.	Feb.	Mar.	Apr.	May	June	July	Aug.	Sep.	Oct.	Nov.	Dec.			
66	Wadi Halfa	0	0	0	0	01.0	0	01.0	0	0	00.5	0	0			02.5
		0	0	0	0	40.0	0	40.0	0	0	20.0	0	0			
67	Port Sudan	06.5	01.0	00.5	00.5	01.0	00.5	08.0	03.5	0	10.0	62.0	28.5			122
		05.3	00.8	00.4	00.4	00.8	00.4	06.6	02.7	0	08.2	50.9	23.5			
68	Abu Hamed	0	0	0	0	01.0	0	02.0	09.5	01.0	0	0	0			13.5
		0	0	0	0	07.4	0	14.8	70.4	07.4	0	0	0			
69	Gebeit	01.5	01.0	02.0	03.5	09.0	07.5	27.0	47.0	13.0	08.0	02.5	01.0			123
		01.2	00.8	01.6	02.8	07.3	06.1	22.0	38.2	10.6	06.5	02.1	08.0			
70	Sinkat	0	0	0	02.5	09.5	08.0	24.5	54.0	17.5	07.0	02.0	0			125
		0	0	0	02.0	07.6	06.4	19.6	43.2	14.0	05.6	01.6	0			
71	Kareima	0	0	0	0	01.0	0	09.0	17.0	03.0	01.0	0	0			31
		0	0	0	0	03.2	0	29.0	54.9	09.7	03.2	0	0			
72	Tokar	21.0	04.0	01.0	01.5	02.5	01.0	05.0	03.0	01.0	08.5	24.0	18.0			90.5
		23.2	04.4	01.1	01.7	02.7	01.1	05.5	03.3	01.1	09.4	26.5	19.9			
73	Tahamiyam	0	0	01.0	02.0	07.0	04.5	26.0	34.5	12.5	06.5	01.0	0			95
		0	0	01.1	02.1	07.4	04.7	27.3	36.3	13.2	06.8	01.1	0			
74	Talguharia	0	0	0	01.0	03.5	02.0	22.0	34.5	14.5	04.0	00.5	0			82
		0	0	0	01.2	04.3	02.4	26.8	42.1	17.7	04.9	00.6	0			
75	Atbara	0	0	0	01.0	02.5	02.0	19.5	36.5	06.0	01.0	0	0			68.5
		0	0	0	01.5	03.5	02.9	28.5	53.3	08.8	01.5	0	0			
76	Zediab	0	0	0	0	03.0	02.5	21.5	31.5	06.0	00.5	0	0			65
		0	0	0	0	04.6	03.8	33.1	48.5	09.2	00.8	0	0			
77	Abu-Deleig	0	0	0	01.0	09.0	09.5	66.0	96.5	31.5	06.0	00.5	0			220
		0	0	0	00.5	04.1	04.3	30.0	43.9	14.3	02.7	00.2	0			
78	Khartoum G.C.	0	0	0	01.0	04.0	08.5	53.5	72.0	20.5	04.5	0	0			164
		0	0	0	00.6	02.5	05.2	32.6	43.9	12.5	02.7	0	0			

APPENDIX C (continued)

No.	Name	Jan.	Feb.	Mar.	Apr.	May	June	July	Aug.	Sep.	Oct.	Nov.	Dec.	Year
79	Khartoum	0	0	0	00.5	05.0	05.5	48.5	68.5	25.0	03.0	0	0	156
		0	0	0	00.3	03.2	03.5	31.1	43.9	16.1	01.9	0	0	
80	Kassala	0	0	01.0	03.5	13.0	28.5	94.0	123.0	56.0	08.0	01.0	0	327
		0	0	00.2	01.1	04.0	09.7	28.7	37.6	17.1	02.4	00.2	0	
81	Jebel Aulia	0	0	0	00.5	04.0	16.0	59.0	85.5	27.0	07.0	0	0	199
		0	0	0	00.3	02.0	08.0	29.6	43.0	13.6	03.5	0	0	
82	Wadi Turabi	0	0	0	02.0	13.0	26.5	80.0	93.0	35.5	08.5	00.5	0	259
		0	0	0	00.8	05.0	10.2	30.9	35.9	13.7	03.3	00.2	0	
83	Kamlin	0	0	0	0	11.0	16.5	79.5	98.0	42.5	06.5	0	0	254
		0	0	0	0	04.3	06.5	31.3	38.6	16.7	02.6	0	0	
84	Khashm el-Girba	0	0	0	01.0	15.0	39.0	146.0	152.0	67.0	09.5	01.5	0	431
		0	0	0	00.3	03.5	09.0	33.9	35.3	15.5	02.2	00.3	0	
85	Geteina	0	0	0	01.0	05.5	09.5	60.5	95.0	26.0	04.5	0	0	202
		0	0	0	00.5	02.7	04.7	30.0	47.0	12.9	02.2	0	0	
86	Rufaa	0	0	0	03.0	09.5	25.5	101.0	118.5	47.0	07.5	0	0	312
		0	0	0	01.0	03.0	08.2	32.3	38.0	15.1	02.4	0	0	
87	Wadi-Shair	0	0	0	01.5	10.0	28.0	93.5	103.0	43.0	07.0	0	0	286
		0	0	0	00.5	03.5	09.8	32.8	36.0	15.0	02.4	0	0	
88	Wad-Medani	0	0	0	03.0	13.0	33.0	128.0	137.0	57.0	13.0	01.0	0	385
		0	0	0	00.8	03.4	08.6	33.2	35.5	14.8	03.4	00.3	0	
89	Managil	0	0	01.0	01.5	11.5	28.0	107.0	125.0	56.5	09.5	0	0	340
		0	0	00.3	00.4	03.4	09.2	31.5	36.8	16.6	02.8	0	0	
90	Gedaref	0	0	01.5	06.0	28.5	97.0	178.0	202.0	103.5	26.5	06.0	0	649
		0	0	00.2	00.9	04.4	14.9	27.4	31.2	15.9	04.2	00.9	0	
91	Dueim	0	0	00.5	02.0	14.0	24.5	93.5	119.0	49.0	12.5	0	0	315
		0	0	00.2	00.6	04.4	07.8	29.7	37.8	15.6	03.9	0	0	

APPENDIX C (continued)

No.	Name	Jan.	Feb.	Mar.	Apr.	May	June	July	Aug.	Sep.	Oct.	Nov.	Dec.	Year
							Month of the year							
92	Hadj Abdalla	0	0	0	04.0	22.5	40.5	120.0	165.0	65.5	17.5	01.0	0	436
		0	0	0	00.9	05.2	09.3	27.6	37.8	15.0	04.0	00.2	0	
93	Wadi Haddad	0	0	01.0	04.5	13.5	42.0	118.0	140.0	65.0	19.0	0	0	403
		0	0	00.2	01.1	03.3	10.5	29.3	34.8	16.1	04.7	0	0	
94	Bara	0	0	0	00.5	10.5	20.5	91.0	116.0	47.5	11.0	0	0	297
		0	0	0	00.2	03.5	06.9	30.6	39.1	16.0	03.7	0	0	
95	El-Fasher	0	0	0	01.0	10.5	15.5	95.0	133.0	36.5	05.5	0	0	297
		0	0	0	00.3	03.5	05.2	32.0	44.8	12.4	11.8	0	0	
96	Mafaza	0	0	0	0	26.0	88.0	167.0	183.0	86.0	21.0	03.0	0	574
		0	0	0	0	04.5	15.3	29.1	31.9	15.0	03.7	00.5	0	
97	Sennar D.S.	0	0	00.5	03.5	25.0	55.5	127.0	163.0	66.5	21.0	01.0	0	463
		0	0	00.1	00.8	05.4	12.0	27.5	35.0	14.4	04.6	00.2	0	
98	Geneina	0	0	01.0	04.5	27.0	39.0	164.5	233.0	72.5	08.5	0	0	550
		0	0	00.2	00.8	04.9	07.1	29.9	42.4	13.2	01.5	0	0	
99	El-Obeid	0	0	00.5	02.5	20.5	33.0	115.5	140.0	79.0	21.0	0	0	412
		0	0	00.1	00.6	05.0	08.0	28.0	34.0	19.2	05.1	0	0	
100	Kosti/Rabak	0	0	00.5	03.0	18.5	41.5	109.5	141.5	59.0	29.0	00.5	0	403
		0	0	00.1	00.8	04.6	10.3	27.2	35.1	14.6	07.2	00.1	0	
101	Singa	0	0	0	04.5	31.0	73.5	163.0	191.5	89.0	26.5	01.0	0	580
		0	0	0	00.8	05.3	12.7	28.1	33.0	15.3	04.6	00.2	0	
102	Tendelti	0	0	0	02.5	16.0	32.0	110.5	147.0	65.0	21.0	0	0	394
		0	0	0	00.6	04.1	08.1	28.0	37.3	16.5	05.4	0	0	
103	Om Rwaba	0	0	0	00.5	14.5	29.5	129.0	142.5	56.5	12.5	0	0	385
		0	0	0	00.1	03.8	07.7	33.5	37.0	14.7	03.2	0	0	
104	Rahad	0	0	0	02.0	21.5	40.5	123.0	131.5	65.5	22.0	0	0	406
		0	0	0	00.5	05.3	10.0	30.3	32.4	16.1	05.4	0	0	

APPENDIX C (continued)

No.	Name	Jan.	Feb.	Mar.	Apr.	May	June	July	Aug.	Sep.	Oct.	Nov.	Dec.	Year
105	El-Nahud	0 / 0	0 / 0	0 / 0	04.0 / 01.0	24.5 / 06.1	42.0 / 10.4	102.0 / 25.3	129.0 / 31.9	82.0 / 20.3	19.5 / 04.8	01.0 / 00.2	0 / 0	404
106	Jebelein	0 / 0	0 / 0	0 / 0	03.0 / 00.7	21.5 / 05.0	48.0 / 11.1	122.0 / 28.3	124.0 / 28.8	81.5 / 18.9	27.0 / 06.3	04.0 / 00.9	0 / 0	431
107	Nyala	0 / 0	0 / 0	01.0 / 00.2	04.0 / 00.8	32.0 / 06.4	60.0 / 12.0	132.0 / 26.3	168.5 / 33.6	85.5 / 17.0	19.0 / 03.7	0 / 0	0 / 0	502
108	Dilling	0 / 0	0 / 0	02.0 / 00.3	10.5 / 01.6	53.5 / 08.0	89.5 / 13.4	168.5 / 25.2	170.5 / 25.5	124.0 / 18.5	49.5 / 07.4	01.0 / 00.1	0 / 0	669
109	Rashad	0 / 0	00.5 / 00.1	01.0 / 00.1	14.5 / 01.9	70.0 / 09.3	98.5 / 13.1	116.0 / 15.4	191.5 / 25.4	162.5 / 21.5	93.5 / 12.4	06.0 / 00.8	0 / 0	754
110	Roseires	0 / 0	00.5 / 00.1	02.0 / 00.3	14.0 / 01.8	58.5 / 07.4	129.5 / 16.4	178.0 / 22.7	216.5 / 27.6	147.5 / 18.8	33.5 / 04.3	05.0 / 00.6	0 / 00	785
111	Renk	0 / 0	0 / 0	0 / 0	06.5 / 01.2	33.5 / 06.2	81.5 / 15.1	133.0 / 24.6	144.0 / 26.6	92.5 / 17.1	46.0 / 08.5	04.0 / 00.7	0 / 0	541
112	Abri	0 / 0	02.0 / 00.3	00.5 / 00.1	10.0 / 01.4	63.0 / 08.7	100.0 / 14.0	163.0 / 22.7	172.0 / 24.0	134.5 / 18.8	67.0 / 09.4	04.0 / 00.6	0 / 0	716
113	Kadugli	0 / 0	01.5 / 00.2	01.5 / 00.2	17.0 / 02.2	83.0 / 10.9	120.0 / 15.7	151.5 / 19.9	158.5 / 20.8	140.5 / 18.4	83.5 / 11.0	05.0 / 00.7	0 / 0	762
114	Talodi	0 / 0	0 / 0	05.0 / 00.6	21.0 / 02.6	85.5 / 10.5	119.0 / 14.6	169.0 / 20.7	173.0 / 21.2	151.0 / 18.5	86.0 / 10.4	07.5 / 00.9	0 / 0	817
115	Kurmuk	0 / 0	01.0 / 00.1	05.5 / 00.6	26.0 / 02.8	112.0 / 11.9	140.5 / 14.9	166.0 / 17.7	203.5 / 21.6	164.0 / 17.5	105.0 / 11.2	15.5 / 01.6	01.0 / 00.1	940
116	Melut	0 / 0	0 / 0	02.5 / 00.4	14.0 / 02.2	55.0 / 08.5	94.5 / 14.7	134.0 / 20.8	154.0 / 23.9	117.5 / 18.2	69.5 / 10.8	03.0 / 00.5	0 / 0	644
117	Kodok	0 / 0	0 / 0	02.5 / 00.4	19.0 / 02.6	65.0 / 08.8	126.5 / 17.0	152.0 / 20.6	174.0 / 23.6	123.0 / 16.7	71.5 / 09.7	04.5 / 00.6	0 / 0	738

Station — Month of the year

APPENDIX C (continued)

No.	Name	Jan.	Feb.	Mar.	Apr.	May	June	July	Aug.	Sep.	Oct.	Nov.	Dec.	Year
118	Malakal	0 / 0	0 / 0	06.0 / 00.7	28.0 / 03.4	90.5 / 11.1	123.5 / 15.1	167.5 / 20.5	181.5 / 22.2	138.0 / 16.8	74.0 / 09.0	09.0 / 01.1	01.0 / 00.1	819
119	Tonga	0 / 0	01.0 / 00.1	05.5 / 00.6	26.0 / 03.0	86.5 / 10.0	135.0 / 15.6	166.5 / 19.1	206.0 / 23.7	152.5 / 17.6	84.0 / 09.7	05.0 / 00.6	0 / 0	868
120	Abwong	0 / 0	0 / 0	06.0 / 00.8	26.0 / 03.4	74.0 / 09.7	112.0 / 14.7	166.0 / 21.8	175.0 / 22.9	121.5 / 15.9	73.0 / 09.6	09.5 / 01.2	0 / 0	763
121	Fangak	0 / 0	01.0 / 00.1	07.5 / 00.7	38.5 / 03.6	115.0 / 10.7	147.5 / 13.7	229.5 / 21.4	242.5 / 22.6	176.0 / 16.5	106.5 / 09.9	09.0 / 00.8	0 / 0	1076
122	Aweil	0 / 0	02.5 / 00.3	10.0 / 01.1	35.5 / 03.7	133.0 / 14.0	156.5 / 16.4	206.0 / 21.6	203.5 / 21.4	149.0 / 15.7	51.5 / 05.4	03.5 / 00.4	0 / 0	951
123	Nasser	0 / 0	01.0 / 00.1	11.0 / 01.3	37.5 / 04.7	119.0 / 14.8	127.5 / 15.8	145.5 / 18.1	177.5 / 22.1	126.0 / 15.7	50.5 / 06.3	08.0 / 01.0	00.5 / 00.1	804
124	Raga	0 / 0	03.0 / 00.3	16.0 / 01.3	54.5 / 04.6	144.0 / 12.1	173.5 / 14.6	223.5 / 18.8	264.0 / 22.2	202.0 / 17.0	96.0 / 08.1	11.0 / 00.9	01.5 / 00.1	1189
125	Meshra er Rek	0 / 0	01.0 / 00.1	05.0 / 00.6	26.5 / 03.2	75.0 / 09.0	135.5 / 16.4	169.0 / 20.4	168.0 / 20.2	154.5 / 18.6	86.5 / 10.4	09.0 / 01.1	0 / 0	830
126	Gambeila	06.5 / 00.5	11.0 / 00.9	35.0 / 02.7	77.0 / 06.0	163.0 / 12.7	171.5 / 13.3	223.0 / 17.3	258.0 / 20.0	182.5 / 14.2	99.5 / 07.7	48.5 / 03.7	12.5 / 01.0	1288
127	Akobo	01.0 / 00.1	03.0 / 00.3	20.0 / 02.1	76.5 / 08.1	130.5 / 13.9	119.0 / 12.7	159.5 / 17.0	194.0 / 20.6	144.0 / 15.3	72.0 / 07.7	18.0 / 01.9	02.5 / 00.3	940
128	Wau	01.0 / 00.1	05.0 / 00.4	23.5 / 02.1	64.0 / 05.7	135.5 / 12.1	167.5 / 15.0	193.5 / 17.3	216.5 / 19.4	171.0 / 15.3	126.5 / 11.3	14.0 / 01.3	0 / 0	1118
129	Tonj	01.0 / 00.1	01.5 / 00.1	21.5 / 02.0	69.5 / 06.6	129.0 / 12.1	171.5 / 16.1	204.0 / 19.3	205.0 / 19.3	171.5 / 16.1	73.0 / 06.9	15.5 / 01.4	0 / 0	1063
130	Ghabe-Shambe	0 / 0	04.0 / 00.5	17.0 / 02.1	47.5 / 06.0	87.5 / 11.0	125.5 / 15.8	149.5 / 18.8	163.0 / 20.5	121.5 / 15.3	68.0 / 08.6	10.0 / 01.3	01.0 / 00.1	794.5

APPENDIX C (continued)

No.	Name	Jan.	Feb.	Mar.	Apr.	May	June	July	Aug.	Sep.	Oct.	Nov.	Dec.	Year
131	Rumbek	0 0	06.5 00.6	26.0 02.6	86.5 08.5	145.0 14.2	155.5 15.3	174.0 17.1	199.5 19.6	138.0 13.5	71.5 07.0	15.5 01.5	01.0 00.1	1019
132	Pibor Post	06.5 00.7	10.5 01.2	34.5 03.8	80.0 08.8	123.5 13.5	117.0 12.9	141.5 15.6	146.0 16.1	115.5 12.7	78.5 08.6	40.0 04.4	15.5 01.7	909
133	Bor	02.0 00.2	08.0 00.9	29.0 03.2	83.5 09.3	124.0 13.7	118.5 13.1	139.0 15.4	136.0 15.1	129.0 14.3	101.5 11.2	26.0 02.9	06.5 00.7	903
134	Amadi	03.0 00.3	18.5 01.5	49.0 04.1	130.0 10.9	175.5 14.6	146.5 12.2	190.0 15.9	170.0 14.2	147.0 12.2	128.0 10.7	35.0 02.9	05.5 00.5	1198
135	Terakekka	00.3 00.3	12.5 01.4	33.5 03.7	92.5 10.2	159.5 17.6	117.0 12.9	136.0 15.0	128.0 14.1	105.0 11.6	81.0 09.0	29.0 03.3	09.0 00.9	905
136	Maridi	10.0 00.7	27.5 02.0	68.0 05.0	162.0 11.8	194.0 14.1	172.5 12.5	183.0 13.3	188.0 13.7	159.0 11.5	135.0 09.8	62.0 04.5	15.0 01.1	1375
137	Juba	04.0 00.4	11.0 01.1	40.5 04.2	108.5 11.2	153.0 15.7	127.5 13.1	135.0 13.9	136.5 14.0	109.0 11.2	99.0 10.2	36.0 03.7	13.0 01.3	973
138	Yambio	14.5 01.0	29.0 02.0	93.5 06.5	151.5 10.6	179.0 12.5	161.5 11.2	171.0 11.9	189.5 13.2	176.0 12.3	177.0 12.3	74.5 05.2	19.0 01.3	1436
139	Loka	10.0 00.8	24.0 01.8	64.0 04.8	135.0 10.3	160.0 12.2	164.5 12.5	178.5 13.5	198.0 15.0	144.5 10.9	151.0 11.4	62.5 04.7	28.0 02.1	1320
140	Yei	10.0 00.7	30.0 02.1	70.5 05.0	154.5 10.9	183.5 12.8	159.5 11.2	179.5 12.7	195.0 13.8	172.0 12.1	167.0 11.8	76.0 05.4	20.5 01.5	1418
141	Kajo Kaji	06.5 00.5	24.5 01.8	68.0 05.1	159.0 12.0	168.0 12.6	161.0 12.1	166.0 12.7	180.5 13.6	151.5 11.4	148.0 11.1	72.0 05.4	23.0 01.7	1328
142	Nimule	07.0 00.6	24.0 03.1	57.5 05.0	122.5 10.6	152.0 13.1	121.5 10.5	150.0 12.9	156.5 13.5	137.5 11.8	129.5 11.1	79.0 06.7	24.0 02.1	1161
143	Kitgum	06.5 00.5	28.5 02.3	77.5 06.1	133.5 10.5	179.0 14.2	136.0 10.8	176.0 13.9	174.5 14.1	128.5 10.2	112.0 09.9	70.5 05.5	37.5 03.0	1265

Station   Month of the year

APPENDIX C (continued)

No.	Name	Jan.	Feb.	Mar.	Apr.	May	June	July	Aug.	Sep.	Oct.	Nov.	Dec.	Year
144	Atua	22.5 / 01.6	49.5 / 03.5	85.0 / 06.1	143.0 / 10.2	141.0 / 10.1	121.5 / 08.7	149.0 / 10.6	192.0 / 13.7	169.0 / 12.1	180.0 / 12.8	102.5 / 07.3	46.0 / 03.3	1401
145	Gulu	11.5 / 00.8	46.5 / 03.0	88.5 / 05.8	173.5 / 11.3	203.0 / 13.3	146.0 / 09.6	162.5 / 10.6	224.0 / 14.6	172.0 / 11.2	161.0 / 10.5	98.5 / 06.4	45.0 / 02.9	1532
146	Moroto	09.0 / 01.0	35.0 / 04.0	71.5 / 08.1	126.0 / 14.3	144.5 / 16.3	83.5 / 09.4	143.5 / 16.2	98.5 / 11.1	55.5 / 06.3	45.0 / 05.1	47.0 / 05.4	25.0 / 02.8	884
147	Negetta Farm/Lira	25.0 / 01.8	37.0 / 02.6	89.0 / 06.3	175.0 / 12.4	190.0 / 13.5	129.5 / 09.2	126.0 / 08.9	216.5 / 15.4	158.5 / 11.3	137.5 / 09.8	82.0 / 05.8	42.0 / 03.0	1408
148	Aduku dispensary	41.0 / 03.2	50.0 / 03.9	87.5 / 06.8	139.0 / 10.7	164.0 / 12.7	103.5 / 08.0	108.0 / 08.3	168.5 / 13.0	139.0 / 10.7	133.0 / 10.3	98.0 / 07.6	62.5 / 04.8	1294
149	Katakawi	18.0 / 01.6	39.0 / 03.3	77.0 / 06.6	150.0 / 12.9	170.5 / 14.6	122.0 / 10.6	136.5 / 11.7	157.5 / 13.5	115.5 / 09.9	86.5 / 07.4	61.0 / 05.2	31.5 / 02.7	1165
150	Butiaba	13.5 / 01.7	31.5 / 04.1	59.0 / 07.6	99.5 / 12.8	97.5 / 12.6	57.5 / 07.4	67.5 / 08.7	87.0 / 11.2	78.0 / 10.1	86.0 / 11.2	70.0 / 09.0	28.0 / 03.6	775
151	Soroti	19.0 / 01.4	52.5 / 04.0	79.0 / 06.0	183.5 / 13.9	194.0 / 14.6	128.0 / 09.7	117.0 / 08.9	180.0 / 13.6	140.0 / 10.6	115.0 / 08.7	83.0 / 06.3	31.0 / 02.3	1322
152	Masindi	28.5 / 02.2	57.0 / 04.4	100.5 / 07.8	159.5 / 12.4	147.5 / 11.5	102.0 / 07.9	110.0 / 08.5	136.5 / 10.7	139.5 / 10.8	143.0 / 11.1	118.0 / 09.2	45.0 / 03.5	1287
153	Ongino	20.5 / 01.7	58.0 / 04.8	101.0 / 08.3	190.5 / 15.6	197.5 / 16.2	113.0 / 09.3	122.0 / 10.0	130.0 / 10.7	88.5 / 07.3	74.5 / 06.1	79.5 / 06.5	43.0 / 03.5	1218
154	Serere Agr. Stat.	22.0 / 01.6	57.5 / 04.2	95.0 / 06.9	207.5 / 15.2	187.0 / 13.7	107.0 / 07.8	117.5 / 08.6	165.0 / 12.1	146.0 / 10.7	113.5 / 08.3	96.0 / 07.0	53.0 / 03.9	1367
155	Kyere	21.0 / 01.6	45.5 / 03.5	91.0 / 07.0	218.0 / 16.9	192.0 / 14.9	114.0 / 08.8	130.0 / 10.2	151.0 / 11.7	115.5 / 08.9	87.0 / 06.7	81.0 / 06.3	45.0 / 03.5	1291
156	Bulindi	33.5 / 02.6	06.5 / 05.0	110.5 / 08.4	176.0 / 13.4	132.5 / 10.1	89.0 / 06.8	87.5 / 06.7	152.0 / 11.6	163.0 / 12.5	138.0 / 10.5	107.0 / 08.2	55.0 / 04.2	1309

Month of the year

APPENDIX C (continued)

No.	Station Name	Jan.	Feb.	Mar.	Apr.	May	June	July	Aug.	Sep.	Oct.	Nov.	Dec.	Year
157	Ngora	24.0 01.8	54.0 04.0	106.5 07.9	185.5 13.8	205.5 15.3	125.5 09.3	121.5 09.1	121.5 09.1	174.0 12.9	122.0 09.1	83.0 06.2	41.5 03.1	1344
158	Nakasangola	18.5 01.8	39.5 03.9	85.5 08.5	162.0 16.1	137.0 13.6	65.0 06.5	77.0 07.7	109.5 10.9	86.5 08.6	89.0 08.9	95.0 09.4	41.5 04.1	1006
159	Bukedia	22.0 01.8	40.5 03.3	109.5 09.0	199.0 16.2	173.5 14.2	119.0 09.7	132.0 10.8	151.0 12.4	87.5 07.2	74.0 06.1	77.0 06.3	37.0 03.0	1222
160	Kachimbala	23.5 02.0	50.5 04.3	103.5 08.9	170.0 14.6	166.0 14.3	128.5 11.0	112.0 09.6	138.0 11.9	99.0 08.5	66.5 05.7	60.5 05.2	46.0 04.0	1164
161	Kibale	35.5 02.6	58.5 04.3	104.0 07.6	213.5 15.6	201.0 14.7	121.5 08.9	111.5 08.2	157.5 11.5	113.5 08.2	99.0 07.2	84.5 06.2	68.0 05.0	1368
162	Bugaya	34.0 02.6	40.5 03.0	103.5 07.8	217.0 16.3	173.5 13.1	91.0 06.8	103.0 07.8	137.0 10.3	113.5 08.5	117.0 08.8	129.0 09.7	70.0 05.3	1329
163	Mbale	28.5 02.4	55.0 04.7	94.0 08.0	162.0 13.7	167.5 14.2	125.5 10.6	113.5 09.6	134.0 11.3	111.0 09.3	85.0 07.2	64.5 05.5	41.5 03.5	1182
164	Namasagali	31.5 02.6	50.5 04.2	124.0 10.3	175.0 14.6	145.0 12.1	74.5 06.2	65.0 05.4	120.5 10.0	119.0 09.9	124.0 10.3	100.5 08.4	71.5 06.0	1201
165	Vukula	39.5 03.0	59.5 04.5	116.5 08.7	197.0 14.7	190.5 14.3	88.5 06.6	101.0 07.6	132.5 09.9	113.0 08.4	111.0 08.3	124.0 09.3	63.0 04.7	1336
166	Kiboga	34.5 02.9	52.0 04.3	93.5 07.8	176.0 14.7	133.5 11.1	54.5 04.5	78.0 06.5	135.5 11.4	121.5 10.1	149.0 12.4	104.0 08.7	67.0 05.6	1199
167	Bulopa	42.5 03.1	74.5 05.5	95.0 07.0	194.0 14.4	165.5 12.3	78.0 05.8	107.0 07.9	143.0 10.6	121.0 09.0	128.5 09.5	138.5 10.3	62.5 04.6	1350
168	Ntenjeru	49.0 04.1	73.5 06.2	101.5 08.5	193.0 16.2	122.0 10.2	55.0 04.6	67.5 05.6	103.0 08.6	111.0 09.3	118.5 09.9	126.0 10.5	75.0 06.3	1195
169	Bukalasa	47.0 03.7	66.0 05.2	129.0 10.3	141.5 11.2	130.0 10.3	80.5 06.4	76.0 06.0	114.5 09.1	128.5 10.2	148.0 11.9	114.0 09.1	84.0 06.7	1259

APPENDIX C (continued)

No.	Name	Jan.	Feb.	Mar.	Apr.	May	June	July	Aug.	Sep.	Oct.	Nov.	Dec.	Year
170	Kahangi Estate	44.0 / 03.2	75.0 / 05.5	122.5 / 08.9	175.5 / 12.8	130.0 / 09.5	75.5 / 05.5	59.5 / 04.3	112.0 / 08.1	159.5 / 11.6	180.0 / 13.1	160.5 / 11.7	81.0 / 05.8	1375
171	Tororo	44.0 / 03.3	67.5 / 05.2	127.0 / 09.5	197.5 / 14.8	207.0 / 15.6	115.0 / 08.7	102.5 / 07.7	105.5 / 07.9	92.0 / 06.9	113.0 / 08.5	97.5 / 07.3	61.5 / 04.6	1330
172	Fort Portal	32.5 / 02.2	74.5 / 05.0	138.0 / 09.3	188.0 / 12.7	143.5 / 09.7	80.5 / 05.4	62.0 / 04.2	116.0 / 07.8	188.0 / 12.7	216.0 / 14.6	167.0 / 11.3	75.0 / 05.1	1481
173	Kalagala Agr. Stat.	50.5 / 04.2	82.5 / 06.8	117.5 / 09.7	151.5 / 12.5	128.5 / 10.6	86.5 / 07.1	62.0 / 05.1	103.0 / 08.5	125.0 / 10.3	119.5 / 09.9	118.0 / 09.7	68.5 / 05.6	1213
174	Iganga	48.5 / 03.7	64.5 / 04.9	143.0 / 10.9	185.0 / 14.0	157.5 / 12.0	81.0 / 06.2	65.5 / 05.0	122.5 / 09.3	123.0 / 09.4	121.5 / 09.2	120.0 / 09.1	83.0 / 06.3	1315
175	Mubende	37.5 / 03.1	65.5 / 05.4	103.0 / 08.5	152.0 / 12.5	104.5 / 08.6	60.5 / 05.0	58.5 / 04.8	121.0 / 10.0	143.5 / 11.8	157.0 / 13.0	138.0 / 11.4	71.0 / 05.9	1212
176	Nawanzu	48.0 / 03.6	56.5 / 04.3	145.0 / 10.9	194.0 / 14.7	150.5 / 11.4	80.0 / 06.0	89.0 / 06.7	116.5 / 08.8	122.5 / 09.2	127.0 / 09.6	120.5 / 09.1	75.5 / 05.7	1325
177	Dabani	64.0 / 04.1	77.0 / 04.9	131.0 / 08.4	232.0 / 14.9	212.0 / 13.6	106.0 / 06.8	99.0 / 06.3	124.0 / 07.9	148.0 / 09.5	139.5 / 08.9	148.0 / 09.5	80.5 / 05.2	1561
178	Nagoje	62.5 / 04.8	72.5 / 05.6	120.5 / 09.3	196.5 / 15.1	164.0 / 12.6	102.0 / 07.9	55.0 / 04.2	90.5 / 07.0	99.5 / 07.7	121.0 / 09.3	120.0 / 09.3	92.0 / 07.2	1297
179	Masafu Dispensary	57.0 / 03.7	66.0 / 04.3	161.0 / 10.5	224.5 / 14.7	219.0 / 14.3	98.0 / 06.4	107.0 / 07.0	117.0 / 07.6	143.5 / 09.4	133.0 / 08.7	128.5 / 08.4	75.5 / 04.9	1530
180	Lugala	61.0 / 04.4	86.5 / 06.3	142.5 / 10.4	190.0 / 13.9	159.5 / 11.7	102.5 / 07.5	70.0 / 05.1	105.0 / 07.7	99.0 / 07.2	134.5 / 09.8	116.5 / 08.5	102.0 / 07.5	1369
181	Monika Estate	60.5 / 04.3	100.5 / 07.1	134.0 / 09.5	189.0 / 13.4	171.5 / 12.3	102.0 / 07.2	76.0 / 05.4	104.0 / 07.4	119.5 / 08.4	131.5 / 09.3	119.0 / 08.6	100.5 / 07.1	1408
182	Namanve	57.5 / 04.3	77.5 / 05.8	126.5 / 09.6	196.0 / 14.8	154.0 / 11.6	78.0 / 05.9	58.0 / 04.4	93.5 / 07.1	110.5 / 08.3	123.0 / 09.3	148.5 / 11.2	102.0 / 07.7	1325

APPENDIX C (continued)

	Station		503

No.	Name	Jan.	Feb.	Mar.	Apr.	May	June	July	Aug.	Sep.	Oct.	Nov.	Dec.	Year
						Month of the year								
183	Mukono Agr. Stat.	83.5 05.8	97.5 06.8	148.5 10.3	211.0 14.7	168.5 11.7	94.5 06.5	58.0 04.0	106.5 07.4	106.0 07.3	130.5 09.0	135.0 09.3	104.5 07.2	1444
184	Bwavu	68.0 05.0	93.5 06.9	136.5 10.1	178.5 13.3	159.0 11.8	91.0 06.7	76.5 05.7	99.0 07.3	116.0 08.6	111.5 08.3	117.5 08.7	103.0 07.6	1350
185	Nsayamba	58.5 05.1	74.0 06.5	103.5 09.0	167.0 14.5	149.5 13.1	70.5 06.2	47.5 04.2	86.0 07.5	90.5 07.9	101.5 08.9	123.5 10.8	72.0 06.3	1144
186	Kabasanda	71.0 05.6	75.0 05.9	112.5 08.9	176.5 13.6	107.5 08.5	69.5 05.5	54.0 04.3	96.5 07.6	119.0 09.4	127.0 10.0	147.0 11.7	113.5 09.0	1264
187	Budo King's College	53.0 04.4	62.5 05.2	115.0 09.5	175.0 14.6	145.0 12.0	76.0 06.3	49.0 04.1	96.0 08.0	110.0 09.1	115.0 09.5	123.5 10.2	85.0 07.1	1205
188	Ngogwe Coffee Nursery	88.0 05.7	102.5 06.6	154.5 10.0	220.5 14.2	205.5 13.3	130.0 08.4	89.5 05.8	95.0 06.1	98.5 06.4	102.0 06.5	147.0 09.5	117.0 07.5	1550
189	Buvuma Isl.	91.0 05.7	113.5 07.1	184.5 11.5	250.0 15.6	272.0 17.0	116.0 07.3	60.5 03.8	71.5 04.5	80.5 05.0	91.5 05.7	128.0 08.0	141.0 08.8	1600
190	Kisubi	54.5 03.8	92.5 06.6	153.0 10.9	243.0 17.3	227.5 16.2	97.5 06.9	63.0 04.5	78.0 05.5	73.0 05.2	97.0 06.9	126.0 09.0	101.0 07.2	1406
191	Entebbe	69.0 04.5	88.5 05.8	164.5 10.8	262.0 17.1	250.5 16.4	119.0 07.8	77.0 05.0	76.5 05.0	76.0 04.9	94.0 06.2	131.0 08.6	120.0 07.9	1528
192	Nkosi	49.0 04.7	48.5 04.6	110.5 10.6	165.5 15.9	142.5 13.7	61.0 05.9	31.0 03.0	60.0 05.8	78.5 07.5	86.5 08.3	122.0 11.7	86.0 08.3	1041
193	Kalungu	44.5 04.5	63.0 06.4	96.0 09.7	159.5 16.0	125.5 12.7	41.5 04.2	26.0 02.6	73.0 07.4	88.0 08.9	91.5 09.2	107.0 10.8	75.5 07.6	991
194	Katigondo	50.5 04.7	73.0 06.8	111.5 10.4	160.0 14.9	156.5 14.6	42.5 04.0	26.5 02.5	61.0 05.7	87.5 08.2	108.5 10.2	108.5 10.2	83.0 07.8	1069
195	Lyantonde Dispensary	53.0 05.8	66.0 07.3	95.5 10.5	121.5 13.3	77.0 08.5	27.0 03.0	17.5 01.9	55.5 06.1	102.0 11.2	115.5 12.7	104.5 11.5	75.0 08.2	910

APPENDIX C (continued)

No.	Name	Jan.	Feb.	Mar.	Apr.	May	June	July	Aug.	Sep.	Oct.	Nov.	Dec.	Year
							Month of the year							
196	Masaka	51.5 04.7	66.5 06.1	115.0 10.5	179.5 16.4	168.0 15.4	46.0 04.1	36.5 03.3	54.5 05.0	86.0 07.9	102.5 09.4	99.0 09.1	88.0 08.1	1093
197	Kiwala Estate	54.5 04.8	65.5 05.8	115.0 10.2	178.0 15.8	214.0 18.9	67.0 05.9	33.5 02.9	51.5 04.6	70.0 06.2	80.0 07.1	100.0 08.9	100.0 08.9	1129
198	Kalangala Dispensary	121.5 05.4	149.0 06.7	244.5 10.9	354.0 15.8	352.5 15.7	208.0 09.3	117.0 05.2	108.5 04.8	105.0 04.7	117.0 05.2	179.5 08.0	186.5 08.3	2243
199	Buunga	51.0 04.4	61.5 05.2	138.5 11.7	205.0 17.3	212.0 17.9	65.5 05.5	36.0 03.0	45.5 03.8	89.5 07.6	86.5 07.3	100.0 08.4	94.0 07.9	1185
200	Kyananw- kaka	58.5 05.1	67.5 05.8	139.0 12.0	197.5 17.1	206.5 17.9	57.0 04.9	30.0 02.6	42.0 03.6	81.5 07.1	81.5 07.1	101.0 08.7	93.0 08.1	1155
201	Busenyi	76.0 06.0	105.0 08.3	124.0 09.8	166.0 13.2	100.0 07.9	40.0 03.2	42.0 03.3	86.0 06.8	118.0 09.3	148.5 11.7	151.0 11.9	109.5 08.6	1266
202	Mbarara	43.5 04.8	66.5 07.4	97.0 10.7	119.0 13.2	79.0 08.8	27.0 03.0	22.5 02.5	59.0 06.5	102.0 11.3	104.0 11.5	110.5 12.2	73.0 08.1	903
203	Bikira	58.5 05.3	66.0 06.0	155.0 14.0	171.5 15.5	185.5 16.7	32.0 02.9	23.0 02.1	37.0 03.3	91.0 08.2	95.5 08.6	109.0 09.8	84.0 07.6	1108
204	Lwasamaire	61.5 06.0	87.0 08.5	108.5 10.6	118.5 11.5	81.0 07.9	27.0 02.6	26.5 02.6	72.0 07.0	104.0 10.1	109.0 10.6	128.0 12.4	105.0 10.2	1028
205	Katera	77.0 06.5	82.0 06.9	154.5 13.1	220.5 18.7	213.5 18.0	36.0 03.0	20.5 01.7	41.5 03.5	57.5 04.9	79.5 06.7	95.5 08.1	105.0 08.9	1183
206	Kabale	56.0 05.3	86.5 08.3	118.0 11.3	145.5 13.9	112.5 10.7	42.0 04.0	37.5 03.6	74.0 07.1	97.5 09.3	96.5 09.2	100.0 09.5	82.0 07.8	1048
207	Kapenguria	12.0 01.0	31.5 02.6	80.5 06.6	149.0 12.4	201.0 16.6	132.5 10.9	188.0 15.5	170.5 14.1	97.5 08.1	70.0 05.8	52.5 04.3	26.0 02.2	1211
208	Endebess	22.5 01.9	39.5 03.3	79.5 06.6	152.5 12.6	152.5 12.7	123.5 10.2	177.0 14.7	174.0 14.4	98.0 08.1	84.0 07.0	68.0 05.6	35.0 02.9	1206

APPENDIX C (continued)

No.	Name	Jan.	Feb.	Mar.	Apr.	May	June	July	Aug.	Sep.	Oct.	Nov.	Dec.	Year
							Month of the year							
209	Kitale Agr. Stat.	18.0 01.6	39.5 03.4	67.0 05.8	154.0 13.2	167.0 14.3	128.0 11.0	161.0 13.8	171.0 14.7	103.0 08.9	64.5 05.5	54.0 04.6	37.0 03.2	1164
210	Turbo	26.0 02.0	43.0 03.4	75.0 05.9	164.5 12.8	166.5 13.0	147.5 11.5	183.0 14.2	248.0 19.3	111.0 08.7	43.0 03.4	49.0 03.8	25.5 02.0	1282
211	Tambach	34.5 02.9	36.5 03.0	91.0 07.6	202.5 16.9	180.0 15.0	104.0 08.7	133.0 11.1	121.0 10.1	58.0 04.8	74.0 06.2	102.5 08.5	63.0 05.2	1200
212	Myanga	37.5 02.7	47.5 03.5	122.5 08.9	223.5 16.3	233.5 17.1	120.5 08.8	94.5 06.9	112.5 08.2	100.5 07.3	119.0 08.7	93.5 06.8	66.0 04.8	1371
213	Bugoma V.S.	47.0 03.0	68.0 04.4	119.0 07.6	222.0 14.2	243.0 15.6	149.0 09.6	129.5 08.3	138.0 08.8	139.0 08.9	116.0 07.4	106.5 06.8	84.0 05.4	1561
214	Mumais	43.0 02.4	95.0 05.3	142.5 07.9	247.0 13.7	279.5 15.5	172.0 09.5	146.0 08.1	165.0 09.2	162.5 09.0	140.5 07.8	114.0 06.3	96.0 05.3	1803
215	Tororo/Naninga	49.5 03.6	69.5 05.1	122.0 09.8	200.5 14.7	192.0 14.0	85.0 06.2	82.0 06.0	103.5 07.6	115.0 08.4	137.0 10.0	140.0 10.2	73.0 05.3	1369
216	Kakamega	79.5 04.1	101.0 05.3	147.0 07.7	255.5 13.3	283.5 14.7	200.0 10.4	166.5 08.7	208.5 10.9	178.0 09.3	122.0 06.3	99.0 05.1	80.5 04.2	1921
217	Kapasabet	47.5 03.1	74.0 04.8	117.5 07.6	194.0 12.6	210.5 13.7	158.0 10.3	173.0 11.2	201.0 13.1	144.5 09.4	87.0 05.6	77.0 05.0	56.0 03.6	1540
218	Rangala	55.0 03.6	77.5 05.0	138.5 09.0	230.0 15.0	214.0 13.9	108.5 07.1	96.0 06.2	116.0 07.6	144.0 09.4	122.5 08.0	140.0 09.1	94.0 06.1	1536
219	Maseno V.S.	61.5 04.0	86.5 05.6	151.0 09.8	230.0 14.9	208.5 13.5	118.5 07.7	92.0 06.0	127.5 08.3	126.5 08.2	114.5 07.4	131.0 08.5	93.5 06.1	1541
220	Equator	34.0 02.2	59.0 03.8	94.0 06.1	206.5 13.4	221.0 14.4	156.5 10.1	185.0 12.0	221.0 14.4	139.5 09.1	80.5 05.2	81.0 05.3	62.0 04.0	1540
221	Miwani	48.0 03.9	76.5 06.3	134.5 11.0	217.0 17.7	159.0 13.0	87.5 07.1	69.0 05.6	92.0 07.5	75.0 06.1	88.0 07.2	100.0 08.2	77.5 06.4	1224

APPENDIX C (continued)

							Month of the year							
No.	Name	Jan.	Feb.	Mar.	Apr.	May	June	July	Aug.	Sep.	Oct.	Nov.	Dec.	Year
222	Kisumu P.C.	48.5 04.3	80.5 07.1	134.5 11.9	188.0 16.7	150.0 13.3	83.0 07.3	58.0 05.1	76.0 06.7	66.0 05.8	56.0 05.0	90.0 08.0	99.5 08.8	1130
223	Chemelil	54.0 04.1	83.0 06.3	126.0 09.6	232.0 17.7	191.0 14.6	94.5 07.2	82.5 06.3	104.0 08.0	76.5 05.8	74.5 05.7	106.5 08.2	84.5 06.5	1309
224	Muhoroni	60.5 03.9	118.0 07.6	160.0 10.3	249.5 16.0	194.0 12.5	119.5 07.7	109.0 07.0	134.0 08.6	103.0 06.6	80.0 05.1	122.5 07.9	106.0 06.8	1556
225	Fort Ternan	38.5 03.1	79.5 06.4	119.0 09.6	177.0 14.3	155.0 12.4	124.5 10.0	134.0 10.8	135.0 10.9	86.0 06.9	63.0 05.1	74.0 06.0	55.5 04.5	1241
226	Londiani	30.0 02.7	46.0 04.1	82.0 07.3	156.0 14.0	141.0 12.6	122.0 10.9	135.5 12.1	165.0 14.8	95.0 08.5	50.5 04.5	55.5 05.0	39.5 03.5	1118
227	Lumbwa	35.0 03.0	63.5 05.4	94.0 08.1	173.5 14.9	162.5 14.0	128.5 11.1	125.0 10.8	134.0 11.5	90.0 07.8	59.0 05.1	53.5 04.6	42.5 03.7	1161
228	Molo	34.5 02.7	52.5 04.1	100.0 07.8	181.5 14.2	149.0 11.7	122.5 09.6	142.5 11.2	175.0 13.7	99.5 07.8	64.5 05.1	89.0 07.0	64.5 05.1	1275
229	Kericho	65.5 03.6	95.5 05.2	157.0 08.6	264.0 14.4	246.5 13.4	164.5 09.0	152.5 08.3	188.0 10.3	150.0 08.2	136.5 07.4	124.0 06.8	89.0 04.8	1833
230	Sotik	68.0 05.0	103.0 07.6	137.0 10.2	183.5 13.6	145.0 10.7	120.5 08.9	92.0 06.8	127.5 09.5	114.0 08.5	86.5 06.4	93.0 06.9	79.0 05.9	1349
231	Kisii	63.0 03.7	99.5 05.8	164.5 09.5	252.5 14.7	211.0 12.2	146.5 08.6	105.0 06.1	150.5 08.7	154.0 08.9	131.5 07.6	143.5 08.3	102.5 05.9	1724
232	Bukoba	152.5 07.7	158.0 08.0	243.0 12.2	337.0 17.0	308.5 15.6	87.0 04.4	45.0 02.3	79.0 04.0	102.5 05.2	124.0 06.3	161.5 08.1	183.0 09.2	1981
233	Tarime	82.0 05.9	99.5 07.2	157.5 11.4	231.5 16.7	159.5 11.5	79.5 05.7	46.0 03.3	78.0 05.6	82.5 05.9	120.5 08.7	129.0 09.3	121.5 08.8	1387
234	Musoma	58.0 07.6	67.0 08.7	117.5 15.3	164.5 21.5	106.5 13.9	23.0 03.0	18.0 02.3	20.5 02.7	23.0 03.0	34.5 04.5	74.0 09.7	59.5 07.8	766

APPENDIX C (continued)

Station								Month of the year							Year
No.	Name	Jan.	Feb.	Mar.	Apr.	May	June	July	Aug.	Sep.	Oct.	Nov.	Dec.	
235	Kagondo	115.5 07.2	133.0 08.2	217.0 13.4	301.0 18.7	237.0 14.7	38.5 02.4	27.5 01.7	53.5 03.3	76.0 04.7	108.0 06.7	157.0 09.7	150.0 09.3	1614
236	Kawalinda	113.5 07.0	135.5 08.4	194.0 12.0	294.0 18.1	237.0 14.6	41.5 02.6	21.5 01.3	60.0 03.7	76.5 04.7	123.5 07.6	161.5 10.0	162.5 10.0	1621
237	Rubya	115.5 08.8	133.0 10.1	172.0 13.1	216.5 16.4	156.5 11.9	27.5 02.1	11.0 00.8	39.0 02.9	73.0 05.5	109.0 08.3	138.0 10.5	127.0 09.6	1318
238	Igabiro	95.5 08.5	103.5 09.3	167.5 15.0	195.5 17.5	123.5 11.0	23.0 02.1	09.0 00.8	24.5 02.2	53.0 04.7	90.5 08.1	106.0 09.5	126.5 11.3	1118
239	Ikizu	86.5 08.3	83.0 08.0	142.5 13.7	154.0 14.8	99.0 09.5	35.5 03.4	09.0 00.9	35.0 03.4	47.5 04.6	106.5 10.3	127.5 12.3	112.0 10.8	1038
240	Kome	95.0 08.7	99.0 09.1	147.5 13.5	157.0 14.3	75.5 06.9	20.5 01.9	16.0 11.5	48.0 04.4	75.0 06.9	125.5 11.5	129.0 11.8	104.0 09.5	1092
241	Ngara	100.0 09.7	110.5 10.6	120.5 11.6	184.0 18.1	102.0 09.8	14.5 01.4	04.0 00.4	21.0 02.0	62.0 06.0	75.5 07.3	129.0 12.5	110.0 10.6	1033
242	Mwanza	95.5 09.2	108.0 10.4	149.5 14.5	185.5 18.0	93.0 09.0	18.0 01.7	12.5 01.2	21.0 02.0	39.5 03.8	54.5 05.3	120.5 11.6	137.5 13.3	1035
243	Biharamulo	97.0 10.0	111.0 11.5	150.5 15.6	190.5 19.7	76.0 07.9	12.5 01.3	03.5 00.4	20.0 02.1	33.0 03.4	62.0 06.4	99.0 10.2	111.0 11.5	966
244	Ukiriguru	95.0 11.4	82.0 09.8	132.5 15.9	139.5 16.7	72.5 08.7	09.5 01.1	01.0 00.1	13.5 01.6	23.0 02.8	41.5 05.0	119.0 14.2	106.0 12.7	835
245	Sumvwe Mission	82.0 10.6	80.5 10.4	120.0 15.5	123.5 15.9	77.5 10.0	16.5 02.1	06.0 00.8	07.0 00.9	16.0 02.1	36.0 04.7	103.0 13.3	106.0 13.7	774
246	Geita	94.5 09.7	106.0 10.9	138.5 14.2	162.5 16.6	80.5 08.3	09.5 01.0	03.0 00.3	20.0 02.1	35.5 03.6	64.5 06.6	130.0 13.3	130.5 13.4	975
247	Ngudu	90.0 10.9	96.0 11.6	135.5 16.4	154.0 18.6	69.0 08.3	10.0 01.2	03.5 00.4	04.0 00.5	11.0 01.3	26.5 03.2	100.5 12.1	128.0 15.5	828

507

APPENDIX C (continued)

No.	Station Name	Month of the year													Year
		Jan.	Feb.	Mar.	Apr.	May	June	July	Aug.	Sep.	Oct.	Nov.	Dec.	
248	Kijima Mission	105.5 13.9	86.0 11.4	134.5 17.8	118.0 15.6	64.0 08.5	09.0 01.2	01.0 00.1	06.0 00.8	14.5 01.9	30.5 04.0	72.5 09.6	115.5 15.2	757
249	Shanwa	116.0 14.7	97.0 12.3	117.0 14.9	147.0 18.7	51.5 06.5	04.0 00.5	0 0	04.0 00.5	08.5 01.1	26.0 03.3	87.0 11.0	130.0 16.5	788
250	Addis Ababa	14.0 01.2	39.5 03.3	65.5 05.4	83.0 06.9	84.0 07.0	130.5 10.8	275.5 22.8	286.5 23.7	188.0 15.6	19.5 01.6	14.0 01.2	06.0 00.5	1206

APPENDIX D - Monthly and annual discharges of the Nile and its tributaries at a number of key stations

TABLE 1   Monthly and annual discharges of the Kagera River at Kyaka Ferry, million m^3 (WMO, 1974)

Year	Jan.	Feb.	Mar.	Apr.	May	June	July	Aug.	Sep.	Oct.	Nov.	Dec.	Total
1940	418	370	434	411	459	486	549	542	477	466	411	385	5.420
1941	385	333	385	390	422	396	418	437	429	428	420	437	4.890
1942	431	398	505	540	651	672	676	648	591	542	471	443	6.570
1943	412	367	397	399	412	390	384	375	354	347	315	322	4.490
1944	322	304	326	336	384	333	350	360	354	360	339	338	4.100
1945	319	288	319	312	263	351	338	332	315	325	306	338	3.900
1946	341	300	332	312	356	318	328	335	312	305	285	301	3.800
1947	298	281	304	330	487	504	657	620	537	515	441	431	5.420
1948	404	360	390	388	433	409	398	382	349	344	324	324	4.510
1949	322	291	325	331	345	325	326	306	280	275	272	282	3.680
1950	287	260	292	319	351	319	325	322	310	304	274	290	3.650
1951	293	266	301	326	360	460	606	550	465	450	405	496	4.980
1952	518	482	592	596	647	604	604	635	591	552	484	454	6.760
1953	410	369	416	418	451	420	421	403	398	388	365	400	4.860
1954	368	325	382	385	503	544	547	512	419	409	378	364	5.140
1955	339	318	369	372	388	362	391	397	368	350	343	355	4.350
1956	335	298	299	316	432	479	520	485	465	452	400	423	4.900
1957	403	343	403	410	556	670	680	660	605	566	504	486	6.290
1958	445	392	450	457	527	506	515	491	432	425	383	402	5.430
1959	391	337	377	379	436	388	406	419	400	419	386	396	4.740
1960	410	379	433	530	631	701	694	586	509	476	437	425	6.210
1961	397	365	400	398	436	424	449	450	397	396	420	494	5.000
1962	434	536	778	824	924	928	960	927	795	724	654	634	9.120
1963	641	636	810	873	1148	1136	1339	1328	1133	931	756	757	11.490
1964	826	844	951	1052	1204	1243	1245	1048	771	665	601	623	11.070
1965	617	555	626	644	746	808	819	715	600	555	502	532	7.720
1966	506	496	410	699	885	901	913	671	584	536	473	464	7.410
1967	497	448	499	496	570	544	566	564	522	513	499	534	6.250
1968	496	508	693	847	1204	1300	1300	1115	873	725	636	678	10.370
1969	656	686	838	882	1082	963	861	730	612	565	527	521	8.920
1970	530	480	562	660	900	1080	1010	850	700	630	545	530	8.480
1971	505	450	515	540	640	620	695	690	630	610	570	565	7.030

Source: Date from 1940-1966 - W.D.D. Uganda
           Date from 1967-1971 - Hydromet. Project

510

TABLE 2   Monthly and annual discharges of the Victoria Nile at Jinja, mlrd m^3 (WMO, 1974)

Year	Jan.	Feb.	Mar.	Apr.	May	June	July	Aug.	Sep.	Oct.	Nov.	Dec.	Total
1946	1.30	1.08	1.16	1.12	1.32	1.36	1.33	1.33	1.30	1.30	1.26	1.42	15.28
1947	1.45	1.35	1.57	1.74	2.23	2.25	2.21	2.18	2.05	2.10	2.94	1.99	23.06
1948	1.92	1.78	1.92	1.86	2.08	2.02	2.01	1.99	1.88	1.86	1.81	1.82	22.95
1949	1.76	1.55	1.61	1.56	1.68	1.61	1.57	1.53	1.43	1.41	1.30	1.32	18.33
1950	1.36	1.20	1.33	1.42	1.58	1.50	1.51	1.44	1.37	1.38	1.32	1.33	16.74
1951	1.32	1.99	1.32	1.47	1.69	1.65	1.59	1.48	1.34	1.35	1.40	1.71	17.51
1952	1.96	1.81	1.98	2.07	2.52	2.50	2.41	2.32	2.21	2.23	2.12	2.11	26.24
1953	2.05	1.74	1.88	1.86	2.12	1.96	1.92	1.79	1.67	1.67	1.61	1.75	22.02
1954	1.72	1.47	1.62	1.74	1.89	1.91	1.84	1.78	1.66	1.66	1.52	1.56	20.27
1955	1.53	1.44	1.56	1.58	1.73	1.56	1.50	1.42	1.36	1.48	1.39	1.46	18.01
1956	1.56	1.52	1.56	1.63	1.86	1.78	1.71	1.63	1.56	1.60	1.53	1.58	19.52
1957	1.58	1.44	1.63	1.75	2.02	2.07	2.04	1.92	1.77	1.67	1.61	1.69	21.16
1958	1.69	1.54	1.71	1.66	1.91	1.86	1.86	1.79	1.67	1.66	1.52	1.58	20.45
1959	1.16	1.14	1.59	1.57	1.72	1.64	1.61	1.50	1.40	1.44	1.44	1.55	17.76
1960	1.54	1.46	1.69	1.88	2.06	1.94	1.84	1.67	1.58	1.61	1.57	1.58	20.42
1961	1.52	1.36	1.52	1.62	1.80	1.71	1.65	1.59	1.55	1.61	1.95	2.69	20.57
1962	3.01	2.69	2.99	3.06	3.52	3.45	3.46	3.35	3.22	3.36	3.20	3.38	38.69
1963	3.48	3.16	3.63	3.56	3.89	4.10	4.15	4.01	3.67	3.63	3.53	3.99	44.80
1964	4.25	3.79	4.14	4.23	4.55	4.40	4.61	4.28	4.12	4.20	3.88	4.01	50.46
1965	4.47	4.12	4.60	4.28	4.40	4.31	3.70	3.55	3.29	3.31	3.31	3.52	46.86
1966	3.52	3.17	3.62	3.66	4.23	4.25	3.65	3.46	3.33	3.37	3.30	3.38	42.94
1967	3.30	2.89	3.10	3.04	3.42	3.28	3.30	3.16	2.92	3.01	2.99	3.35	37.76
1968	3.27	3.01	3.46	3.57	4.11	4.07	4.00	3.80	3.49	3.46	3.36	3.71	43.31
1969	3.79	3.54	4.03	4.01	4.35	4.14	4.03	3.79	3.51	3.47	3.32	4.02	46.00
1970	2.90	3.20	3.65	3.84	4.25	4.08	4.01	3.84	3.67	3.72	3.51	3.60	44.27

TABLE 3   Monthly and annual mean discharges of the Kyoga Nile at Paraa, m^3/sec (WMO, 1974)

Year	Jan.	Feb.	Mar.	Apr.	May	June	July	Aug.	Sep.	Oct.	Nov.	Dec.	Year
1948	718	682	667	655	671	687	708	635	737	725	714	685	699
1949	656	615	572	539	549	552	545	549	569	571	524	475	560
1950	450	420	404	417	429	450	465	484	525	532	505	456	461
1951	417	398	390	417	465	515	544	571	575	581	590	642	509
1952	719	705	715	671	674	693	718	667	718	720	671	627	687
1953	581	538	503	522	551	551	545	537	534	516	498	458	528
1954	428	437	448	434	435	584	624	660	676	708	669	627	561
1955	579	558	543	540	523	530	534	601	720	650	595	556	577
1956	524	492	471	464	483	541	569	595	664	692	698	657	571
1957	613	588	573	569	622	694	768	788	706	666	643	608	653
1958	550	524	506	496	528	541	581	620	640	626	555	508	557
1959	528	508	498	492	520	536	546	560	604	640	636	630	558
1960	640	631	631	644	695	820	870	820	890	772	756	604	739
1961	700	620	574	580	610	640	654	732	816	884	1100	1318	769
1962	1442	1340	1320	1123	1177	1210	1237	1273	1350	1408	1500	1527	1326
1963	1405	1352	1322	1357	1627	1757	1740	1735	1694	1537	1487	1515	1544
1964	1457	1345	1675	1683	1707	1688	1691	1815	1882	1040	1927	1830	1728
1965	1700	1618	1557	1742	1742	1668	1630	1558	1482	1472	1498	1490	1596
1966	1403	1347	1363	1408	1435	1468	1510	1473	1490	1485	1498	1412	1441
1967	1322	1228	1165	1217	1235	1262	1285	1305	1293	1360	1368	1402	1287
1968	1327	1287	1312	1330	1463	1607	1640	1635	1585	1547	1543	1530	1483
1969	1458	1447	1538	1530	1600	1623	1625	1586	1550	1482	1450	1443	1527
1970	1373	1300	1280	1338	1462	1508	1545	1578	1662	1664	1657	1532	1492

TABLE 4   Monthly* and annual discharges of the River Semliki for the period
1948-1970 (WMO, 1974)

Year	Jan.	Feb.	Mar.	Apr.	May	June	July	Aug.	Sep.	Oct.	Nov.	Dec.	Total mlrd m³
1948	104	100	109	107	125	135	162	169	184	156	153	130	4.29
1949	68	95	82	94	118	126	124	127	141	150	113	109	3.53
1950	90	78	86	97	100	96	124	136	153	156	125	123	3.60
1951	102	100	102	163	155	158	140	193	133	157	148	276	5.08
1952	223	182	163	185	216	176	173	192	189	165	153	129	5.65
1953	114	93	99	99	118	110	95	133	113	101	123	112	3.44
1954	89	81	77	105	112	126	127	116	133	130	112	121	3.50
1955	100	100	97	104	102	89	107	111	126	164	122	120	3.53
1956	110	103	96	122	132	104	118	136	143	171	146	131	3.97
1957	126	107	123	143	158	179	141	151	136	121	130	156	4.38
1958	122	106	96	104	134	126	169	138	131	131	112	115	3.91
1959	160	103	84	92	118	90	79	103	100	114	127	113	3.37
1960	103	101	102	112	105	91	103	100	107	115	118	91	3.28
1961	72	74	63	83	84	87	95	127	140	168	–	–	–
1962	207	171	162	165	222	200	183	265	239	–	280	275	6.78
1963	278	273	284	317	500	388	351	324	286	271	278	286	10.09
1964	268	229	213	247	284	231	210	243	280	271	231	232	7.73
1965	210	179	170	170	166	151	158	138	131	158	194	155	5.20
1966	128	135	138	210	166	145	143	149	172	160	165	144	4.89
1967	119	104	98	106	147	132	138	131	110	125	143	145	3.94
1968	117	116	146	147	151	157	145	147	126	131	151	179	4.51
1969	153	146	151	138	171	150	145	132	136	123	136	123	4.48
1970	119	111	124	210	184	172	187	207	215	194	183	153	5.42

TABLE 5   Monthly and annual discharges of the Albert Nile at Panyango for the
period 1948-1970 (WMO, 1974)

Year	Jan.	Feb.	Mar.	Apr.	May	June	July	Aug.	Sep.	Oct.	Nov.	Dec.	Total mlrd m³
1948	918	878	844	794	821	841	862	911	972	997	1010	963	28.41
1949	874	818	769	752	743	736	724	724	736	750	729	694	23.78
1950	638	587	552	551	556	541	560	571	601	624	625	568	18.32
1951	526	488	467	486	492	501	500	519	540	564	617	717	16.87
1952	728	703	661	733	780	799	792	816	845	870	868	840	24.79
1953	780	719	661	644	683	664	657	661	660	653	664	646	21.25
1954	590	558	541	540	568	575	571	612	648	680	683	650	18.95
1955	642	624	594	600	640	520	475	505	630	545	525	500	17.88
1956	540	580	630	575	685	645	645	510	476	525	600	590	17.91
1957	550	492	480	494	514	580	577	592	662	622	655	655	18.05
1958	620	537	482	412	480	462	530	582	620	625	590	575	17.10
1959	625	541	480	417	485	460	540	580	600	650	610	590	17.30
1960	435	390	370	387	427	439	437	510	627	697	815	825	16.70
1961	754	623	547	530	489	482	510	620	760	932	1373	1694	24.44
1962	1765	1790	1829	1845	1886	1916	1932	1933	1982	2018	2073	2097	60.54
1963	2071	2049	2046	2050	2123	2146	2141	2142	2143	2111	2096	2100	66.20
1964	2069	2027	2997	2981	1980	1964	1964	1983	2019	2068	2081	2033	63.43
1965	1903	1874	1845	1831	1821	1799	1778	1768	1750	1740	1770	1780	56.50
1966	1766	1728	1702	1704	1728	1720	1726	1729	1743	1759	1782	1763	54.20
1967	1630	1550	1460	1360	1380	1360	1380	1390	1400	1430	1515	1560	45.30
1968	1490	1450	1385	1395	1390	1480	1500	1550	1610	1600	1660	1680	48.40
1969	1645	1635	1625	1613	1621	1628	1608	1626	1610	1590	1597	1597	50.96
1970	1522	1437	1357	1406	1501	1523	1532	1597	1664	1678	1673	1644	48.70

*monthly discharges given in million m³/month

TABLE 6   Monthly and annual discharges of the Bahr el Jebel at Mongalla for the period 1912-1973 (Cairo University - Massachusetts Institute of Technology, 1977)

Year	Measured flows in million m^3												
	Jan.	Feb.	Mar.	Apr.	May	June	July	Aug.	Sep.	Oct.	Nov.	Dec.	Year
1912	1670	1380	1380	1430	1610	1550	2300	2920	3000	2200	2040	1950	23430
1913	1480	1370	1470	1660	2340	2320	2480	2440	1880	1810	1930	1820	23000
1914	1680	1400	1530	1480	1860	1700	2040	2810	2590	2660	3270	2500	25520
1915	2010	1710	1870	1900	2330	2340	2260	2620	2820	2970	2790	2270	27890
1916	1990	1730	1780	1910	2490	2920	3300	4080	5250	4810	4040	3590	37890
1917	3200	2850	3060	3060	4380	4910	4990	5420	6430	7350	5300	4850	55800
1918	4850	4080	4410	4110	4320	3970	3950	4010	3610	3650	3180	2990	47130
1919	2770	2330	2390	2360	2660	2380	3010	2780	2860	2700	2570	2380	31190
1920	2320	1760	1660	1860	2200	2310	2340	2560	2190	2480	2140	1990	25910
1921	1530	1220	1200	1120	1190	1200	1630	1830	1580	1650	1290	1180	16620
1922	1070	891	999	1050	1250	1160	1240	1560	1950	1550	1460	1080	15260
1923	983	801	853	913	1430	1330	2090	2850	1930	2200	2170	1780	19330
1924	1640	1440	1420	1650	1910	1550	1600	1720	1930	2020	1920	1660	20460
1925	1540	1300	1430	1450	1670	1520	1560	1860	1610	1520	1750	1650	18860
1926	1390	1180	1300	1470	1880	1660	2320	3130	2750	3020	2400	2350	24850
1927	2230	1940	2090	2150	2240	2230	2260	2340	2220	2260	2090	1990	26040
1928	1850	1620	1640	1960	3830	2620	2520	2330	2060	2380	2000	1830	26640
1929	1700	1440	1480	1510	2240	1760	1780	1950	1930	2000	1880	1660	21330
1930	1540	1320	1500	1750	1940	1800	1820	2040	2120	2300	2440	2060	22630
1931	1960	1680	1880	1940	2300	2180	2760	3300	3130	3020	2470	2410	29030
1932	2250	1940	2180	2060	2680	2380	2980	3720	3570	3480	2760	2650	32650
1933	2480	2180	2350	2260	2500	2280	2570	2680	3410	3040	2470	2350	30570
1934	2180	1810	1910	1980	2420	2130	2480	3080	2460	1990	1960	1890	26290
1935	1760	1500	1590	1670	2340	2170	2280	2130	2260	2240	1820	1720	23480
1936	1600	1420	1560	1600	1860	2040	2210	2490	2450	2310	1940	1900	23380
1937	1810	1630	1730	1890	2500	2250	2910	3170	2350	2650	2730	2510	28130
1938	2310	2000	2100	2060	2430	2550	2570	3470	3110	2810	2470	2270	30150
1939	2140	1830	1940	2120	2190	2020	2220	2300	2140	2010	2120	1900	24930
1940	1720	1530	1610	1710	2230	1730	2050	2500	2100	1760	1630	1680	22250
1941	1580	1320	1560	1520	2150	2760	2100	1980	1970	1970	1920	2040	22870
1942	1890	1650	1940	1960	2600	2650	3100	3780	3990	3150	2720	2730	32160
1943	2610	2210	2270	2170	2460	2530	2760	2750	2640	2270	1950	1840	28460
1944	1710	1450	1510	1540	2140	1610	1950	1850	1970	1950	1600	1450	20730
1945	1320	1070	1070	986	1510	1550	1790	2300	2200	1920	1520	1500	18736
1946	1280	1100	1120	1190	1600	2100	1870	3260	2890	2280	1900	1620	22210
1947	1510	1340	1500	1850	2210	2000	2590	3300	3350	3310	2440	2510	27910
1948	2340	2090	2150	2060	2350	2540	2770	3130	3430	3620	2940	2480	31900
1949	2220	1890	1960	1860	2160	2050	2600	2890	2890	2480	1910	1770	26680
1950	1630	1350	1410	1530	1660	1560	1880	2720	2580	2880	1710	1450	22360
1951	1350	1130	1190	1310	1510	1560	1510	1890	1450	1870	2120	2100	19990
1952	1870	1640	1680	2400	2200	2180	2420	3530	3180	3150	2400	2250	28610
1953	1990	1660	1690	1610	1910	1950	2130	2340	1880	1920	1900	1650	22630
1954	1510	1280	1380	1520	1910	1800	2020	2890	3010	2230	1870	1700	22920
1955	1640	1440	1510	1500	1740	1530	1690	2200	2860	3060	2370	1780	23320
1956	1620	1430	1490	1640	1950	1850	1940	2500	3060	3010	2100	1860	24450
1957	1780	1570	1780	1940	2380	2690	2170	2480	2090	2080	1960	1910	24830
1958	1810	1570	1700	1730	2080	2100	2770	2870	2520	2470	1940	1880	25440
1959	1760	1490	1580	1550	2120	1910	1900	2440	2390	2220	2030	1820	23210
1960	1630	1490	1680	1830	2090	2860	2210	2500	2640	2780	2350	2040	25100

TABLE 6   (continued)

| Year | Measured flows in million m^3 | | | | | | | | | | | | |
	Jan.	Feb.	Mar.	Apr.	May	June	July	Aug.	Sep.	Oct.	Nov.	Dec.	Year
1961	1900	1620	1770	1790	2010	2040	2590	3550	2730	4250	4730	4080	34060
1962	3450	3040	3560	3660	4330	4090	4700	4890	5000	5040	4520	4370	50650
1963	4420	3890	4280	4790	6050	5640	3500	5540	5310	4970	5010	5070	60470
1964	4390	3740	3860	4510	5290	4940	5720	6320	6860	7340	5760	5270	64000
1965	5260	4530	4780	4570	4740	4430	4580	4810	4440	5040	4900	4840	56920
1966	4290	3780	4130	4130	4360	4060	4250	4510	4690	4750	4750	4310	51980
1967	3990	3410	3620	3370	3780	3710	4070	5084	4790	5000	5120	4560	50504
1968	3720	3190	3503	3480	4030	4140	4557	5115	4620	4774	4620	4650	50399
1969	4309	3944	4309	4020	4650	4320	4495	5363	5040	4526	4410	4340	53726
1970	4061	3813	3720	3810	4185	4200	4557	5766	6600	6200	5250	4557	56719
1971	4402	4185	4030	3900	4185	3960	4216	4557	4920	4898	4140	3844	51237
1972	3596	3565	3286	3060	3317	3390	3503	3534	3570	3968	4320	4247	43356
1973	3844	3658	3193	3240	3937	3630	3689	4278	4140	4050	3960	4844	45443

TABLE 7   Monthly and annual discharges of the White Nile at Malakal for the period 1912-1973 (Cairo University - Massachusetts Institute of Technology, 1977)

| Year | Measured flows in million m^3 | | | | | | | | | | | | |
	Jan.	Feb.	Mar.	Apr.	May	June	July	Aug.	Sep.	Oct.	Nov.	Dec.	Year
1912	1880	1460	1370	1220	1190	1580	2310	2920	3160	3310	3070	2710	26180
1913	1810	1390	1460	1330	1710	1720	2250	2620	2730	2960	2220	1630	23830
1914	1450	1230	1310	1260	1310	1630	2170	2730	3230	3590	3380	3340	26630
1915	2600	1460	1390	1300	1510	1850	2440	2820	2980	3270	3270	3050	27940
1916	1950	1410	1360	1300	1460	1840	2500	2910	3400	3990	4080	4280	30480
1917	4280	3240	2220	1860	2010	2390	2880	3230	3520	4140	4400	4810	38980
1918	4970	4620	4840	2880	2450	2910	3330	3800	4010	4000	3610	2930	44350
1919	2190	1810	1800	1580	1760	2260	2810	3160	3220	3500	3460	2880	30430
1920	1870	1490	1420	1240	1450	1990	2380	2630	2740	3040	2960	2760	25970
1921	1750	1240	1230	1130	1430	1720	2210	2500	2580	2870	2710	1600	26840
1922	1500	1040	963	860	1040	1560	2110	2440	2540	2860	2860	2820	23593
1923	1710	1040	1060	1300	1410	2010	2460	2940	3160	3360	3070	2960	26480
1924	1830	1320	1270	1310	1540	1680	2180	2490	2620	2940	2840	2810	24830
1925	1860	1270	1290	1210	1470	1830	2360	2660	2750	2990	2880	2590	25160
1926	1650	1180	1340	1240	1590	1970	2310	2680	2880	2330	3150	3090	26410
1927	2500	1400	1390	1310	1340	1700	2250	2560	2650	2850	2710	1810	24470
1928	1400	1170	1200	1320	1820	2070	2490	2850	2980	3300	3240	3080	26920
1929	1920	1410	1410	1360	1960	2300	2680	2870	2990	3310	3150	3080	28440
1930	2180	1430	1400	1340	1530	1830	2340	2630	2670	2830	2670	2040	24890
1931	1550	1190	1250	1200	1270	1690	2310	2640	2870	3180	3160	2810	25120
1932	1680	1360	1360	1310	1510	2000	2470	2840	3120	3690	3720	3870	28930
1933	3420	1820	1670	1600	1670	1960	2420	3130	3450	3730	3660	3420	31950
1934	2300	1510	1540	1460	1680	1970	2590	2940	3150	3420	3380	3410	29350
1935	2420	1520	1520	1510	1720	2230	2700	2930	2970	3180	3120	3070	28890
1936	1980	1570	1470	1310	1570	2030	2390	2620	2750	2980	2880	2400	25950
1937	1640	1320	1340	1240	1540	1980	2320	2820	2960	3320	3080	2750	26310
1938	1800	1350	1430	1380	1530	1900	2430	2710	2860	3300	3290	3400	27380

TABLE 7 (continued)

Year	Measured flows in million m^3												
	Jan.	Feb.	Mar.	Apr.	May	June	July	Aug.	Sep.	Oct.	Nov.	Dec.	Year
1939	3170	1780	1600	1480	1750	2160	2540	2710	2790	3030	3000	2530	28540
1940	1650	1350	1370	1280	1370	1730	2190	2490	2590	2790	2630	1900	23320
1941	1460	1190	1240	1160	1340	1940	2360	2620	2710	2880	2840	2910	24650
1942	2050	1330	1440	1210	1480	1940	2380	2690	2960	3140	3000	2730	26350
1943	1640	1259	5340	1310	1510	1730	2210	2590	2740	2950	2900	2510	28689
1944	1600	1320	1320	1320	1600	2040	2420	2670	2810	3080	2860	2400	25440
1945	1670	1270	1270	1070	1230	1810	2260	2570	2810	3240	3150	3110	25460
1946	2290	1310	1220	1060	1220	1700	2260	2950	3510	3780	3740	3880	28920
1947	3780	2340	1460	1420	1560	1930	2410	2720	2920	3270	3210	3350	30370
1948	2700	1510	1440	1320	1480	2000	2450	2720	2890	3140	3150	3340	28140
1949	2940	1730	1520	1450	1450	1920	2370	2720	3010	3440	3280	3340	29170
1950	3080	1710	1480	1420	1710	1960	2430	2820	3090	3420	3300	3180	29600
1951	2210	1350	1350	1160	1190	1630	2110	2330	2460	2690	2650	2620	23750
1952	1780	1270	1220	1180	1420	1740	2190	2470	2570	2780	2750	2560	23930
1953	1670	1260	1280	1250	1430	1750	2200	2600	2770	3040	2970	2450	24670
1954	1710	1250	1290	1290	1360	1780	2320	2770	3130	3490	3360	3220	26970
1955	2610	1550	1400	1400	1500	1930	2350	2590	2790	3080	3040	3090	27330
1956	2840	1970	1580	1540	1820	2080	2480	2750	3000	3320	3180	3240	29800
1957	3040	1780	1650	1790	1600	2060	2450	2740	2790	2960	2800	2340	28000
1958	1600	1280	1290	1180	1380	1750	2340	2670	2840	3110	2960	3010	25410
1959	2220	1360	1380	1240	1560	1910	2300	2530	2680	2920	2860	2940	25900
1960	2320	1430	1500	1320	1610	1950	2390	2640	2680	2840	2780	2730	26090
1961	1820	1320	1380	1380	1380	1730	2310	2850	3220	3730	3430	3400	27950
1962	3300	2700	2420	1810	1900	2260	2720	3060	3240	3530	3500	3730	34170
1963	3760	2950	2270	1880	2350	2610	3020	3450	3880	4630	4680	4760	40240
1964	3930	3100	2890	2480	2440	2560	3170	4150	5200	6090	6210	6420	48640
1965	6060	4460	3800	3070	2800	2800	3500	4050	4200	4560	4400	4130	47830
1966	3200	2320	2190	1990	2320	2780	3250	3610	3920	4410	4520	4400	38920
1967	3560	2390	2140	1900	1870	2170	2740	3250	3560	4210	4090	4060	35940
1968	3720	2674	2266	1848	1888	2214	2706	3097	3330	4689	3720	3255	35407
1969	2592	2158	2189	2040	2130	2508	3063	3286	3420	3813	3870	3813	34882
1970	2806	2216	2167	1929	1975	2373	2861	3255	3570	3968	3990	4154	35264
1971	3472	2500	2372	2067	1993	2181	2775	3255	3570	3968	3960	4092	36205
1972	3255	2248	2068	1836	2306	2451	3013	3255	3210	3317	3030	2492	32481
1973	2043	1682	1693	1581	1916	2283	2691	2979	3150	3379	3360	3286	30043

TABLE 8   The monthly and annual outflows from Lake Tana in the period 1920-33
(Hurst, H.E., and Philips, P., 1933)

Year	Monthly mean discharges in million m³/day												Total mlrd m³
	Jan.	Feb.	Mar.	Apr.	May	June	July	Aug.	Sep.	Oct.	Nov.	Dec.	
1920	-	-	-	-	-	-		12.4	25.0	26.3	18.0	11.4	-
1921	7.3	4.8	3.1	1.5	0.7	0.3	1.2	11.8	25.7	25.8	16.7	10.6	3.34
1922	6.7	4.3	2.4	1.1	0.7	0.4	2.3	15.8	36.2	31.8	20.0	12.1	4.08
1923	7.8	5.0	3.1	1.7	1.1	1.1	3.2	19.1	38.4	31.5	20.1	13.4	4.44
1924	8.6	5.5	3.3	2.2	1.0	0.9	4.1	21.9	48.4	34.1	23.8	15.2	5.16
1925	9.2	5.9	3.7	2.0	1.1	0.7	1.4	11.4	21.3	18.7	12.3	7.9	2.94
1926	5.3	3.1	-	-	-	-	-	-	-	-	-	-	-
1927	-	-	-	-	-	-	-	-	-	-	-	-	-
1928	6.6	4.0	1.6	1.0	0.8	0.7	3.5	14.8	25.9	19.4	14.3	8.5	3.08
1929	4.8	3.5	2.0	1.0	1.1	2.1	8.6	33.6	55.0	38.7	20.6	13.2	5.63
1930	7.1	5.2	3.5	1.8	1.0	1.5	4.7	16.0	30.3	19.3	12.5	7.7	3.36
1931	4.7	3.4	1.8	0.9	0.6	0.5	1.3	10.1	25.2	23.1	13.8	8.3	2.85
1932	5.4	3.7	2.0	1.0	0.7	0.6	2.9	18.1	39.4	29.3	16.8	12.6	4.04
1933	6.9	4.1	2.7	1.4	0.7	0.6	1.5	10.1	28.7	24.5	17.3	11.6	3.35

TABLE 9   Monthly and annual discharges of the Blue Nile at Roseires for the
period 1912-1973 (Cairo University - Massachusetts Institute of
Technology, 1977)

Year	Measured flows in million m³												
	Jan.	Feb.	Mar.	Apr.	May	June	July	Aug.	Sep.	Oct.	Nov.	Dec.	Year
1912	856	573	358	238	248	1450	5930	14200	8690	3440	1620	871	38474
1913	576	352	272	260	604	396	2320	6620	6090	2140	743	319	20692
1914	190	143	146	232	187	975	5660	16900	11400	8850	4610	1740	51033
1915	873	468	337	227	560	1250	3640	8190	11100	6940	2960	1350	37895
1916	726	410	240	219	490	1390	7450	19400	16800	10900	4500	2210	64735
1917	1190	670	500	372	632	1760	8210	18400	21100	10900	3910	2030	69674
1918	1160	753	649	540	949	1980	5850	12600	8670	3860	1720	883	39614
1919	700	464	318	162	500	1600	7200	15100	13100	4280	1800	942	46164
1920	625	393	396	227	882	2210	7230	12700	9910	7350	2860	1320	46103
1921	778	468	306	204	428	1290	4330	15000	12300	5690	2190	1100	44084
1922	640	384	268	184	371	1460	5510	14200	12200	6850	2330	1220	45617
1923	707	428	412	478	1080	1930	6890	18000	13100	5160	2380	1530	52095
1924	804	535	385	530	564	1830	6960	15100	13900	5930	3540	1630	51708
1925	922	524	399	291	701	2030	4890	12900	9900	5280	2380	1160	41377
1926	722	427	398	354	1850	1990	7320	17300	14000	7060	2670	1550	55691
1927	797	436	400	242	266	1850	6130	12500	9080	5220	1840	990	39751
1928	573	315	254	413	1540	2250	8900	17600	11900	5760	2560	1330	53395
1929	752	467	334	373	1880	4130	11000	19300	16600	9890	3230	1890	69846
1930	1070	622	459	526	608	1610	7090	14300	11300	4280	2020	1080	44965
1931	630	344	257	192	221	1450	4890	15300	12700	7520	2570	1200	47274
1932	675	383	259	203	747	1520	6260	16000	15500	7140	2240	1210	52137
1933	707	419	327	248	512	1210	4510	13300	13400	7130	2900	1510	46173
1934	805	452	317	280	452	1700	7620	17900	12700	7390	2750	1590	53936
1935	878	485	360	380	1100	2590	10900	18900	16700	8100	2830	1590	45813
1936	999	746	471	445	535	1510	8410	16400	14700	5760	2330	1360	53606

TABLE 9   (continued)

Year													Measured flows in million m^3
	Jan.	Feb.	Mar.	Apr.	May	June	July	Aug.	Sep.	Oct.	Nov.	Dec.	Year
1937	824	478	371	236	598	1370	7360	17000	13100	4810	2280	1270	49697
1938	696	383	384	217	370	1660	9120	19000	16500	10000	3230	1610	63170
1939	932	541	405	364	650	1630	5620	11800	10800	6500	2910	1390	43542
1940	792	470	334	243	345	1090	3950	15100	10200	3870	1580	816	38790
1941	475	291	208	126	733	2350	6170	12200	10100	7040	3070	1350	44123
1942	661	370	715	288	556	1420	8070	16600	12900	6870	2210	1210	51314
1943	734	500	438	222	419	864	5070	14800	13800	5940	2410	1230	46427
1944	684	400	281	246	765	1580	6370	14400	11000	4190	1900	1940	43756
1945	616	344	225	180	756	1390	5660	12700	13700	8490	3530	1690	49281
1946	945	502	326	292	335	1820	9480	25200	15200	7227	3030	1580	65937
1947	906	517	461	688	415	1040	4920	17100	14900	6580	2250	1380	51162
1948	734	500	438	219	348	2580	7730	14500	14000	10200	3530	1590	56369
1949	893	500	419	358	439	2070	7740	16000	13900	6600	2490	1680	52985
1950	937	474	359	508	802	1780	5940	15100	13900	5570	2090	1190	48650
1951	718	405	350	243	382	1170	4340	15500	9670	7440	3160	1540	45118
1952	763	425	321	245	361	1150	6020	15300	11500	6510	2170	1110	45875
1953	616	336	280	245	569	912	6850	17700	11200	5840	2330	1290	48071
1954	768	431	320	252	304	1630	8690	18100	15000	8570	3040	1620	58725
1955	1100	599	379	459	669	1500	7500	16800	15300	8750	3050	1640	57746
1956	911	514	382	467	440	2330	7160	15000	12000	14100	4320	1950	59574
1957	1090	605	1100	1290	706	1780	5660	17200	10500	3720	1720	1000	46371
1958	590	398	256	319	382	1870	7780	19000	13800	9670	3410	1700	59175
1959	1050	642	456	283	610	1100	5380	16200	16100	9290	3850	1980	56941
1960	1190	714	536	405	547	1310	7040	16300	13600	6980	2450	1400	52472
1961	794	509	366	502	361	1410	8990	17300	16800	11100	3550	2330	63912
1962	1150	602	489	289	558	1700	5350	15100	14200	8940	2560	1460	52608
1963	881	483	400	416	1250	1520	6140	16900	12500	4860	2620	2300	50270
1964	944	565	340	396	443	1670	8870	16900	14600	10600	3720	1950	60998
1965	1150	666	448	483	277	1200	4760	13400	9420	7200	3070	1750	43824
1966	880	563	467	376	608	2190	5170	12300	11300	2260	2230	2410	40754
1967	1600	512	725	489	487	1420	5920	14200	14200	9052	3120	2020	53745
1968	1321	1122	1057	771	353	1647	7440	12990	5460	7409	2166	1485	31530
1969	1293	1299	1528	978	753	1944	6479	18200	10470	3100	1929	1804	33397
1970	1575	1491	741	327	302	1095	4650	16957	11190	6200	2541	1262	48331
1971	1448	1183	1085	537	412	1290	6851	15314	10680	4526	2646	1500	47472
1972	1494	966	970	654	409	1404	5084	9021	4800	3441	1674	927	30844
1973	899	707	415	447	638	1926	5532	15655	10800	6138	2346	1274	46577

TABLE 10   Monthly and annual discharges of the Blue Nile at Sennar for the period 1912-1973 (Cairo University - Massachusetts Institute of Technology, 1977)

Year	\multicolumn{13}{c}{Measured flows in million m^3}
	Jan.	Feb.	Mar.	Apr.	May	June	July	Aug.	Sep.	Oct.	Nov.	Dec.	Year
1912	1100	677	463	293	199	1280	6590	14100	9940	4340	2290	1260	42532
1913	765	450	326	201	697	472	2090	6750	6750	2140	752	374	21767
1914	254	148	142	189	158	689	5620	18000	12400	10300	5540	2260	55700
1915	1240	705	472	334	607	1350	3650	9390	11300	7520	3160	1640	41368
1916	862	476	262	186	384	935	7160	20500	19400	13700	5600	2850	72315
1917	1600	915	727	505	668	1630	8390	19900	24500	13700	4620	2620	79575
1918	1450	947	719	584	982	2010	180	14100	10600	4720	2150	1130	39572
1919	654	429	276	146	343	1450	6850	14500	13000	3970	1790	966	44374
1920	606	416	370	233	497	1860	7130	13300	10800	8330	3030	1490	48032
1921	788	439	295	164	307	1050	3730	14100	12300	6250	2360	1230	43013
1922	702	389	265	173	245	1020	5090	15600	13700	7840	2680	1300	49006
1923	731	445	403	463	811	1920	6800	14500	13200	6030	2800	1540	49643
1924	783	511	369	489	518	1280	6490	15100	13300	6010	3450	1650	53950
1925	953	527	393	261	460	1730	4350	12900	9670	5120	1810	1100	39274
1926	619	405	580	510	1240	1810	6630	17200	14000	7030	2250	1370	53644
1927	706	455	465	480	331	1700	6240	14300	10700	5240	1310	804	42731
1928	426	318	393	627	1470	2240	8440	16700	11400	5140	2040	1180	50374
1929	669	505	483	572	1860	3960	10500	18700	16600	10000	2920	1660	68429
1930	952	658	594	539	722	1520	6580	14000	11100	3940	1480	844	42929
1931	440	289	307	395	226	1140	4360	15100	13000	7520	2010	1010	45797
1932	473	361	331	222	779	1390	5920	15800	14900	6770	1810	1020	49776
1933	537	387	374	335	573	1170	3890	13000	13800	7320	2540	1340	45266
1934	630	362	424	360	496	1540	7150	18000	13100	7050	2200	1340	52652
1935	719	440	420	529	1020	2600	10100	17900	15500	8010	2340	1460	61038
1936	843	639	540	542	603	1330	7940	16300	14400	5640	1820	1040	51637
1937	573	400	426	353	560	1320	7170	17200	12800	4460	1710	1050	48022
1938	506	354	421	294	421	1460	8560	18700	16400	9910	2750	1430	61206
1939	749	486	489	424	687	1530	4980	11400	10700	6140	2490	1180	41245
1940	621	417	371	300	388	1050	3490	14900	9900	3440	984	572	36433
1941	267	235	233	152	626	2420	5610	11700	9520	6420	2540	1100	40823
1942	501	310	687	363	660	1330	7450	16400	12100	6340	1610	875	48626
1943	434	360	348	242	498	614	4910	15100	13300	5560	1880	998	44324
1944	471	415	289	280	820	1410	5180	13700	10500	4210	1280	801	39356
1945	369	304	244	216	776	1380	5398	12900	13900	8140	3210	1570	48407
1946	782	439	430	362	391	1800	9060	24000	15000	7060	2730	1380	63434
1947	671	465	516	701	502	922	4230	15990	13200	5320	1670	1150	45337
1948	490	388	544	260	357	2310	7150	13600	13100	9310	3140	1340	51989
1949	648	423	413	446	417	1870	6650	15600	13900	6400	1910	1240	49917
1950	712	411	333	498	852	1570	5970	15000	13800	5190	1490	859	44685
1951	410	347	368	342	368	1000	4000	14400	9150	7180	2760	1370	41695
1952	648	476	429	324	438	939	5530	14500	11000	5890	1560	805	42469
1953	471	302	309	230	554	1010	6290	16400	10400	5070	1700	998	44034
1954	649	408	363	284	343	120	7950	18900	16200	8690	2330	1200	57347
1955	852	554	404	660	688	1310	6930	16500	15500	8280	2190	1250	53918
1956	720	434	380	398	510	1540	6660	14700	11600	12500	3510	1530	54482
1957	862	479	866	770	719	1580	5400	16200	10400	3200	1090	740	42306
1958	486	400	252	330	398	1780	7320	18200	13200	9020	2680	1340	53426
1959	939	590	492	294	673	1030	4830	14700	15100	8040	3150	1530	51268
1960	939	563	427	486	607	1180	6160	15300	13000	6350	1760	862	47634

TABLE 10   (continued)

Year	Jan.	Feb.	Mar.	Apr.	May	June	July	Aug.	Sep.	Oct.	Nov.	Dec.	Year
					Measured flows in million m³								
1961	434	331	266	457	437	951	6800	14900	15900	9710	2980	1980	55146
1962	869	344	360	338	477	1700	4350	13400	12800	6180	1740	818	44376
1963	376	213	166	318	1200	1420	5090	16000	11300	4220	1320	1050	43269
1964	533	281	84	155	365	1350	8090	18200	15600	9430	2580	1290	57958
1965	687	330	117	401	360	973	3880	12400	8570	5870	1780	1020	36388
1966	411	257	210	200	433	1960	4320	11300	10900	1230	887	1380	33488
1967	1128	310	254	357	589	1062	4681	14198	12540	6417	2103	1345	44984
1968	796	777	675	537	254	1050	6727	14477	5550	4247	1074	644	36808
1969	458	713	1023	726	635	1296	5332	17515	8940	1754	723	709	39248
1970	657	658	322	246	280	822	3720	16771	10320	3875	1362	313	39346
1971	709	727	520	399	381	735	5890	14415	9450	3162	1539	647	38574
1972	725	429	362	420	282	843	3999	8742	3360	2046	510	272	21990
1973	254	220	232	240	381	1239	3689	15717	9480	4216	1056	403	37127

TABLE 11   Monthly and annual discharges of the Blue Nile at Khartoum for the period 1912-1973 (Cairo University - Massachusetts Institute of Technology, 1977)

Year	Jan.	Feb.	Mar.	Apr.	May	June	July	Aug.	Sep.	Oct.	Nov.	Dec.	Year
					Measured flows in million m³								
1912	1340	824	576	324	207	484	5910	15900	11400	4390	2210	1230	44805
1913	870	473	351	217	566	466	1880	7520	8620	3000	1140	390	25693
1914	340	203	166	158	246	734	5250	19000	14300	11300	5720	2260	59677
1915	1250	680	478	318	461	971	3320	8060	11300	8430	3510	1840	40618
1916	942	537	358	250	368	828	6760	20200	21500	16100	6050	2960	76853
1917	1730	1020	835	551	645	1410	7300	17700	22900	15800	4820	2500	77211
1918	1290	780	625	537	400	1170	5370	18900	11300	5000	2000	1150	48522
1919	726	438	321	175	241	1120	6200	16800	17300	5180	2220	1230	51951
1920	646	368	358	308	313	1650	6170	13200	11400	8750	3170	1870	48203
1921	888	433	347	202	259	865	3410	14600	14500	7350	2310	1180	46344
1922	727	388	297	171	175	801	4570	16000	16100	8990	2560	1350	52129
1923	782	507	375	454	569	2000	5890	16500	13900	7290	2850	1700	52817
1924	848	572	420	414	497	1000	5890	15000	15600	7200	3800	1830	53071
1925	860	493	436	304	417	1260	3840	13300	10700	6010	2070	1110	40800
1926	644	454	507	535	1090	1710	5590	18900	15100	7950	2440	1480	56400
1927	813	456	464	416	391	1290	5450	16200	12900	6900	1660	860	47800
1928	500	299	301	420	1110	1720	8100	17100	12800	6090	2170	1180	51790
1929	632	386	348	355	1380	3610	10300	18600	18400	11300	3240	1700	70231
1930	1020	622	535	374	578	1370	6330	14700	12200	4750	1680	859	52749
1931	460	273	282	350	215	868	3570	14800	15100	8630	2440	1100	48088
1932	577	345	316	235	516	1240	4890	15600	15900	8070	2090	1040	45319
1933	649	349	384	251	495	1030	3220	13200	15100	8400	2940	1470	47489
1934	721	361	395	351	142	1260	7020	18200	14900	8530	2620	1470	55966
1935	891	483	440	464	756	2190	9450	20300	18400	9140	2760	1540	66814
1936	914	621	583	513	547	913	7170	17400	16700	7580	2200	1050	56191

TABLE 11   (continued)

Year	Jan.	Feb.	Mar.	Apr.	May	June	July	Aug.	Sep.	Oct.	Nov.	Dec.	Year
							Measured flows in million m^3						
1937	611	387	464	365	433	1050	6470	17500	14800	5600	1950	1100	50730
1938	588	397	426	316	373	995	7380	18800	18200	11700	3290	1460	63925
1939	771	496	489	397	550	1170	4680	12100	11800	7160	2920	1280	43813
1940	655	393	401	302	353	851	2940	16000	12700	4070	1120	697	40482
1941	383	261	302	163	359	1940	4830	11400	10900	7110	2900	1100	41648
1942	557	330	537	409	511	1010	6560	18000	14200	8210	1860	912	53096
1943	477	352	327	257	422	439	4050	15800	15200	6750	2290	974	47338
1944	509	377	280	239	542	778	4880	13800	11400	4910	1310	634	39659
1945	398	266	276	210	392	841	3810	13400	14800	9600	3490	1380	48863
1946	656	387	329	338	380	1250	7750	23000	13000	7970	2950	1500	59510
1947	788	507	533	747	570	799	3630	16400	16300	7280	1990	1270	56814
1948	597	424	591	343	332	1760	7520	14600	14400	10800	3670	1440	56477
1949	777	483	453	438	268	1750	6190	15900	15200	7910	2310	1430	53109
1950	824	446	368	436	781	1330	5390	16500	16300	8270	1880	916	53664
1951	518	348	394	356	289	889	3570	14400	11000	8180	3170	1450	44558
1952	738	461	428	293	410	770	4860	15400	13500	7050	1810	860	46560
1953	533	321	338	233	486	1100	6230	18000	12600	6260	1930	1070	48301
1954	710	440	389	310	278	1090	7800	20000	18200	10500	2600	1300	63617
1955	925	635	425	457	613	1160	5980	17600	16200	10200	2570	1450	58215
1956	896	467	401	365	541	1230	6500	15700	13100	14500	4490	1690	59880
1957	997	587	797	1140	759	1500	4270	16400	11800	3570	1270	758	43848
1958	450	389	215	275	344	1330	6600	19000	14400	9370	2960	1410	56743
1959	887	482	475	297	540	810	3970	14200	17800	9210	3890	1740	54301
1960	950	563	462	438	498	807	5190	17000	14300	7930	1940	966	61044
1961	450	324	354	402	460	850	7340	17400	18300	11000	2920	1990	61890
1962	883	480	392	383	427	1280	3870	15300	14500	8850	1830	978	49153
1963	450	280	168	300	870	1120	5180	16700	12900	4780	1320	1040	45108
1964	658	284	136	168	453	984	7460	19300	17300	11200	3380	1340	62663
1965	748	433	205	204	466	637	3120	12400	9690	6980	2020	1100	38003
1966	486	285	207	272	265	1400	3580	12300	12200	2050	1070	1410	35525
1967	1110	384	170	443	472	825	4320	15000	14500	8210	2270	1500	49204
1968	850	740	536	495	450	1038	5766	14167	6420	5952	1386	1063	38869
1969	388	722	1020	921	651	1071	4743	17887	10230	2037	906	626	41202
1970	704	580	487	711	419	582	2142	15407	12360	4619	1773	527	40311
1971	691	693	446	606	391	591	5084	14260	11310	3968	2106	725	40871
1972	896	447	347	641	239	462	3286	9641	3780	2858	549	242	23388
1973	198	168	229	444	384	1041	3224	14539	10710	6014	1425	564	38940

TABLE 12   Monthly and annual discharges of the Main Nile at Tamaniat for the
period 1912-1973 (Cairo University - Massachusetts Institute of
Technology, 1977)

Year	Measured flows in million m^3												
	Jan.	Feb.	Mar.	Apr.	May	June	July	Aug.	Sep.	Oct.	Nov.	Dec.	Year
1912	3320	2280	1830	1450	1380	2010	6500	16800	11800	7820	5030	3900	64120
1913	2960	1810	1650	1410	1870	2050	3270	8740	10800	5710	3470	2370	46110
1914	1710	1280	1310	1230	1420	1820	5970	20100	17500	14900	9170	5520	81930
1915	4240	2830	2100	1470	1820	2650	4960	10000	12800	11000	6270	4600	64740
1916	3360	2140	1690	1510	1630	2460	8430	21800	23800	20500	10300	6440	104060
1917	4940	4090	3380	2140	2100	2910	8280	18100	25400	21000	8980	6200	107520
1918	5050	3920	4110	4390	4480	4490	7780	14800	16500	9690	5820	4390	85420
1919	3220	2270	2100	1700	1710	3020	7620	17900	19700	9780	5110	3750	77880
1920	2690	1890	1720	1410	1440	3150	7890	15900	15200	13100	6490	4580	75460
1921	3410	2010	1860	1540	1500	2440	5100	15800	16400	11500	5500	4140	71200
1922	2930	1750	1340	950	1050	2060	5920	15100	18800	13400	5980	4330	76610
1923	3360	1670	1420	1330	1700	3440	7320	18300	18100	10900	5410	4410	77360
1924	3320	1980	1530	1560	1810	2360	6990	16600	18900	10300	6450	4410	76210
1925	3370	1980	1600	1280	1680	2840	5820	14500	13300	9310	4860	3790	64330
1926	2800	1720	1700	1630	2230	3320	6410	18700	17500	11300	5810	4510	77630
1927	3600	2340	1900	1680	1660	2490	6440	16300	14400	9700	4350	3300	68160
1928	2070	1510	1520	1670	2660	3380	9460	18200	17000	9690	5110	4120	76390
1929	3050	1890	1700	1590	2980	5510	11800	20200	21200	15300	6620	4630	96470
1930	3670	2330	2020	1750	2090	2830	7630	18900	17100	8210	4110	3190	73830
1931	2130	1460	1410	1350	1300	2150	5580	16500	18300	11600	5500	3890	71170
1932	2780	1710	1610	1390	1740	2880	6260	17300	18000	11700	5530	4400	75500
1933	4070	3150	2300	1860	2060	2660	4960	14600	17800	12400	6600	4940	77400
1934	3730	2180	1890	1720	1980	2830	8180	18800	18500	12100	5840	4620	82370
1935	3790	2420	1980	1840	2170	3760	10200	20000	20000	13800	5940	4350	90250
1936	3350	2300	2160	1740	1840	2440	7960	18000	20100	11400	5200	3770	80260
1937	2640	1720	1770	1520	1710	2580	7460	18300	18000	9050	4650	3810	73210
1938	2780	2080	2070	1600	1610	2200	7740	19600	21300	15200	5990	4320	86490
1939	3460	2630	2170	2380	2300	2840	5230	13100	14600	9630	5430	3900	67690
1940	2580	2410	2500	1620	1620	2150	4110	15900	14300	7090	3380	2580	60180
1941	1960	2110	2180	1370	1420	3160	5310	12900	12100	10000	5880	3370	61760
1942	2730	2120	2620	2430	1710	2580	7120	18200	14800	11200	4360	3510	73380
1943	2310	2340	2590	2120	1720	1860	4950	15100	16500	8970	4450	3140	66050
1944	2010	2610	2800	2380	1930	2710	6190	14800	12800	7590	3860	2810	62490
1945	2000	2270	2580	2140	1550	2670	5250	13800	15700	12300	6370	4220	70850
1946	3180	2120	2930	2520	1860	2570	8840	24100	20400	11100	6130	4200	89950
1947	3860	3040	3030	2950	3090	3030	4660	16900	17600	10100	4580	3900	76140
1948	3310	2570	2950	2660	2170	3170	7890	16600	15900	13700	6460	4020	81400
1949	3200	2110	2830	2840	2450	3030	7120	16600	16300	10800	5290	4070	76640
1950	3460	2620	2840	2850	2980	2750	6160	17100	17000	10700	4660	3600	76720
1951	3060	1880	2940	2790	1660	2120	4460	14600	12200	10200	5240	3680	74830
1952	2530	2650	2640	2080	1590	2160	5780	15400	14100	8910	3940	3070	62690
1953	2270	1980	2830	2160	1740	2400	7040	17800	13900	8810	4280	3320	66530
1954	2150	2350	2760	2420	1510	2460	8450	20700	20600	13500	5430	4160	86490
1955	3400	2360	2650	2920	3120	2740	7190	18100	18000	12900	5070	3920	82370
1956	3310	2480	2810	2710	2950	3290	6900	18000	15900	17000	7510	4370	87230
1957	3620	2660	2850	3090	3350	3970	5790	17100	13700	5800	3510	2720	68160
1958	2140	2060	2380	2380	1620	2700	7560	20200	16100	12000	5470	3720	78330
1959	3160	2100	2730	2690	2230	2410	5510	15800	19900	11400	5990	3910	77830
1960	2960	2240	2820	2820	2520	2370	6240	16300	15600	10600	4090	3050	72470

TABLE 12   (continued)

Year												Year
	Jan.	Feb.	Mar.	Apr.	May	June	July	Aug.	Sep.	Oct.	Nov.	Dec.	Year

Measured flows in million m^3

Year	Jan.	Feb.	Mar.	Apr.	May	June	July	Aug.	Sep.	Oct.	Nov.	Dec.	Year
1961	2400	2000	2890	2800	1840	2110	8160	18900	20200	14100	5770	4590	85760
1962	3700	2770	3060	3310	3440	4080	5560	16000	16500	11200	4700	3920	78240
1963	3220	2720	2640	3050	3830	3660	8010	19000	15300	8590	4140	4460	78620
1964	4150	3360	3100	3370	4240	4490	9330	21200	20800	15000	8050	5910	103000
1965	5800	4830	4270	3660	4910	4570	5860	13600	12300	10900	5640	4680	81020
1966	3710	2900	2650	3870	3510	3820	6200	13300	15400	5750	4610	5200	70900
1967	4670	2900	1990	3490	3450	3270	5850	16400	16800	11800	5580	5090	81290
1968	3844	3538	3193	3630	3689	3360	7192	6262	9270	9145	4230	4061	71434
1969	2635	2436	3224	4740	3968	2934	5797	17949	13020	5394	3620	3503	69320
1970	5022	2477	2691	3870	3072	2451	4061	15531	14070	8122	5220	3255	69842
1971	3565	3190	2883	4080	3379	2544	6572	15097	13410	7223	4770	3751	70646
1975	3844	2807	2920	3840	3056	2568	5177	9455	6900	6200	3120	2508	51395
1973	2136	1668	2027	3060	3224	3360	4557	13702	12840	8122	4110	3091	61897

TABLE 13   Monthly and annual discharges of the Main Nile at Hassanab for the period 1912-1973 (Cairo University - Massachusetts Institute of Technology, 1977)

Measured flows in million m^3

Year	Jan.	Feb.	Mar.	Apr.	May	June	July	Aug.	Sep.	Oct.	Nov.	Dec.	Year
1912	4350	2970	2440	1980	1820	2250	6780	18500	16700	9890	6260	4810	78750
1913	3710	2370	2090	1830	1830	1860	2670	8030	10500	5650	3360	2370	46270
1914	1710	1070	1050	949	1110	1460	4610	17100	16300	14700	7290	5110	72459
1915	4100	2810	2010	1520	1690	2250	4470	11200	14100	12200	6670	4470	67490
1916	3300	2000	1520	1130	1240	2020	7150	19000	21500	18800	10100	6190	93950
1917	4710	3620	3150	1890	1680	2560	7670	18200	23100	18900	8990	6000	100470
1918	4790	3790	4070	4000	3560	3450	7230	13800	14600	8900	5210	3970	77370
1919	2800	1930	1740	1420	1270	2090	6520	16100	18800	9250	5020	3860	70800
1920	2580	1630	1460	1170	1130	2720	7320	16300	14300	12500	6700	4330	72140
1921	3260	2120	1530	1030	1070	1890	4570	15200	18100	12000	5630	4030	70430
1922	2970	1800	1270	857	862	1700	5220	17400	18100	13200	6130	4160	73669
1923	3200	1840	1520	1470	1620	3230	6780	19000	18400	12600	5910	4520	80090
1924	3610	2290	1750	1560	1780	2260	6660	16400	18300	11900	7016	4660	77286
1925	3420	2030	1690	1380	1640	2440	5390	14400	12900	9260	5120	4060	63730
1926	2950	1810	1680	1560	2040	3390	5970	18900	18400	11900	6010	4350	78960
1927	3520	2390	1860	1620	1620	2160	5300	15500	14100	9770	4390	3260	65490
1928	2060	1520	1460	1600	2560	3120	8550	16600	16200	9820	5540	4250	73280
1929	3240	2050	1850	1640	2440	1800	10900	20100	21400	16200	7240	4720	86340
1930	3800	2490	2000	1650	2010	2520	6410	16800	15200	8930	4450	3340	69600
1931	2260	1490	1420	1340	1330	1880	4530	15200	17200	12700	5950	3920	69220
1932	2940	1770	1580	1400	1630	2460	6040	17800	18600	12800	5860	4320	77200
1933	3940	3050	2240	1690	1800	2350	4340	14200	17100	12700	7120	4860	75390
1934	3840	2350	2000	1600	1830	2560	8300	18400	19100	13200	6310	4550	84040
1935	3860	2410	1960	1860	2040	3480	9610	20000	20600	14700	6310	4450	91280
1936	3610	2350	2160	1810	1820	2320	7850	17600	19700	11800	5490	3780	80290

TABLE 13   (continued)

Year	Measured flows in million m³												
	Jan.	Feb.	Mar.	Apr.	May	June	July	Aug.	Sep.	Oct.	Nov.	Dec.	Year
1937	2730	1770	1750	1430	1580	2370	6620	18200	18100	9740	4860	3920	73070
1938	2920	2080	2080	1600	1580	2060	6640	19800	20600	16100	6500	4590	86610
1939	3689	2880	2210	2370	2300	2690	5030	13100	15100	10300	5890	3890	69450
1940	2620	2300	2500	1620	1510	1940	3780	15400	14400	7390	3610	2700	59770
1941	1850	2090	2180	1460	1300	2950	5050	12000	12200	10200	6010	3480	60770
1942	2880	2200	2690	2580	1700	2550	6570	18300	15800	11500	4560	3560	74890
1943	2580	2340	2560	2180	1650	1710	4560	13600	16400	10000	4920	3210	65710
1944	2130	2380	2650	2410	1880	2560	5480	14500	13400	8820	3960	2810	62780
1945	2090	2150	2590	2030	1300	2430	4393	13800	15900	12900	6050	4340	69973
1946	3210	2020	2840	2390	1810	2310	7340	20800	21300	12100	6920	4840	87860
1947	3930	3070	3060	2900	2970	2960	4120	15900	18200	11100	4750	3970	76930
1948	3300	2550	2920	2690	2200	2840	7130	16000	16100	13600	6760	4200	80290
1949	3390	2490	2720	2750	2550	2660	6630	17000	17300	11300	5600	4160	78550
1950	3560	2590	2750	2770	2960	2770	5430	16300	16800	10100	4890	3670	74590
1951	3010	1860	2810	2650	1740	1950	4440	14100	12600	10500	5750	3750	65160
1952	2680	2670	2750	2250	1610	2040	4460	11500	14700	9670	4440	3160	61930
1953	2460	2060	2810	2300	1750	2390	6290	16600	14400	9990	4760	3560	69370
1954	2360	2330	2840	2590	1670	2260	7440	18500	18500	14000	5860	4390	82710
1955	3620	2510	2560	2820	2910	2660	5530	15600	16500	12800	5300	3980	76790
1956	3490	2570	2800	2740	2980	3060	6750	16200	14900	14400	7740	4880	82510
1957	3630	2760	2800	3030	3250	3840	5440	14100	13400	6100	3640	2920	64030
1958	2350	2060	2380	2560	1840	2560	6510	16900	15000	12400	3650	3900	72110
1959	3090	2210	2760	2630	2190	2250	4920	16100	20900	11900	6390	3950	79190
1960	2960	2260	2540	2630	2390	2120	5470	16300	15800	10900	4110	3010	70480
1961	2300	1820	2600	2690	1890	1840	7140	18000	21300	14400	6050	4580	84730
1962	3630	2740	2950	3200	3300	3740	5540	15000	16900	11700	4940	3960	78600
1963	3180	2700	2630	3140	3800	3660	2710	17300	15800	9150	4680	4780	73530
1964	4380	3460	3150	3300	4050	4280	8540	19400	20600	15000	7980	5310	99450
1965	5710	4880	4380	3640	4180	4060	5340	12400	13400	10400	5800	4800	79110
1966	3920	2950	2580	3650	3550	3610	5480	11900	15200	6300	4650	5050	68840
1967	4690	3030	2110	3480	3510	3170	5540	15700	17800	12200	5790	4900	81920
1968	4120	3570	3010	3300	3320	3210	6570	15250	10020	9610	4290	4000	70670
1969	2800	2330	2970	4380	3910	2870	5740	17580	14280	5450	3810	3600	69720
1970	3130	2290	2290	3450	2950	2260	3970	16710	15480	8150	5340	3840	69860
1971	3780	3330	2590	3960	3440	2480	5920	16180	15060	7500	5040	3910	73190
1972	3840	2820	2660	3360	3010	2960	5050	8770	7050	6450	3390	2550	51910
1973	2210	3330	2590	2630	2610	2550	4280	14380	14370	8340	4320	3130	62070

TABLE 14   Monthly and annual discharges of the River Atbara at Atbara, near mouth, for the period 1912-1973 (Cairo University - Massachusetts Institute of Technology, 1977)

Year	\multicolumn{13}{c}{Measured in flow in million m^3}
	Jan.	Feb.	Mar.	Apr.	May	June	July	Aug.	Sep.	Oct.	Nov.	Dec.	Year
1912	30	5	0	0	0	126	1990	6990	3330	407	136	46	13060
1913	34	3	0	0	0	0	593	2350	1690	210	32	5	4717
1914	0	0	0	0	0	10	1570	7180	3280	1030	331	82	13453
1915	25	3	0	0	0	136	858	3060	2750	684	146	35	7697
1916	5	0	0	0	0	79	5160	13200	6440	1540	454	173	27051
1917	68	10	0	0	0	221	1240	1430	7410	1210	270	95	14954
1918	35	9	0	0	3	22	1050	3620	1200	390	67	8	6404
1919	0	0	0	0	0	100	1790	4650	2910	457	79	2	9988
1920	0	0	0	0	0	219	1860	6040	2600	922	200	43	11884
1921	0	0	0	0	0	50	1150	5520	3430	776	147	12	11085
1922	0	0	0	0	0	43	1990	7900	7470	1000	169	17	18389
1923	0	0	0	0	0	72	1280	6100	3680	725	139	45	12041
1924	12	0	0	0	0	64	2480	6080	4460	774	259	58	14187
1925	0	0	0	0	0	48	1120	3950	2020	466	74	0	7678
1926	0	0	0	0	0	0	2020	4580	3340	624	142	0	10706
1927	0	0	0	0	0	55	1610	4350	2750	615	75	0	9455
1928	0	0	0	0	0	173	1300	4670	2350	476	105	0	7074
1929	0	0	0	0	0	236	2580	7240	3760	974	165	36	14991
1930	0	0	0	0	0	41	2370	4170	2510	449	110	19	9669
1931	0	0	9	0	0	22	1060	5520	3550	931	186	54	11332
1932	0	0	0	0	0	7	1930	6330	4360	813	145	49	13634
1933	0	0	0	0	0	0	874	3640	4190	994	290	119	10107
1934	8	0	0	0	0	333	2830	7420	3300	1040	171	66	15168
1935	0	0	0	0	6	232	1460	5040	4850	1130	192	66	12976
1936	0	0	0	0	0	0	1740	5580	4750	810	176	64	13100
1937	0	0	0	0	0	9	1560	6400	3780	780	154	47	12730
1938	0	0	0	0	0	0	1590	7360	4820	1380	260	93	15503
1939	0	0	0	0	0	13	1270	3970	2580	820	190	39	8882
1940	0	0	0	0	0	0	1070	4700	2230	390	78	28	8496
1941	0	0	0	0	0	37	789	3090	1710	562	274	60	6522
1942	0	0	0	0	0	0	1600	5420	3220	807	150	59	11256
1943	0	0	0	0	0	0	1270	6010	5640	998	234	69	14221
1944	25	8	0	0	0	0	2360	5390	2720	593	123	50	11269
1945	17	0	0	0	123	295	1210	4510	3810	1390	290	82	11727
1946	16	0	0	0	0	316	2440	10680	3750	901	294	102	18499
1947	37	17	1	0	0	34	893	4940	3350	826	145	70	10313
1948	39	13	0	0	0	231	1650	4580	2990	990	191	89	10763
1949	50	13	0	0	0	92	1090	4110	3180	834	183	81	9633
1950	59	0	0	0	0	0	2180	6550	4930	832	182	92	14825
1951	54	2	0	0	0	31	1080	4570	2860	856	314	115	9882
1952	46	0	0	0	0	0	1120	4950	3260	637	137	44	10238
1953	18	1	0	0	0	212	2060	7350	3710	950	216	112	14629
1954	62	19	2	0	0	171	2310	9020	6880	2140	384	163	21049
1955	0	0	0	0	0	0	1030	5250	5010	1860	298	76	13524
1956	0	0	0	0	0	0	1940	6020	3790	1940	607	158	14455
1957	112	61	29	16	9	72	774	5270	3140	410	108	71	10072
1958	57	37	5	0	0	97	1540	7360	4020	1110	207	81	14514
1959	43	14	2	0	0	0	1740	8240	5560	1100	337	129	17165
1960	59	23	5	0	0	8	1510	4200	3330	805	170	55	10165

TABLE 14   (continued)

Year	Measured flow in million m^3												
	Jan.	Feb.	Mar.	Apr.	May	June	July	Aug.	Sep.	Oct.	Nov.	Dec.	Year
1961	22	4	0	0	0	47	4440	6130	4830	1200	253	155	17081
1962	67	25	6	0	0	0	1180	6050	4490	1000	230	158	13206
1963	44	18	0	0	15	17	1450	5050	3350	812	149	103	11008
1964	37	23	7	0	0	0	2040	7230	4760	917	191	18	15223
1965	165	135	0	0	15	40	1120	3960	2260	324	35	0	8054
1966	6	0	0	126	117	415	692	3490	1820	217	42	31	6956
1967	8	3	3	78	190	111	1580	5960	4000	934	90	14	12971
1968	0	0	0	0	0	0	1670	3880	680	370	10	3	6613
1969	0	0	0	0	0	0	1250	5210	1440	80	10	4	7994
1970	0	0	0	0	0	0	1070	6320	2990	310	70	31	10791
1971	0	0	0	0	0	0	1180	3690	2370	280	30	0	7550
1972	0	0	0	0	0	0	2510	3500	590	150	0	0	6750
1973	0	0	0	0	0	0	0	6080	1540	400	42	0	8062

TABLE 15   Monthly and annual discharges of the Main Nile at Dongola for the
period 1912-1973 (Cairo University - Massachusetts Institute of
Technology, 1977)

Year	Measured flows in million m^3												
	Jan.	Feb.	Mar.	Apr.	May	June	July	Aug.	Sep.	Oct.	Nov.	Dec.	Year
1912	3880	2540	1920	1450	1230	1080	3700	20800	19600	11100	6100	4580	77980
1913	3650	2310	1870	1480	1360	1700	2230	7680	13400	7860	4140	3020	50800
1914	2100	1420	1290	1100	1180	1140	2900	22900	22300	18000	10500	6260	91090
1915	4190	2930	2050	1380	1290	1600	3080	12300	16400	15000	7840	4910	72970
1916	3510	2170	1500	1180	1130	1320	6100	27100	31700	23200	12200	7060	118170
1917	5340	4130	3790	2410	1760	2080	5830	19600	31500	24200	11600	6880	119120
1918	5350	4350	4810	4540	4340	3460	6300	15600	18800	11900	6670	5010	91130
1919	3700	2390	2070	1620	1470	1680	5070	19200	22400	13000	6260	4500	83360
1920	3430	2050	1590	1310	1220	1970	6230	20700	19100	15800	8710	5180	87290
1921	3800	2180	1680	1300	1120	1380	3300	17900	21500	15200	7140	4580	81080
1922	3420	1960	1380	1050	880	1000	3820	19300	25500	17000	8130	5060	88500
1923	3650	1970	1260	1130	1520	2370	5000	22500	23100	16700	6970	4790	90960
1924	3730	2400	1650	1240	1400	1550	5730	19600	23300	14500	7900	5290	88290
1925	3760	2470	1820	1370	1290	1630	4440	15700	17900	12800	5940	4080	73210
1926	3390	2230	1820	1690	1760	2700	4960	21600	23100	14900	7500	4810	90580
1927	3850	2740	1960	1590	1570	1500	4750	18300	20300	13700	5510	3670	79440
1928	2590	1690	1460	1330	1720	2710	7520	19900	21000	12200	6610	4350	83080
1929	3480	2230	1810	1480	1790	3770	10000	25400	25800	18700	9320	5210	108990
1930	4210	2870	2210	1730	1740	1910	5590	19700	18500	11500	5230	3660	78850
1931	2720	1640	1470	1260	1340	1200	3370	16900	21600	14800	7920	4420	78640
1932	3440	2010	1679	1360	1390	1980	5070	20200	23100	15900	7300	4680	88109
1933	4200	3380	2730	1670	1760	1930	3810	14400	22800	15100	8690	5660	86130
1934	4190	2710	2010	1670	1660	1820	7600	22100	24400	16500	7400	5020	97080
1935	4200	2850	2140	1700	1840	2680	8000	22900	24200	17200	7520	4950	100180
1936	3950	2640	2230	1770	1730	1800	6230	20000	25500	15600	6890	4200	92640

TABLE 15   (continued)

Year	Measured flows in million m^3												
	Jan.	Feb.	Mar.	Apr.	May	June	July	Aug.	Sep.	Oct.	Nov.	Dec.	Year
1937	3170	1900	1650	1450	1400	1800	4840	21600	23800	13500	5790	4310	85210
1938	3350	2090	2210	1670	1470	1600	4520	23800	27000	19800	8850	5260	101620
1939	4120	3160	2340	2140	2220	2170	4230	14000	18400	12400	7580	4360	77120
1940	3160	2080	2500	1830	1350	1550	3430	16200	18200	9760	4280	2990	67350
1941	4040	1820	2150	1670	1150	1930	4430	13300	15100	11100	7430	3990	66110
1942	2990	2080	2540	2520	1810	1920	5450	22400	21000	14500	5520	3810	86540
1943	3030	1850	2500	2220	1610	1560	3280	12000	24300	12900	6090	5300	76640
1944	2670	1700	2570	2450	1800	2130	5190	18300	18500	12100	5040	3640	76090
1945	2610	1880	2750	2280	1500	2010	3890	16000	19600	15900	8000	4890	81310
1946	3700	2299	2510	2470	2130	1610	7370	26900	27700	14600	8230	5200	104719
1947	3990	3080	3060	2600	2860	2920	3480	16400	22900	13900	5760	4060	85010
1948	3410	2680	2500	2560	2200	1830	5860	17500	19200	15500	8690	4560	86490
1949	3490	2590	2410	2520	2550	2010	4630	18500	20200	13200	6960	4380	83440
1950	3750	2720	2540	2510	2720	2270	5430	22400	23300	14100	5910	3780	91430
1951	3110	1890	2320	2420	1990	1300	3400	16500	18200	11900	7610	4420	75060
1952	3040	2320	2760	2250	1590	1490	3930	15790	20450	11400	5300	3280	73600
1953	2490	1830	2320	2420	1510	1740	5200	23000	20400	12700	5430	3620	82660
1954	2560	1790	2550	2350	1650	1440	6870	25200	27600	19000	7370	4490	102870
1955	3520	2580	2150	2480	2530	2390	5030	19900	22900	17200	6400	4090	91170
1956	3410	2730	2470	2590	2720	2630	6660	20100	20800	17900	11100	5260	98370
1957	3750	3020	2480	2920	3070	3390	4770	18600	19500	8530	4360	3020	77410
1958	2320	1880	2170	2400	1770	1610	5920	25200	22700	15200	7150	4540	92860
1959	3220	2310	2280	2540	2270	1860	3970	19500	27800	14800	8470	4350	93370
1960	3120	2220	2230	2500	2500	1800	4480	17400	20400	13500	5370	3140	78660
1961	2550	2780	2310	2600	2030	1510	6550	23200	26800	17400	7560	4920	99210
1962	3670	2760	2590	2990	2980	3220	5000	17400	22600	15500	5830	4020	88560
1963	3350	2820	2560	2820	3250	3590	6990	22100	19700	10600	4880	4560	87220
1964	4300	3240	2860	2700	3630	4100	7920	24900	25700	16400	9450	5850	111050
1965	5750	5080	4390	3580	4170	4520	5820	14800	15600	11000	6600	4910	86220
1966	3930	3020	2540	3210	3550	3110	5570	12800	17500	7290	4470	4530	71520
1967	4650	3306	2108	2985	3317	3030	5890	19902	22050	13919	6240	5270	92667
1968	4092	3422	2883	3150	3255	2988	7006	17825	11400	9951	4590	3689	74251
1969	2954	2294	2635	3840	3782	2571	5890	19778	17400	6262	3960	3348	74714
1970	3255	2575	2145	3480	3193	2199	3441	19499	20760	9114	4860	3751	78272
1971	3472	3422	2297	3660	3503	2352	5487	19096	20160	9300	5550	4185	82484
1972	3472	3103	2378	3660	3348	2490	5084	11129	8820	7378	3630	2558	57050
1973	2192	1494	1559	2568	2858	2769	4092	17174	16560	10199	4860	3131	69456

TABLE 16   Monthly and annual discharges of the Main Nile at Aswan for the
period 1912-1973 (Cairo University - Massachusetts Institute of
Technology, 1977)

Year	Measured flows in million m^3												
	Jan.	Feb.	Mar.	Apr.	May	June	July	Aug.	Sep.	Oct.	Nov.	Dec.	Year
1912	3380	2470	1890	1410	1180	984	3170	18500	18200	11100	5690	4170	72140
1913	3260	2040	1480	1280	1170	1470	1740	6500	12200	7540	4120	2830	45630
1914	1720	1150	1070	947	998	975	2010	19400	20000	16500	10700	6930	82480
1915	4510	3090	2130	1360	1170	1460	2850	10700	14900	14700	8090	5270	70230
1916	3850	2400	1610	1180	1090	1340	5010	25000	26600	22400	13300	7880	111660
1917	5370	3910	3580	2320	1680	1860	4830	17500	27800	22800	11600	7460	110710
1918	5270	3990	4330	4090	3960	3230	5190	13600	17700	11100	6210	4580	83250
1919	3440	2180	1910	1520	1350	1520	3680	16600	20800	12600	6140	4410	76350
1920	3320	2090	1600	1370	1220	1710	5370	18200	17800	15000	8960	5570	82410
1921	3820	2370	1760	1330	1180	1300	2980	15400	19600	14800	7590	4700	76830
1922	3530	2090	1460	996	801	900	3280	18800	23600	15900	8240	4880	84480
1923	3030	2350	1540	1200	1380	2050	4440	20100	21400	16200	7250	5170	86110
1924	4000	2650	1850	1320	1470	1630	5050	18000	22300	14300	7990	5590	86150
1925	3840	2540	1830	1370	1260	1580	3910	13400	16600	12500	6450	4480	69760
1926	3440	2150	1640	1340	1330	2390	4360	19200	20600	14500	8330	5230	84510
1927	3940	2910	2030	1440	1300	1290	4180	15600	18300	13300	6090	4010	74390
1928	2600	1630	1310	1160	1320	2560	6290	18200	20000	12400	7300	4820	79590
1929	3670	2400	1770	1340	1390	3250	8770	23200	24500	18000	9910	5580	103780
1930	4250	2950	2120	1540	1530	1660	4400	18300	17600	11800	5640	4020	75810
1931	2350	1790	1500	1220	1090	1140	2850	15900	21300	15000	8620	4840	77600
1932	3580	2260	1680	1320	1270	1700	4280	18900	22900	13800	7720	5050	84460
1933	4170	3470	2850	1680	1600	1800	3280	13100	22100	15300	9150	5940	84440
1934	4340	2960	2110	1550	1440	1700	6050	20500	23700	16400	8370	5250	94370
1935	4170	2990	2170	1530	1520	2330	7100	21300	24100	17700	8220	5200	98330
1936	4000	2800	2180	1560	1450	1530	5490	19000	23900	15800	7420	4550	89680
1937	3340	2100	1740	1360	1130	1500	3950	19800	22300	14300	6050	4590	82160
1938	3540	2160	2160	1560	1280	1450	3710	23200	27800	19500	9520	5520	101400
1939	4120	3240	2430	1960	2220	1960	3920	13100	18400	13000	7980	4770	77100
1940	3420	2130	2430	1810	1310	1330	3050	15300	18000	10900	4630	3370	67080
1941	2260	1830	2070	1630	1100	1550	4320	12000	14400	11100	8040	4540	64840
1942	3200	2330	2460	2310	1930	1710	4530	20300	20300	14900	6360	4240	84570
1943	3230	1910	2570	2080	1670	1480	2670	15600	24700	14200	6840	4270	81220
1944	3070	2000	2660	2270	1890	1810	4180	16700	18100	12600	5240	3850	74370
1945	2590	1340	2550	2150	1580	1610	3580	14900	18900	16400	8410	5360	79370
1946	3840	2550	2360	2310	2120	1340	6130	25100	27800	15800	9100	5980	104430
1947	4420	3480	3180	2520	2770	2850	3170	14900	22500	15000	6550	4640	85980
1948	3710	2950	2510	2370	2330	1530	5430	16600	19300	15700	9790	5190	87410
1949	3610	2830	2440	2400	2510	1900	4080	18100	20400	14300	7490	4840	84900
1950	3860	2900	2500	2390	2530	2280	4610	20800	22500	14800	6610	4460	90240
1951	3390	2240	2270	2340	2150	1180	2850	15400	17800	12000	7790	4940	74250
1952	3280	2380	2680	2100	1610	1690	3210	14900	21100	12500	6240	3710	75400
1953	2790	1890	2180	2410	1600	1460	3860	22000	20800	13900	6330	4120	83340
1954	2870	1830	2550	2270	1820	1180	5630	24800	28300	20500	8550	5050	105350
1955	3820	2710	2080	2280	2490	2180	4320	19200	22700	18000	7410	4580	91770
1956	3580	2800	2350	2310	2250	2370	5780	19700	20700	18400	12400	5760	98700
1957	3990	3110	2380	2630	3010	2960	4450	17600	21400	9920	4890	3380	79720
1958	2560	1850	2080	2150	1700	1300	5000	24600	24500	15700	8410	4870	94720
1959	3430	2420	2250	2420	2340	1620	3370	18700	29600	16600	9670	5090	97510
1960	3650	2420	2220	2460	2370	1730	3790	17100	20900	14500	6610	3740	81490

TABLE 16   (continued)

Year													
	Jan.	Feb.	Mar.	Apr.	May	June	July	Aug.	Sep.	Oct.	Nov.	Dec.	Year
1961	2940	1980	2210	2540	2220	1310	5400	22600	27000	18600	9410	5680	101890
1962	4140	3080	2780	2900	2840	3140	4690	15800	22500	26700	7260	4730	90560
1963	3710	2960	2700	2550	2880	3310	5310	20400	21800	13200	6150	5110	90080
1964	4770	3570	3040	2550	3480	3840	5840	24900	27000	18300	12300	6980	116570
1965	5880	5250	4940	3660	3660	4540	5600	13400	17300	12400	9770	5750	92150
1966	4380	3140	2510	2580	3570	2330	5480	12300	16900	11400	5020	4870	74480
1967	2745	3390	3888	3557	4480	5869	6618	8921	7096	12610	5598	4670	69540
1968	3281	4228	4063	3700	4785	6350	6065	6105	4180	3795	3590	3240	53800
1969	2900	3700	3970	3660	4870	6570	6710	5890	4200	3810	3640	3390	53510
1970	3060	3980	4090	3920	5420	6520	6720	6110	4285	3750	3395	3245	54700
1971	3554	3785	4275	3940	5475	6490	6965	6225	4445	3775	3670	3295	55890
1972	3450	4020	4240	4040	5290	6535	6990	6290	4245	3745	3590	3025	55470
1973	2610	3505	4380	4000	5360	6580	7005	6380	4235	3855	3920	3580	55430

Measured flows in million m^3

528

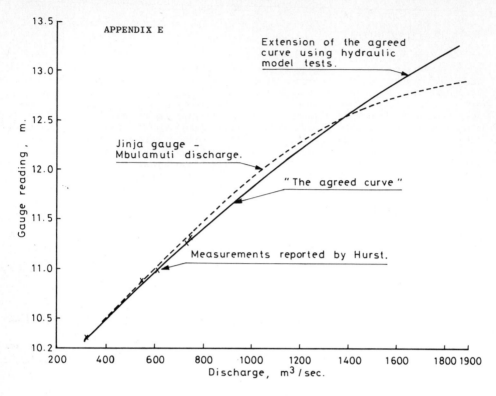

Fig. 1   Rating curve of the Victoria Nile at Jinja

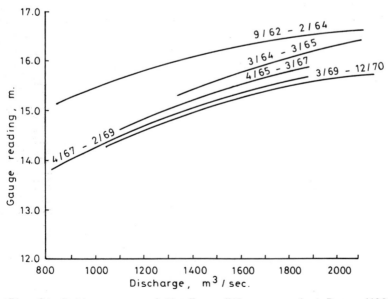

Fig. 2   Rating curves of the Kyoga Nile measured at Paraa (1962-1970)

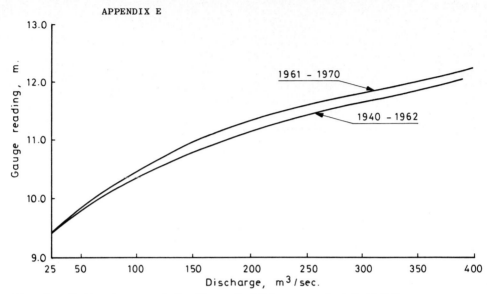

Fig. 3   Rating curves of the River Semliki measured at Bweramul

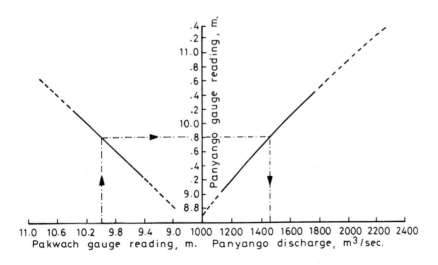

Fig. 4   Rating curve of the Albert Nile at Panyango

APPENDIX E

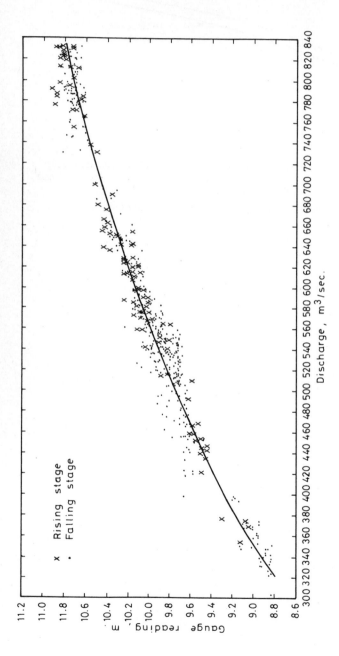

Fig. 5   Rating curve of the Bahr el Jebel and its tributaries measured at Nimule (1923-1927)

APPENDIX E

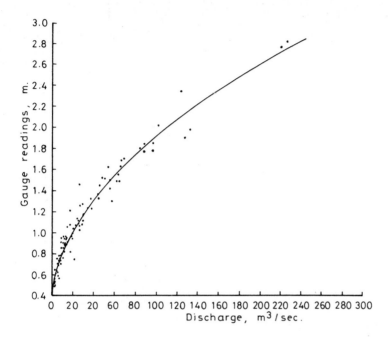

Fig. 6   Rating curve of the River Assua (1923-1927)

APPENDIX E

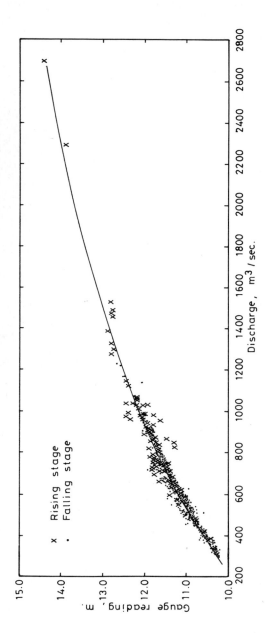

Fig. 7 Rating curve of the Bahr el Jebel measured at Mongalla Station (1911–1927)

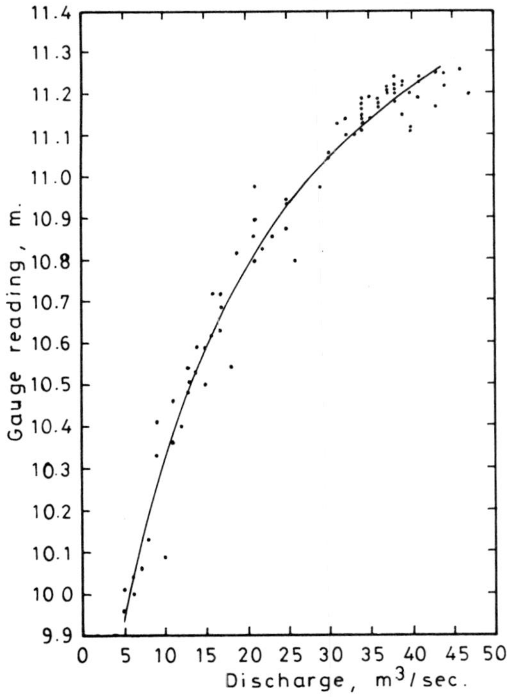

APPENDIX E

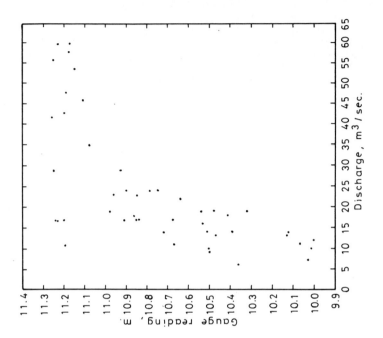

Fig. 9 Rating curve of the River Aliab measured at tail 1,
16 km north of Bor (1925-1927)

Fig. 8 Rating curve of Khor Unyam Kojie
measured 2.5 km south of Bor (1925-1927)

APPENDIX E

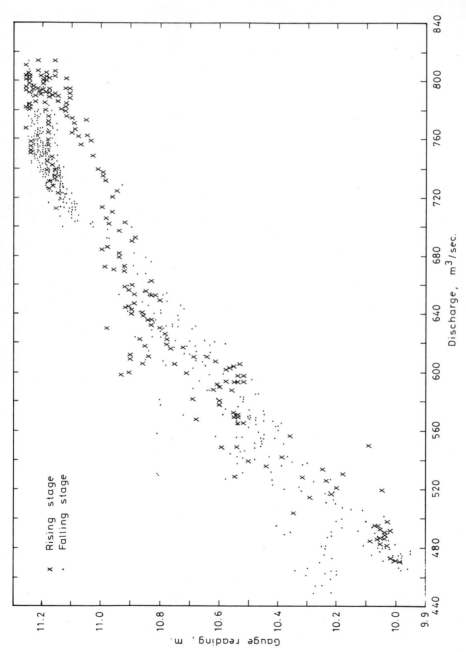

Fig. 10   Rating curve of the Bahr el Jebel measured at Bor (1925–1927)

APPENDIX E

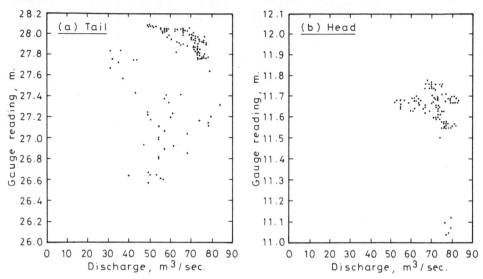

Fig. 11   Rating curve of the Jebel-Zaraf Cut 1 at (a) tail and (b) head (1926-1927)

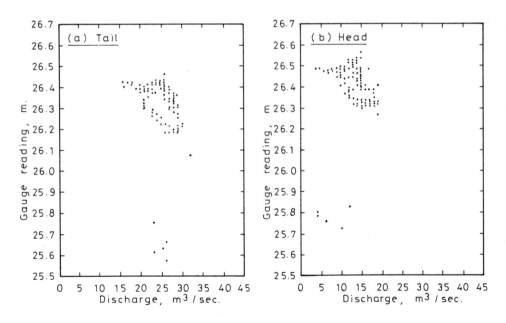

Fig. 12   Rating curve of the Jebel-Zaraf Cut 2 at (a) tail and (b) head (1926-1927)

536

APPENDIX E

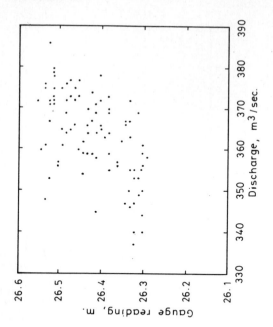

Fig. 14 Rating curve of the Bahr el Jebel measured
at a distance of 281 km from Lake No D.S. tail of
Peake's channel (1926-27)

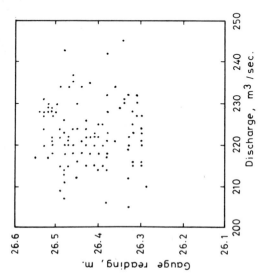

Fig. 13 Rating curve of the regular discharges of the Bahr el Jebel measured D.S.
of the head of Jebel-Zeraf Cut 2 (1926-27).

APPENDIX E

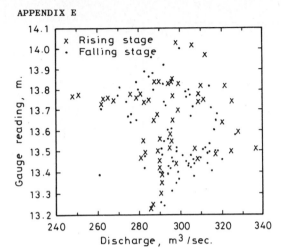

Fig. 15   Rating curve of the White Nile measured in the neighboured of Lake No
(1922-1927)

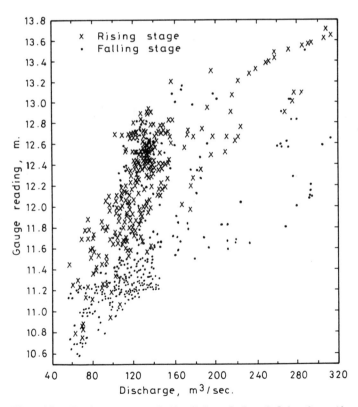

Fig. 16   Rating curve of the Bahr el Zaraf 3 km from the mouth (1908-27)

APPENDIX E

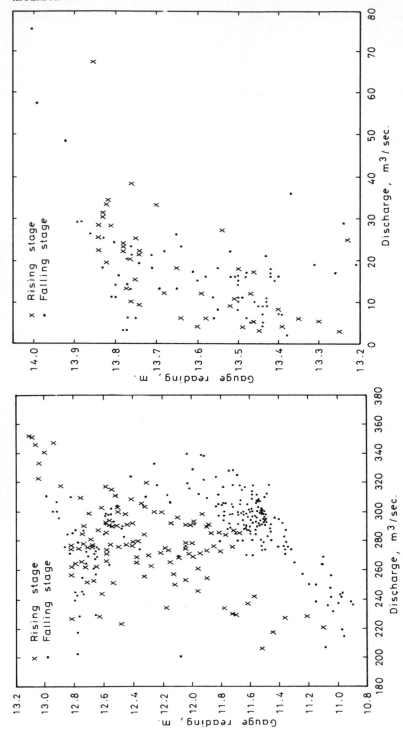

Fig. 17 Rating curve of the White Nile and some of
the Khors measured at Abu Tong in the neighbourhood
of the Zeraf mouth (Tonga Station 1922-26)

Fig. 18 Rating curve of the Bahr el Ghazal and its tributaries measured near Lake No (1921-27)

APPENDIX E

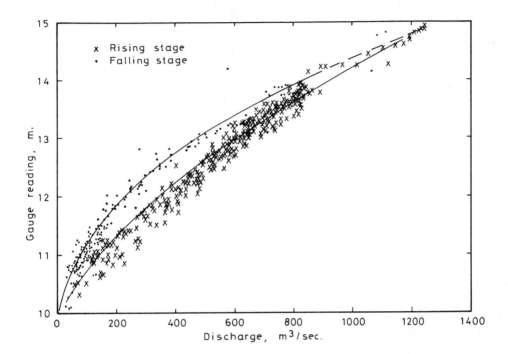

Fig. 19   Average rating curves of the Sobat at Hillet Doleib 9 km above the river mouth (1911-27)

APPENDIX E

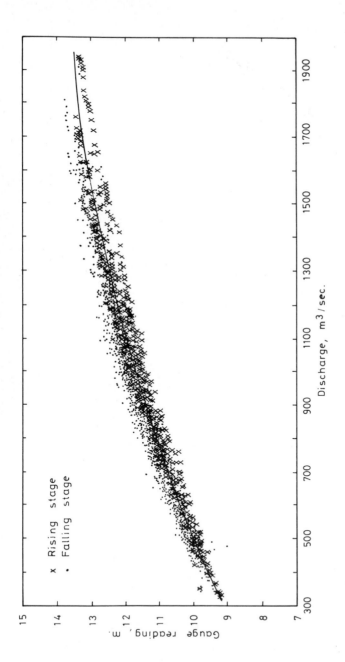

Fig. 20   Average rating curve of the White Nile at Malakal Station (1908-72)

APPENDIX E

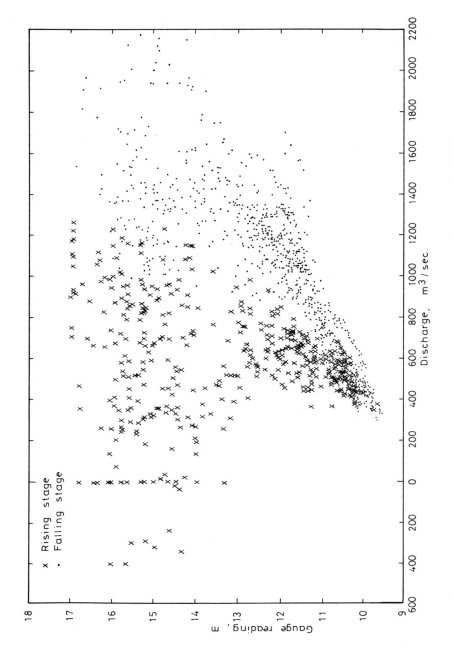

Fig. 21   Rating curve of the White Nile at Mogren Station near Khartoum (1904-27)

APPENDIX E

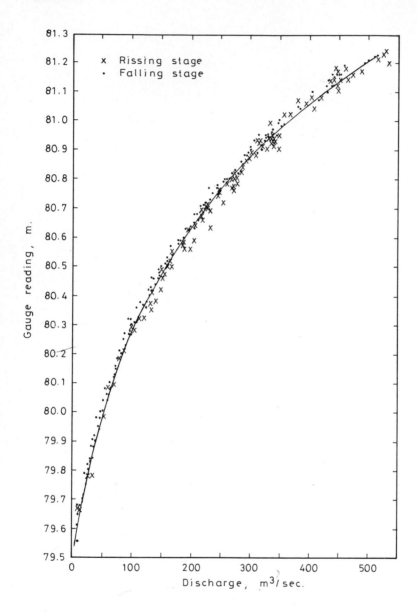

Fig. 22   Average rating curve of the Blue Nile measured at its exit from Lake Tana (Burifasas Station 1920-24)

APPENDIX E

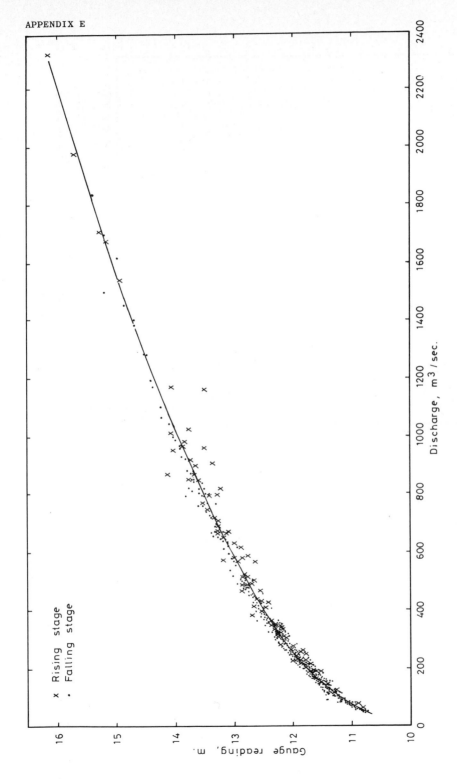

Fig. 23   Average rating curve of the Blue Nile at Roseires Station (1911-27)

544

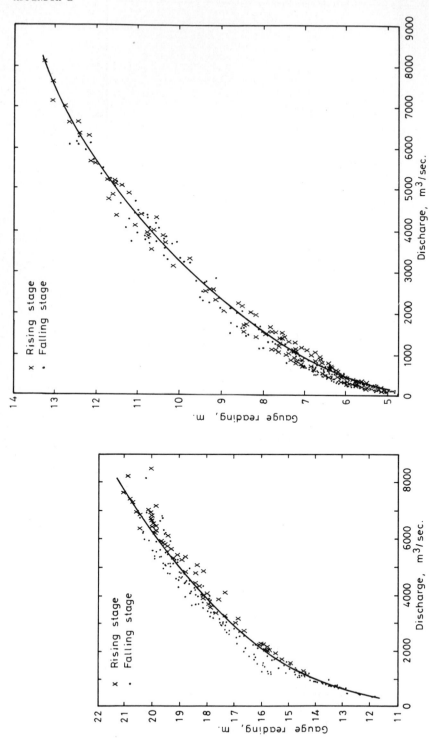

Fig. 25 Average rating curve of the Blue Nile at Sennar Station
(Makwar Gauge 1923-27)

Fig. 24 Average rating curve of the Blue Nile
measured at Singa (Wad el Aies Station 1922-26)

APPENDIX E

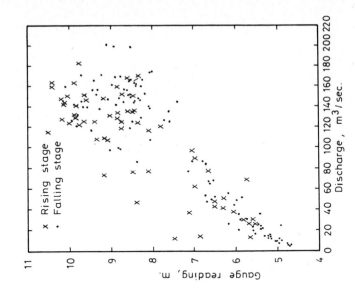

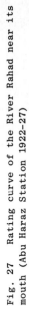

Fig. 27 Rating curve of the River Rahad near its
mouth (Abu Haraz Station 1922–27)

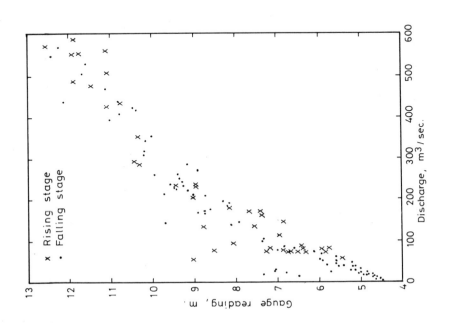

Fig. 26 Rating curve of the River Dinder near its
mouth (Hillet Idris Station 1924–27)

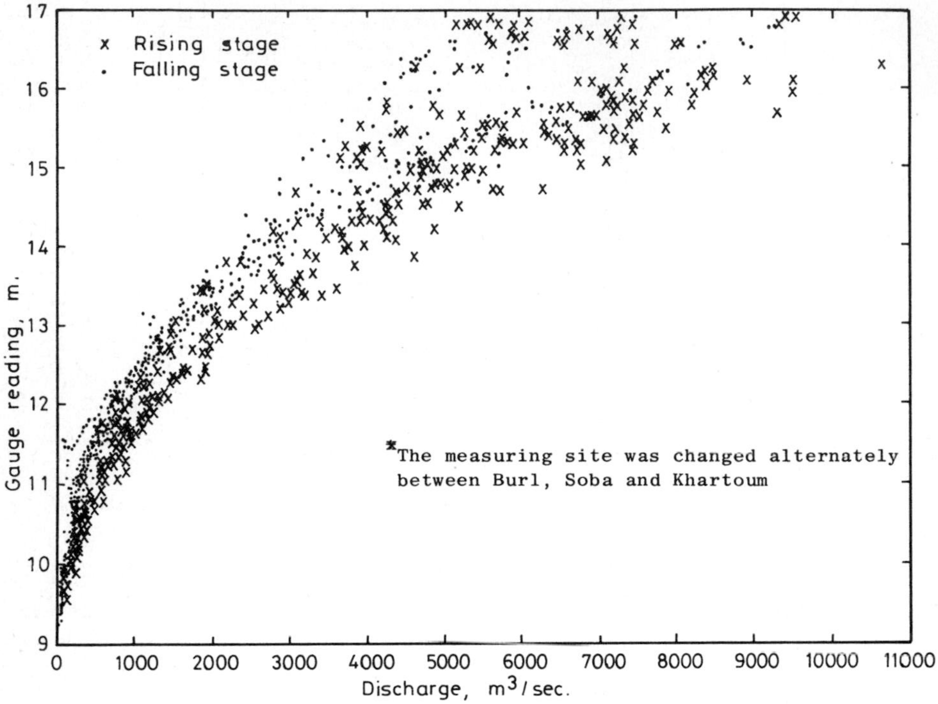
x Rising stage
• Falling stage
*The measuring site was changed alternately between Burl, Soba and Khartoum
Gauge reading, m.
Discharge, m3/sec.
9
10
11
12
13
14
15
16
17
0
1000
2000
3000
4000
5000
6000
7000
8000
9000
10000
11000

Fig. 28 Rating curve of the Blue Nile at Khartoum* (1902-27)

APPENDIX E

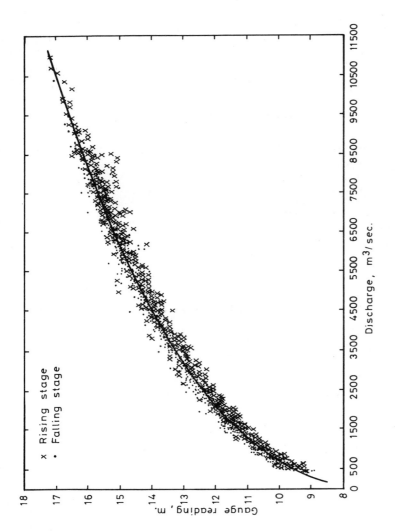

Fig. 29   Average rating curve of the Nile at Tamaniat Station (1911-72)

APPENDIX E

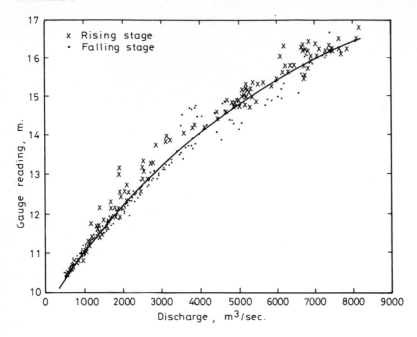

Fig. 30   Rating curve of the Main Nile at Hassanab 5 km upstream the mouth of the Atbara (1924-27)

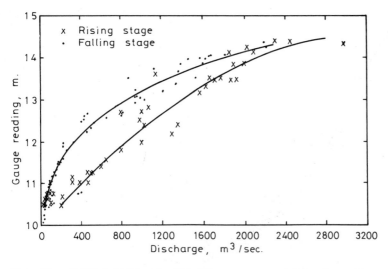

Fig. 31   Rating curve of the Atbara measured 3 km from its mouth (1926-27)

549

APPENDIX E

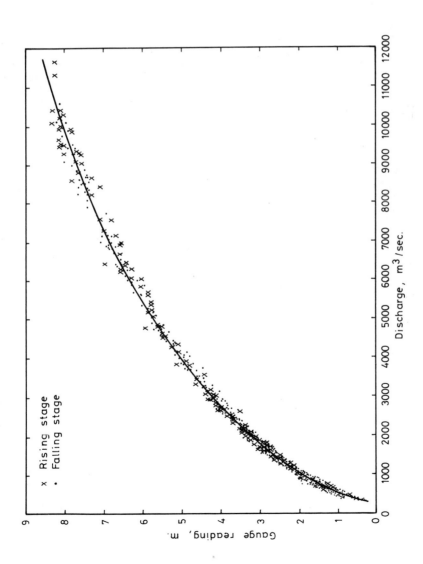

Fig. 32   Average rating curve of the Main Nile at Wadi-Halfa Station (1911–27)

APPENDIX E

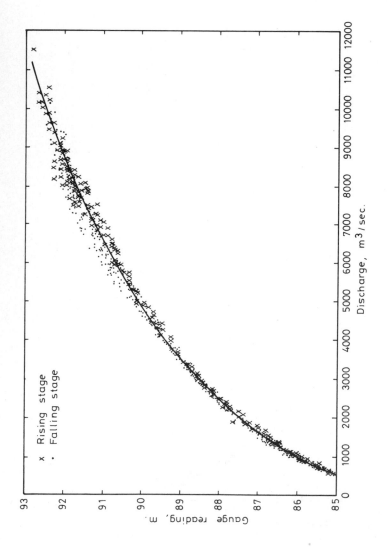

Fig. 33 Average rating curve of the Main Nile at Aswan (Khannaq Station (1918-27)

APPENDIX E

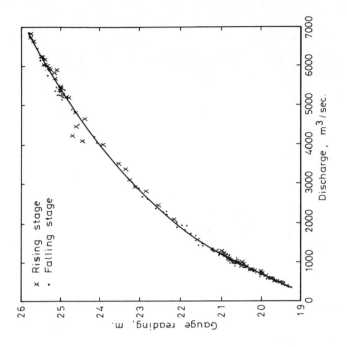

Fig. 35   Rating curve of the Main Nile measured at Koraimat (Beleida Station 1920-22)

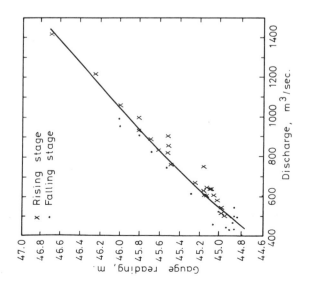

Fig. 34   Rating curve of the Main Nile measured near Assiut (Hawatka Station 1926-27)

APPENDIX F

ASWAN DAM

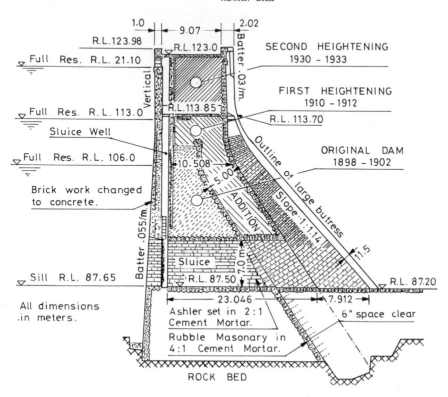

Fig. 1   Cross-section of the pierced part set 2 of the Aswan Dam

Some Technical Data

Location   : at Aswan on the Main Nile, Egypt
Period of construction: 1898-1902
First heightening   : 1912
Second heightening   : 1934

Item	Original dam	1st heightening	2nd heightening
Storage capacity, mlrd m^3	1.06	2.30	5.10
Total length of dam, m	1950.00	1982.00	2129.00
Height above bed, m	21.50	26.50	35.50
Height above foundation, m	38.80	43.80	52.80
Maximum storage level	106.00	113.00	121.00

APPENDIX F

Aswan Dam (continued)

Total number of vents: 180 having levels and dimensions as follows:

No. of vents	sill level	height, m	width, m
65	87.65	7.0	2.0
75	92.00	7.0	2.0
18	96.00	3.5	2.0
22	100.00	3.5	2.0

Number of lock chambers: 5, each 80.0 m long x 9.5 m wide.

The hydro-electric generating plant comprises seven 47-MW and two 11.5-MW vertical Kaplan turbines and alternators operating over a range of 10-33 m. The output of the main turbines varies from 6500 HP at 6 m head to 65000 HP at a head of 27-31 m.

APPENDIX F

SENNAR (MAKWAR) DAM

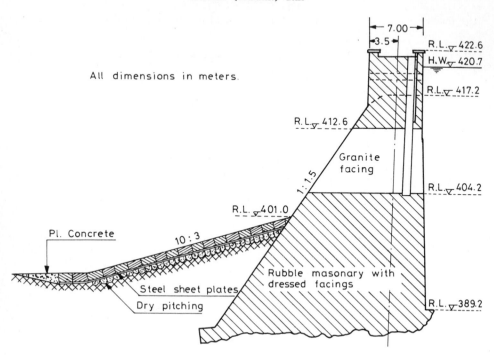

All dimensions in meters.

Fig. 2   Cross-section of Sennar (Makwar) Dam, The Sudan

Some Technical Data

Location   : on the Blue Nile about 350 km southeast of
                                   Khartoum
Year of construction   : 1925
Storage capacity of reservoir   : 0.8 mlrd m^3 (between the levels of 420.7 and
                                   417.2)
Length of pierced section of dam: 600 m
Number of sluices   : 80
Sluice dimensions   : 2.0 m in width x 8.4 m in height
Number of spillways   : 112
Width of spillway   : variable, from 3 to 5 m
Total length of dam   : 3000 m

APPENDIX F

JEBEL EL-AULIA DAM

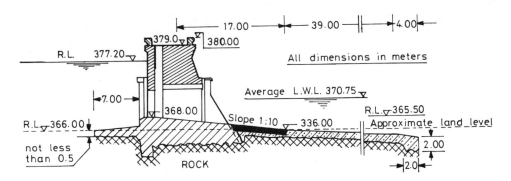

Fig. 3   Cross-section of the Jebel el-Aulia Dam at kilo, 1004

Some Technical Data

Location   : on the White Nile about 40 km upstream of its
                                    junction with the Blue Nile

Period of construction   : 1934-1937

Reservoir dimensions   : 314 km in length x 4.2 km in width

Initial storage capacity   : 3.5 mlrd m^3

Length of solid part of dam: 1126 m

Length of pierced section   :   454 m

Number of sluices   : total = 60, operating = 50, blind (reserve) = 10

Sluice dimensions   : 3.0 m in width x 4.5 m in height

Maximum head difference   : 6.45 m

Lock section   : 60 m

Length of earthern section : 3360 m

Total length of dam   : 5000 m

APPENDIX F

KHASHM EL-GIRBA DAM

Fig. 4

Some Technical Data

Location   : on the Atbara River at Khashm el-Girba some
                                       400 km south-east the river mouth at Atbara
Period of construction   : 1960-1964
Length of concrete section   : 466 m
Length of earthern section   : 3380 m
Total length of dam   : 3846 m
Total number of sluices   : 12
Lower sluices   : 5, each with a capacity of 200 m^3/sec.
Upper sluices   : 7 (for passing flood), each with a capacity of
                                       1100 m^3/sec.
Maximum release capacity   : 8700 m^3/sec., or 750 x 10^6 m^3/day
Dimensions of storage reservoir: 80 km in length x 1.5 km in width
Initial storage capacity (1964): 1.30 mlrd m^3
Actual storage capacity (1971) : 0.97 mlrd m^3
Estimated rate of sedimentation: 40 million m^3/yr

APPENDIX F

EL-ROSEIRES DAM

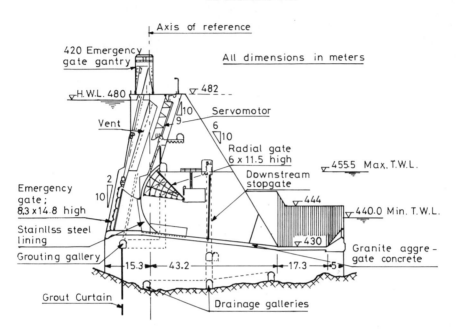

Fig. 5a Typical section deep sluices

Some Technical Data

Location   : on the Blue Nile at Damazin rapids, approximately
                                    120 km from the Ethiopian borders

Period of construction   : 1961-66

Total length of dam   : 16 km

Length of concrete section   : 1 km

Number of buttresses   : 69

Spacing of buttresses   : 14 m increased to 18 m at the intake of power house.
                                    The standard buttresses are modified in various
                                    parts of the concrete section to accommodate the
                                    intakes for irrigation canals on the two river
                                    banks

Number of sluices   : 5 low-level sluices, 7 high-level spillways and the
                                    intakes for 7 turbines in the hydro-electric power
                                    station

Length of earth embankments: 15 km

558

APPENDIX F

El-Roseires Dam (continued)

Maximum storage level in
the first stage   : 480 m above mean sea level

Maximum storage capacity in
the first stage (gross)   : 3 mlrd m^3

Maximum storage level in
the second stage   : 490 m above mean sea level

Maximum storage capacity in
the second stage (gross)   : 7.6 mlrd m^3

Minimum reservoir level to
command irrigation by
gravity   : 467 m above mean sea level

Mean net head for hydro-
electric power development
in the first stage   : 27.8 m

Mean net head for hydro-
electric power development
in the second stage   : 30.6 m

Type of turbines   : vertical Kaplan turbines of 41500 HP under a head
                               of 29 m

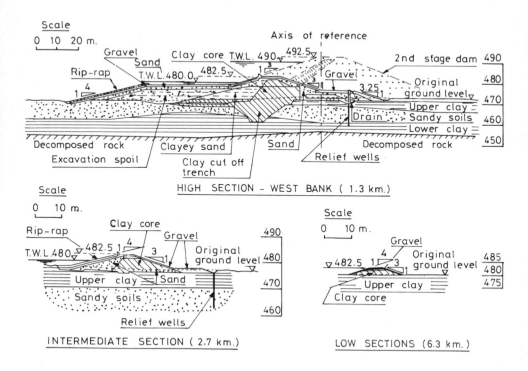

Fig. 5b   Typical cross-sections of earth embankments

APPENDIX F

OWEN FALLS DAM

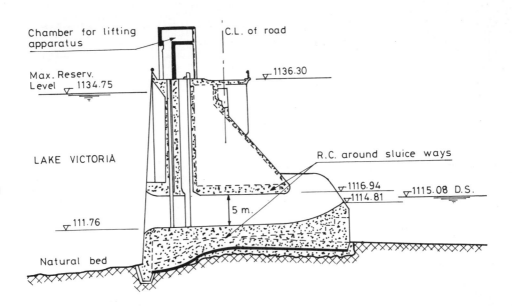

Fig. 6   Cross-section of the Owen Falls Dam

Some Technical Data

Location   : on Victoria Nile about 3 km downstream of the exit at
                               the foot of Owen Falls below the Rippon Falls

Period of construction: 1948-1950

Type of dam   : mass concrete gravity type with some steel reinforcement
                               around the sluices

Dimensions   : about 30 m in height and 760 m in length

Sluices   : 6 in number, each 5 m in height and 3 in width

Outflow   : average is about 620 m^3/sec. reduced to about 500 m^3/
                               sec. The difference is used to build up storage in the
                               lake until required by Egypt

Available head   : between 18 and 22 m, with an average of 20 m

Turbines   : 10 Kaplan turbines each coupled to a 15 000 KW alter-
                               nator. The plant commenced generating electricity in
                               1954 with one unit and since then the other units have
                               been installed progressively

APPENDIX F

THE HIGH DAM AT ASWAN

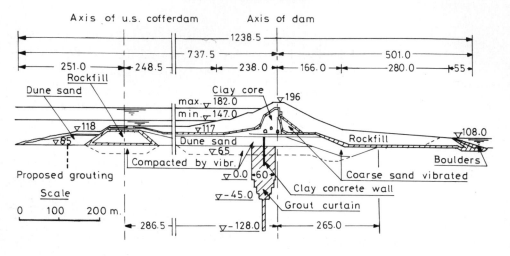

Fig. 7   Cross-section of the High Dam at Aswan

Some Technical Data

Location   : on the Main Nile near Aswan, Egypt, 6.5 km upstream of
                                 the old dam

Periods of construction: First phase   1960-64

                               : Second phase 1965-68

                               : Hydro-electric power plant 1967-72

Type of dam, material
and dimensions   : rock-fill 111 m in height, 40 m top width and 980 m
                                 base width. The dam has a total length of 3600 m of
                                 which 520 m span the Nile. Depth of grout curtain is
                                 213 m (see the cross-section)

Diversion canal   : located on the eastern bank of the river. Designed for
                                 a maximum discharge of 11000 m^3/sec. It consists of
                                 open reaches in the upstream and the downstream parts
                                 joined in the middle by the main control tunnels under
                                 the dam. The upstream tail of the canal is 1150 m in
                                 length, the downstream tail is 485 m and the tunnels
                                 and powerhouse 315 m, i.e. total length = 1950 m. The
                                 upstream canal width = 250 metres narrowing down to
                                 50 m in a distance of 630 m. The narrowest width is
                                 kept constant then increases gradually 230 m at the

APPENDIX F

The High Dam at Aswan (continued)

inlet of the tunnel. The downstream reach begins with
a width of 278.5 m which narrows gradually to 40 m in
a distance of 330 m from the powerhouse.

Tunnels : six tunnels each 282 m in length serve in connecting
the upstream reach of the diversion canal with its
downstream tail across the powerhouse. The tunnel is
circular in the major part of its length, the diameter
being about 15 m and the maximum velocity is nearly
10 m/sec. To resist this high velocity the tunnels have
a reinforced concrete lining more than 1 m thick.
Before joining the powerhouse, each tunnel branches off
into two rectangular conduits, each 7.5 x 2.2 m; one
delivers water to the generating unit and the other
carries the excess water needed for irrigation directly
to the downstream canal without passing through the
turbine.

Power station : The station is located at the downstream end of the
spillway tunnels. It consists of 12 generating units
each fed from one of the 12 branches. The generating
unit is a Francis type turbine directly connected to a
generator. Range of head = 35-77 m, capacity at design
head = 180 00 KW and discharge = 346 m^3/sec. A trans-
former station helps in raising the voltage from
15 700 to 500 000 volts.

I - Edfina barrage (1948 -1951)

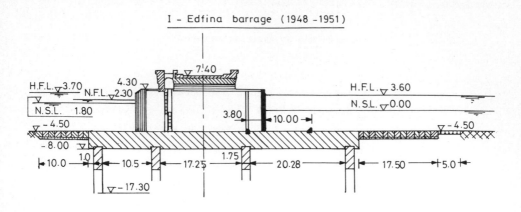

II - Zifta barrage (1901-1903 and 1953-1955)

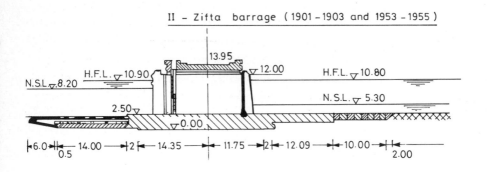

III - New Rosetta barrage (1937-1939)

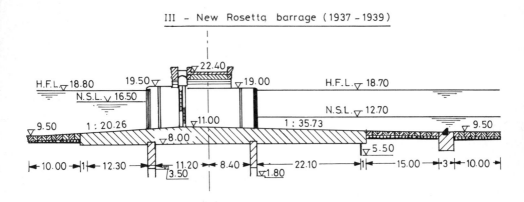

Fig. 8   Sections of large barrages on the Nile between Aswan and the Mediterranean Sea

IV - Assiut barrage (1898 - 1902 and 1934 - 1938)

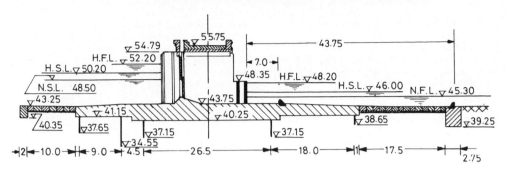

V - Nag - Hammadi barrage (1927 - 1930)

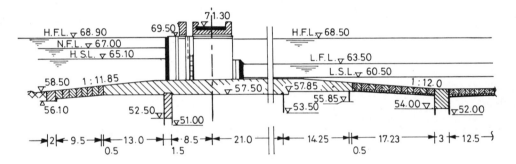

VI - Esna barrage (1945 - 1947)

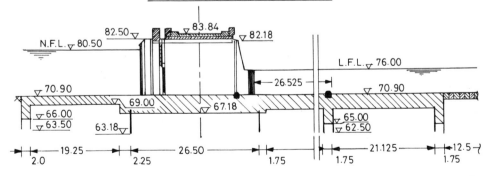

Fig. 8   continued

INDEX

[+]R.G.S. = rain gauging station, [++]G.S. = river gauging station, [*]M.S. = meteorological station, [**]A.S. = agricultural station

---

[+]R.G.S. = rain gauging station, [++]G.S. = river gauging station, [*]M.S. = meteoro-logical station, [**]A.S. = agricultural station

---

[+]R.G.S. = rain gauging station, [++]G.S. = river gauging station, [*]M.S. = meteorological station, [**]A.S. = agricultural station

---

[+]R.G.S. = rain gauging station, [++]G.S. = river gauging station, [*]M.S. = meteorological station, [**]A.S. = agricultural station

Hargreave's formula, 181-183, 187, 191, 198, 200, 205, 208, 263, 274, 276, 277, 279
Harmonic: parameter, coefficient. 152, 342, 346, 366, 370
El-Hassana: R.G.S.,[+] 124, 470, 491, 548
Hassanab: G.S.,[++] 198, 395-398, 403, 406, 521
Heliopolis, 203
Helwan: M.S.,[*] 64, 69, 74, 79, 87, 91, 96, 98, 124, 167, 470, 492
Heat index, 261
High-Aswan Dam, 12, 13, 50, 54, 251, 272, 413, 419, 428, 432, 439, 441, 443, 460, 466, 560
Hillet Doleib: G.S., on Sobat, 373, 410, 539
Hillet Idris: G.S., on Dinder, 386, 545
Hillet Nuer (Adok), 359
Humid: climate, 61
Humidity: absolute, relative, 64, 77, 78, 182, 200, 203, 209, 212, 218, 259, 267, 277
Hurghada: M.S., 64, 69, 74, 91, 124, 200, 201, 216, 472, 493
Hydrograph, 338, 343, 351, 353, 366, 373, 387, 400, 407, 409, 411
Hydrometeorological Survey of the Catchments of Lakes Victoria, Kyoga and Albert, 13, 61, 95, 113, 120, 121, 133, 152, 157, 172, 317, 345

Igabiro: R.G.S., 132, 153, 155, 480, 507
Igneous rocks, 299, 300
Ikizu: R.G.S., 132, 480, 507
Imatong: mountain, 30
Index of raininess, 157
Inganga: R.G.S., 129, 476, 502
Inselberg: 306
Integration method: 271
Intertropical Convergence Zone (ITCZ), 105
Irrigation cycle, 229
Ishasha: River, 28
Isohyet: 61, 274
Isotherm: 68

Jebel Aulia
  dam, reservoir, G.S., 10, 12, 186, 379, 430, 555
  M.S., 65, 472, 495
Jebelein: R.G.S., 127, 150, 474, 497

Jensen and Haise's formula, 266
Jimma: M.S., 65, 70, 87, 92, 94, 95, 101
Jinja: M.S., G.S., on Victoria Nile, 86, 215, 285, 326, 330, 331, 510, 528
Jonglei Canal, 13, 362, 451
Jordan Valley, 19
Juba
  on Bahr el Jebel, 30, 67, 96, 107, 181, 182, 216, 287
  M.S., 65, 70, 75, 80, 87, 92, 101, 128, 476, 499
Jur: River, 34, 364
Jurassic: geologic time, 293

Kabale: M.S., 65, 76, 86, 88, 92, 94, 95, 101, 103, 478, 504
Kabanyolo: M.S., 285
Kabasande: R.G.S., 130, 478, 503
Kachimbala: R.G.S., 129, 476, 501
Kadugli: R.G.S., 127, 474, 497
Kafr el-Dawar: R.G.S., 123, 470, 490
  el-Sheikh: R.G.S., 123, 470, 490
  el-Zayyat: R.G.S., 124, 470, 491
Kafu: River, 23, 25
Kagera
  Basin, 63, 116, 318, 321
  River, 6, 19, 20, 23, 318, 340
Kagondo: R.G.S., 480, 507
Kahangi Estate: R.G.S., 129, 476, 502
Kaja: River, 30
Kajinarty: G.S. on Main Nile, 427
Kajo-Kaji: R.G.S., 128, 354, 476, 499
Kakamega: R.G.S., 131, 478, 505
Kakitumba: River, 19
Kalagala: A.S.,[**] 129, 153, 155, 476, 502
Kalangala: M.S., 504
Kalangassa: River, 19
Kalungu: R.G.S., 130, 478, 503
Kamlin: R.G.S., 126, 387, 472, 495
Kampala: M.S., 65, 88, 215
Kangen: River, 39
Kapasabet: R.G.S., 131, 478, 505
Kapenguria: R.G.S., 131, 478, 504
Karagwe-Ankolean: geological formation, 293
Karima (Kareima): M.S., 65, 87, 125, 472, 494
El-Kasr: M.S., 204, 209, 212
Kassala
  M.S., 65, 70, 75, 79, 87, 92, 95, 101, 167, 194, 195, 216, 472, 495
  Province, 306
Katakawi: R.G.S., 128, 476, 500

+R.G.S. = rain gauging station, ++G.S. = river gauging station, *M.S. = meteorological station, **A.S. = agricultural station

+R.G.S. = rain gauging station, ++G.S. = river gauging station, *M.S. = meteorological station, **A.S. = agricultural station

---

[+]R.G.S. = rain gauging station, [++]G.S. = river gauging station, [*]M.S. = meteorological station, [**]A.S. = agricultural station

[+]R.G.S. = rain gauging station, [++]G.S. = river gauging station, [*]M.S. = meteorological station, [**]A.S. = agricultural station

---

[+]R.G.S. = rain gauging station, [++]G.S. = river gauging station, [*]M.S. = meteorological station

---

+R.G.S. = rain gauging station, *M.S. = meteorological station